PROTEIN BIOCHEMISTRY, SYNTHESIS, STRUCTURE AND CELLULAR FUNCTIONS

FIBRONECTIN

CURRENT CONCEPTS IN STRUCTURE, FUNCTION AND PATHOLOGY

PROTEIN BIOCHEMISTRY, SYNTHESIS, STRUCTURE AND CELLULAR FUNCTIONS

Additional books in this series can be found on Nova's website
under the Series tab.

Additional e-books in this series can be found on Nova's website
under the e-books tab.

PROTEIN BIOCHEMISTRY, SYNTHESIS, STRUCTURE AND CELLULAR FUNCTIONS

FIBRONECTIN

CURRENT CONCEPTS IN STRUCTURE, FUNCTION AND PATHOLOGY

JAMES BEATTIE
EDITOR

New York

For permission to use material from this book please contact us:
Telephone 631-231-7269; Fax 631-231-8175
Web Site: http://www.novapublishers.com

Library of Congress Cataloging-in-Publication Data

Fibronectin : current concepts in structure, function, and pathology / editor, James Beattie.
p. cm.
Includes index.
ISBN 978-1-62257-298-4 (hardcover)
1. Fibronectins. I. Beattie, James.
QP552.F53F532 2012
572.8'4--dc23

2012021409

Published by Nova Science Publishers, Inc. † New York

Contents

Preface		**vii**
Chapter I	Fibronectin Functions: Mechanisms and Applications *Preety Sharma, Miles F. Beaux II and Nam Hoon Kwon*	**1**
Chapter II	FN Modulation of Growth Factor Activity *James Beattie*	**21**
Chapter III	Fibronectin: The N-Terminal Region and Its Role in Cell Migration - Implications for Disease and Healing *Ian R Ellis, Mohammad R. Islam, Lateef E. Aljorani and Sarah J. Jones*	**35**
Chapter IV	Fibronectin Assembly and Function: A Multi-Dimensional View *Wing S. To and Kim S. Midwood*	**71**
Chapter V	Vesicular Trafficking of the Fibronectin Receptor $\alpha 5\beta 1$ Integrin *Chuan Hu, Nazarul Hasan and Krista A. Riggs*	**125**
Chapter VI	Methods to Study Fibronectin Secretion and Matrix Assembly by Fibroblasts In Vitro and In Vivo *Anna Goc, Maha Abdalla and Payaningal R. Somanath*	**141**
Chapter VII	Pathways Regulating the Secretion and Assembly of Fibronectin Matrix: An Update Methods to Study Fibronectin Secretion and Matrix Assembly by Fibroblasts In Vitro and In Vivo *Maha Abdalla, Anna Goc and Payaningal R. Somanath*	**155**
Chapter VIII	Modulation of Fibronectin Expression in Astrocytes Affects Neuritogenesis: Relevance in Developmental Neurotoxicity *Lucio G. Costa, Marina Guizzetti, Daniella Pizzurro and Gennaro Giordano*	**191**

Chapter IX Evidence for Fibronectin Role on Uterine Natural Killer Cells Migration, Proliferation, Differentiation and Survival during Pregnancy **207**
Bruno Zavan, Andréa Mollica do Amarante Paffaro, Paulo P. Joazeiro, Áureo. T.Yamada and Valdemar A.Paffaro Jr

Chapter X Fibronectin: Structure, Biosynthesis and Roles in Disease **213**
Kátia da Silva Calabrese, Luiz Otávio Pereira Carvalho, Tânia Zaverucha do Valle, Flávia de Oliveira Cardoso and Ana Lúcia Abreu-Silva

Chapter XI Fibronectin in Renal Diseases and Peritoneal Dialysis **227**
Susan Yung and Tak Mao Chan

Chapter XII Role of Fibronectin in Tissue Remodeling during Kidney Disease Development **265**
Séin O'Connell, Stephen Nolan, Craig Slattery, Michael P Ryan and Tara McMorrow

Chapter XIII Role of Fibronectin in Neoplasm **283**
Zhenggang Xiong and Liqiong Liu

Chapter XIV Fibronectin and Brain Tumors **301**
Mariella Caffo, Gerardo Caruso, Valeria Barresi, Marcello Passalacqua, Filippo F. Angileri and Francesco Tomasello

Index **321**

Preface

The multidomain extracellular protein fibronectin (Fn) is assuming an increasingly important role in cell biology. Interactions between Fn and juxtaposed cells (mainly via cell surface heterodimeric integrin receptors) play an important role in the regulation of cell migration, adhesion, growth, differentiation and apoptosis. This classical "outside-in" signalling leads to the activation of intra-cellular signalling pathways and results in the transient formation of protein-protein complexes and alterations in the states of protein phosphorylation which ultimately lead to changes in gene expression. It is apparent that in many instances Fn-integrin complexes are able to "cross-talk" to other growth factor – receptor complexes (e.g. insulin-like growth factor-I (IGF-I) and IGF-IR. Such interaction may be required for growth factors to display the full range and appropriate temporal aspects of their activity. In addition to cellular interactions, Fn also associates with many other extracellular glycosaminoglycans (GAGS) - heparin, fibrin and collagen to regulate such processes as blood clotting and wound healing. These interactions are often controlled by events within different types of cells in the affected tissues – so called "inside-out" signalling. Finally, the role of Fn in the aetiogy and regulation of different disease processes is an increasingly popular research area. The role of Fn in the establishment and spread of tumour cells is one such topic. This volume draws on the expertise of many laboratories to provide up to date research findings for those interested in this field.

James Beattie, PhD,
Deartment of Oral Biology
University of Leeds
January 2012

In: Fibronectin
Editors: James Beattie

ISBN 978-1-62257-298-4

Chapter I

Fibronectin Functions: Mechanisms and Applications

***Preety Sharma*[1], *Miles F. Beaux II*[2,*] *and Nam Hoon Kwon*[3]**

[1]Center of radiological research, College of Physicians and Surgeons, Columbia University, New York, NewYork
[2]MPA Division, Los Alamos National Laboratory, Los Alamos, New Mexico
[3]Medicinal Bioconvergence Research Center, Seoul National University, Seoul, Korea

Abstract

Fibronectin (Fn) is a high molecular weight multidomain glycoprotein present in extracellular matrix (ECM) interacting with specific adhesion cell surface receptors. These interactions regulate various cellular functions including cell adhesion, migration, morphology, growth, differentiation, wound repair, carcinogenesis and bacterial infiltration. The purpose of this chapter is to provide a brief overview of various *in vivo* functions of Fn, and its research and clinical applications. We begin with highlighting the key structural features and forms that account for its multiple functions. Next, we describe briefly its important *in vivo* functions. Lastly we focus on how these functional mechanisms have been exploited for research, clinical and drug delivery applications of Fn including tissue culture, adhesion assays, disease diagnosis and wound repair.

* Email: mbeaux@lanl.gov.

1. Introduction

Since its discovery in the early 1970s [1], the prominence of fibronectin (Fn) protein in scientific literature has persisted and will likely continue to do so into the foreseeable future. Its discovery came about as researchers searched for differential surface chemistry between cancer and healthy cells, and it was originally thought to be a cancer protein [2]. Fn has since been shown to play a role in a broad range of cellular functions unrelated to cancer [3-8]. In fact the role of Fn proteins have been shown to be vital to the development and normal physiological state of higher organisms. All of the functions of Fn can be described by its structure, specifically the many binding sites [9]. In this chapter, a brief summary is given of the relevant structure and properties of Fn proteins. A review of various applications of Fn proteins which exploit these functions is also provided.

2. FN Structure and Forms

Advancement in the understanding of structure, function, and distribution of Fn protein has been possible through the use of cell culture, proteolytic studies, genetic engineering techniques, immunocytochemistry, knock out, and gene targeted mouse models. Fn protein is the product of a 90 kbp gene composed of 47 exons localized on chromosome 2 [10]. Expression of the Fn gene is regulated by a set of different growth factors and hormones that regulate its developmental specificity. A complex vascular system has evolved in higher organisms such as vertebrates and Fn is a critical component of blood vasculature [11]. It is a highly conserved protein with greater than 90% homology within mammals and more than 80% homology between chicken and mammalian sequences. The highly conserved structural complexity of Fn in different animals, as well as the presence of interspecies and tissue specific multi-variant isoforms indicates that this protein carries out important functions in maintaining a normal physiological state. Among the ECM proteins, Fn is a critical protein in higher organisms as demonstrated by lethality of Fn-null mice caused by early vascular defects in embryonic development [12]. Clinically, Fn has emerged as a protein which is associated with a variety of disease conditions [13-16]. The significant role of Fn can be better appreciated with a closer understanding of its structural details including interaction sites, isoforms, and also tissue distribution.

2.1. Structure

Fn exists in multiple forms with about 20 splice variants which have different functional and physical properties. The polypeptide regions within the globular domains are susceptible to proteolysis producing Fn fragments of defined size and properties. Studies carried out in the early 1980s on these Fn fragments have been key to providing insights into their physiological role [17, 18]. Fn exists in a dimeric soluble form in plasma (300μg/ml) and other body fluids. It is also present as a multimeric insoluble fibrillar protein in extracellular matrix (ECM) in multiple tissues and on cell surfaces of a large number of cell types [19]. Dimeric Fn molecules consist of two identical or near identical 220-250 kDa monomer subunits. Each monomer consists of three different modules: type I, type II and type III domains. Type I and II domains have 12 and 2 repeating units, respectively, each of about 40 and 60 amino acids, respectively. Nine of the 12 type I repeating units are present at the amino terminal of the Fn polypeptide and the remaining three are present at the carboxyl terminal. The two type II repeating units are located immediately following the first six type I repeats on the amino terminal.

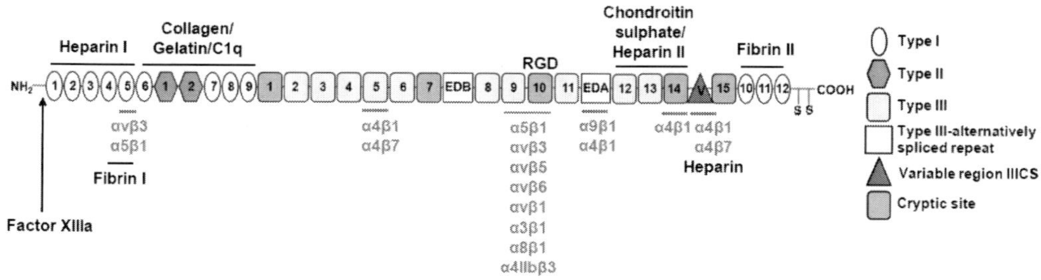

Figure 1. Schematic representation of the Fn monomer consisting of three modules- I, II and III with repeating units, major binding and interacting sites, alternative splicing regions, amino and carboxyl terminal and cryptic sites.

The type III domain has 15 centrally located modules each of about 90 amino acids that constitute the largest part of the Fn polypeptide (Figure 1). Fn beta strands are cross linked together with two disulphide bonds present only in type I or type II modules. Structurally type III modules exhibit a seven beta stranded sandwich motif [19-22].

2.2. Type I and II Modules

Type I modules play an important role in cellular matrix assembly, blood clotting and pathogen recognition. The amino-terminal type I domain has a strong affinity binding site for fibrin, a plasma protein critical for the blood clotting process. Also present are sites for cell surface heparan sulfate proteoglycans, glycosphingolipids, and microbial surface components recognizing adhesive matrix molecule (MSCRMMS) on gram positive bacteria. Immediately following the amino- terminal domain is the highly glycosylated collagen binding domain consisting of type I and type II modules. Fn binds *in vitro* to collagen but more effectively to denatured collagen (gelatin) [19-22]. The C1q component of the complement system has the same binding site as collagen on Fn. This interaction of C1q complement with Fn helps in the clearance of immune complexes and cellular debris during immune responses [23].

2.3. Type III Modules

Within the central type III domains is the major heparin binding site consisting of type III repeating domains and a variable region that is tissue specific. Besides the heparin binding domain, an Arg-Gly-Asp (RGD) sequence within the type III domain is identified as the main cell binding site. This site, in the cellular form of Fn (cFn), interacts with a type of cell surface receptor called integrins.

Integrins are heterodimeric αβ transmembrane receptors. These activate multiple signaling pathways that mediate various biochemical processes involved in cell-cell, cell-matrix interaction, adhesion and other functions. Various cell types such as platelets, megakaryocytes, endothelial cells, lymphocytes, leukocytes variably express integrins such as $\alpha_5\beta_1$ $\alpha_v\beta_3$, $\alpha_v\beta_5$, $\alpha_{IIb}\beta_3$, $\alpha_4\beta_1$, and $\alpha_4\beta_7$ that interact with the ECM and cellular adhesion molecules through known binding sites on Fn [24]. The RGD motif in the type III Fn domain is the major integrin binding site for all integrins except $\alpha_4\beta_1$ which binds to RGD homologues present in different regions of the Fn protein. Besides the Fn RGD integrin binding domain in type III repeating units 9 and 10, elements within the C-terminal heparin binding domain regulate the formation of cytoskeleton-ECM connections called focal contacts.

Although the integrin-Fn interactions are largely responsible for cell adhesion mechanisms, coordinated interactions of Fn with other proteins in focal contacts are required for the entire cellular response. Studies on cultured cells have shown that events such as actin organization and formation of filipodia and lamellipodia occur within minutes of cell contact with the Fn type III module 13 segment of the heparin-binding region [25]. The heparin binding domain also plays a role during acute responses of clotting, and platelet adhesion and aggregation.

Within the type III domain are three alternative splicing regions that give rise to multiple isoforms of Fn. These include the connecting sequence (IIICS) present between type III units 14 and 15 and extra domains (ED) A and B. EDA and EDB are also referred to as EIIIA and EIIIB in the literature. The EDA is present between type III units 11 and 12 and EDB is located between units 7 and 8 [19-22]. Studies using genetically engineered mice have looked closely at the *in vivo* functions for the EDA and EDB domain expressing Fn peptides. EDA domain knock-out mice were observed to have wound healing defects, altered behavior, impaired motor coordination, and a shorter life span. In contrast no distinct phenotype was observed in EDB knock-out mouse models indicating that EDA may compensate functionally in EDB knock outs. EDA and EDB double knock-out mice on the other hand have lethal defects [26-28].

2.4. Splicing Isotypes

Fn gene functions are regulated by alternative mRNA splicing that generates multiple isoforms of the protein. The mRNA splicing events result in inclusion or exclusion of EDA and EDB exons and the IIICS region. Depending on the species however the IIICS also known as the V (Variable in length) element can be partially included in the Fn mRNA. The complexity of the IIICS (V) region splicing and total Fn isoforms present increases with increasing species complexity. Up to five V forms and 20 Fn isoforms are present in humans [22, 29-31].

The V region is involved in Fn secretion and binding to integrin and heparin [32]. The Fn isoforms present during embryonic development contain both EDA and EDB domains. In adults this isoform is increased in concentration only temporally during tissue repair, angiogenesis and during pathological conditions such as tissue fibrosis and tumorigenesis [33-35].

The soluble form of Fn (sFn) is synthesized predominantly by hepatocytes and secreted directly into the circulation. These forms almost always lack the alternatively spliced EDA and EDB domains. cFn, however, is produced locally in tissues, predominantly by resident fibroblasts and endothelial cells, and is deposited in the pericellular matrix. cFn consists of variable proportions of EDA and EDB domains and is a highly heterogeneous group of Fn isoforms. This heterogeneity is the result of splicing patterns that are tissue and species-specific, generating Fns with different cell-adhesive, ligand binding and solubility properties [22, 36].

2.5. Fibrillar Matrix Formation

Fn produced and secreted by fibroblasts and endothelial cells is assembled into a fibrillar matrix ECM composing interstitial matrices and basement membranes around the cells and tissues [32, 37]. Soluble Fn binds to $\alpha_5\beta_1$ integrin via its RGD and synergy sites but not to other RGD-dependent integrins [38]. This binding is required for *de novo* Fn fibril formation. Fn binding induces integrin clustering which promotes Fn-Fn interactions by providing locally high concentrations of Fn at the cell surface. This process exposes buried binding sites called cryptic sites which induce further conformational change in Fn, resulting in insoluble fibrillar matrix formation. The cryptic sites have novel biological functions and expose regions that serve as nucleation sites for fibrillogenesis-assembly of Fn into fibrils [39], which in turn stimulates cell growth and contractility [40].

Fn matrix assembly is a dynamic process and several studies suggest that Fn turnover occurs continuously, resulting in fibril formation and solubilization in equilibrium [41, 42]. The $\alpha_5\beta_1$ integrin which plays an important role in Fn matrix assembly is also shown to be involved in the turnover [32]. Caveolin-1 regulates $\alpha_5\beta_1$ endocytosis, and β_1 integrin seems to be involved in the endocytosis of soluble Fn, although the exact mechanisms are not clear [43, 44].

3. FN Functions

3.1. Wound Repair

The extracellular matrix components such as collagen, proteoglycans, hyaluronic acid, Fn, laminin, tenascin and thrombospondin along with growth factors, cytokines and various cell types collectively participate in the normal wound healing process. Fn has many important functions particularly in the initial stages of healing. Following tissue injury, fibroblasts migrate into the wound site, are activated, and produce an early matrix mainly composed of fibrin, Fn, and hyaluronic acid. Fn is crosslinked to the fibrin clot and to itself by activated coagulation factor XIII at the Gln residue close to the Fn amino terminus. This crosslinking in the matrix acts as a scaffold that directs fibroblast cell movement mediating cell attachment, spread, and migration [25]. The cell adherence step is critical for cell migration into the wound site. Structured cell-ECM interactions followed by cell migration, angiogenesis, and integrin signaling is required for a smooth transition from an initial Fn rich matrix to a more mature collagen rich vascular matrix called granulation tissue. The amount of available Fn at the wound site is indicated to be a rate limiting step in the formation of new granulation tissue. Thus if the process of matrix formation is compromised it can result in an abnormal healing process by virtue of the formation of chronic granulation tissue [42, 45-49].

3.2. Organ Specific Functions

Multiple Fn splice variants and forms dictate tissue specific functions of Fn. During early embryonic development Fn is expressed differentially in various body tissues such as brain, kidney, heart, and vessels. Based on immunofluorescence staining, Fn and collagen IV were localized around fetal blood vessels and in the stroma of placental villi highlighting the important roles in tissue organization of placental villi [50]. Fn protein plays a protective role in cerebral ischemic disease. In ischemic brain infarction, there is an increase in permeability of the blood brain barrier and plasma Fn (pFn) enters the brain. This Fn has been proposed to bind to cell surface $\alpha_5\beta_1$ and/or $\alpha_V\beta_3$ integrins causing upregulation of anti-apoptotic protein Bcl-2 preventing caspase-3 mediated apoptosis [51].

Within the eye, the trabecular meshwork (TM) regulates the aqueous humor (AH) outflow and maintains the intraocular pressure. The cells lining the trabecular meshwork express various ECM components such as Fn, laminin, vitronectin, proteoglycans and collagen. The ECM is proposed to regulate flow of AH through TM by acting as a filter and providing physical resistance to the movement of AH. Increase in ECM content is often observed in glaucomatous patients. Excess Fn deposition in TM occurs in corticosteroid induced glaucoma, primary angle open glaucoma (POAG) and in aging eyes. Various Fn domains actively interact with growth factors and integrins activating signaling pathways that control permeability, cellular attachment, matrix metalloproteinase (MMP) expression, and cell survival. [14, 52-54].

In many disorders associated with elevated levels of cFn, a primary feature is predominant production and deposition of ECM proteins in affected tissue. The buildup of scar tissue eventually interferes with the normal functioning of the organ system gradually resulting in organ failure. Pathological effects of fibrosis have been observed in hepatic, pulmonary, and renal systems and reflect the role of underlying tissue specific factors unique to each disorder resulting in complex chronic injury [55]. Changes in the types of Fn isoforms expressed are also observed in tissue lesions. Marked increase in EDA and EDB positive Fn isoforms not normally expressed in a normal state is observed. A correlation between upregulation and accumulation of these isoforms, thickening of tissue, and phenotypic changes has been indicated [15]. Moreover increased pFn was also found to be significantly elevated in a study on patients with ischemic heart disease as compared to the control group. The increased pFn levels were positively correlated with elevated lipid component levels in plasma and blood pressure in these patients [56].

3.3. Bacterial Infiltration

In 1993 it was demonstrated that *Mycobacterium leprae* binds to β_1 integrin receptors after binding to soluble Fn [57]. It was then proposed that this interaction with Fn and β_1 integrin might be responsible for *M. leprae* invasion of nasal epithelial cells. *Staphylococcus aureus* will be used in this section as an example to discuss the topic of bacterial infiltration because of its interaction with Fn to infiltrate cells, its prevalence as a pathogenic cause of chronic infections in both humans and animals, and its relation to drug delivery discussed later in this chapter. As we will see in this section, it is just such an interaction as described above that enables *S. aureus* to infiltrate cells. The binding of *S. aureus* to Fn was demonstrated as early as 1978 [58].

S. aureus is documented as being able to enter nonprofessional phagocytes such as endothelial cells and epithelial cells, to escape from the endosomal vesicle, multiply within the cytoplasm, and induce apoptosis as illustrated by [8, 59] and references therein. Its ability to infiltrate nonprofessional phagocytes suggests that *S. aureus* must form high-affinity bonds with the host cell receptors. Four specific molecules were identified by Dziewanowska *et. al.* that play a key role in the high-affinity binding necessary for *S. aureus* to infiltrate nonprofessional phagocytes: Fn, β_1 integrin, heat shock protein 60 (Hsp60), and Fn binding protein (FnBP) anchored in the bacterial cell wall [60]. In this study Hep-2 cells deficient in Fn synthesis, and MAC-T cells expressing Fn were used. It was demonstrated that an addition of exogenous Fn stimulated uptake of the bacteria by the Fn deficient Hep-2 cells, while introduction of antibodies against Fn or β_1 integrin reduced internalization of the bacteria into MAC-T cells. Furthermore, introduction of Hsp60 antibodies blocking an interaction between FnBP and Hsp60 significantly reduced internalization of the organism [60]. In a supporting study it was shown that FnBP deficient mutants of the organism would not enter the cell while FnBP expression in non-invasive strains resulted in their internalization [61]. This supporting study specified $\alpha_5\beta_1$ integrin as the β_1 integrin involved, given that antibodies against α_5 integrin block cellular adhesion to Fn, and that $\alpha_5\beta_1$ integrin is known as the classic Fn receptor [9, 62].

From these results, it was determined that *S. aureus* achieves invasion by simultaneous binding of its FnBP to both Hsp60 and Fn/β_1 integrin, resulting in the strong binding to the host cell necessary for invasion of nonprofessional phagocytes. Two possible binding configurations are suggested by Dziewanowska *et. al.* for the bridging between the bacterium and host cell: one in which a single FnBP binds to Hsp60 and Fn simultaneously and another in which two distinct FnBPs interact with the two receptors [60]. The binding of *S. aureus* to host cell receptors results in signal transduction initiating the endocytotic internalization of the organism. The exact adhesion factors that are responsible for initiating the internalization mechanism is still an open question, however as we will see in the drug delivery section of this chapter there is evidence to suggest that the Fn/integrin interaction is responsible.

3.4. Cancer Metastasis

In addition to promoting cell adhesion to the extracellular matrix, Fn and integrins may also function in the control of proliferation, tumor cell migration, invasion, and metastasis [63]. Metastasis is a complex series of steps in which malignant cells leave the original tumor site and migrate to other parts of the body via the bloodstream or the lymphatic system. To do so, malignant cells break away from the primary tumor first. The epithelial to mesenchymal transition (EMT) which is characterized by the disruption of intercellular contacts and the enhancement of cell motility is an important feature in tumor metastasis. The loss of epithelial-like phenotype and acquisition of mesenchymal-like phenotype is critical for tumor invasion and dissemination in the earlier stages of metastasis. However, the reverse phenotype called MET (mesenchymal to epithelial transition) is more favorable for the disseminating cells to arrive in and settle down at their targeted new organs in the later stages. Regulated expression of cell surface molecules is needed for the EMT and MET processes, therefore several membranous or secreted proteins such as E-cadherin, cytokeratins, vimentin, MMP-3, MMP-9 and Fn are down- or up-regulated during metastasis [64].

The exact role of Fn in metastasis is still unclear; however there is increasing evidence for a correlation between Fn and cancer metastasis, with a few discrepancies. Fn protein was found to promote the haptotactic migration of murine melanoma cells via attachment to the subendothelial matrix [65, 66]. In a new metastatic cell line, SPC-A-1sci isolated from mouse pulmonary metastatic lesions, overexpression of Fn was associated with highly metastatic and aggressive behaviors of the cell line [67]. Downregulation of Fn decreased the metastatic abilities of SPC-A-1sci *in vitro* and *in vivo*. In B16F1 melanoma and 3LL lung carcinoma cell lines, elimination of pFn reduced lung metastasis by reversing the prometastatic effects of blood clotting and tumor cell integrin $\alpha_v\beta_3$ [68]. All these reports suggest that Fn overexpression may help cancer metastasis. In the same context, peptide and antibody inhibitors of Fn and integrin or Fn-mimicking peptides acting as a competitor of Fn successfully inhibited metastasis, suggesting Fn as an important target for cancer metastasis [63, 69].

In other studies, silencing of Fn expression was indicated to be one of the key mechanisms for the enhancement of metastasis [70-72]. When two murine mammary adenocarcinomas, M3 and MM3 were analyzed, the highly metastatic MM3 tumor (95% lung incidence) had small mRNA transcript and Fn protein levels compared to moderately metastatic M3 (40% incidence to lung) [70]. More studies focusing on the relationship between Fn and metastasis stages (EMT or MET) and on the underlying mechanism is necesary to better understand the role of Fn in cancer metastasis.

4. Applications of FN

4.1. Research Applications

Tissue Culture

Synthetic fragments of Fn can duplicate cell attachment activity [73]. This *in vivo* adhesion function has been exploited for routine *in vitro* cell culture, biomaterials and tissue engineering applications. Numerous studies conducted during the 1970s identified the effects of sera and various sera components on cell adhesion to surfaces for *in vitro* culturing [74-76]. Tissue cultureware coated with purified soluble Fn protein or vitronectin serve as anchorage for uniform attachment and spread of a variety of cell types such as fibroblasts, hepatocytes, hematopoietic stem/progenitor cells, and keratinocytes [77-80]. Some cell culture procedures replace serum containing media with serum free media either for economical reasons or to fulfill specific experimental requirements. In such instances, one of the approaches to maintain the optimal growth conditions for adherent cultures is to include certain additives such as adhesive factors, growth factors, or hormones in the medium to compensate for absence of the rich serum components. For many cell types, the presence of Fn as a coating on the tissue cultureware surface or as soluble protein in the media may be pertinent for initial adhesion of the cells to the growth surface without which they do not attach.

Other cell types may grow without Fn protein but overall morphology, physiology, and tissue specific gene expression may be optimally improved in the presence of Fn [79]. ECM proteins coated on cultureware not only adsorb to the biomaterial surfaces that are used to culture cells but they also act as ligands interacting with cell surface receptors activating signaling pathways for cell differentiation, growth, and survival [81-84].

Tissue culture techniques have played an immense role in advancing the understanding of the structure and function of the genetic and physiological aspects of cells and their components. Yet it is important to recognize that growth of cells on artificial substrates such as plastic or glass force the cells to grow as two dimensional (2D) sheets on flat and rigid surfaces. The cells within 2D culture system are devoid of the complex, dynamic, multi-cellular and cell-matrix three dimensional (3D) interactions naturally present in living tissue.

The components of extracellular matrices, scaffolds and proteins if included in the laboratory cell culture techniques provide *in vivo* like morphology and physiologically relevant environments with dynamic cell-cell and cell-matrix interactions. Moreover the types of integrins and adhesion proteins expressed on the cell surface and the *in vitro* signaling pathways that are activated are strongly affected by the type of surface the cells are cultured on; therefore growth of cells in 3D gels or matrices such as collagen, Fn, vitronectin and gelatin alone or as a mixture has been recognized as an important advancement towards mimicking normal cell behavior in various cell systems. Numerous examples of the use of 3D matrices and gels to culture a variety of cell types can be found in the literature [85-90]. Affinity chromatography purified Fn derived from human and bovine plasma is commercially available either alone or in combination with other ECM and basement components for cell culture applications. Fn in 3D gels has found important applications in the study of Fn fibrillogenesis [52] and cancer biology [91].

Adhesion Assays

Cell adhesion assays are commercially available for rapid, quantitative and reliable assessment of factors that influence cell-matrix interactions. The delineation of these interactions is integral to understanding not only the basic events involved in maintaining normal cell physiology but also the complex dynamics of tumor invasion and metastasis in cancer biology. For determination of adherent cell interaction with Fn, cells are seeded onto the Fn coated plates and cultured for a desired time period followed by determination of cell adhesion using a fluorometric or colorimetric assay.

4.2. Clinical Applications

4.2.1. Biomarker for Disease Diagnosis

Renal Conditions

Glomerular extracellular matrix consists largely of Fn. Variation in the levels of Fn excreted in urine is a useful clinical marker for assessment of pathologic renal conditions such as fibrosis in nephritis patients. High amounts of Fn is reported to reflect severity of kidney fibrogenesis [92, 93].

Cancer

Urinary Fn fragmented forms are known to be increased in patients with prostate cancer as compared to normal subjects. Urinary Fn is observed to be present as variable sized fragments containing the central cell-binding domain. Monoclonal antibodies specific against this Fn domain are used for *in vitro* antibody assays for quantitative measurements [94, 95]. Studies have also been done to evaluate the clinical usefulness of measurements of pFn in sera from patients with malignant disease [96, 97].

Commercial *in vitro* enzyme immunoassay (EIA) kits designed for specific quantitative determination of soluble human Fn in serum, urine, cell culture supernatants and other biological fluids are available for such measurements.

Preterm Labor

Fetal Fn is found in amniotic fluid, placental tissue and the extracellular substance of deciduas basalis. Mechanical or inflammatory damage to the membranes or placenta before birth can release Fn, an important marker to clinically predict spontaneous preterm birth. Birth before 37 weeks gestation is regarded as preterm and occurs in 7-11% of pregnancies and in 3-4% of pregnancies before 34 weeks gestation. Infants born at 34 weeks or earlier often have life threatening medical conditions [98, 99]. It is proposed that the use of fetal Fn detection can identify those women that are not at immediate risk and thereby treatment measures can be focused on the ones that seriously need attention.

Cervicovaginal fetal Fn testing involves taking ectocervix swabs and performing an enzyme-linked immunosorbent assay (ELISA) containing FDC-6 monoclonal antibody to detect fetal Fn. The samples are processed to detect the presence or absence of $\geq$ 50ng/mL of fetal Fn and is interpreted to predict the risk of preterm labor [98, 100].

Alcoholic Liver Disease

Fn is the most abundant matrix protein in the liver given that it is synthesized by hepatocytes. cFn primarily existing as bundles connected to the microvilli of hepatocytes in the space of Disse is present at very low levels throughout the liver.

Under normal physiological conditions, cFn and pFn are present at similarly balanced levels and unusually elevated levels of cFn are indicative of underlying pathological tissue condition. Studies on patients with chronic liver damage report a considerable increase in patient blood plasma levels of cFn. Higher levels of cFn may have a role in the onset of progressive damage. This has initiated considerable attention in the use of cFn as a potential biomarker for diagnostic tests [55].

Cardiovascular Disease

Increased expression of specific Fn isoforms normally absent in adult cardiovascular tissue is associated with development and progression of various cardiovascular pathologies.

One of the clinically useful treatment strategies is to use therapeutic antibodies and peptide targeting drugs that are highly specific and effective. Abciximab, a humanized mouse Fab monoclonal antibody against integrins such as $\alpha_{IIb}\beta_3$, $\alpha_v\beta_3$ that efficiently blocks platelet aggregation is clinically utilized during percutaneous coronary intervention (PCI) procedures for the treatment of myocardial infarction and acute coronary syndromes. Additional treatment strategies are suggested specifically targeting intervention of Fn-cell adhesion and Fn deposition pathways in lesion tissues [15, 24, 101].

4.2.2. Wound Repair

As mentioned earlier in the chapter, Fn plays an important role during the initial stages of wound repair and the healing process. Wounds with decreased healing rate extending over a period of 3 months or longer and those that do not follow the normal wound healing process despite optimum wound care are described as chronic wounds. Difficult-to-heal wounds such as venous stasis, pressure ulcers, and diabetic wounds may over time evolve to become a chronic wound [102-104]. Roughly 5 million people in US alone are afflicted with pressure and leg ulcers [105]. Considerable efforts are focused on developing treatment strategies to combat the defective healing process. One of the significant factors contributing to defective chronic wounds is considered to be an extracellular matrix deficient state. As mentioned earlier in the chapter, formation of a functional Fn matrix at an initial stage is critical for subsequent events in the normal wound healing process. In a pathological state at the wound site ECM components may be present at decreased levels or in a defective state. This may be due to prolonged inflammation resulting from increased pro-inflammatory cytokines that produce MMP. These MMPs cleave ECM components such as Fn, collagen and hyaluronic acid into peptide fragments. These accumulated fragments at the wound site are hypothetically unavailable for matrix formation. Moreover these fragments can also act as chemotactic peptides further increasing the inflammation interfering with the healing process [106-108].

One approach to stimulate the healing process is to replenish the defective or reduced levels of ECM components at the wound site. Besides collagen which is available for topical application on wounds, the following have also been under investigation for stimulating wound repair: Fn in eye drops, wound dressings soaked in Fn, synthetic polymeric substrates containing biologically active Fn fragments functioning as adhesive and cell migration substrates, and high affinity Fn peptides [105, 109]. Fn has about 12 integrin binding sites referred to as "cell binding sites" that can be exploited clinically to act as biologically active short peptides functioning as cell migration substrates, activating integrin signaling [110] and cell adhesion pathways. These peptides either alone or in combination can be useful for wound repair and tissue engineering [105].

4.3. Drug Delivery

The ability of *S. aureus* to infiltrate nonprofessional phagocytes via a Fn bridge between the organism and host cell receptors inspired a study into the use of Fn to internalize nanomaterial drug carriers into human and bovine epithelial cells [111, 112]. In the initial study, silica-based nanowires (NWs) were coated with Fn in an attempt to internalize them into epithelial cells [111]. It was determined that Fn would bind to the nanowires via an interaction with surface hydroxyls, however a specific binding site of the Fn protein involved in this binding was never identified [113]. The role of FnBP in the infiltration of *S. aureus* is important; however, internalization of Fn-coated NWs seems to be mediated by endocytosis of soluble Fn in turn mediated by $\alpha_5\beta_1$ integrin during the Fn turnover process. Thus this approach makes the assumption that the interaction between FnBP and Hsp60 is not critical for bacterial internalization. The successful internalization of Fn coated NWs would suggest that the interaction between FnBP and Hsp60 serves more to strengthen the binding of *S. aureus* to host cell receptors or to induce signal transduction for bacteria-favorable circumstance, rather than to induce endocytosis. After demonstrating that Fn coated NWs could be internalized by MAC-T cells, they were utilized as carriers to deliver a subunit of *Escherichia Coli* Shiga-like toxin into both MAC-T and Hep-2 cells. [112, 114] Shiga toxin is an AB_5 molecule in which the five B subunits are responsible for receptor binding and internalization of the toxin, and the A subunit is a toxin that inhibits eukaryotic protein synthesis by catalytically inactivating the 60S ribosomal subunits, resulting in cell death. By isolating the toxic A subunit from the B subunit counterparts prior to internalization, the toxic A subunit was unable to enter cells on its own. Only when attached to the Fn coated NWs was the toxin able to infiltrate the cells.

This demonstrated ability to deliver a toxic agent into cells with target ability could have potential applications for immunotherapy of cancer. However it was later pointed out that the Fn coating does not provide the level of specificity needed to target specific cell types, and replacement of Fn with a more specific target ligand should be identified for targeted drug delivery applications [115]. Replacing Fn with an RGD sequence could maintain the cell binding function while reducing volume and saving space for antibody coatings to enhance cell specificity. As an example, mRNA gene delivery [116] might find advantages in an RGD/antibody mediated delivery since, like the Shiga-like toxin I A subunit, mRNA is required to reach the ribosomes in the cytoplasm to be effective.

Conclusion

Since the discovery of Fn in the 1970s as a cold insoluble globulin [117], great progress has been made in the understanding of the structural features of Fn and its isoforms that enable this protein to participate in diverse *in vivo* functions. The multiple binding sites on the Fn polypeptide are excluded or included in variant isoforms of the protein. These isoforms are differentially expressed depending on the species, developmental stage and tissue type, and primarily govern the function of this protein. The expression of specific Fn variants is tightly regulated at genetic levels and by expression of various growth factors, hormones and cytokines. Deficiency or expression of defective Fn forms or emergence of forms that are not usually expressed at a developmental stage disrupts normal tissue functioning. These observations strongly suggest that Fn is indispensible for maintaining normal cell functions such as ECM assembly, blood vessel formation, wound healing, cell adhesion, migration, and growth, and highlight its role in pathological conditions. Years of intense research into Fn function has shed light on its potential as an important disease marker for diagnostic and therapeutic purposes and as a ligand for drug delivery. Although much yet remains to be explored in the arena of Fn targeted therapeutics current strategies developed in the last two decades have proven to be clinically useful. Continued investigation into the mechanisms and functions of this vastly versatile protein may lead to a better understanding of its role in metastasis and potentially a number of other physiological conditions and processes.

References

[1] Hynes, R.O., Alteration of cell-surface proteins by viral transformation and by proteolysis. *Proceedings of the National Academy of Sciences of the United States of America*, 1973. 70(11): p. 3170-4.

[2] Longtin, R., Birthday of a breakthrough: fibronectin research proves important, but not as originally expected. *Journal of the National Cancer Institute,* 2004. 96(1): p. 6-8.

[3] Ruoslahti, E., Fibronectin in Cell-Adhesion and Invasion. *Cancer and Metastasis Reviews*, 1984. 3(1): p. 43-51.

[4] Yamada, K.M., Fibronectin peptides in cell migration and wound repair. *The Journal of clinical investigation*, 2000. 105(11): p. 1507-9.

[5] Grinnell, F., J.R. Head, and J. Hoffpauir, Fibronectin and cell shape in vivo: studies on the endometrium during pregnancy. *The Journal of cell biology,* 1982. 94(3):p.597-606.

[6] Sottile, J., D.C. Hocking, and P.J. Swiatek, Fibronectin matrix assembly enhances adhesion-dependent cell growth. *Journal of cell science*, 1998. 111 (Pt 19): p. 2933-43.

[7] Moursi, A.M., et al., Fibronectin regulates calvarial osteoblast differentiation. *Journal of cell science*, 1996. 109 (Pt 6): p. 1369-80.

[8] Bayles, K.W., et al., Intracellular Staphylococcus aureus escapes the endosome and induces apoptosis in epithelial cells. *Infection and Immunity,* 1998. 66(1): p. 336-42.

[9] Milam, S.B., et al., Cell adhesion proteins in oral biology. *Critical reviews in oral biology and medicine : an official publication of the American Association of Oral Biologists,* 1991. 2(4): p. 451-91.

[10] Zardi, L., et al., Species-specific monoclonal antibodies in the assignment of the gene for human fibronectin to chromosome 2. *EMBO J., 1982*. 1(8): p. 929-33.

[11] Hynes, R.O. and Q. Zhao, The evolution of cell adhesion. *J. Cell Biol.,* 2000. 150(2): p. F89-96.

[12] George, E.L., et al., Defects in mesoderm, neural tube and vascular development in mouse embryos lacking fibronectin. *Development,* 1993. 119(4): p. 1079-91.

[13] Castelletti, F., et al., Mutations in FN1 cause glomerulopathy with fibronectin deposits. *Proc. Natl. Acad. Sci. U S A,* 2008. 105(7): p. 2538-43.

[14] Faralli, J.A., et al., Functional properties of fibronectin in the trabecular meshwork. *Exp. Eye Res.,* 2009. 88(4): p. 689-93.

[15] Magnusson, M.K. and D.F. Mosher, Fibronectin: structure, assembly, and cardiovascular implications. *Arterioscler. Thromb. Vasc. Biol.,* 1998. 18(9): p. 1363-70.

[16] Polin, R.A., Role of fibronectin in diseases of newborn infants and children. *Rev. Infect. Dis.,* 1990. 12 Suppl 4: p. S428-38.

[17] Sekiguchi, K. and S. Hakomori, Functional domain structure of fibronectin. *Proc. Natl. Acad. Sci. U S A,* 1980. 77(5): p. 2661-5.

[18] Wagner, D.D. and R.O. Hynes, Topological arrangement of the major structural features of fibronectin. *J. Biol. Chem.*,1980.255(9): p.4304-12.
[19] Hynes, R.O., *Fibronectins.* 1st ed. 1990: Springer-Verlag: New York.
[20] Carnemolla, B., et al., The inclusion of the type III repeat ED-B in the fibronectin molecule generates conformational modifications that unmask a cryptic sequence. *J. Biol. Chem.*, 1992. 267(34): p. 24689-92.
[21] Potts, J.R. and I.D. Campbell, Structure and function of fibronectin modules. *Matrix Biol.,* 1996. 15(5): p. 313-20; discussion 321.
[22] White, E.S., F.E. Baralle, and A.F. Muro, New insights into form and function of fibronectin splice variants. *J. Pathol.,* 2008. 216(1): p. 1-14.
[23] Rimoldi, M.T., et al., Complement component C1q enhances invasion of human mononuclear phagocytes and fibroblasts by Trypanosoma cruzi trypomastigotes. *J. Clin. Invest.*, 1989. 84(6): p. 1982-9.
[24] Millard, M., S. Odde, and N. Neamati, Integrin targeted therapeutics. *Theranostics*, 2011. 1: p. 154-88.
[25] Bloom, L., K.C. Ingham, and R.O. Hynes, Fibronectin regulates assembly of actin filaments and focal contacts in cultured cells via the heparin-binding site in repeat III13. *Mol. Biol. Cell,* 1999. 10(5): p. 1521-36.
[26] Muro, A.F., et al., Regulated splicing of the fibronectin EDA exon is essential for proper skin wound healing and normal lifespan. *J. Cell Biol.,* 2003. 162(1): p. 149-60.
[27] Tan, M.H., et al., Deletion of the alternatively spliced fibronectin EIIIA domain in mice reduces atherosclerosis. *Blood,* 2004. 104(1): p. 11-8.
[28] Chauhan, A.K., et al., Impaired motor coordination in mice lacking the EDA exon of the fibronectin gene. *Behav. Brain Res.,* 2005. 161(1): p. 31-8.
[29] Norton, P.A. and R.O. Hynes, Alternative splicing of chicken fibronectin in embryos and in normal and transformed cells. *Mol. Cell Biol.,* 1987. 7(12): p. 4297-307.
[30] Schwarzbauer, J.E., et al., Three different fibronectin mRNAs arise by alternative splicing within the coding region. *Cell,* 1983. 35(2 Pt 1): p. 421-31.
[31] Umezawa, K., A.R. Kornblihtt, and F.E. Baralle, Isolation and characterization of cDNA clones for human liver fibronectin. *FEBS Lett.,* 1985. 186(1): p. 31-4.
[32] Singh, P., C. Carraher, and J.E. Schwarzbauer, Assembly of fibronectin extracellular matrix. *Annu. Rev. Cell Dev. Biol.,* 2010. 26: p. 397-419.
[33] Dean, D.C., Expression of the fibronectin gene. *Am. J. Respir. Cell Mol. Biol.,* 1989. 1(1): p. 5-10.
[34] Ffrench-Constant, C., et al., Reappearance of an embryonic pattern of fibronectin splicing during wound healing in the adult rat. *J. Cell Biol.,* 1989. 109(2): p. 903-14.
[35] Magnuson, V.L., et al., The alternative splicing of fibronectin pre-mRNA is altered during aging and in response to growth factors. *J. Biol. Chem.,* 1991. 266(22): p. 14654-62.
[36] Pankov, R. and K.M. Yamada, Fibronectin at a glance. *J. Cell Sci.,* 2002. 115(Pt 20): p. 3861-3.
[37] McDonald, J.A., Extracellular matrix assembly. *Annu. Rev. Cell Biol.,* 1988. 4: p. 183-207.
[38] Huveneers, S., et al., Binding of soluble fibronectin to integrin alpha5 beta1 - link to focal adhesion redistribution and contractile shape. *Journal of cell science*, 2008. 121(Pt 15): p. 2452-62.

[39] Zhong, C., et al., Rho-mediated contractility exposes a cryptic site in fibronectin and induces fibronectin matrix assembly. *J. Cell Biol.,* 1998. 141(2): p. 539-51.

[40] Hocking, D.C. and K. Kowalski, A cryptic fragment from fibronectin's III1 module localizes to lipid rafts and stimulates cell growth and contractility. *J. Cell Biol.,* 2002. 158(1): p. 175-84.

[41] Wierzbicka-Patynowski, I., Y. Mao, and J.E. Schwarzbauer, Continuous requirement for pp60-Src and phospho-paxillin during fibronectin matrix assembly by transformed cells. *J. Cell Physiol.,* 2007. 210(3): p. 750-6.

[42] Sottile, J. and D.C. Hocking, Fibronectin polymerization regulates the composition and stability of extracellular matrix fibrils and cell-matrix adhesions. *Mol. Biol. Cell,* 2002. 13(10): p. 3546-59.

[43] Shi, F. and J. Sottile, Caveolin-1-dependent beta1 integrin endocytosis is a critical regulator of fibronectin turnover. *Journal of cell science,* 2008. 121(Pt 14): p. 2360-71.

[44] Sottile, J. and J. Chandler, Fibronectin matrix turnover occurs through a caveolin-1-dependent process. *Mol. Biol. Cell,* 2005. 16(2): p. 757-68.

[45] Juliano, R.L. and S. Haskill, Signal transduction from the extracellular matrix. *J. Cell Biol.,* 1993. 120(3): p. 577-85.

[46] El Ghalbzouri, A., E. Lamme, and M. Ponec, Crucial role of fibroblasts in regulating epidermal morphogenesis. *Cell Tissue Res.,* 2002. 310(2): p. 189-99.

[47] Badylak, S.F., The extracellular matrix as a scaffold for tissue reconstruction. *Semin. Cell Dev. Biol.,* 2002. 13(5): p. 377-83.

[48] Wysocki, A.B. and F. Grinnell, Fibronectin profiles in normal and chronic wound fluid. *Lab. Invest.,* 1990. 63(6): p. 825-31.

[49] Nakagawa, S., P. Pawelek, and F. Grinnell, Extracellular matrix organization modulates fibroblast growth and growth factor responsiveness. *Exp. Cell Res.,* 1989. 182(2): p. 572-82.

[50] Isemura, M., et al., Distribution of fibronectin and other connective tissue components in human placenta. *Tohoku. J. Exp. Med.,* 1985. 145(4): p. 373-9.

[51] Sakai, T., et al., Plasma fibronectin supports neuronal survival and reduces brain injury following transient focal cerebral ischemia but is not essential for skin-wound healing and hemostasis. *Nat. Med.,* 2001. 7(3): p. 324-30.

[52] Mao, Y. and J.E. Schwarzbauer, Stimulatory effects of a three-dimensional microenvironment on cell-mediated fibronectin fibrillogenesis. *J. Cell Sci.,* 2005. 118(Pt 19): p. 4427-36.

[53] Li, A.F., N. Tane, and S. Roy, Fibronectin overexpression inhibits trabecular meshwork cell monolayer permeability. *Mol. Vis.,* 2004. 10: p. 750-7.

[54] Zhou, L., et al., Expression of integrin receptors in the human trabecular meshwork. *Curr. Eye Res.,* 1999. 19(5): p. 395-402.

[55] Aziz-Seible, R.S. and C.A. Casey, Fibronectin: Functional character and role in alcoholic liver disease. *World J. Gastroenterol.,* 2011. 17(20): p. 2482-99.

[56] Song, K.S., et al., Plasma fibronectin levels in ischemic heart disease. *Atherosclerosis,* 2001. 154(2): p. 449-53.

[57] Byrd, S.R., R. Gelber, and L.E. Bermudez, Roles of Soluble Fibronectin and Beta-1-Integrin Receptors in the Binding of Mycobacterium-Leprae to Nasal Epithelial-Cells. *Clinical Immunology and Immunopathology,* 1993. 69(3): p. 266-271.

[58] Kuusela, P., Fibronectin binds to Staphylococcus aureus. *Nature,* 1978. 276(5689): p. 718-20.

[59] Dziewanowska, K., et al., Fibronectin binding protein and host cell tyrosine kinase are required for internalization of Staphylococcus aureus by epithelial cells. *Infection and Immunity*, 1999. 67(9): p. 4673-4678.

[60] Dziewanowska, K., et al., Staphylococcal fibronectin binding protein interacts with heat shock protein 60 and integrins: Role in internalization by epithelial cells. *Infection and Immunity*, 2000. 68(11): p. 6321-6328.

[61] Sinha, B., et al., Fibronectin-binding protein acts as Staphylococcus aureus invasin via fibronectin bridging to integrin a5b1. *Cellular Microbiology*, 1999. 1(2): p. 101-117.

[62] Pistone, M., et al., Integrin synthesis and utilization in cultured human osteoblasts. *Cell Biology International*, 1996. 20(7): p. 471-479.

[63] Akiyama, S.K., K. Olden, and K.M. Yamada, Fibronectin and integrins in invasion and metastasis. *Cancer Metastasis Rev.,* 1995. 14(3): p. 173-89.

[64] Zhang, W., et al., Four-and-a-half LIM protein 2 promotes invasive potential and epithelial-mesenchymal transition in colon cancer. *Carcinogenesis,* 2010. 31(7): p. 1220-9.

[65] Nicolson, G.L., et al., The role of fibronectin in adhesion of metastatic melanoma cells to endothelial cells and their basal lamina. *Exp. Cell Res.,* 1981. 135(2): p. 461-5.

[66] McCarthy, J.B. and L.T. Furcht, Laminin and fibronectin promote the haptotactic migration of B16 mouse melanoma cells in vitro. *The Journal of cell biology*, 1984. 98(4): p. 1474-80.

[67] Jia, D., et al., Development of a highly metastatic model that reveals a crucial role of fibronectin in lung cancer cell migration and invasion. *BMC Cancer*, 2010. 10: p. 364.

[68] Malik, G., et al., Plasma fibronectin promotes lung metastasis by contributions to fibrin clots and tumor cell invasion. *Cancer Res*., 2010. 70(11): p. 4327-34.

[69] Humphries, M.J., K. Olden, and K.M. Yamada, A synthetic peptide from fibronectin inhibits experimental metastasis of murine melanoma cells. *Science,* 1986. 233(4762): p. 467-70.

[70] Werbajh, S.E., et al., Downregulation of fibronectin transcription in highly metastatic adenocarcinoma cells. *FEBS Lett.,* 1998. 440(3): p. 277-81.

[71] Urtreger, A., et al., Expression of RGD minus fibronectin that does not form extracellular matrix fibrils is sufficient to decrease tumor metastasis. *Int. J. Cancer*, 1998. 78(2): p. 233-41.

[72] Akamatsu, H., et al., Suppression of transformed phenotypes of human fibrosarcoma cells by overexpression of recombinant fibronectin. *Cancer Res.*,1996.56(19): p.4541-6.

[73] Pierschbacher, M.D. and E. Ruoslahti, Cell Attachment Activity of Fibronectin Can Be Duplicated by Small Synthetic Fragments of the Molecule. *Nature,* 1984. 309(5963): p. 30-33.

[74] Hynes, R.O., Cell surface proteins and malignant transformation. *Biochim. Biophys. Acta,* 1976. 458(1): p. 73-107.

[75] Weston, J.A. and K.L. Hendricks, Reversible transformation by urea of contact-inhibited fibroblasts. *Proc. Natl. Acad. Sci. U S A,* 1972. 69(12): p. 3727-31.

[76] Yamada, K.M. and K. Olden, Fibronectins--adhesive glycoproteins of cell surface and blood. *Nature*, 1978. 275(5677): p. 179-84.

[77] Gilchrest, B.A., R.E. Nemore, and T. Maciag, Growth of human keratinocytes on fibronectin -coated plates. *Cell Biol. Int. Rep.*, 1980. 4(11): p. 1009-16.

[78] Yokota, T., et al., Growth-supporting activities of fibronectin on hematopoietic stem/progenitor cells in vitro and in vivo: structural requirement for fibronectin activities of CS1 and cell-binding domains. *Blood,* 1998. 91(9): p. 3263-72.

[79] Sanchez, A., et al., Fibronectin regulates morphology, cell organization and gene expression of rat fetal hepatocytes in primary culture. *J. Hepatol.,* 2000. 32(2): p. 242-50.

[80] Zijlstra, A. and M.E. Schelling, Detection of multiple fibronectin isoforms in fetal fibroblast monolayer culture: a novel method for the qualitative and quantitative detection of multiple antigens. *Histochem. Cell Biol.,* 1999. 111(2): p. 163-9.

[81] Detrait, E., et al., Fibronectin-pluronic coadsorption on a polystyrene surface with increasing hydrophobicity: relationship to cell adhesion. *J. Biomed. Mater Res.,* 1999. 45(4): p. 404-13.

[82] Steele, J.G., et al., Adsorption of fibronectin and vitronectin onto Primaria and tissue culture polystyrene and relationship to the mechanism of initial attachment of human vein endothelial cells and BHK-21 fibroblasts. *Biomaterials,* 1995. 16(14): p. 1057-67.

[83] Steele, J.G., G. Johnson, and P.A. Underwood, Role of serum vitronectin and fibronectin in adhesion of fibroblasts following seeding onto tissue culture polystyrene. *J. Biomed. Mater Res.,* 1992. 26(7): p. 861-84.

[84] Rainaldi, G., et al., Fibronectin facilitates adhesion of K562 leukemic cells normally growing in suspension to cationic surfaces. J. *Biomed. Mater Res.,* 2001. 55(1): p. 104-13.

[85] Grinnell, F., Fibroblast biology in three-dimensional collagen matrices. *Trends Cell Biol.,* 2003. 13(5): p. 264-9.

[86] Li, M.L., et al., Influence of a reconstituted basement membrane and its components on casein gene expression and secretion in mouse mammary epithelial cells. *Proc. Natl. Acad. Sci. U S A,* 1987. 84(1): p. 136-40.

[87] Weaver, V.M., et al., Reversion of the malignant phenotype of human breast cells in three-dimensional culture and in vivo by integrin blocking antibodies. *J. Cell Biol.*, 1997. 137(1): p. 231-45.

[88] Barcellos-Hoff, M.H., et al., Functional differentiation and alveolar morphogenesis of primary mammary cultures on reconstituted basement membrane. *Development,* 1989. 105(2): p. 223-35.

[89] Lang, S.H., et al., Prostate epithelial cell lines form spheroids with evidence of glandular differentiation in three-dimensional Matrigel cultures. *Br. J. Cancer,* 2001. 85(4): p. 590-9.

[90] Cukierman, E., R. Pankov, and K.M. Yamada, Cell interactions with three-dimensional matrices. *Curr. Opin. Cell Biol.,* 2002. 14(5): p.633-9.

[91] Winters, B.S., et al., Three-dimensional culture regulates Raf-1 expression to modulate fibronectin matrix assembly. *Mol. Biol. Cell,* 2006. 17(8): p. 3386-96.

[92] Gwinner, W., I. Jackle-Meyer, and H. Stolte, Origin of urinary fibronectin. *Lab. Invest.,* 1993. 69(2): p. 250-5.

[93] Kozlovskaia, L.V., et al., [Urinary fibronectin as an indicator of kidney fibrosis in nephritis]. *Ter. Arkh.,* 1999. 71(6): p. 34-8.

[94] Katayama, M., et al., Urinary fibronectin fragments (a potential tumor marker) measured by immunoenzymometric assay with domain-specific monoclonal antibodies. *Clin. Chem*., 1991. 37(3): p. 466-71.
[95] Katayama, M., et al., Increased fragmentation of urinary fibronectin in cancer patients detected by immunoenzymometric assay using domain-specific monoclonal antibodies. *Clin. Chim. Acta*,1993.217(2): p.115-28.
[96] Choate, J.J. and D.F. Mosher, Fibronectin concentration in plasma of patients with breast cancer, colon cancer, and acute leukemia. *Cancer,* 1983. 51(6): p. 1142-7.
[97] Ruelland, A., et al., Level of plasma fibronectin in patients with breast cancer. *Clin. Chim. Acta,* 1988. 178(3): p. 283-7.
[98] Honest, H., et al., Accuracy of cervicovaginal fetal fibronectin test in predicting risk of spontaneous preterm birth: systematic review. *BMJ,* 2002. 325(7359): p. 301.
[99] Stewart, A.L., et al., Brain structure and neurocognitive and behavioural function in adolescents who were born very preterm. *Lancet,* 1999. 353(9165): p. 1653-7.
[100] Matsuura, H., et al., The oncofetal structure of human fibronectin defined by monoclonal antibody FDC-6. Unique structural requirement for the antigenic specificity provided by a glycosylhexapeptide. *J. Biol. Chem.,* 1988. 263(7): p. 3314-22.
[101] Coller, B.S., Platelet GPIIb/IIIa antagonists: the first anti-integrin receptor therapeutics. *J. Clin. Invest*., 1997. 100(11 Suppl): p. S57-60.
[102] Loots, M.A., et al., Differences in cellular infiltrate and extracellular matrix of chronic diabetic and venous ulcers versus acute wounds. *J. Invest. Dermatol.,* 1998. 111(5): p. 850-7.
[103] Tarnuzzer, R.W. and G.S. Schultz, Biochemical analysis of acute and chronic wound environments. *Wound Repair Regen.,* 1996.4(3): p.321-5.
[104] Trengove, N.J., et al., Analysis of the acute and chronic wound environments: the role of proteases and their inhibitors. *Wound Repair Regen.,* 1999. 7(6): p. 442-52.
[105] Yamada, K.M., Fibronectin peptides in cell migration and wound repair. *J. Clin. Invest.,* 2000. 105(11): p. 1507-9.
[106] Robson, M.C., D.L. Steed, and M.G. Franz, Wound healing: biologic features and approaches to maximize healing trajectories. *Curr. Probl. Surg.,* 2001. 38(2): p. 72-140.
[107] Cook, H., et al., Defective extracellular matrix reorganization by chronic wound fibroblasts is associated with alterations in TIMP-1, TIMP-2, and MMP-2 activity. *J. Invest. Dermatol.,* 2000. 115(2): p. 225-33.
[108] Trengove, N.J., H. Bielefeldt-Ohmann, and M.C. Stacey, Mitogenic activity and cytokine levels in non-healing and healing chronic leg ulcers. *Wound Repair Regen*., 2000. 8(1): p. 13-25.
[109] Livant, D.L., et al., The PHSRN sequence induces extracellular matrix invasion and accelerates wound healing in obese diabetic mice. *J. Clin. Invest.,* 2000. 105(11): p. 1537-45.
[110] Giancotti, F.G. and E. Ruoslahti, Integrin signaling. *Science,* 1999. 285(5430): p. 1028-32.
[111] Beaux, M.F., et al., Fibronectin Bonding to Nanowires and Their Internalization by Epithelial Cells. *Journal of Biomedical Nanotechnology,* 2006. 2(1): p. 23-28.

[112] Kwon, N.H., et al., Nanowire-based delivery of Escherichia coli O157 shiga toxin 1 A subunit into human and bovine cells. *Nano letters*, 2007. 7(9): p. 2718-23.
[113] Beaux, M.F., et al., X-ray photoelectron spectroscopic analysis of the surface chemistry of silica nanowires. *Applied Surface Science*, 2011. 257(13): p. 5766-5771.
[114] Hovde, C.J., et al., Evidence That Glutamic Acid-167 Is an Active-Site Residue of Shiga-Like Toxin-I. *Proceedings of the National Academy of Sciences of the United States of America*, 1988. 85(8): p. 2568-2572.
[115] Beaux, M.F., 2nd, D.N. McIlroy, and K.E. Gustin, Utilization of solid nanomaterials for drug delivery. *Expert opinion on drug delivery*, 2008. 5(7): p. 725-35.
[116] Yamamoto, A., et al., Current prospects for mRNA gene delivery. *European Journal of Pharmaceutics and Biopharmaceutics*, 2009. 71(3): p. 484-489.
[117] Dessau, W., B.C. Adelmann, and R. Timpl, Identification of the sites in collagen alpha-chains that bind serum anti-gelatin factor (cold-insoluble globulin). *Biochem. J.,* 1978. 169(1): p. 55-9.

In: Fibronectin
Editors: James Beattie

ISBN 978-1-62257-298-4

Chapter II

FN Modulation of Growth Factor Activity

James Beattie*
Dept of Oral Biology. Leeds Dental Institute,
University of Leeds, Leeds, UK

Abstract

Fibronectin (FN) is a glycoprotein constituent of the extra-cellular matrix (ECM) which has an important role in regulation of many aspects of cell physiology. These functions are mediated through interaction with a diverse range of biomolecules including cell surface integrin receptors, growth factors/receptors and ECM components such as heparin and fibrin. In this capacity FN can modulate cell responses to changes in the pericellular environment (outside-in signalling). In addition increasing evidence suggests that intracellular signalling can affect the organisation and function of ECM components –including FN (inside-out signalling). This chapter aims to summarise recent findings in this area and to suggest some future directions that research may take. In doing so we will focus on the area of growth factor/receptor interaction with FN bound integrins and the signalling events associated with formation of such complexes.

FN Gene and Protein

The human FN gene is approximately 82 kb in length and contains 46 exons located on chromosome 2 (2q34). Processing of FN mRNA by differential splicing at exons denoted EDA, EDB and the variable region IICSresults in 20 potential isoforms of the protein [1]. Soluble plasma FN (pFN) is secreted mainly by hepatocytes and does not contain the EDA and EDB exons which exist in the insoluble cell associated FN (cFN) secreted by fibroblasts in various tissues. It is this latter form which signals through surface integrin receptors. FN

* E-mail: J.Beattie@leeds.ac.uk.

monomers are linked through a disulphide bond at the C-terminus to form dimers. Post-translational N- and O-linked glycosylation, together with serine and threonine phosphorylation [2-4] effects the function of FN. This includes the interaction including interactions of this ECM protein with normal and cancerous cells [5, 6]. A FN network is formed through isopeptide bonds between glutamic acid and the ε amino groups of lysine residues in the N-terminal region of the molecule. The association of cFN with integrins is important for this process and is reviewed in [7].

FN and Integrins

FNcontains Type I, II and III modules which provide binding sites for other ECM components[8-13].Probably the most important targets for FN are the cell surface integrins.and Tthis binding serves to provide a physical and biological link between cells and ECM.Integrins are cell surface hetero-dimeric proteins which are associated non-covalently. They comprise α and β subunits whose extracellular domainsprovide a binding site for FN. ManySstudies have identified a "synergy site" and an RGD peptide loop in the 9th and 10th Type III FN domains (FNIII 9-10) respectively as important motifs for integrin binding. This region of FN has been termed the cell binding domain (CBD) [14-21] .]. Evidence also suggests that the FNIII 8 domain may also contribute to the structural integrity of FN-CBD [22].Although α5β1 and αvβ3 integrins are the best characterised receptors, FN can bind to at least 11 integrin heterodimers in a cell and tissue specific manner. This interaction of FN with integrin receptors plays a crucial role in the process of "cross-talk" between integrin and GF receptors and the description of some of the mechanisms involved in this process forms the main part of this review. We describe first the interaction of FN with growth factors themselves and then conclude with some comments on intracellular signalling pathways involved in subsequent GF receptor - integrin signalling.

FN and Growth Factors

In common with other components of the extracellular matrix FN binds to growth factors. This serves to partition GFs between cell surface receptors, ECM and interstitial fluid and has important implications for GF activity. FN binds GFs from thePDGF, CTGF, HGF, VEGF, TGF, FGF and IGF families although specificity of binding exists amongst individual members of these families [23-30]. Recent evidence suggests that FN domains III 12-14 forms an important and promiscuous binding surface for many GFs [31, 32]. The functional consequences of GF-FN association are diverse and dependant on GF, tissue/cell type and biological endpoint. Association of growth factors with ECM can also affect the qualitative outcomes of subsequent signalling events. One recent study has reported that matrix association of VEGF confers altered biological properties for the growth factor when compared with soluble VEGF [33]. Thus ECM bound VEGF was associated with a differential and prolonged tyrosine phosphorylation (at Y1214) of VEGFR-2 in transfected porcine aortic endothelial cells. This resulted in activation of different signalling pathways when compared to soluble VEGF and the effect of ECMgel bound VEGF was dependent on the presence of β1 integrin expression. Although in this instance the growth factor was associated with Type I collagen collagengels, as indicated above, VEGF also binds to fibronectin and it would be of interest to examine whether such effects were reproducible with an alternative ECM substrate. A very recent study has analysed in detail the binding of PDGF-BB to different FN domains [34] located both N-terminal and C-terminal to the CBD of FN. Furthermore,Tthese domains along with the CBD were required for the survival and PDGF-BB stimulated proliferation of mouse FN-null fibroblasts. The PDGF-AA isoform of this growth factor also binds to FN, but in this instance binding is only apparent after heparin treatment of FN and.tThese authors proposed an important role for fibronectin PDGF-AA binding in the migration of mesodermal cells during embryogenesis. On a similar developmental theme, the heparin binding domain (HBD-FNIII12-14) was shown to induce angiogenesis both in vivo and in vitro. In addition by sequesteringthe most active form of VEGF-A (non-poly ADP ribosylated) HBD was able to enhance angiogenesisand increase the expression of the angiogenic marker VEGF receptor-2 (VEGFR-2) [35]. A new and growing research area concerns the growth factor regulation of FN gene expression. For example in a well differentiated mouse hepatocyte cell line (mhAT3F), TGF-β caused up regulation of FN along with α5 integrin subunit which in combination with the β1 subunit forms a main fibronectin receptor. ThisSuch alteration in gene expression maintained cell migration when plated onto fibronectin enriched Matrigel surfaces[surfaces [36].

FN and GF Receptors

Direct evidence for a binary interaction between integrins and growth factor receptors has not been reportedis somewhat lacking. However GF receptors can be co-immunoprecipitated with integrins following ligation of the latter with ECM constituents. Indeed ligand independent activation of GF receptors under these conditions is well documented. For example in pancreatic cells, the pro-survival effect of FN is mediated via the IGF-IR in an IGF-independent manner by sequestration of the tyrosine phosphatase SHP-2 by integrin β3 subunit (integrin β1 subunit does not cause this effect). This causes sustained activation of the IGF-IR and downstream activation of Akt. Interestingly, soluble FN also stimulates IGF-I receptor phosphorylation in non-adherent cells.

In the human intestinal epithelial cell line Caco-2 which had been stably transfected with α5 integrin subunit, growth on FN coated plates resulted in ligand independent activation of the EGF-R. However this alone was not sufficient to stimulate cell proliferation in the absence of growth factor despite the fact that FN alone activated signalling pathways downstream from the EGF-R. It was suggested by these authors that the action of integrins was to potentiate the subsequent activity of growth factors [37].

Direct evidence of the importance of FN-integrin association for IGF-I activity was obtained by simple but elegant experiments which compared the effects of IGF-I on cells cultured on plastic with those cultured on FN. Althoughculture of cells on FN alone was sufficient to promote cellattachment and spreading [38], it was found that the presence of FN enhanced the activity of IGF-I with respect to both these biological endpoints. In addition, culture on FN enhanced cell migratory and differentiation activity in response to IGF-I when compared with the same cells grown on plastic. Such effects were typically associated with enhanced tyrosine phosphorylation of IGF-IR and association of FN with αvβ3 integrin [39-41]. Specificity of FN-integrin interactions was demonstrated by growth of cells on FN coated surfaces together with the use of integrin specific antibodies and/or disintegrin inhibitors of integrin-FN association to inhibit integrin dependent IGF signalling, demonstrating the requirement for activation of both IGF and integrin signalling pathways to stimulate processes such as cell adhesion and migration [42-45]. Cell proliferation, adhesion and migration are important for cancer cell growth and metastasis and this has prompted further studies of the molecular basis for interaction between the IGF axis and integrin receptors. As IGF-IR is the immediate route for IGF signalling, efforts initially focussed on investigating if a physical link existed between IGF-IR and integrin heterodimers thus forming a direct basis for signalling crosstalk. Traditionally these interactions are demonstrated by techniques such as co-immunoprecipitation of target proteins and direct binding can be investigated by such techniques as yeast two hybrid analysis (Y2H), isothermal titration (ITC) calorimetry or surface plasmon resonance (SPR). Although no evidence exist for direct contact between IGF-IR and integrin subunits, studies have shown co-immunoprecipitation of IGF-IR with β-integrin subunits [46]. This suggested that other protein(s) may be involved in the IGF-IR-integrin complex and initiated a search for such proteins.

A potential candidate for this role is the "7-WD domain" protein RACK1 (receptor for activated Ckinase). RACK1 is a member of the G protein β subunit family and was clonedtwenty years ago [47]. The WD domains of RACK1 form a 7-bladed propeller structure composed mainly of β sheets which caninteract with other signalling proteins [48].Following the observation thatRACK1 bound to activated PKC isozymes [49]it was shown that RACK1 was an adaptor for several proteins involved in cell signalling and that different proteins interacted specifically with different WD domains in RACK1 - reviewed in [50]. see. Therefore RACK1 regulates the activity ofseveralsignalling pathways [51].

Significantly, Y2H and co-immunoprecipitation studies showed RACK1 binding to a membrane proximal region of β1, 2 and 5 integrin cytosolic domains via WD motifs 5,6 and 7[52, 53]. Studies showed that tyrosine Y302 in the WD7 domain of RACK1 was essential for this interaction [54]. In the context of cell motility, IGF-I acting through PI3 kinase stimulates RACK1 – integrin association.this function. The authors of this study also quantified the strength of binding cells to ECM and indicated that an intermediate affinity of binding (i.e. neither too high nor too low) was required for optimal cell binding [55]. An IGF-I dependant association of RACK1 with the cytosolic domain of the IGF-IR β subunit was also demonstrated by Y2H [56] and a tripartite complex of IGF-IR-RACK1-β1 integrin was reported in breast cancer cells and fibroblasts [46]. Tyrosine residues (Y1250 and Y1251) of the IGF-IR were essential for formation of these tri-partite complexes [57]. Further studies reported an IGF-I dependant dissociation of RACK1 and the serine/threonine phosphatase PP2A causing decreased phosphatase activity [58].Growth of fibroblasts or breast cancer cells on FN enhancedrelease of PP2A from RACK1 and simultaneously increased β1 integrin recruitment to the RACK1/IGF-IR complex.

Thereforecomplex. Therefore PP2A and β1 integrin may compete for the same binding surface on RACK1 and IGF-Imay cause displacement of PP2Aand promote integrin association with RACK1 [54] (see Figure 1). As FN binding to integrins is required for IGF-I stimulated dissociation of the RACK1/PP2A complex in non-adherent cell cultures this mechanism provides a direct link between IGF, FN and integrin signalling. Kuemmerle, working with primary cultures of smooth muscle cells showed that SHP-1 recruitment to the IGF-IR occurs following inhibition of FN association with $\alpha_V\beta3$ integrin and results in dephosphorylation and attenuation of IGF-IR signalling [40].

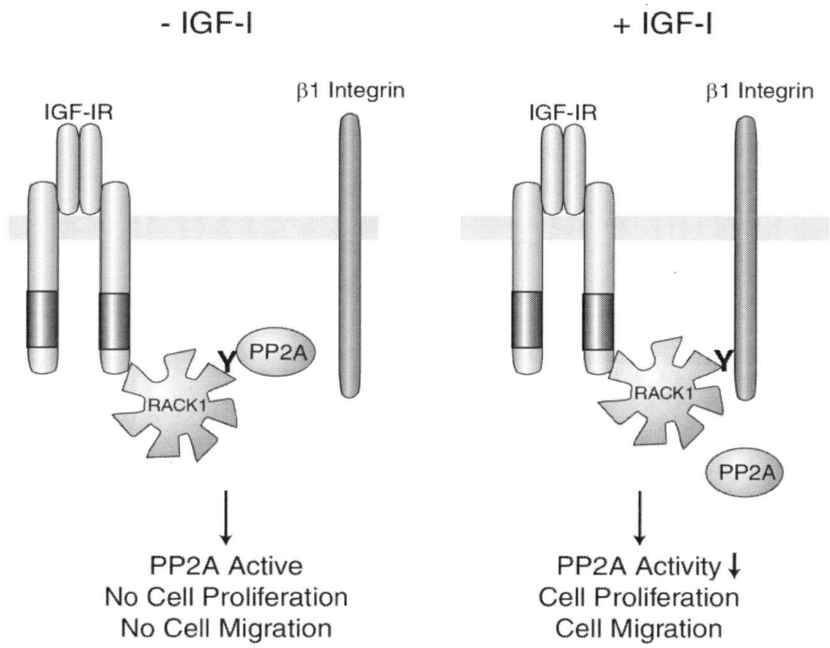

Figure 1. Regulation of cell proliferation and migration by PP2A (adapted from ref 58).

In an analogous fashion, pro-survival effects of FN in pancreatic cells are mediated by sequestration of SHP-2 by β3 integrin resulting in activation of IGF-IR. The requirement of IGF-I for this effect appears to be cell type specific [39].

In mesenchymal human bone marrow cells, adhesion to FN causes ligand independent phosphorylation of PDGFR-β. This effect is dependent on α5β1 integrin and results in co-immunoprecipitation of PDGR-β with α5 integrin and co-localises PDGFR-β with α5β1integrin in FAK complexes.FN also potentiates PDGF-BB stimulation of PDGFR-β and FAK in an integrin dependent manner resulting in downstream activation PI3K/Akt, actin reorganisation and cell migration. Howevercell surface heparan sulphate proteoglycans are also involved in regulation of FN-PDGFR-β- α5β1integrin complexes and evidence also exists that that PDGF-BB binds directly to α5β1integrin [59] . Other signalling pathways may be involved in PDGF –integrin regulation of cell migration. A very recent study has highlighted the involvement of the serine-threonine integrin-linked kinase (ILK) in PDGF stimulated migration in primary cultures of mouse smooth muscle cells [59].

The link between growth factor and fibronectin in the stimulation of various aspects of cell biology has been so widely confirmed that efforts to synthesise and test hybrid molecules consisting of growth factor fused to the collagen binding domain of fibronectin has been reported for HGF, VEGF and EGF. It is envisioned that these molecules will be concentrated in the vicinity of the cell membrane - ECM junction and that the effects of growth factors will be subsequently potentiated [60-62].

IGFBPs and Integrin Crosstalk

There are 6 soluble IGF binding proteins (IGFBP1-6) which bind IGF-I and IGF-II with high affinity and act as transporters and possibly tissue storage depots for the growth factors.The majority of IGFBPs are able to bind specifically to integrins, ECM constituents or in some instances both ECM and integrins. As such, IGFBPs provide an alternative route through which members of the IGF axis may influence integrin activity. Several groups have postulated that direct binding of IGFBPs to integrin heterodimers leads to biological effects although these appear to be cell type dependant with both inhibitory [63-67] and stimulatory [68-71] effects described.IGFBP-1 and -2 possess an RGD sequence in their C-terminal domains and IGFBP-1 binds in an RGD-dependant manner to α5β1 integrins expressed in CHO cells [72].IGFBP-1 association with α5β1 integrin on endometrial stromal cells has been associated with the process of decidualisation in human endometrium [73] and there are reports of IGFBP-1 stimulation of cell migration through direct interaction with α5β1 integrin [69, 72]. Conversely, some studies reported an inhibitory effect of IGFBP-1 on FAK phosphorylation, cell attachment and migration in human breast cancer cell lines possibly by direct binding to α5β1 integrins or by disruption of the α5β1 interaction with FN [65, 66]. Despite also containing an RGD sequence early studies suggested that IGFBP-2 did not bind integrin heterodimers [70]. However more recent work has clearly shown that IGFBP-2 binding to integrins on the surface of breast cancer cellscan up regulate genes involved in apoptosis (including NFκB and TGF beta-1).and that this inhibits tumour cell migration [64] .In contrast, over expression of IGFBP-2 in a human glioma cell line resulted in increased cell mobility and lamellipodia formation by a mechanism which involved IGFBP-2 interaction with α5 integrin and stimulation of the GTPase activating protein Rac1 [68]. This study is a rare example of direct activation of a signalling pathway by IGFBP-integrin interaction, as usually details of such pathways remain unknown. We believe this maybe a promising area for further study.

Although there are no reports of other IGFBPs binding directly to integrin proteins, IGFBP-3 and -5 are able to influence activity through the integrin axis indirectly by binding to extrac ellular proteins which themselves bind to integrin heterodimers. Binding of IGFBP-3 and -5 to both FN and VN has been described although contradictory reports exist on the biological effects of these interactions [74-78]. From a different perspective it is interesting that recent reports suggest that IGFBP-5 is able to up regulate FN expression and that increased FN concentrations may be involved in the development of fibrotic lesions in both skin and lung [79-82]. Recent Eevidence indicates that this effect may be mediated via MAPK up regulation and subsequent binding of early growth response (Egr-1) Zn finger transcription factor to nuclear transported IGFBP-5 [83]. Regulation of gene expression by IGFBPs is a relatively new field and we expect further interesting developments in this area.

Direct Effects of FN

There is some evidence that intact FN or peptide fragments have direct effects in some cell culture systems. For example FNIII (6-8) repeats are able to directly stimulate the differentiation of neural progenitor stem cells into neurons and inhibit differentiation to oligodendrocyte lineages. Although histologically the effect of FNIII(6-8) do not appear small to be large the observation that this FN fragment also stimulates expression of both NGF and BDNF from cultures of microglia cellsargues for a role in differentiation of neural stem cells [84].However these very recent observations in primary rat CNS cell cultures require further validation and extension into the putative role of FN in regulating directly the differentiation of stem cells.

Summary

FN, integrins, GFs and GFRs interact to regulate signalling events through both integrins and GFsS. The mechanisms by which this occurs are diverse and space limitations mean we can only discuss a fewlimited amount of these pathways. However, we anticipate that in this activeexpanding research area more information on integrin-GFR crosstalk will become available in the near future. Such studies will make a valuable contribution to understanding such processes as cell differentiation, migration and growth, events closely associated with the development and dissemination of tumour cells.

References

[1] White, E. S., Baralle, F. E. and Muro, A. F. (2008) New insights into form and function of fibronectin splice variants. *The Journal of pathology.* 216, 1-14.

[2] Liu, T., Qian, W. J., Gritsenko, M. A., Camp, D. G., 2nd, Monroe, M. E., Moore, R. J. and Smith, R. D. (2005) Human plasma N-glycoproteome analysis by immunoaffinity subtraction, hydrazide chemistry, and mass spectrometry. *Journal of proteome research.* 4, 2070-2080.

[3] Olsen, J. V., Blagoev, B., Gnad, F., Macek, B., Kumar, C., Mortensen, P. and Mann, M. (2006) Global, in vivo, and site-specific phosphorylation dynamics in signaling networks. *Cell.* 127, 635-648.

[4] Tressel, T., McCarthy, J. B., Calaycay, J., Lee, T. D., Legesse, K., Shively, J. E. and Pande, H. (1991) Human plasma fibronectin. Demonstration of structural differences between the A- and B-chains in the III CS region. *Biochem. J.* 274 (Pt 3), 731-738.

[5] Mandel, U., Hamilton Therkildsen, M., Reibel, J., Sweeney, B., Matsuura, H., Hakomori, S., Dabelsteen, E. and Clausen, H. (1992) Cancer-associated changes in glycosylation of fibronectin. Immunohistological localization of oncofetal fibronectin defined by monoclonal antibodies. *Apmis.* 100, 817-826.

[6] Jones, G. E., Arumugham, R. G. and Tanzer, M. L. (1986) Fibronectin glycosylation modulates fibroblast adhesion and spreading. *The Journal of cell biology.* 103, 1663-1670.

[7] Mao, Y. and Schwarzbauer, J. E. (2005) Fibronectin fibrillogenesis, a cell-mediated matrix assembly process. *Matrix Biol.* 24, 389-399.

[8] Balbona, K., Tran, H., Godyna, S., Ingham, K. C., Strickland, D. K. and Argraves, W. S. (1992) Fibulin binds to itself and to the carboxyl-terminal heparin-binding region of fibronectin. *J. Biol. Chem.* 267, 20120-20125.

[9] Balian, G., Click, E. M. and Bornstein, P. (1980) Location of a collagen-binding domain in fibronectin. *J. Biol. Chem.* 255, 3234-3236.

[10] Hayashi, M., Schlesinger, D. H., Kennedy, D. W. and Yamada, K. M. (1980) Isolation and characterization of a heparin-binding domain of cellular fibronectin. *J. Biol. Chem.* 255, 10017-10020.

[11] Richter, H., Seidl, M. and Hormann, H. (1981) Location of heparin-binding sites of fibronectin. Detection of a hitherto unrecognized transamidase sensitive site. *Hoppe-Seyler's Zeitschrift fur physiologische Chemie.* 362, 399-408.

[12] Rostagno, A. A., Schwarzbauer, J. E. and Gold, L. I. (1999) Comparison of the fibrin-binding activities in the N- and C-termini of fibronectin. *Biochem. J.* 338(Pt 2),375-386.

[13] Sekiguchi, K. and Hakomori, S. (1980) Identification of two fibrin-binding domains in plasma fibronectin and unequal distribution of these domains in two different subunits: a preliminary note. *Biochemical and biophysical research communications.*97,709-715.

[14] Aota, S., Nomizu, M. and Yamada, K. M. (1994) The short amino acid sequence Pro-His-Ser-Arg-Asn in human fibronectin enhances cell-adhesive function. *J. Biol. Chem.* 269, 24756-24761.

[15] Kubomura, S., Obara, M., Karasaki, Y., Taniguchi, H., Gotoh, S., Tsuda, T., Higashi, K., Ohsato, K. and Hirano, H. (1987) Genetic analysis of the cell binding domain region of the chicken fibronectin gene. *Biochim. Biophys. Acta.* 910, 171-181.

[16] Obara, M., Kang, M. S., Rocher-Dufour, S., Kornblihtt, A., Thiery, J. P. and Yamada, K. M. (1987) Expression of the cell-binding domain of human fibronectin in E. coli. Identification of sequences promoting full to minimal adhesive function. *FEBS letters.* 213, 261-264.

[17] Obara, M., Kang, M. S. and Yamada, K. M. (1988) Site-directed mutagenesis of the cell-binding domain of human fibronectin: separable, synergistic sites mediate adhesive function. *Cell.* 53, 649-657.

[18] Pierschbacher, M. D. and Ruoslahti, E. (1984) Cell attachment activity of fibronectin can be duplicated by small synthetic fragments of the molecule. *Nature.* 309, 30-33.

[19] Waite, K. A., Mugnai, G. and Culp, L. A. (1987) A second cell-binding domain on fibronectin (RGDS-independent) for neurite extension of human neuroblastoma cells. *Experimental cell research.* 169, 311-327.

[20] Grant, R. P., Spitzfaden, C., Altroff, H., Campbell, I. D. and Mardon, H. J. (1997) Structural requirements for biological activity of the ninth and tenth FIII domains of human fibronectin. *J. Biol. Chem.* 272, 6159-6166.

[21] Ruoslahti, E. and Pierschbacher, M. D. (1987) New perspectives in cell adhesion: RGD and integrins. *Science* (New York, N.Y. 238, 491-497.

[22] Altroff, H., van der Walle, C. F., Asselin, J., Fairless, R., Campbell, I. D. and Mardon, H. J. (2001) The eighth FIII domain of human fibronectin promotes integrin alpha5beta1 binding via stabilization of the ninth FIII domain. *J. Biol. Chem.* 276, 38885-38892.

[23] Bossard, C., Van den Berghe, L., Laurell, H., Castano, C., Cerutti, M., Prats, A. C. and Prats, H. (2004) Antiangiogenic properties of fibstatin, an extracellular FGF-2-binding polypeptide. *Cancer research.* 64, 7507-7512.

[24] Gui, Y. and Murphy, L. J. (2001) Insulin-like growth factor (IGF)-binding protein-3 (IGFBP-3) binds to fibronectin (FN): demonstration of IGF-I/IGFBP-3/fn ternary complexes in human plasma. *The Journal of clinical endocrinology and metabolism.* 86, 2104-2110.

[25] Hoshijima, M., Hattori, T., Inoue, M., Araki, D., Hanagata, H., Miyauchi, A. and Takigawa, M. (2006) CT domain of CCN2/CTGF directly interacts with fibronectin and enhances cell adhesion of chondrocytes through integrin alpha5beta1. *FEBS letters.* 580, 1376-1382.

[26] Mooradian, D. L., Lucas, R. C., Weatherbee, J. A. and Furcht, L. T. (1989) Transforming growth factor-beta 1 binds to immobilized fibronectin. *Journal of cellular biochemistry.* 41, 189-200.

[27] Rahman, S., Patel, Y., Murray, J., Patel, K. V., Sumathipala, R., Sobel, M. and Wijelath, E. S. (2005) Novel hepatocyte growth factor (HGF) binding domains on fibronectin and vitronectin coordinate a distinct and amplified Met-integrin induced signalling pathway in endothelial cells. *BMC cell biology.* 6, 8.

[28] Sahni, A., Khorana, A. A., Baggs, R. B., Peng, H. and Francis, C. W. (2006) FGF-2 binding to fibrin(ogen) is required for augmented angiogenesis. *Blood.* 107, 126-131.

[29] Schultz, G. S. and Wysocki, A. (2009) Interactions between extracellular matrix and growth factors in wound healing. *Wound Repair Regen.* 17, 153-162.

[30] Wijelath, E. S., Rahman, S., Murray, J., Patel, Y., Savidge, G. and Sobel, M. (2004) Fibronectin promotes VEGF-induced CD34 cell differentiation into endothelial cells. *J. Vasc. Surg.* 39, 655-660.

[31] Martino, M. M. and Hubbell, J. A. The 12th-14th type III repeats of fibronectin function as a highly promiscuous growth factor-binding domain. *Faseb. J.* 24, 4711-4721.

[32] Wijelath, E. S., Rahman, S., Namekata, M., Murray, J., Nishimura, T., Mostafavi-Pour, Z., Patel, Y., Suda, Y., Humphries, M. J. and Sobel, M. (2006) Heparin-II domain of fibronectin is a vascular endothelial growth factor-binding domain: enhancement of VEGF biological activity by a singular growth factor/matrix protein synergism. *Circulation research.* 99, 853-860.

[33] Chen, T. T., Luque, A., Lee, S., Anderson, S. M., Segura, T. and Iruela-Arispe, M. L. Anchorage of VEGF to the extracellular matrix conveys differential signaling responses to endothelial cells. *The Journal of cell biology.* 188, 595-609.

[34] Lin, F., Ren, X. D., Pan, Z., Macri, L., Zong, W. X., Tonnesen, M. G., Rafailovich, M., Bar-Sagi, D. and Clark, R. A. Fibronectin growth factor-binding domains are required for fibroblast survival. *The Journal of investigative dermatology.* 131, 84-98.

[35] Viji, R. I., Kumar, V. B., Kiran, M. S. and Sudhakaran, P. R. (2008) Angiogenic response of endothelial cells to heparin-binding domain of fibronectin. *The international Journal of biochemistry and cell biology.* 40, 215-226.

[36] Biname, F., Lassus, P. and Hibner, U. (2008) Transforming growth factor beta controls the directional migration of hepatocyte cohorts by modulating their adhesion to fibronectin. *Mol. Biol. Cell.* 19, 945-956.

[37] Kuwada, S. K. and Li, X. (2000) Integrin alpha5/beta1 mediates fibronectin-dependent epithelial cell proliferation through epidermal growth factor receptor activation. *Mol. Biol. Cell.* 11, 2485-2496.

[38] Charo, I. F., Nannizzi, L., Smith, J. W. and Cheresh, D. A. (1990) The vitronectin receptor alpha v beta 3 binds fibronectin and acts in concert with alpha 5 beta 1 in promoting cellular attachment and spreading on fibronectin. *The Journal of cell biology.* 111, 2795-2800.

[39] Edderkaoui, M., Hong, P., Lee, J. K., Pandol, S. J. and Gukovskaya, A. S. (2007) Insulin-like growth factor-I receptor mediates the prosurvival effect of fibronectin. *J. Biol. Chem.* 282, 26646-26655.

[40] Kuemmerle, J. F. (2006) Occupation of alphavbeta3-integrin by endogenous ligands modulates IGF-I receptor activation and proliferation of human intestinal smooth muscle. *American journal of physiology.* 290, G1194-1202.

[41] Maile, L. A., Imai, Y., Clarke, J. B. and Clemmons, D. R. (2002) Insulin-like growth factor I increases alpha Vbeta 3 affinity by increasing the amount of integrin-associated protein that is associated with non-raft domains of the cellular membrane. *J. Biol. Chem.* 277, 1800-1805

[42] Ando, Y. and Jensen, P. J. (1993) Epidermal growth factor and insulin-like growth factor I enhance keratinocyte migration. *The Journal of investigative dermatology.* 100, 633-639.

[43] Aplin, J. D., Lacey, H., Haigh, T., Jones, C. J., Chen, C. P. and Westwood, M. (2000) Growth factor-extracellular matrix synergy in the control of trophoblast invasion. *Biochemical Society transactions.* 28, 199-202.

[44] Nakasaki, M., Yoshioka, K., Miyamoto, Y., Sasaki, T., Yoshikawa, H. and Itoh, K. (2008) IGF-I secreted by osteoblasts acts as a potent chemotactic factor for osteoblasts. *Bone.* 43, 869-879.

[45] Tai, Y. T., Podar, K., Catley, L., Tseng, Y. H., Akiyama, M., Shringarpure, R., Burger, R., Hideshima, T., Chauhan, D., Mitsiades, N., Richardson, P., Munshi, N. C., Kahn, C. R., Mitsiades, C. and Anderson, K. C. (2003) Insulin-like growth factor-1 induces adhesion and migration in human multiple myeloma cells via activation of beta1-integrin and phosphatidylinositol 3'-kinase/AKT signaling. *Cancer research.* 63, 5850-5858.

[46] Kiely, P. A., Sant, A. and O'Connor, R. (2002) RACK1 is an insulin-like growth factor 1 (IGF-1) receptor-interacting protein that can regulate IGF-1-mediated Akt activation and protection from cell death. *J. Biol. Chem.* 277, 22581-22589.

[47] Guillemot, F., Billault, A. and Auffray, C. (1989) Physical linkage of a guanine nucleotide-binding protein-related gene to the chicken major histocompatibility complex. *Proceedings of the National Academy of Sciences of the United States of America.* 86, 4594-4598.

[48] Sondek, J. and Siderovski, D. P. (2001) Ggamma-like (GGL) domains: new frontiers in G-protein signaling and beta-propeller scaffolding. *Biochemical pharmacology.* 61, 1329-1337.

[49] Ron, D., Chen, C. H., Caldwell, J., Jamieson, L., Orr, E. and Mochly-Rosen, D. (1994) Cloning of an intracellular receptor for protein kinase C: a homolog of the beta subunit of G proteins. *Proceedings of the National Academy of Sciences of the United States of America.* 91, 839-843.

[50] McCahill, A., Warwicker, J., Bolger, G. B., Houslay, M. D. and Yarwood, S. J. (2002) The RACK1 scaffold protein: a dynamic cog in cell response mechanisms. *Molecular pharmacology*. 62, 1261-1273.

[51] Chang, B. Y., Conroy, K. B., Machleder, E. M. and Cartwright, C. A. (1998) RACK1, a receptor for activated C kinase and a homolog of the beta subunit of G proteins, inhibits activity of src tyrosine kinases and growth of NIH 3T3 cells. *Molecular and cellular biology*. 18, 3245-3256.

[52] Besson, A., Wilson, T. L. and Yong, V. W. (2002) The anchoring protein RACK1 links protein kinase Cepsilon to integrin beta chains. Requirements for adhesion and motility. *J. Biol. Chem.*277, 22073-22084.

[53] Liliental, J. and Chang, D. D. (1998) Rack1, a receptor for activated protein kinase C, interacts with integrin beta subunit. *J. Biol. Chem.* 273, 2379-2383.

[54] Kiely, P. A., Baillie, G. S., Lynch, M. J., Houslay, M. D. and O'Connor, R. (2008) Tyrosine 302 in RACK1 is essential for insulin-like growth factor-I-mediated competitive binding of PP2A and beta1 integrin and for tumor cell proliferation and migration. *J. Biol. Chem.*283, 22952-22961.

[55] Lynch, L., Vodyanik, P. I., Boettiger, D. and Guvakova, M. A. (2005) Insulin-like growth factor I controls adhesion strength mediated by alpha5beta1 integrins in motile carcinoma cells. *Mol. Biol. Cell.*16, 51-63.

[56] Hermanto, U., Zong, C. S., Li, W. and Wang, L. H. (2002) RACK1, an insulin-like growth factor I (IGF-I) receptor-interacting protein, modulates IGF-I-dependent integrin signaling and promotes cell spreading and contact with extracellular matrix. *Molecular and cellular biology*. 22, 2345-2365.

[57] Kiely, P. A., Leahy, M., O'Gorman, D. and O'Connor, R. (2005) RACK1-mediated integration of adhesion and insulin-like growth factor I (IGF-I) signaling and cell migration are defective in cells expressing an IGF-I receptor mutated at tyrosines 1250 and 1251. *J. Biol. Chem.*280, 7624-7633.

[58] Kiely, P. A., O'Gorman, D., Luong, K., Ron, D. and O'Connor, R. (2006) Insulin-like growth factor I controls a mutually exclusive association of RACK1 with protein phosphatase 2A and beta1 integrin to promote cell migration. *Molecular and cellular biology*. 26, 4041-4051.

[59] Esfandiarei, M., Yazdi, S. A., Gray, V., Dedhar, S. and van Breemen, C. Integrin-linked kinase functions as a downstream signal of platelet-derived growth factor to regulate actin polymerization and vascular smooth muscle cell migration. *BMC cell biology*. 11, 16.

[60] Ishikawa, T., Eguchi, M., Wada, M., Iwami, Y., Tono, K., Iwaguro, H., Masuda, H., Tamaki, T. and Asahara, T. (2006) Establishment of a functionally active collagen-binding vascular endothelial growth factor fusion protein in situ. *Arteriosclerosis, thrombosis, and vascular biology*.26, 1998-2004.

[61] Ishikawa, T., Terai, H. and Kitajima, T. (2001) Production of a biologically active epidermal growth factor fusion protein with high collagen affinity. *Journal of biochemistry*. 129, 627-633.

[62] Kitajima, T., Terai, H. and Ito, Y. (2007) A fusion protein of hepatocyte growth factor for immobilization to collagen. *Biomaterials*.28, 1989-1997.

[63] Schutt, B. S., Langkamp, M., Rauschnabel, U., Ranke, M. B. and Elmlinger, M. W. (2004) Integrin-mediated action of insulin-like growth factor binding protein-2 in tumor cells. *J. Mol. Endocrinol.* 32, 859-868.

[64] Pereira, J. J., Meyer, T., Docherty, S. E., Reid, H. H., Marshall, J., Thompson, E. W., Rossjohn, J. and Price, J. T. (2004) Bimolecular interaction of insulin-like growth factor (IGF) binding protein-2 with alphavbeta3 negatively modulates IGF-I-mediated migration and tumor growth. *Cancer research.*64, 977-984.

[65] Zhang, X. and Yee, D. (2002) Insulin-like growth factor binding protein-1 (IGFBP-1) inhibits breast cancer cell motility. *Cancer research.* 62, 4369-4375.

[66] Perks, C. M., Newcomb, P. V., Norman, M. R. and Holly, J. M. (1999) Effect of insulin-like growth factor binding protein-1 on integrin signalling and the induction of apoptosis in human breast cancer cells. *J. Mol. Endocrinol.* 22, 141-150.

[67] Gockerman, A., Prevette, T., Jones, J. I. and Clemmons, D. R. (1995) Insulin-like growth factor (IGF)-binding proteins inhibit the smooth muscle cell migration responses to IGF-I and IGF-II. *Endocrinology.*136, 4168-4173.

[68] Wang, G. K., Hu, L., Fuller, G. N. and Zhang, W. (2006) An interaction between insulin-like growth factor-binding protein 2 (IGFBP2) and integrin alpha5 is essential for IGFBP2-induced cell mobility. *J. Biol. Chem.* 281, 14085-14091.

[69] Gleeson, L. M., Chakraborty, C., McKinnon, T. and Lala, P. K. (2001) Insulin-like growth factor-binding protein 1 stimulates human trophoblast migration by signaling through alpha 5 beta 1 integrin via mitogen-activated protein Kinase pathway. *The Journal of clinical endocrinology and metabolism.* 86, 2484-2493.

[70] Galiano, R. D., Zhao, L. L., Clemmons, D. R., Roth, S. I., Lin, X. and Mustoe, T. A. (1996) Interaction between the insulin-like growth factor family and the integrin receptor family in tissue repair processes. Evidence in a rabbit ear dermal ulcer model. *The Journal of clinical investigation.*98, 2462-2468.

[71] Clemmons, D. R., Jones, J. I., Busby, W. H. and Wright, G. (1993) Role of insulin-like growth factor binding proteins in modifying IGF actions. *Ann. N Y Acad. Sci.*692, 10-21.

[72] Jones, J. I., Gockerman, A., Busby, W. H., Jr., Wright, G. and Clemmons, D. R. (1993) Insulin-like growth factor binding protein 1 stimulates cell migration and binds to the alpha 5 beta 1 integrin by means of its Arg-Gly-Asp sequence. *Proceedings of the National Academy of Sciences of the United States of America.* 90, 10553-10557.

[73] Matsumoto, H., Sakai, K. and Iwashita, M. (2008) Insulin-like growth factor binding protein-1 induces decidualization of human endometrial stromal cells via alpha5beta1 integrin. *Molecular human reproduction.*14, 485-489.

[74] Beattie, J., Kreiner, M., Allan, G. J., Flint, D. J., Domingues, D. and van der Walle, C. F. (2009) IGFBP-3 and IGFBP-5 associate with the cell binding domain (CBD) of fibronectin. *Biochemical and biophysical research communications.* 381, 572-576.

[75] Xu, Q., Yan, B., Li, S. and Duan, C. (2004) Fibronectin binds insulin-like growth factor-binding protein 5 and abolishes Its ligand-dependent action on cell migration. *J. Biol. Chem.*279, 4269-4277.

[76] Martin, J. A., Miller, B. A., Scherb, M. B., Lembke, L. A. and Buckwalter, J. A. (2002) Co-localization of insulin-like growth factor binding protein 3 and fibronectin in human articular cartilage. Osteoarthritis and cartilage / OARS, *Osteoarthritis Research Society.* 10, 556-563.

[77] Kricker, J. A., Towne, C. L., Firth, S. M., Herington, A. C. and Upton, Z. (2003) Structural and functional evidence for the interaction of insulin-like growth factors (IGFs) and IGF binding proteins with vitronectin. *Endocrinology.*144, 2807-2815.

[78] Nam, T., Moralez, A. and Clemmons, D. (2002) Vitronectin binding to IGF binding protein-5 (IGFBP-5) alters IGFBP-5 modulation of IGF-I actions. *Endocrinology.*143, 30-36.

[79] Sureshbabu, A., Okajima, H., Yamanaka, D., Shastri, S., Tonner, E., Rae, C., Szymanowska, M., Shand, J. H., Takahashi, S., Beattie, J., Allan, G. J. and Flint, D. J. (2009) IGFBP-5 induces epithelial and fibroblast responses consistent with the fibrotic response. *Biochemical Society transactions.*37, 882-885.

[80] Yasuoka, H., Jukic, D. M., Zhou, Z., Choi, A. M. and Feghali-Bostwick, C. A. (2006) Insulin-like growth factor binding protein 5 induces skin fibrosis: A novel murine model for dermal fibrosis. *Arthritis and rheumatism.* 54, 3001-3010.

[81] Yasuoka, H., Zhou, Z., Pilewski, J. M., Oury, T. D., Choi, A. M. and Feghali-Bostwick, C. A. (2006) Insulin-like growth factor-binding protein-5 induces pulmonary fibrosis and triggers mononuclear cellular infiltration. *The American journal of pathology.* 169, 1633-1642.

[82] Pilewski, J. M., Liu, L., Henry, A. C., Knauer, A. V. and Feghali-Bostwick, C. A. (2005) Insulin-like growth factor binding proteins 3 and 5 are overexpressed in idiopathic pulmonary fibrosis and contribute to extracellular matrix deposition. *The American journal of pathology.* 166, 399-407.

[83] Yasuoka, H., Yamaguchi, Y. and Feghali-Bostwick, C. A. (2009) The pro-fibrotic factor IGFBP-5 induces lung fibroblast and mononuclear cell migration. *American journal of respiratory cell and molecular biology.*41, 179-188.

[84] Liao, H., Huang, W., Niu, R., Sun, L. and Zhang, L. (2008) Cross-talk between the epidermal growth factor-like repeats/fibronectin 6-8 repeats domains of Tenascin-R and microglia modulates neural stem/progenitor cell proliferation and differentiation. *Journal of neuroscience research.*86, 27-34.

In: Fibronectin
Editors: James Beattie

ISBN 978-1-62257-298-4

Chapter III

Fibronectin: The N-Terminal Region and Its Role in Cell Migration Implications for Disease and Healing

*Ian R Ellis**, *Mohammad R. Islam*, *Lateef E. Aljorani and Sarah J. Jones*
Unit of Cell and Molecular Biology, Dundee Dental School, University of Dundee, Park Place, Dundee, Scotland, UK

Abstract

The smallest truncated member of the fibronectin (Fn) family, migration stimulating factor (MSF), contains the heparin/fibrin domain (HepI/FibI), the gelatin binding domain (GBD), part of the first type III module of the cell binding domain (CBD) and a unique ten amino acid sequence, the result of intronic read through. MSF and fragments of Fn have been studied in our laboratory for a number of years in terms of cell motility and their expression in various diseases. The active sequence for cell motility contains the amino acids: isoleucine, glycine and aspartic acid (IGD), which is conserved in four of the type I repeats, ^{3}FnI, ^{5}FnI, ^{7}FnI and ^{9}FnI.

Computational analysis of the structure of these domains predicts that the IGD motifs are at the end of 'finger-like' projections. Mutation of these IGD sequences has allowed identification of the residues necessary for cell migration (although in other cell types the HEEGH sequence found in 8FnI is necessary for motility).

The type of assay and the matrix, upon which the cells are grown, influences the cellular response to the Fn fragments. Using a 3D collagen migration assay, we have identified the most bioactive amino acid sequences and which cell surface receptors are important for cell motility. Here, we present additional data identifying the signalling

* Author for correspondence: Dr Ian Ellis email: irellis @dundee.ac.uk. Telephone: +44 (0)1382 660111 ext 35826. Fax: +44 (0) 1382 665995.

pathways important in initiating cell motility in response to both GBD and IGDS. This response appears to be transmitted via an integrin as focal adhesion kinase (FAK) becomes phosphorylated. The PI3kinase/Akt pathway is the most important signalling pathway for motility, induced by the N-terminus of Fn. Addition of MSF to fibroblasts causes a de-phosphorylation of Akt and inhibitors of this pathway block motility of these cells.

One of the most obvious effects of the N-terminal region of Fn is increased extracellular matrix production. Both MSF and GBD act by increasing Hyaluronan (HA) synthesis, specifically high molecular weight HA, which appears to be important for increased cell motility.

The control of MSF and Fn expression appears to be influenced by a number of growth factors such as TGFβ-1. These growth factors lead to an increase in MSF expressed by fibroblasts grown in tissue culture. Bioactivity and consequently cell motility are controlled in part, by the size of the Fn protein, for example, zebrafish produce a slightly larger inactive form of truncated Fn. A recombinant mammalian Fn variant containing up to the ^{3}FnIII module can be activated by the binding of bacterial proteins, which has important implications for disease.

We will discuss the N-terminal domain of Fn and its role in cell motility, disease progression and wound healing. Our work in this area has shown that MSF is over-expressed in cancers and can be used as a marker of some forms of this disease. MSF and IGD-like compounds have also been used in wound healing assays and are potent stimulators of healing and may be important in developing novel wound healing therapies.

Keywords: Fibronectin, cell migration, HA, MSF

Introduction

The extracellular matrix (ECM) is the substance that binds the cells of all multicellular organisms together. The ECM has an influence on everything that happens in homeostasis and diseases, such as cancer. The ECM is both instructive in terms of development (Caplan, 1986) and important in adult tissue in terms of tissue homeostasis and can also influence progression of disease (Lukashev and Werb, 1998). The ECM is made up of an array of proteins, glycoproteins and glycosaminoglycans. Each matrix has a slightly different composition and this controls how a tissue develops and in turn, the differentiation of the cells in that tissue. Each matrix has the ability to allow cell migration or obstruct (physically) or inhibit (chemically) the motility of cells. The major components of any extracellular matrix are collagens, hyaluronan and fibronectin. These three molecules have been the focus of our research over the past few decades.

Type I collagen is the most common protein found within the body. The primary feature of collagen is a long stiff, triple stranded helical structure, in which three collagen polypeptide chains called alpha chains are wound around one another in a rope-like superhelix. In Type I collagen, the molecules are remarkably similar; each chain consists of 1050 amino acids. They are the same length and thickness, 300nm long and 1.5nm thick. Each fibril has 3 alpha chains coiled together in the form of a superhelix (Kielty and Grant, 1992). These fibres give collagen a high tensile strength and also a low susceptibility to proteolytic breakdown. As tissues age, the type of collagen produced changes with a greater amount of Type III collagen being produced compared with Type I (Lovell et al, 1987). During wound healing collagen synthesis is increased and it is laid down in a manner which is mainly down to speed of repair as opposed to being in the architecturally correct orientation (Ashcroft et al, 1991). Collagens and their fragments are able to stimulate the migration of fibroblasts as demonstrated by using a chemotaxis assay (Postlethwaite et al, 1987). Woodley et al, (1991) reported that collagen type I and III induced the motility of keratinocytes. Therefore collagen is an important substrate in cell motility. Cells have to be able to breakdown or manipulate collagen fibres in order to move through a matrix. Collagen and its fragments are able to be utilised in repair and regeneration, as well as disease, (Bellows et al, 1981).

Hyaluronan (HA) is the only non-sulphated glycosaminglycan (GAG) and is made up of repeating sugars, N-acteyl-glucosamine and glucuronic acid. The "sugar" has potential to build long chains which are found in the ECM and allow for greater flexibility of a tissue. Hyaluronan molecules are able to bind a large volume of water and have the ability to promote cell motility. Hyaluronan has been implicated in promoting cell migration in both development and in tumour progression (Toole, 2001; Toole et al, 2002). Fibroblasts are stimulated to migrate by addition of high molecular weight HA to 3D collagen gels (Ellis et al, 1992).

Fibroblasts also respond to growth factors, such as EGF, by elevating HA synthesis and specifically by synthesis of high molecular weight HA (Ellis et al, 1997; 2007). HA of different sizes has been implicated in the migration of other cell types, for example endothelial cells are stimulated to undergo angiogenesis by low molecular weight HA fragments (West et al, 1985). However, high molecular weight HA has been shown to inhibit angiogenesis (West and Kumar, 1989). The role of HA in tumour cell migration has also been reported (Turley, 1992). Tumour cells appear to have the ability to utilize HA as a substratum or by using it to "push through" the ECM (Bourguignon et al, 1995).

Fibronectin (Fn) is a major glycoprotein found in an insoluble form in the ECM of most tissues. The protein has a molecular weight of around 250kDa (Hynes, 1990), but is usually found as a homodimer joined by a disulphide bridge at the C-terminus. However, an even larger structure known as superfibronectin has been reported, which contains a large number of Fn molecules joined together to form a large complex (Morla et al, 1994). These are joined at the first Fn-III module which induces spontaneous disulphide cross-linking to form multimers. Fn has a wide range of activities including adhesion and migration. Cell migration is important in development of the foetus; wound healing in the adult and in more pathological events such as metastases (Duband et al, 1986). Full length Fn has been implicated as being important for neural crest motility in the developing blastocyst (Newgreen and Thiery, 1980). Fn is important for fibroblast and epithelial cell migration in wound healing and cancer (Schor et al 1981, Woodley et al, 1991). Fn affects cell migration both in its full length form and when it is broken down (Schor et al, 2003), these smaller fragments have so-called cryptic activity (Ellis et al, 2010A).

These activities are cell type dependent, and some activities are only seen when fibronectin fragments are produced. These include migratory activities that are not found in the full length molecule (Fukai et al, 1991; 1995; Clark et al, 1988). Degradation of Fibronectin into fragments may also be a sign of disease and a number of papers have indicated that N terminal fragments are found in rheumatoid arthritis as well as in wound fluid (Wysocki and Grinnell, 1990; Homandberg et al, 1991).

A number of isoforms of fibronectin have been isolated most of which are larger than "normal" fibronectin. Only two truncated forms of fibronectin have been isolated, these are short around 77kDa and 100kDa (Figure 1A; Schor et al, 2003, Vakonakis et al, 2009; Zhao et al, 2001). They are truncated in the first and third type III domains and therefore they cannot dimerise or produce larger super Fn molecules (Vakonakis et al, 2009).

A)

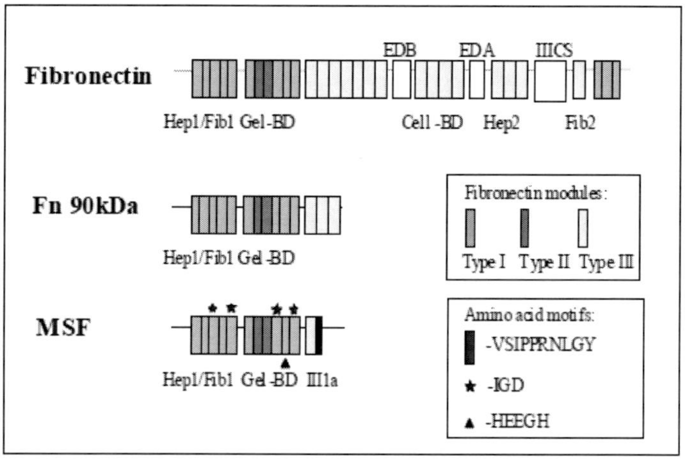

B)
MSF

MLRGPGPGLLLLAVQCLGTAVPSTGASKSKRQAQQMVQPQSPVAVSQSKPGCYDNGKH
YQINQQWERTYLGNALVCTCYGGSRGFNCESKPEAEETCFDKYTGNTYRVGDTYERPK
DSMIWDCTCIGAGRGRISCTIANRCHEGGQSYK**IGD**TWRRPHETGGYMLECVCLGNGK
GEWTCKPIAEKCFDHAAGTSYVVGETWEKPYQGWMMVDCTCLGEGSGRITCTSRNRCN
DQDTRTSYR**IGD**TWRKKDNRGNLLQCICTGNGRGEWKCERHTSVQTTSSGSGPFTDVR
AAVYQPQPHPQPPPYGHCVTDSGVVYSVGMQWLKTQGNKQMLCTCLGNGVSCQETAVT
QTYGGNSNGEPCVLPFTYNGRTFYSCTTEGRQDGHLWCSTTSNYEQDQKYSFCTDHTV
LVQTRGGNSNGALCHFPFLYNNHNYTDCTSEGRRDNMKWCGTTQNYDADQKFGFCPMA
AHEEICTTNEGVMYR**IGD**QWDKQHDMGHMMRCTCVGNGRGEWTCIAYSQLRDQCIVDD
ITYNVNDTFHKR**HEEGH**MLNCTCFGQGRGRWKCDPVDQCQDSETGTFYQ**IGD**SWEKYV
HGVRYQCYCYGRGIGEWHCQPLQTYPSSSGPVEVFITETPSQPNSHPIQWNAPQPSHI
SKYILRWRPVSIPPRNLGY

Figure 1. Fibronectin structure: Fibronectin is made up of a domain structure of type one (FnI), type two (FnII) and type 3 (FnIII) repeats. Each of these modules is highly conserved throughout the molecule and throughout evolution. A) Shows the modular structure of Fn and the related molecules Fn90kDa and MSF. B) The amino acid sequence of MSF with the 'active' sequences in bold.

Fn interacts with molecules on the cell surface to influence cell behaviour. The main cell surface interaction is via receptors known as integrins. Fibronectin can bind to nine different integrins (Johannsonn et al, 1997), which are made up of two subunits, alpha and beta. The integrins that can bind fibronectin are $\alpha_D\ \beta_2$, $\alpha_4\ \beta_1$, $\alpha_5\ \beta_1$, $\alpha_8\ \beta_1$, $\alpha_V\ \beta_1$, $\alpha_V\ \beta_3$, $\alpha_{II}\ \beta_3$, $\alpha_V\ \beta_6$, and $\alpha_4\ \beta_7$ (Johannsonn et al, 1997). Binding of fibronectin to the integrin utilises the amino acid sequence RGDS for most of its interactions. This sequence is found in fibronectin in the FnIII-10 domain and is also present in most other ECM molecules (Ruoslahti, 1996; Carson and Barker, 2009).

The inactive integrin subunits only loosely interact with each other and are active only once a ligand is in place at the cell surface (Giancotti, 2003). Fn binding to any of the integrin subunits triggers a cascade of signalling molecules to the nucleus which directly influence transcription and translation (Morgan et al, 2009). The Fn -integrin complex also interacts with the cell cytoskeleton and this may in turn affect cell migration (Morgan et al, 2009). Fn may also interact with the other cell surface receptors (Ellis et al, 2010A) using a binding site other than RGDS. The N- terminal region of Fn does not contain an RGD sequence but it may bind to a number of cell surface molecules (McKeown-Longo and Mosher, (1985); Sottile, and Mosher, (1997); Hocking, et al, (1998)). Hocking et al, (1998) reported that both of the integrin subunits α_5 and β_1 can bind to the 70kDa N terminus of fibronectin, which in turn causes the phosphorylation of FAK and changes in cell shape around focal contacts. Other binding proteins for the N-terminal region of fibronectin have also been described, including transglutaminase 2, and endosialin (Akimov, and Belkin, 2001; Tomokowicz et al, 2007).

Cell Migration

Cell movement is an essential part of eukaryotic cellular behaviour (Ellis et al, 2010B). This event is required in both tissue development and homeostasis; including morphogenesis, wound healing and inflammation. However, cell motility has a role in more sinister events such as neoplasia.

For simple or single cell eukaryotes to get cells from one place to another is in essence, all about survival. However, whether they are simple or more complex groups of eukaryotic cells, a system for moving cells has developed and appears conserved over the millennia. Dictyostelium is an amoeba with the ability to migrate as a single cell or as a group of cells (Affolter and Weijer, 2005). This simple organism has all the proteins required for cell migration making it a suitable experimental model for investigating cell motility. Eukaryotic cell movement requires interplay of the cell's receptors and cytoskeleton and this involves a number of different signalling pathways. Eukaryotic cells have the ability to migrate towards a stimulus (chemotaxis). Chemotaxis is defined as the directional migration towards an external signal, where some form of chemoattractant is detected by cell surface receptors which direct the cells towards the stimuli (Zigmond and Hirsch, 1976). Genetic manipulation of dictyostelium has lead to some insights into how most eukaryotic cells actually move. Dictyostelium has a very small genome and manipulation of this has lead to some interesting results indicating which proteins and pathways are essential for cell motility (Kolsch et al, 2007) including; Akt, PI3 kinase, PTEN, phospholipase A2 and C. These proteins and their associated pathways are able to regulate the actin cytokskeleton. Cell migration is essential for all multicellular organisms to develop normally into an adult with the ability to heal when wounded, but it also allows tumours to metastasise.

One of the most fundamental questions after why cells move, is how do cells move? A number of models for cell motility have been put forward. However, cells will migrate in different ways depending on the substratum upon which they are plated or surrounded by. Initial work focussed on cells moving across a 2D substratum, but we have now moved towards investigating and understanding migration in the third dimension. Fibroblasts migration in 3D involves a minimum of three interlinked functional events: Attachment of the leading edge of the cell, contraction of the cell and detachment of the back end of the cell. The attachment of cells to the leading edge is via receptors to the ECM and the pulling of the cell across the matrix, detaching from the matrix as they go (Reviewed in: Friedl and Brocker 2000). In a 3D collagen system these events are under the influence of integrins β1 and α2 (Schmidt and Friedl, 2010). Blocking of either integrin, with specific antibodies, results in cells being unable to migrate through a 3D matrix (Maaser et al, 1999). Attachment of the edge of the cell is via focal complexes which in turn become focal adhesions and finally fibrillar adhesions (Morgan et al, 2009). These complexes have a turnover of integrin receptors and phosphotyrosine, leading to the force of motility being transmitted through the actin cytoskeleton. Cell receptors are thus linked from the ECM to the cell and then inside the cell to the cytoskeletal elements required for motility (Morgan et al, 2009). For a cell to move, a large number of cell-matrix contacts must be maintained and broken. The cell receptors must maintain an intricate balance between the matrix on the outside, the membrane where it is located and the cytoskeletal and signal transduction pathways inside the cells. Certain signalling pathways and integrins appear to cluster at specific sites within a migrating cell (Morgan et al, 2009).

Signalling Pathways in Migrating Cells

For a cell to move towards (or away from) a stimulus it must be capable of reacting to an external source. How a cell does this is via cell surface receptors and intracellular signalling pathways. A large number of apparently diverse pathways have been implicated in the migration of cells (Schmidt and Friedl, 2010). The pathways reviewed here are those relevant to integrin and fibronectin related motility. Once an integrin has been activated, a cascade of molecules are phosphorylated, one of the first proteins to be phosphorylated is focal adhesion kinase (FAK) (Akiyama, 1996), activating the MAP-kinase via ras (Schlaepfer and Mitra, 2004). Blocking of this pathway with specific inhibitors reveals its importance in cell motility in response to fibronectin (Anand-Apte et al, 1997).

In chemotaxis one of the most important pathways for cell motility is via phospholipid signalling (Kolsch, Charest and Firtel, 2008). Initial experiments found that Dictyostelium migration was controlled by the production and breakdown of phosphatidylinositol (3, 4, 5) -triphosphate (PIP3). This is controlled, in part, by the actions of two enzymes; phosphorinositide 3-kinase (PI3K) and the phosphatase PTEN (phosphatase and tensin homologue) (Huang et al, 2003; Iijima and Devortes, 2002). By utilising PIP3, cells can activate the actin cytoskeleton necessary for cell motility. PIP3 also interacts with Akt to activate myosin assembly which is also necessary for cell migration (Kolsch, Charest and Firtel, 2008). There are at least three mammalian isoforms of Akt (Zhou et al, 2006) each with a different role in the control of cell migration, with one being responsible for stimulation and one for inhibition (Zhou et al, 2006).To investigate motility of cells a number of cell migration assays have been developed.

Cell Migration Assays

Cell migration assays have been developed over the last few decades. These assays can be split into 2 main types; motility of cells across 2D surfaces and migration in 3-dimensions.

These assays have been employed to investigate everything from growth factors and cytokines to ECM compounds and their proteolytic fragments. The two most popular cell migration assays are the Boyden chamber type assays and the wound or gap closure assays. Our group has developed an assay to investigate motility into a 3D collagen gel; similar assays have substituted Matrigel, made up of basement membrane proteins, to investigate cell motility.

The Boyden chamber type assay was first described in the early 1960s and through various modifications has become one of the most widely used. The early experiments were carried out in chambers requiring a large number of cells. However, microchemotaxis chambers have been developed and these require fewer cells to obtain statistically valid data. The assay in essence consists of plating cells on to the top of a filter with a specific pore size. The chamber is incubated with an agonist for a specified time and then fixed, and stained. The cells that have migrated in response to the agonist, to the lower side of the filter are counted - usually by microscopy (Zigmond and Hirsch, 1973; Figure 2a). These assays can be easily modified and can be used to investigate the migration of a number of different cells types across the membrane. Treatment of the membrane can have important consequences for the motility of the cells through the pores e.g. coating of the membrane with either collagen or denatured collagen leads to different fragments of Fn being active in stimulating migration (Schor et al, 1996 and Figure 6a).

The other popular 2D migration assay is the Scratch assay or monolayer wound healing assay (Valster et al, 2005). This assay involves growing cells on a dish or 24-well plate and once confluent, a parallel scratch wound of approximately 400µm width is made with a plastic pipette tip. The wounds are observed and photographed using phase contrast on an inverted microscope. Photomicrographs are taken at regular intervals over the course of 12-24 hours. Figure 2b shows an oral epithelial tumour line migrating across a scratch in response to the growth factor VEGF. After 24 hours the scratch in the presence of growth factor is closed but not the control.

A number of 3D assays have been developed in the last thirty years. These include the chick allantoic membrane assay used to investigate angiogenesis and invasion (Ribatti et al, 2001). This assay has lead to the designation of large number of cytokines and matrix macromolecules as being angiogenic (Viji et al, 2008). Another assay is the migration of cells into or through matrigel. Matrigel is a basement membrane mixture initially isolated from Engelbreth-Holm-Swarm (EHS) tumour cells injected into mice (Kleinmann and Martin, 2005), migration using this matrix has lead to a growing number of reports of how tumour cells can move in 3 dimensions and to the development of an *in vitro* angiogenesis assay.

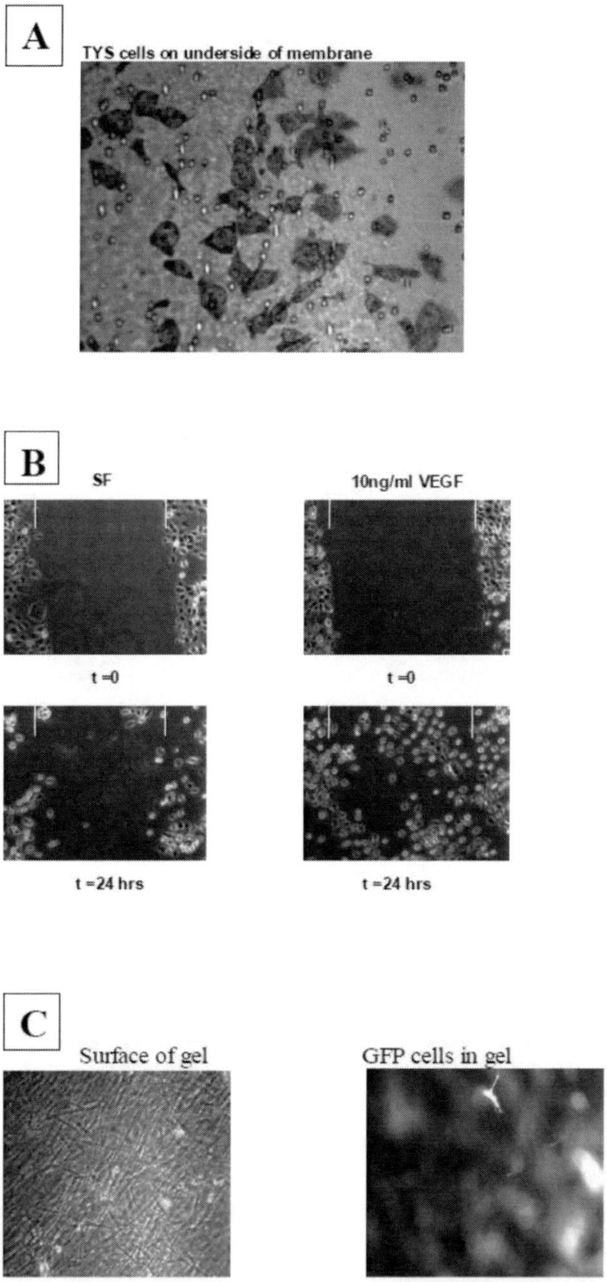

Figure 2. Examples of 2D and 3D cell migration assays: A) Boyden or micro-chemotaxis assay: Movement across or through a filter involves cells being plated onto the top surface of a semi-permeable filter. After a specified length of time, the cells on the top surface are wiped off and the cells underneath stained and counted. B) Scratch assay: Cell motility is measured in terms of cell response to a stimulus. The confluent cell monolayer is scratched and the migration of cells into the denuded area is recorded by photography over a 12-24 hour period. Added growth factors or matrix molecules can be compared in this assay, as well as inhibitors. C) Collagen gel assay: The 3D collagen gel was developed to give a more tissue-like environment to visualise cell migration. The assay involves plating cells onto the surface of a pre-set collagen gel incubated with the motogen. The number of cells that have migrated into the collagen gel after 96 hours is then quantified.

The experimental system used in our laboratory has involved a 3D collagen gel system (Schor, 1980- Figure 2c). This assay involves the plating of cells onto the surface of a 3D matrix of collagen isolated from rat tails (Schor, Ellis and Schor, 2001). A large number of parameters have been tested in the last thirty years. Early experiments lead to differences being observed in the effect fibronectin had on the migration of fibroblast cells and melanoma cells. Full length fibronectin stimulated the migration of melanoma cells into the 3D collagen matrix but had no effect on adult fibroblasts (Schor, Bazill and Schor, 1981). Further research investigated the migration of fibroblasts from different sources into 3D collagen gels. Fibroblasts were plated at confluent (high density, $2x10^5$ cells per 35ml gel) and sub-confluent (low density, $1x10^4$ cells per 35ml gel) onto the surface of the gel and the migration was determined after 96 hours (Schor et al, 1985). Data showed that although migration at low plating density was similar, high plating density migration depended on the origin of the fibroblasts. Foetal fibroblasts had high levels of migration at a confluent density when compared with adult fibroblasts (Schor et al, 1985). Using a mathematical formula, an index of migration was determined, known as the Cell Density Migration Index or CDMI. Cells with an index above +0.4 were described as “adult-like”. Studies of fibroblasts from various cancers indicated that these cells had a migratory phenotype of less than +0.4 (“foetal-like”).

The next step was to determine what the cause of this increased migration in foetal and cancer patients was. Serum free-conditioned medium from foetal, cancer patients and adult fibroblasts was collected and tested on adult fibroblasts (Schor et al, 1988). Conditioned medium from cancer patients and foetal fibroblasts was found to stimulate the migration of adult fibroblasts into the 3D collagen gel. Further biochemical purification of this factor, called migration stimulating factor (MSF), followed and it was found to be related to the N-terminus of fibronectin (Grey et al, 1989, Schor et al, 1996 and Schor et al, 2003). The following results sections will describe the investigation of the localisation and identification of the bio-active motif in fibronectin and the mechanism that fibroblasts use to migrate.

Results

1. Structural Domains of Fibronectin and Their Effect on Migration

Fn has a very consistent domain structure made up of components known as Fn type I (FnI), Fn type II (FnII) and Fn type III (Fn III) modules, containing ~45, 60 and 90 residues, respectively. These modules are highly conserved throughout the Fn family and are able to form different types of structures that enable Fn to have a large and variable number of functions (Hynes, 1990). Historically, Fn has been described by its ability to bind to various cell and extracellular matrix molecules. Due to its susceptibility to protease digestion, Fn fragments have been isolated and purified by their ability to bind to ECM components and cell surface receptors. From the N-terminal region: the HepI/FibI a 29kDa fragment that binds to both heparin and fibrin, the gelatin binding domain (GBD) a 45kDa fragment that binds to gelatin, the cell binding domain (CellBD) a 110kDa fragment that binds to various integrins, but primarily the alpha 5 beta 1 (α_5 β_1), the Hep II a 35kDa fragment that binds to heparin and the Fib II a 25kDa fragment that binds to fibrin.

Fn has the potential to elicit a variety of effects depending on the target cell type, the cell density and the matrix surrounding the cells (Schor et al, 2003). Fn also has the potential to release a large number of cryptic activities when it is broken down into smaller subunits (Ellis et al 2010A). Fn is a very important molecule in normal development, in fact Fn-/- homozygous null embryonic mice die very early in development (Watt and Hodivala, 1994), due to a number of disorders especially in the mesoderm. Fn has a role in various disease states, these include tumours and arthritis (Schor et al, 1996).

Cell migration can be measured using a number of different assays, including scratch assays, Boyden or microchemotaxis assays, the collagen gel assay and collagen gel sandwich assay. Each assay has some merits and some obvious drawbacks. We have used the microchemotaxis, collagen gel and collagen sandwich assays to study the effect of either growth factors or matrix macromolecules on cell migration (Schor et al, 2006; Ellis et al, 2010A). Choice of assay depends upon the cell type being investigated in the experiment. For experiments investigating integrins and signalling pathways it was easier to pre-incubate cells in either antibody or inhibitor prior to the experiment.

Fn was proteolytically digested into smaller domains or recombinant fragments were synthesised using prokaryotic or eukaryotic expression plasmids. Initial experiments using the collagen gel assay and the well characterised Fn fragments; Hep I, Gel-BD, Cell, Hep II and FibI were performed on fibroblasts:. The matrix upon which cells are plated is important for cell motility. Using our 3D collagen gel assay we have shown that full length Fn fibronectin does not stimulate migration of fibroblasts. However, smaller proteolytically derived fragments are able to stimulate migration; *e.g.* the gelatin binding domain stimulates fibroblast motility (Figure 3A). Zebrafish have been reported to have a truncated form of Fn whose molecular weight is around 100kDa. This protein "runs" from the N-terminus of Fn until the third type III repeat. Two different forms of this 100kDa protein were made and were shown not to stimulate fibroblast migration when analysed using the 3D collagen gel assay (Vakonakis et al, 2009, Kay et al unpublished; Figure 3B). We were able to generate an *in silico* model of protein folding (Vakonakis et al, 2009). Analysis of the 100kDa protein indicated that the third FnIII module bound to the fourth FnI module removing the motogenic activity. A point mutation (R222A) in the protein was able to open up these sites and allowed for migration stimulating activity (Figure 3B) to be restored (Vakonakis et al, 2009).

Other recombinant proteins were made including the removal of the third FnIII domain (Fn90kDa) and the removal of all the FnIII domains (Fn70kDa). The removal of the modules produced proteins that have migration stimulating activity (Figure 3C and 3D) suggesting that the residues responsible for creating the inactive confirmation are found in third type III repeat.

Work from our laboratory has indicated that the migration stimulating activity produced by foetal and cancer patient fibroblasts had an amino acid sequence from within the gelatin binding domain of Fn (Grey et al, 1989; Schor et al, 1996). However, cloning of the gene responsible for the migration stimulating activity suggested a protein of around 75kDa (Schor et al, 2003). Sequencing of the gene for MSF found that it was a made up of a sequence from the N-terminus of Fn to the first type III repeat. The first type III repeat is made up of two exons split around an intron. MSF was found to contain a read-through into the intron of 10 amino acids followed by a stop codon. This means that MSF stops at the end of FnIII 1a (Schor et al, 2003). MSF therefore is made up of two of the proteolytic fragments of Fn, the

GBD (made up of four FnI repeats and the two FnII repeats) and the HepI domains (made up of the first five FnI repeats).

We used the collagen gel assay in order to investigate full-length MSF and the proteolytic fragments. Full-length MSF and the proteolytic fragment containing the GBD had migration stimulating activity, whilst in this form the Hep 1 had no activity (Figure 4A, Ellis et al, 2010A).

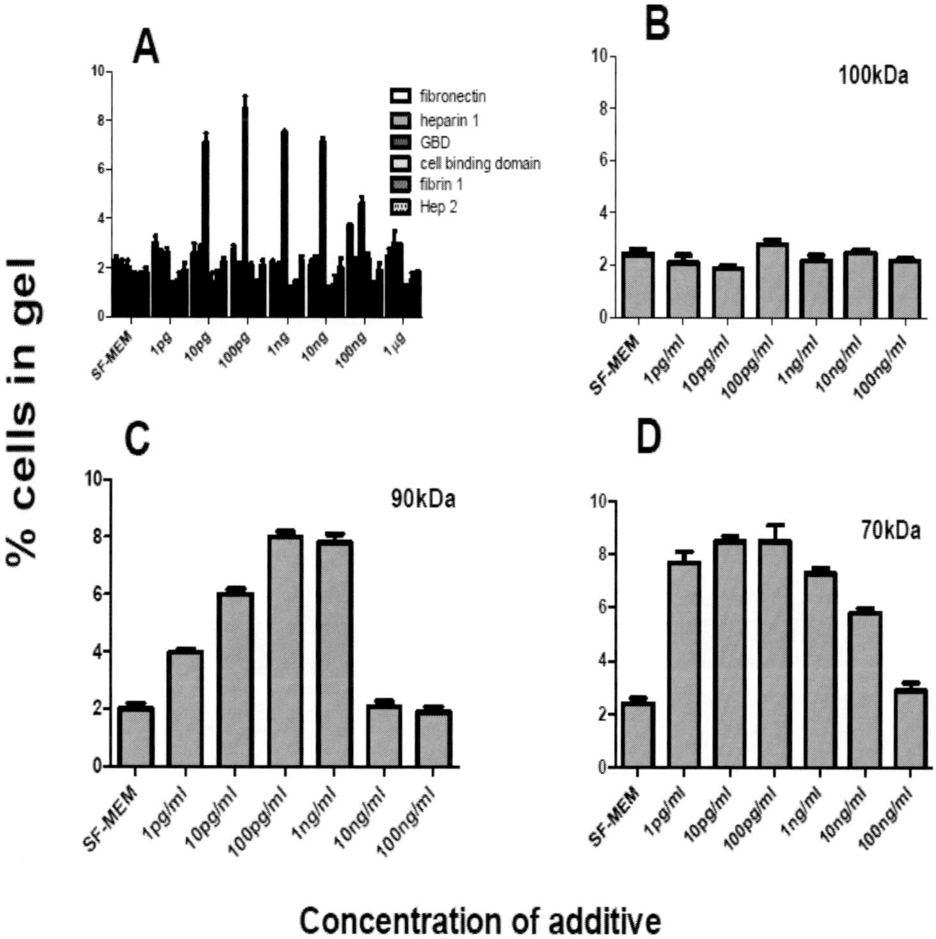

Figure 3. Collagen gel assay: FSF44 fibroblasts were plated onto the surface of a 3D collagen gel, in the presence of added factor. Migration from the surface into the gel was determined microscopically after 96 hours. A) The graph shows the dose response range (1pg/ml-1µg/ml) of added Fn and its proteolytically derived fragments on migratory response. The data indicates that only the gelatin binding domain of Fn stimulated the migration of fibroblasts. The assay produced similar results when performed in SF-MEM, 1% (v/v) DCS, 5% (v/v) DCS and fibronectin stripped-5% (v/v) DCS. B) The graph shows the dose response range (1pg/ml-100ng/ml) of a recombinant truncated form of Fn (100kDa). This protein is made up of the N-terminal region up to the end of the third FnIII domain. C) The graph shows the dose response range (1pg/ml-100ng/ml) of a recombinant truncated form of Fn (90kDa). This protein is made up of the N-terminal region up to the end of the second FnIII domain. D) The graph indicates the dose response range (1pg/ml-100ng/ml) of a recombinant truncated form of Fn (70kDa). This protein is made up the N-terminal region up to the end of the ninth FnI domain.

It was predicted by Hynes (1990) that the IGD sequences found in the type FnI repeats would have some biological role to play. Of the nine FnI repeats four of these have an IGD sequence, two in the Hep1 domain and the other two in the GBD. In looking for the active motif of MSF we made small peptides containing the IGD motif and tested them in the migration assay (Schor et al, 1999). The Hep 1 domain proteolytically derived from Fn has no migratory stimulating activity. In order to investigate this we used recombinantly expressed FnI 1-5 domains. Protein sequencing of this and the Hep1 domain showed that the Hep 1 had extended N-terminal and C terminal sequences. This added about 90 amino acids to the Hep 1 protein (Ellis et al, 2010A). Using the collagen gel migration assay we tested the biological activity of recombinant forms of the protein that make up the Hep 1 domain. The full length recombinant protein stimulates migration of fibroblasts as did the two smaller units of the domain (Figure 4B). This is contrary to the proteolytically derived protein and is probably due to folding of the native protein which masks the IGD motifs. The two IGD containing repeats are the third FnI and the fifth FnI. In order to investigate whether or not the IGD sequences were responsible for migration stimulating activity we used fragments containing two type FnI repeats; the FnI 2-3 and the FnI 4-5 which are able to stimulate migration, whereas the FnI 1-2 did not (Figure 4C), indicating that the IGD containing repeats were probably responsible for migration stimulating activity.

At the C-terminus of MSF the IGD sequences are in the 7 FnI and the 9 FnI. Investigation of these recombinant fragments showed that the FnI 7-8 and the FnI 8-9 were both active and the FnI 7-9 was also active (Figure 4D). Mutation of the IGD motif knocked out bioactivity (Millard et al, 2007). It should be noted that these smaller units have less biological activity than full length MSF or the 70kDa fragment of Fn. This suggests that although the active sequences have bioactivity, larger molecules aid stability and possibly have further interactions with the cell surface receptor. Peptides of around 20 amino acids in length were synthesised around the IGD motifs and tested for migration stimulating activity; the IGDS and IGDQ containing peptides stimulated migration at around 1ng/ml (Figure 4E). However, when the peptide was circularised this lead to a 10 fold enhancement of activity (Figure 4E and Ellis et al, 2010A).

A number of short amino acid sequences within Fn have been implicated in cell migration and adhesion. These include the RGD motif, the IGD motif (Ellis et al, 2010A) and the HEEGH motif (Houard et al, 2005). The VGD motif is found within the FnI repeats but has no reported function (Ellis et al 2010A). These peptides were analysed using the 3D-collagen gel assay; only the IGDS and IGDT stimulated fibroblast migration in this assay.

The concentration of these peptides needed to stimulate migration was found to be very high compared with full length MSF (Figure 4F). This suggests that when binding to the receptor, IGD sequences need to be stabilised by other, as yet undetermined, sequences. The relative concentrations needed for biological activity in the collagen gel assay are summarised in Table 1.

In order to determine whether the IGD sequences were in fact needed for biological activity we made a number of mutants of the MSF protein. As described in Schor et al, (2003) the migration of fibroblasts requires an intact IGD in either module 7 or 9. MSF stimulates migration into 3D collagen gels and that mutation of IGD at both the 7 and 9 modules removes the ability of the protein to stimulate migration (Figure 5 A and B).

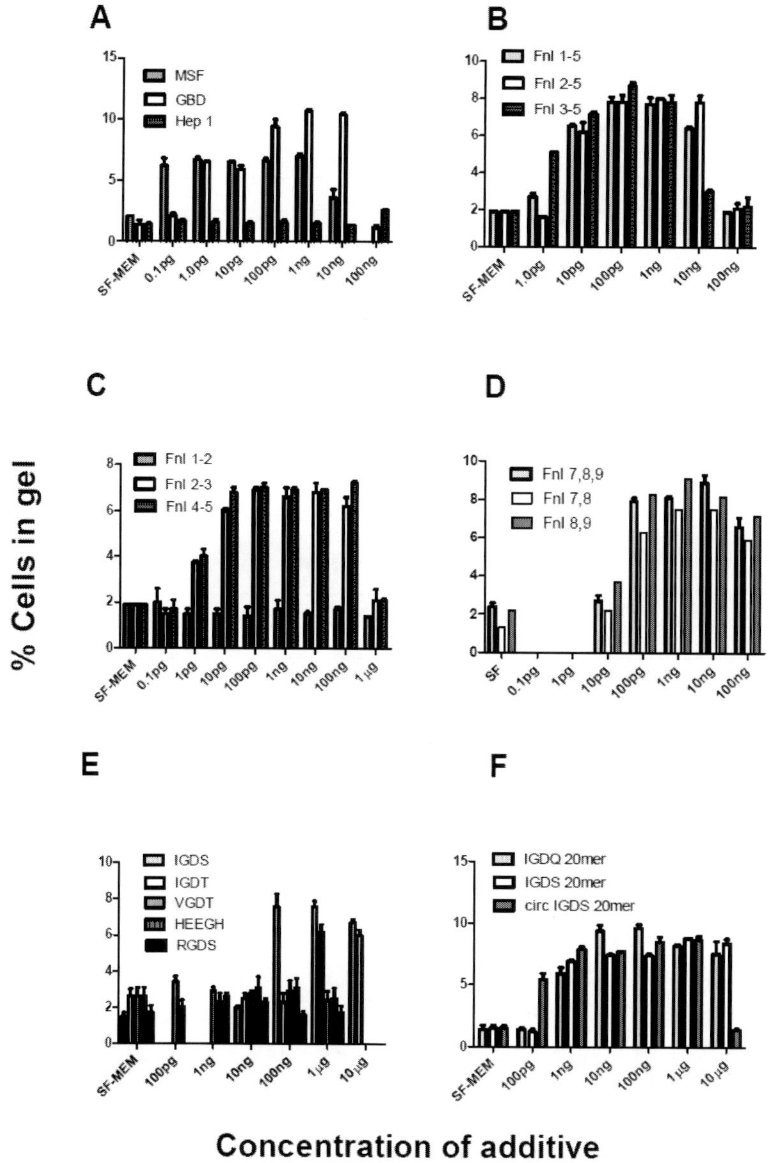

Figure 4. Collagen gel assay: FSF44 fibroblasts were plated onto the surface of a 3D collagen gel, in the presence of added factors. Migration from the surface into the gel was determined microscopically after 96 hours. A) The graph shows a dose response range (0.1pg/ml-100ng/ml) of added recombinant MSF, GBD and HepI both proteolytically derived fragments (Sigma) on migratory response. The data indicates that only the MSF and GBD stimulated the migration of fibroblasts, above background levels (SF). B) The graph shows the dose response range of (1pg/ml-100ng/ml) recombinant proteins FnI 1-5, FnI 2-5 and FnI 3-5. These proteins were expressed in HEK293T cells. C) The graph shows the dose range (0.1pg/ml-1µg/ml)of a recombinant FnI 1-2, FnI 2-3 and FnI 4-5 modules. These proteins were expressed eukaryotically in HEK293T cells. D) The graph indicates the dose response range (10pg/ml-100ng/ml) of recombinant proteins FnI 7-9, FnI7-8 and FnI 8-9. These proteins were expressed in HEK293T cells. E) The graph indicates the dose response (100pg/ml- 10µg/ml) range of chemically synthesised proteins; IGDS (20mer), IGDQ (20mer) and circularised IGDS (20mer). F) The graph indicates the dose range of (100pg/ml-10µg/ml) chemically synthesised proteins; IGDS, IGDT, VGDT, HEEGH and RGDS.

Table 1. Summary of the effect of proteins and peptides added to FSF44 fibroblasts in the collagen gel assay

Protein/peptide	Active range in collagen gel assay (pg or ng/ml)	Active range in collagen gel assay (molar concentration)	Type of response
MSF	0.1pg/ml-10ng/ml	1.3fM-130pM	Bell shaped
GBD	1pg/ml-10ng/ml	22fM-220pM	Bell shaped
Hep 1	Not active	Not active	Not applicable
$^{1\text{-}5}$FnI	10pg/ml-10ng/ml	0.42pM-420pM	Bell shaped
$^{2\text{-}5}$FnI	10pg/ml-10ng/ml	0.53pM-530pM	Bell shaped
$^{3\text{-}5}$FnI	1pg/ml-1ng/ml	71fM-71pM	Bell shaped
$^{1\text{-}2}$FnI	Not active	Not active	Not applicable
$^{2\text{-}3}$FnI	10pg/ml-100ng/ml	1.1pM-11nM	Bell shaped
$^{4\text{-}5}$FnI	10pg/ml-100ng/ml	1.1pM-11nM	Bell shaped
IGDQ (20mer)	1ng/ml-10μg/ml	480pM-4.8μM	plateau
IGDS (20mer)	1ng/ml-10μg/ml	3.8pM-3.8μM	plateau
IGDS (20mer) circularised	100pg/ml-10μg/ml	48pM-4.8μM	Bell shaped
IGDS	100ng/ml-10μg/ml	0.25μM-25μM	plateau
IGDT	1μg/ml-10μg/ml	0.25μM-25μM	plateau
VGDT	Not active	Not active	Not applicable
HEEGH	Not active	Not active	Not applicable
RGDS	Not active	Not active	Not applicable

Table 1; summary of migration using the 3D collagen gel assay. Each protein/peptide used in the paper is listed and how it affected migration with the range it was active.

Other peptide motifs have also been implicated in cell migration. Houard et al, (2005) suggested the important motif for migration stimulating activity for MCF7 cells in the Boyden chamber assay was the HEEGH sequence. This motif has been reported to have protease activity (Houard et al, 2005). In our 3D collagen gel assay a FEEGF mutant had no inhibitory effect on fibroblast migration (Figure 5C) whereas MSF with all four IGD sequences mutated to DGI and wild-type HEEGH had no stimulatory activity (Figure 5D).

The activity of MSF and GBD appears to be very assay and cell type dependent. In order to compare our experimental results with other methods, we investigated cell migration using the micro-chemotaxis or Boyden chamber assay. This is a quick assay, usually complete in less than 6 hours. The assay is also performed in serum free conditions, making the effector molecule the sole stimulator of motility.

Using this type of assay and different cell types other groups have reported migration stimulating activity in the Cell binding domain (Postlethwaite et al, 1981; Clark et al, 1988; Fukai et al, 1991) and the Hep II domain on neural crest cells (Perris et al, 1989). Initial work using gelatin coated membranes suggested that the GBD end of the molecule was not active in stimulating migration. However, when the filter was coated with native collagen rather than gelatin, the GBD was active in stimulating the migration of fibroblasts (Figure 6A). At around 100μg to 1mg Fn and CBD were also active.

Full length MSF stimulates motility of fibroblasts in this assay and at lower concentrations around 10pg/ml-10ng/ml (Table 2, Figure 6B). Mutations of the MSF sequence to knock out the IGD, in FnI modules 3, 5, 7 and 9, removed migration stimulating activity as measured by this assay (Figure 6C). However, mutations in the HEEGH sequence had no effect on motility of fibroblasts as measured by this assay (Figure 6D). Mutation of the IGD sequences in FnI 7 and 9 removed the ability of these MSF mutants to stimulate fibroblast migration.

However, as with the collagen gel assay, we investigated whether migration stimulating activity was cryptic in the N-terminal region, especially around the IGD motifs in FnI 3 and 5. Recombinant FnI1-5 and FnI 2-5 and FnI 3-5 were all tested in the micro-chemotaxis assay; the full length recombinant protein had no activity but the smaller peptides did (Figure 6E).

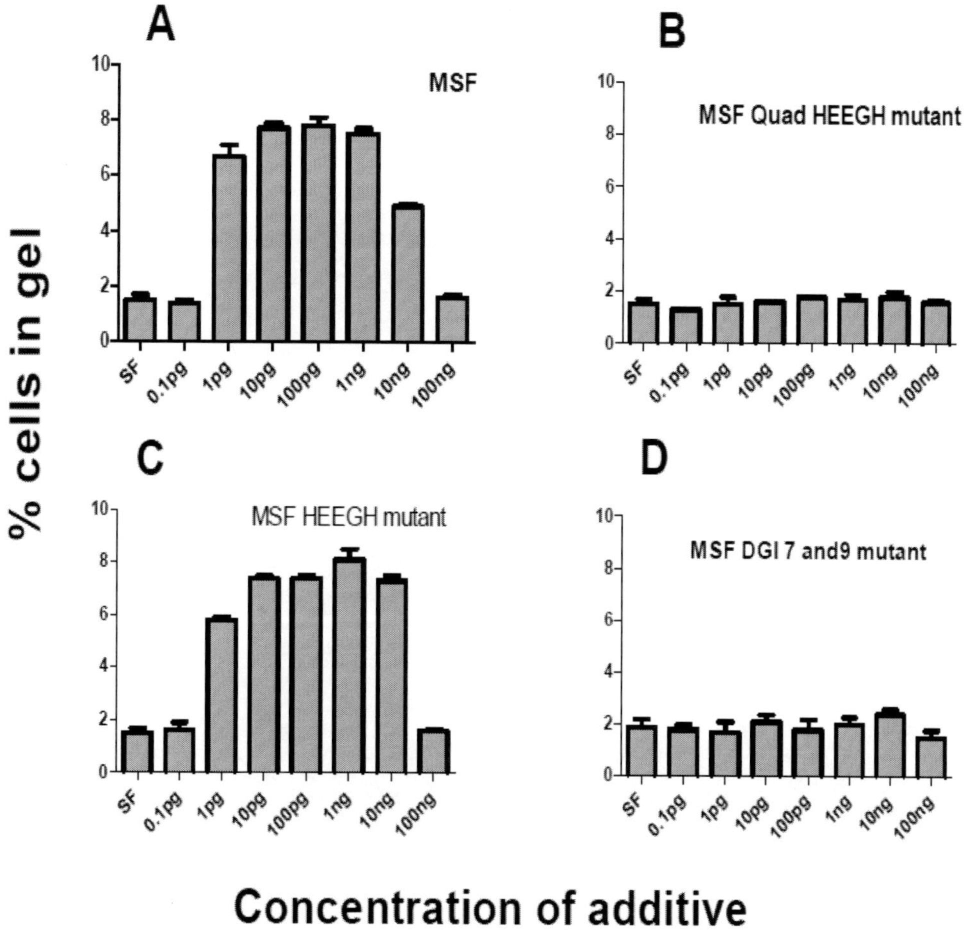

Figure 5. Collagen gel assay: FSF44 fibroblasts were plated onto the surface of a 3D collagen gel, in the presence of added factor (0.1pg/ml-100ng/ml). Migration from the surface into the gel was assessed microscopically after 96 hours. A) The graph indicates a dose response range of recombinant MSF. The protein was expressed as a recombinant protein in bacteria and purified as described in Schor et al, (2003). B) MSF mutant with the IGD sequences in the FnI 7 and FnI 9 modules mutated to DGI. These mutations were made and expressed as a recombinant protein in bacteria and purified as described in Schor et al, (2003). C) MSF mutant with the HEEGH mutated sequence in the FnI8 module. The protein was expressed as a recombinant protein in bacteria and purified. D) MSF mutant with IGD sequences in the FnI 3, FnI 5, FnI 7 and FnI9 mutated to DGI as well as the HEEGH mutated to FEEGF sequence in FnI 8. (See B).

However, activity was fully recovered with the addition of vitronectin to the recombinant FnI 1-5 protein (Ellis et al, 2010a). It is not clear whether there is an interaction between the FnI 1-5 and vitronectin or whether an interaction at the cell surface is responsible. Smaller fragments of FnI (1-2 domains, 2-3 and 4-5 domains) were also expressed as recombinant proteins. Only proteins with IGD sequences within them were found to be active (Figure 6F). It was noticeable that these proteins were all active in the low to high nanogram range. Different lengths of peptides containing IGD sequences were chemically synthesised; 20 amino acids being the largest and 4 amino acids the smallest and were all tested in the micro-chemotaxis assay. All were found to be active but highest activity was found in the high nanogram to low microgram range. It should be noted that all IGD containing motifs have a wide concentration range of activities in the micro-chemotaxis assay (Table 2).

One of the main areas of study has been the effect of Fn and its fragments on fibroblasts. These cells have an important role to play in wound healing and in stromal influence of tumour cells. One of the most fascinating series of experiments was the investigation of the effect of MSF on the migration of other cell types. Oral tumour lines grown in our laboratory have little ability to migrate into collagen gels and therefore to analyse this type of cell we have used the micro-chemotaxis assay. TYS (an oral tumour cell line) cells were investigated for their ability to migrate in response to MSF and its mutants.

Table 2. Summary of the effect of proteins and peptides added to FSF44 fibroblasts in the micro-chemotaxis assay

Protein/peptide	Active range in microchemotaxis assay (pg or ng/ml)	Active range in microchemotaxis assay (molar concentration)	Type of response
MSF	10pg/ml-10ng/ml	0.13pM-130pM	Bell shaped
GBD	100pg/ml-100ng/ml	2.2pM-220pM	Bell shaped
Hep 1	Not active	Not active	Not applicable
$^{1\text{-}5}$FnI (SF-MEM)	Not active	Not active	not applicable
$^{1\text{-}5}$FnI with vitronectin	100pg/ml-10ng/ml	4.2pM-420pM	Not full range
$^{2\text{-}5}$FnI	100pg/ml-100ng/ml	5.3pM-5.3nM	Bell shaped
$^{3\text{-}5}$FnI	100pg/ml-100ng/ml	7.1pM-7.1nM	Bell shaped
$^{1\text{-}2}$FnI	Not active	Not active	Not applicable
$^{2\text{-}3}$FnI	1ng/ml-1μg/ml	0.11nM-110nM	plateau
$^{4\text{-}5}$FnI	1ng/ml-100ng/ml	0.11nM-11nM	Bell shaped
IGDS (20mer)	10ng/ml-10μg/ml	3.8nM-3.8μM	Bell shaped
IGDS (20mer) circularised	10ng/ml-10μg/ml	4.8nM-4.8μM	Bell shaped
IGDT	100ng/ml-100μg/ml	0.25μM-250μM	Bell shaped (?)

Table 2: indicates a summary of migration in the transmembrane assay. Each protein/peptide used in the paper is listed and how it affected migration with the range it was active.

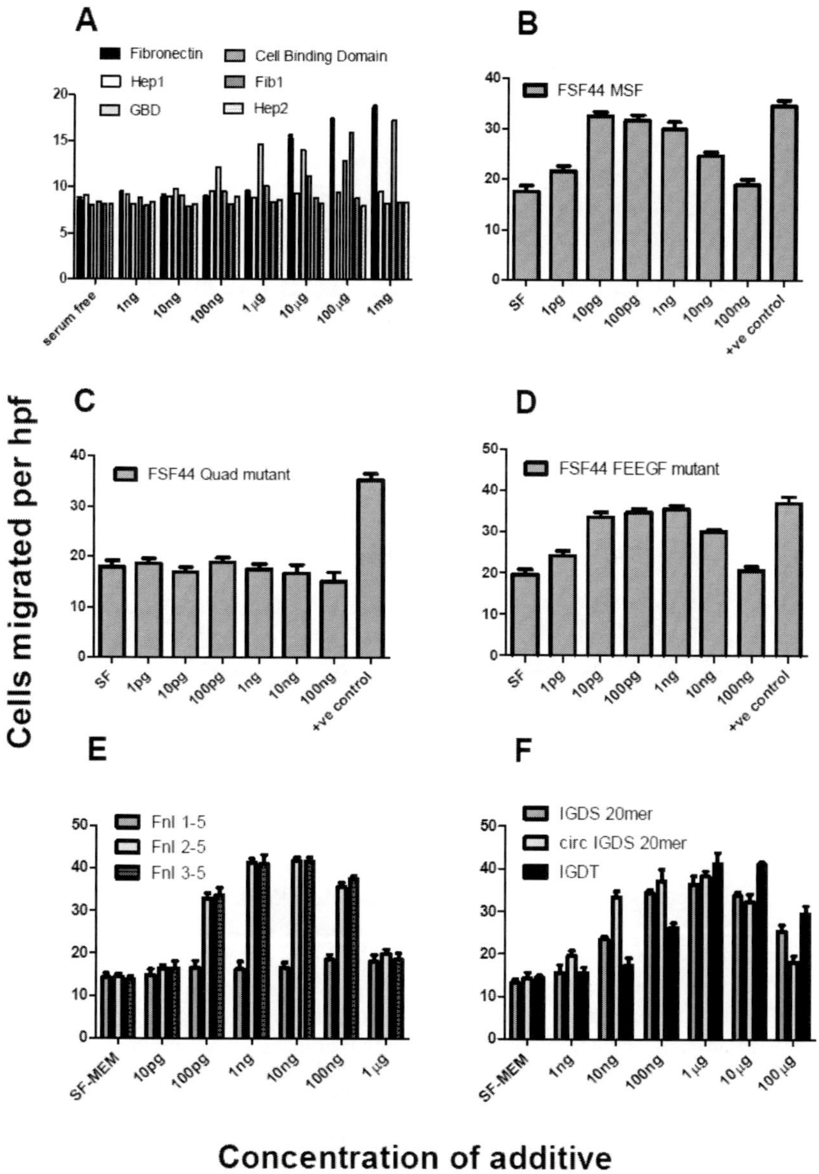

Figure 6. Micro-chemotaxis or Boyden chamber assay: FSF44 fibroblasts were plated onto the upper surface of a native collagen coated 8µm PVP membrane (as described in Ellis et al, 2010A). After 5 hours, migration of cells from the upper surface to the lower surface of the membrane via the pores was determined microscopically after fixation and staining. A) The graph shows the dose response range (1ng/ml-1mg/ml) of added Fn and its proteolytically derived fragments Hep 1, GBD, CBD, Fibrin 1 and Hep 2 (Sigma) on migratory response. The data indicates that GBD stimulated the migration of fibroblasts at lower concentrations than Fn and CBD. B) The graph shows the doseresponse range of a recombinant MSF. The protein was made from a bacterial expression plasmid and purified as described in Schor et al, (2003). C) The graph shows the dose response range of recombinant MSF Quad IGD mutant. These mutations were made as described in Schor et al, (2003). The protein was expressed in a bacterial expression system and purified. D) MSF mutant with the HEEGH sequencemutated to FEEGF in the FnI8 module (Houard et al, (2005)). The protein was made from a bacterial expression plasmid and purified. E) The graph shows the dose response range of recombinant FnI 1-5, FnI 2-5 and FnI 3-5. These proteins were expressed eucaryotically in HEK293T cells. F) The graph indicates the dose range of the chemically synthesised proteins; IGDS (20mer), IGDQ (20mer) and circularised IGDS (20mer).

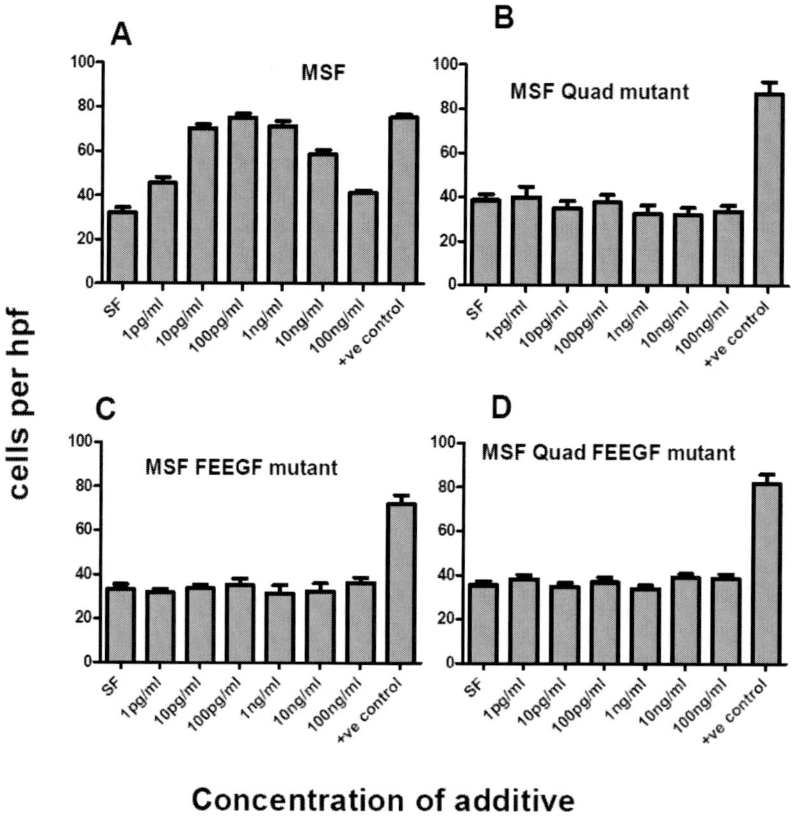

Figure 7. Micro-chemotaxis or Boyden chamber assay: FSF44 fibroblasts were plated onto the upper surface of a native collagen coated 8µ PVP membrane (as described in Ellis et al, 2010A). After 5 hours migration of cells from the upper surface to the lower surface of the membrane via the pores was determined microscopically after fixation and staining. The y axis represents cells per high powered field (hpf). A) The graph shows a dose response range of prokaryotic recombinant MSF as described in Schor et al, (2003). B) MSF mutant with the IGD sequences in the FnI 3, FnI 5, FnI 7 and FnI 9 modules. These mutations were made as described in Schor et al, (2003). The protein was expressed in a bacterial expression system and purified. C) MSF mutant with the HEEGH sequence mutated to FEEGF in the FnI8 module (Houard et al, (2005). The protein was made from a bacterial expression plasmid and purified. D) MSF mutant with IGD sequences in the FnI 3, FnI 5, FnI 7 and FnI9 mutated as well as the HEEGH sequence in FnI 8. The mutations were made and the construct cloned into a bacterial expression plasmid. The protein was expressed and purified.

These cells respond to the addition of MSF, although they form a plateau with respect to motility rather than a bell shaped curve (Figure 7A). However, these cells differ in their response to mutants of MSF. Mutation of all four of the IGD sequences to DGI, in this protein, had no effect on the motility of the tumour cells. However, if the HEEGH sequence was mutated to FEEGF this protein also has no stimulatory activity (Figure 7C) and mutation of all four IGD and the HEEGH sequences also had no stimulatory effect on migration (Figure 7D).

Other cell types will also migrate into 3D collagen gels. The original work with fibroblasts indicated that they migrate to different extents depending on cell density and cell type. The original assay plated cells at both low and high density. Adult fibroblasts plated at low density migrated to a high degree into collagen gels.

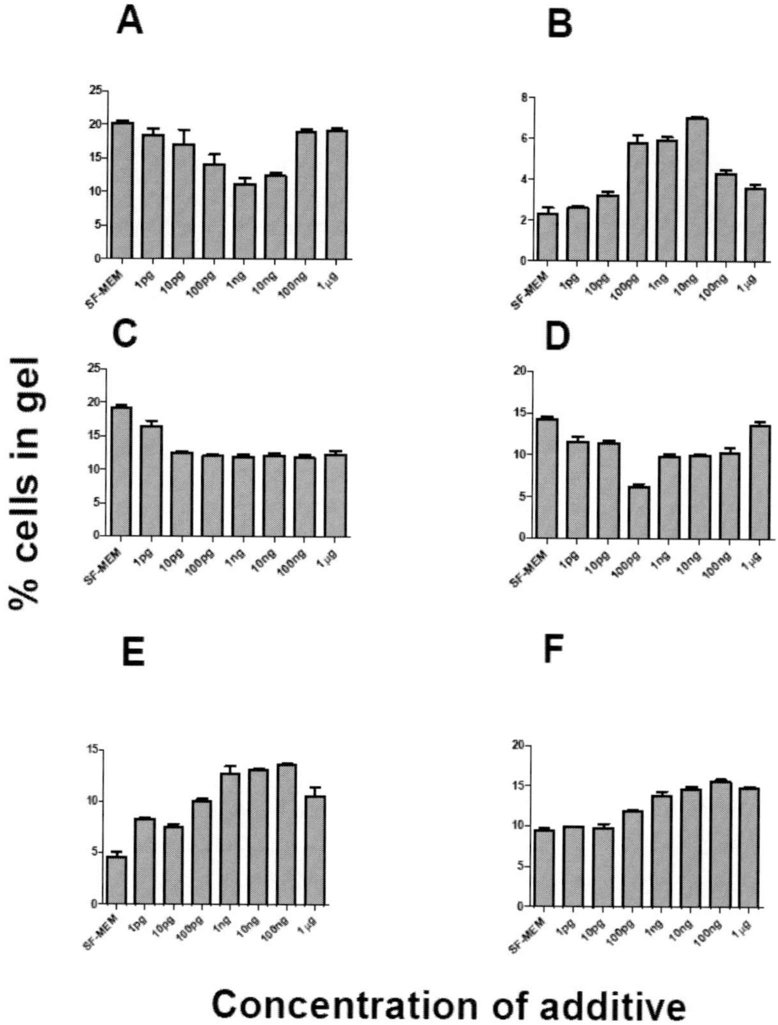

Figure 8. Collagen gel assay: Cells from different origins were plated onto the surface of 3D collagen gels, in the presence of added factor (1pg-1µg/ml). Migration from the surface into the gel was determined microscopically after 4 days. A) The graph shows the effect of different concentrations of GBD on skin fibroblasts plated at low density; in terms of migratory response. The data indicates that GBD inhibited the migration of fibroblasts in a dose dependant manner. It should be noted this data is consistent for at least 4 cell lines. B) The graph shows the effect of different concentrations of MSF on adult oral mucosa fibroblasts plated at high density; in terms of migratory response. The data indicates that MSF stimulated the migration of fibroblasts in a dose dependant manner. It should be noted this data is consistent for at least 4 cell lines. C) The graph shows the effect of different concentrations of GBD on foetal skin fibroblasts plated at low density; in terms of migratory response. The data indicates that GBD inhibited the migration of fibroblasts in a dose dependant manner. It should be noted this data is consistent for at least 4 cell lines. D) The graph shows the effect of different concentrations of GBD on fetal skin fibroblasts plated at high density; in terms of migratory response. The data indicates that GBD inhibited the migration of fibroblasts in a dose dependant manner. It should be noted this data is consistent for at least 4 cell lines. E) The graph shows the effect of different concentrations of GBD on melanoma cell line T8 (a kind gift from Prof I Hart, London) plated at high density; in terms of migratory response. The data indicates that GBD stimulated the migration in a dose dependant manner. F) The graph shows the effect of different concentrations of GBD on breast carcinoma cell line MDA-435 plated at high density; in terms of migratory response. The data indicates that GBD stimulated migration in a dose dependant manner.

Addition of GBD inhibited the ability of adult skin fibroblasts to migrate into 3D collagen gels (Figure 8A), displaying a bell-shaped dose response. However, addition of GBD to adult skin fibroblasts plated at high density produced a stimulation of migration (Figure 8B). This stimulation followed a bell shaped dose response. We have used at least ten different adult skin fibroblast cell lines and all of them display this pattern of migration. Foetal fibroblasts migrate into 3D collagen gels differently to adult fibroblasts (Schor et al, 1985) which is due to their response to endogenous MSF (Grey et al, 1989). GBD added to foetal fibroblasts at both low and high density inhibits cell migration in a dose dependant manner (Figure 8C and 8D).

The effect on low density cells produces an inhibition of 40% migration which is a plateau. This effect was observed with four foetal fibroblast cell lines. The effect at high density was more unpredictable. The inhibition produced a bell shaped response and indicated that at high concentration GBD had no effect. The effect of MSF has also been tested on adult oral mucosa fibroblast cell lines in the migration assay; MSF stimulates the migration of these cells in the 3D collagen gel assay (Figure 8E). This migration follows a bell shaped dose-response curve.

We have also investigated the effects of MSF on cancer cell lines. Human melanoma cells migrate into collagen gels (Schor et al, 1982 and unpublished results), addition of GBD to a human melanoma cell line (T8) stimulated the migration of these cells into the gel (Figure 8F). The response of a number of breast cancer cell lines to MSF has also been investigated; the cell line MDA-MB231 is stimulated by addition of MSF to migrate into 3D collagen gels (Figure 8G). This effect appears to form a plateau rather than a bell shaped dose-response curve. The significance of the difference between a bell shaped dose response and a plateau appears to be a phenomenon associated with cancer cells. Normal cell migration takes place at an optimal concentration and if this is exceeded migration ceases. Some tumour cells do not appear subject to this level of control and therefore migration can occur unchecked potentially leading to metastasis.

2. The Mechanism by Which the N-Terminal Region of Fn Stimulates Cell Migration

Cells migrate in response to a number of stimuli. These stimuli can induce normal cell migration, which is essential in both development and wound healing. However, the same stimuli can also cause cells to metastasise around the body. The understanding of the control of these conserved mechanisms is essential for both wound healing and metastasis. This may lead to the design of therapeutic agents which will be able to target either improving wound healing or blocking of metastasis.

We have determined, using variations to our assay system, the integrin subunits that appear to be important for both binding of the N-terminal region of Fn to cells and those which are important in the migration stimulated by this fragment. We determined the signalling pathways involved by an initial screen of a group of known inhibitors and focussing on the PI3 kinase/Akt pathway.

Early work indicated that HA synthesis was an important aspect of cell migration for fibroblasts in the 3D collagen gel assay. Chen et al, (1989) indicated that foetal and adult fibroblasts produce different amounts of HA. These experiments and the ones described in

Schor et al, (1989) were all performed on plastic tissue culture dishes and therefore did not take into account the influence of the matrix on HA synthesis. Ellis et al, (1992) revisited this work but changed the substrata on which the cells were plated. Fibroblasts were plated on collagen and their migration into collagen assessed on day 4 as well as HA synthesis. Addition of exogenous TGFβ-1 interacted with MSF and modulated fibroblast cell migration and HA synthesis (Ellis et al, 1992 and Ellis and Schor 1995). Foetal and adult fibroblasts plated at high density on collagen gels migrate to different amounts. A higher number of foetal fibroblasts migrate into the collagen gel and were shown to produce more HA. Their migration is susceptible to Streptomyces hyaluronan lyase activity in the collagen gel assay (Ellis, Banyard and Schor, 1997). Adult fibroblasts in the presence of MSF and other growth factors such as aFGF (FGF1) are also susceptible to this enzyme (Ellis et al, 1992 and Ellis, Banyard and Schor, 1997).

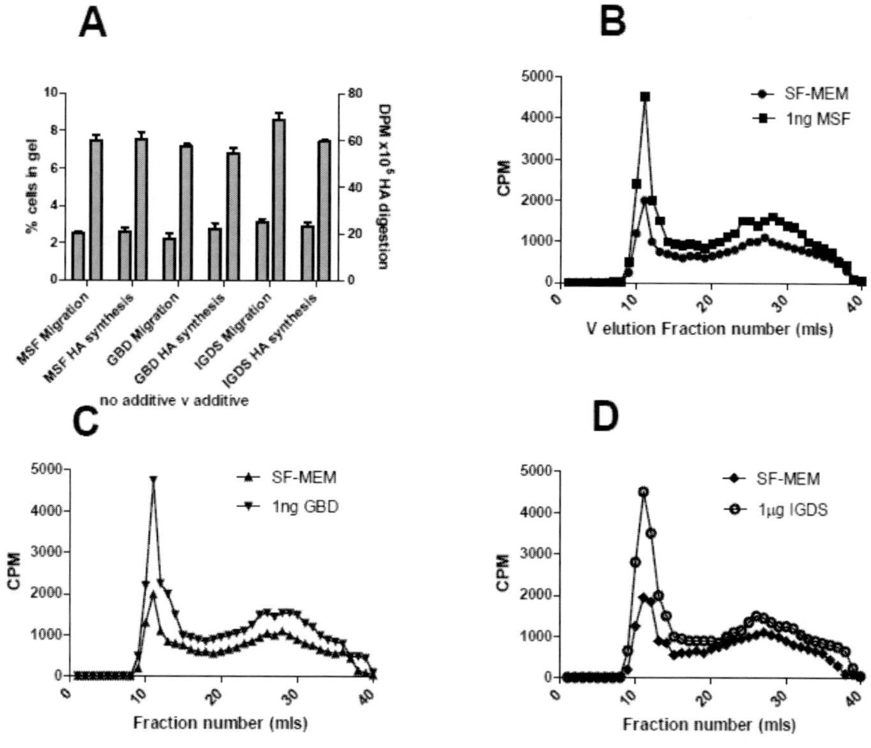

Figure 9. Collagen gel assay and hyaluronan assay: The active concentrations of MSF, GBD and IGD all stimulate HA synthesis. A) Fibroblasts were plated onto the surface of a collagen gel in the presence or absence of 1ng/ml MSF, 1ng/ml GBD and 1µg/ml IGDS. All the gels were incubated with 1µCi of ^{3}[H]-glucosamine. Cells migrated into the gels were counted microscopically. The conditioned medium was collected from each assay and analysed for HA content by Streptomyces hyaluronan lyase digestion. B) Gel filtration chromatography using Sepharose CL-2B was performed to investigate the size distribution of HA made by fibroblasts in response to the addition of 1ng/ml MSF. The data indicates an increase in high molecular weight HA synthesis. The High molecular weight HA is eluted at V0 (void volume, peak on graph). This is equivalent to a molecular weight around 2x10^6 Da. C) Gel filtration chromatography using Sepharose CL-2B was performed to investigate the size distribution of HA made by fibroblasts in response to the addition of 1ng/ml GBD. The data indicates an increase in high molecular weight HA synthesis. D) Gel filtration chromatography using Sepharose CL-2B was performed to investigate the distribution of HA made by fibroblasts in response to the addition of 1µg/ml IGDS. The data indicates an increase in high molecular weight HA synthesis.

This enzyme digests HA only; other hyaluronidases break down other GAGs making it impossible to determine which GAGs are important for migration (Ellis et al, 1997). Adult fibroblasts respond to MSF and its smaller active motifs by increased synthesis of HA. To allow us to directly compare HA synthesis and migration, we plated fibroblasts onto the surface of 3D collagen gels in the presence of 3[H]-glucosamine. Migration was measured microscopically after 4 days and HA synthesis was determined by 3[H]-glucosamine incorporation into HA by digestion with Hyaluronan lyase and ethanol precipitation.

The total incorporation and the digested fraction's incorporation of 3[H] glucosamine were determined by scintillation counting. The difference between the two values was deemed to by total HA. The addition of MSF, GBD and IGDS to fibroblasts was found to stimulate both migration and HA synthesis (Figure 9A). The effect of HA on cell motility and behaviour is determined by its molecular weight. High molecular weight HA has been implicated in inhibiting angiogenesis, whereas small fragments of HA have been reported to stimulate endothelial cell proliferation and migration (West et al, 1987). Fibroblasts synthesize different sizes of HA which can be separated by gel filtration chromatography. Adult fibroblast cells, under control conditions, produce a wide range of different sized HA molecules and respond to the addition of MSF by increasing the amount of high molecular weight HA produced (Figure 9B). A similar response to the smaller active components of MSF was observed (Figure 9C and 9D).

The N-terminal region of Fn/MSF stimulates fibroblasts to migrate into 3-D collagen gels or through the pores of a collagen coated filter. Initially it was not obvious how the cells were stimulated to move either through the collagen gel or through the pores of the filter. The Arginine -Glutamine- Aspartate-Serine (RGDS) sequence is not present in the N-terminal region of Fn and therefore is not a "true" integrin binding protein. We decided to subdivide the movement of the cells into two events: 1) Activation of the receptor/binding protein by the ligand. 2) Manifestation of this activation in terms of cell motility. Activation was analysed by pre-incubation of confluent fibroblasts plated onto plastic, gelatin, collagen or fibrin coated dishes with the cytokine or matrikine of choice. This methodology permitted activation of the cells by the cytokine/matrikine and allowed us to study the receptors and pathways involved in this step (Figure 10A). The cells were then plated into the collagen gel assay in the absence of any further added cytokine and the migration determined as normal. This experimental design allowed us to investigate the receptors and pathways important for migration. It was necessary to pre-incubate the cells with IGDS for around 7 hours to produce an increase in motility (Figure 10B). This pre-incubation event has been investigated for a large number of cytokines in our laboratory and always requires approximately 4 hours before an increase in motility are observed (Ellis et al, 1997, Ellis et al, 2010A and Schor et al, 2011).

These experiments allowed us to investigate activation and migration as separate events, as well as a whole. By using antibodies to specific receptors and to signal transduction pathways we have been able to determine how fibroblasts migrate in response to MSF into 3D collagen gels. It must be noted that the way fibroblasts migrate in response to MSF appears to be very assay specific and the pathways utilised in motility in the collagen assay are different from those in the micro-chemotaxis assay.

An investigation into the effect of IGDS on migration both in assay and in terms of pre-incubation was performed (similar results have been shown for MSF- data not shown). The basis of this type of experiment was to determine which integrin receptors were important for binding of the ligand and then in cell motility. When IGDS is present in the assay it stimulates migration of fibroblasts into 3D collagen gels. When IGDS was incubated in the assay with blocking antibodies to specific integrins, migration was inhibited by antibodies to integrins α2, αv, β1, β2, β3, which reduced migration to around background levels (Figure 11A). Antibodies to α4β1 and αVβ3 inhibited fibroblast migration but antibodies against the classic Fn receptor, α5β1, did not.

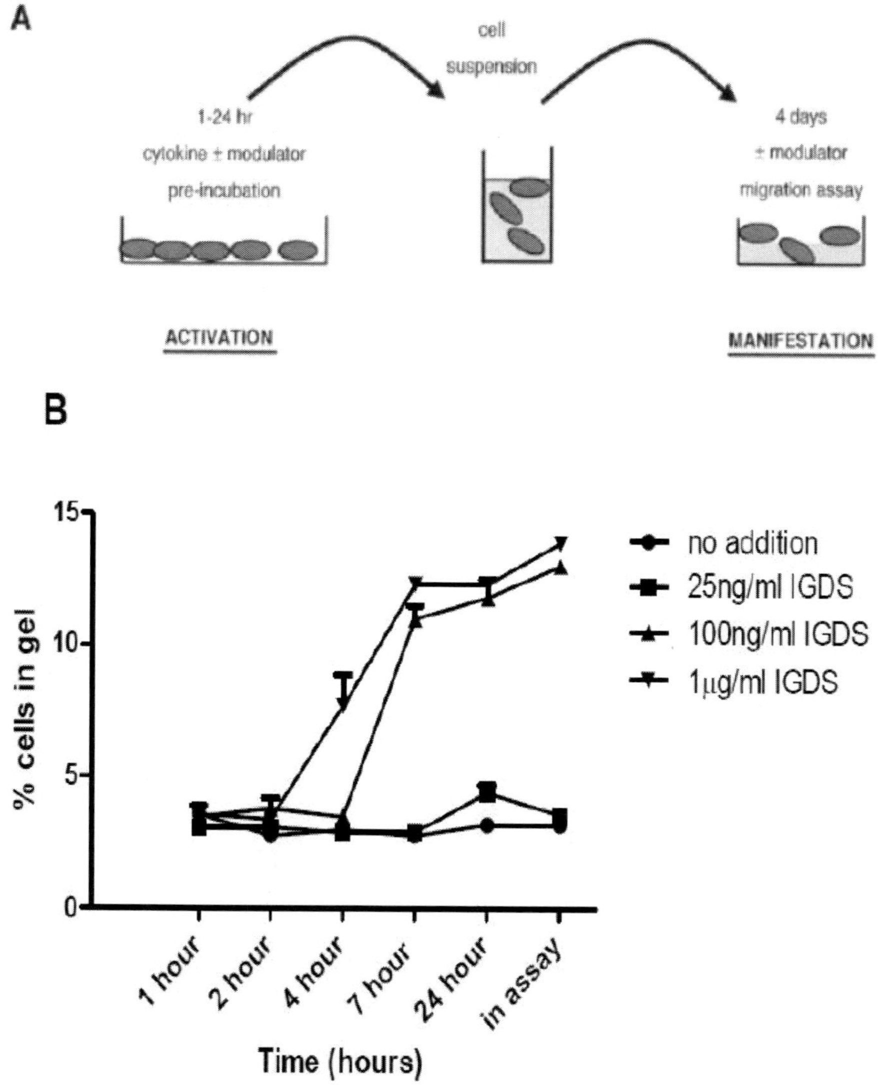

Figure 10. Investigating the activation and manifestation of MSF and cytokines. A) The cytokine/matrikine (and inhibitor/antibody) was pre-incubated on either a plastic tissue culture dish (or a dish coated with a known matrix). After a designated period of time the cells were removed from the dish. Removal of cells from coated dishes is by using trypsin; however cells plated on collagen gels need incubation with collagenase prior to the addition of trypsin. After the cells have been pelleted they were plated in the 96 hours collagen gel assay. The assay may include inhibitors or activators. B) FSF44 fibroblasts were pre-incubated for various times in the presence of 1μg/ml IGDS. After the allotted time period the cells were washed trypsinised and plated onto the surface of a 3D collagen gel. After 96 hours their migration was determined microscopically.

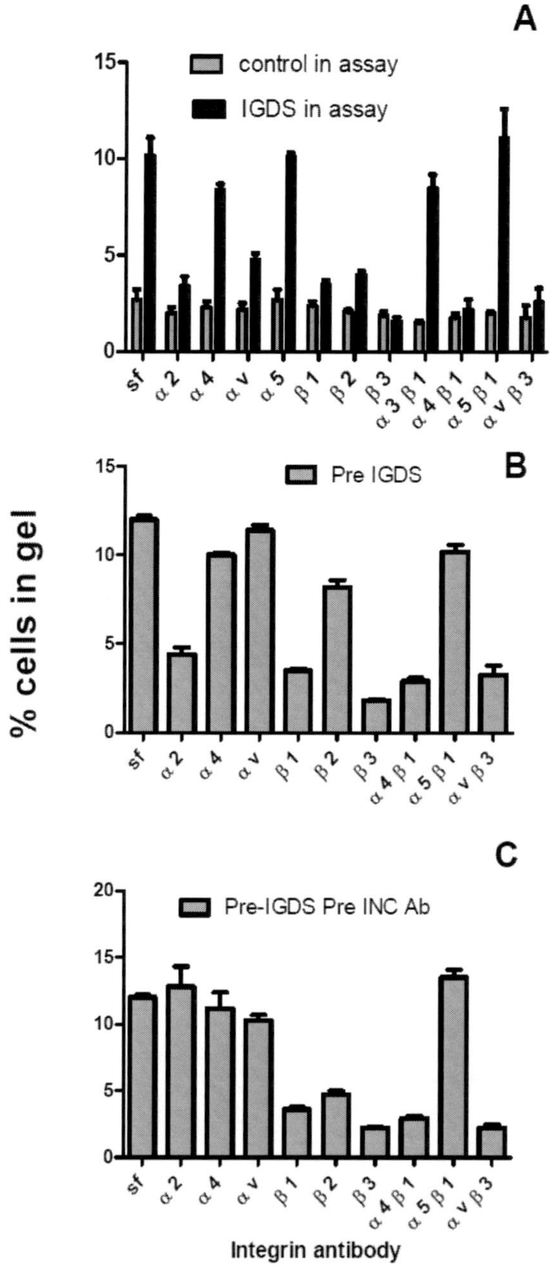

Figure 11. Investigation of activation and manifestation of IGDS on migration of fibroblasts. A) FSF44 fibroblasts were plated onto the surface of a 3D collagen gel. Cells were plated in the presence and absence of 1μg/ml IGDS and 1μg/ml of various anti- integrin antibodies (as listed in Schor et al, 1999). Migration was determined microscopically on Day 4. B) IGDS was pre-incubated (Pre) with FSF44 fibroblasts for 17 hours on plastic tissue culture dishes. After removing the cells from the dishes the cells were plated onto the surface of a 3D collagen gel in the presence of various anti-integrin antibodies. Migration was determined microscopically after 96 hours. C) IGDS and Integrin antibodies were pre-incubated with FSF44 fibroblasts for 17 hours on plastic tissue culture dishes. After removing the cells from the dishes the cells were plated onto the surface of a 3D collagen gel, in the absence of additional integrin antibodies. Migration was determined microscopically on Day 4.

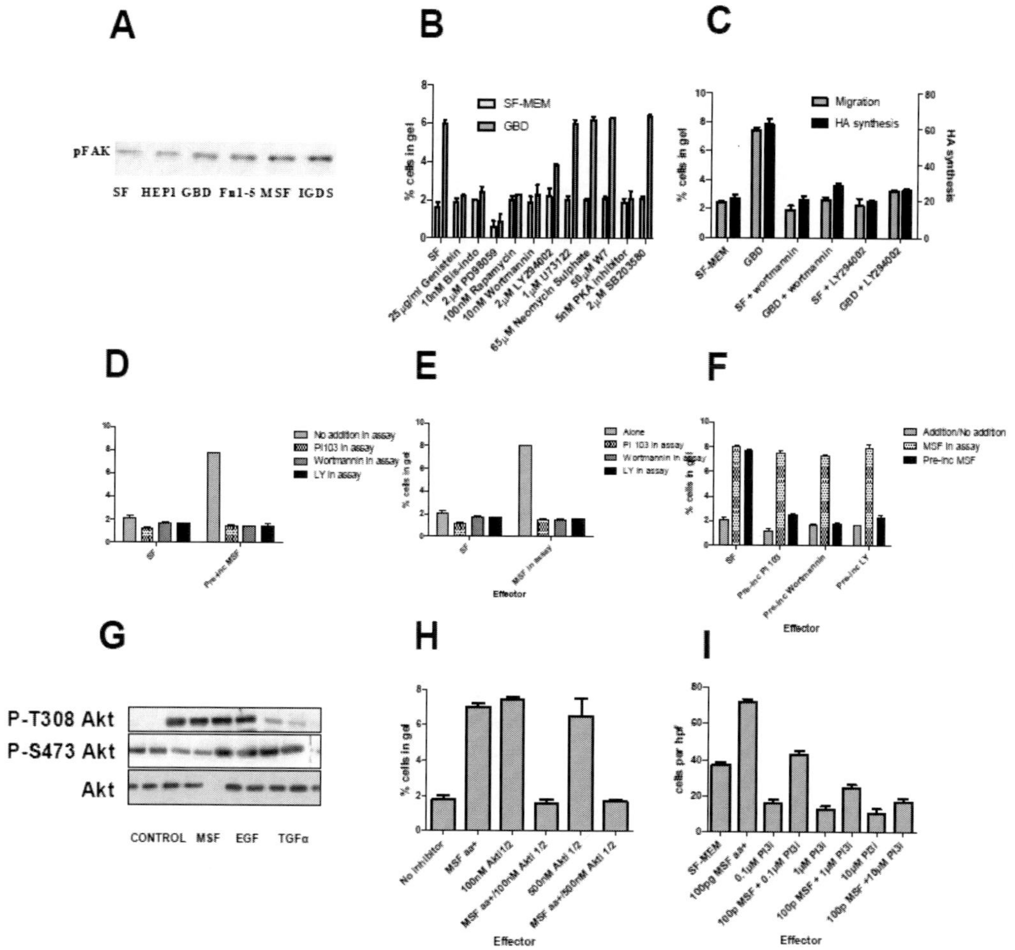

Figure 12. The effect of signal transduction pathway inhibitors on fibroblasts in terms of phosphorylation, motility (using 3D-collagen migration assay) and HA synthesis. A) IGD containing proteins/peptides were added to serum starved fibroblasts plated on plastic tissue culture dishes for 15 minutes. The cells were washed and lysed with RIPA buffer and the lysates fractionated by SDS-PAGE, Western blotted and probed with an anti-pFAK antibody (experimental details Ellis et al, 2010A). B) GBD was used in the 3D collagen gel assay with a range of signal transduction pathway inhibitors. These inhibitors were used at the concentrations indicated on the graph. Inhibitors were used at suitable concentrations for the assay (method and sources of inhibitors Schor et al, 1999). C) Effect of GBD and PI3-kinase pathway inhibitors on HA synthesis and migration. Fibroblasts were plated onto the surface of a collagen gel in the presence or absence of 1ng/ml GBD and Wortmannin or LY294002. All the gels were incubated with 1μCi of ^{3}[H]-glucosamine. Afterwards migration of the cells into the gel was counted microscopically. The medium was collected and analysed for HA content by Streptomyces hyaluronan lyase digestion. D) The effect of in assay MSF and the PI3-kinase pathway inhibitors on migration of FSF44 fibroblasts. Cells were plated onto the surface of a collagen gel and migration determined on day 4. E) The effect of PI3-kinase inhibitors on pre-incubated MSF (17 hours) in terms of migration on FSF44 fibroblasts. Cells were plated onto the surface of a collagen gel and migration determined on day 4. F) Pre-incubation of cells with MSF and PI-3 kinase inhibitors for 17 hours on plastic tissue culture dishes. The cells were trypsinised and plated onto the surface of collagen gels in the presence or absence of added MSF and migration determined on day 4. G) Effect of EGF, TGFα and MSF incubated on FSF44 fibroblasts; in terms of Serine 473 phosphorylation of the protein Akt. Cells were plated on plastic tissue culture dishes and once confluent were serum starved overnight.

They were then incubated with the added growth factor for 15 minutes, lysed and then fractionated by SDS-PAGE. The proteins were Western blotted and probed with an anti-phospho 473 Akt antibody (Ellis et al, 2010B). H) Effect of Akt inhibitors on FSF44 fibroblast migration into 3D collagen gels. MSF and/ or the inhibitor were added in assay. The cell migration into the gel was determined microscopically on Day 4. I) Effect of PI3-kinase inhibitors on TYS migration using the micr-chemotaxis assay. MSF and the inhibitor PI-103 were placed in the lower chamber and TYS cells into the upper chamber. After 5 hours, migration of cells from the upper surface to the lower surface of the membrane via the pores was determined microscopically after fixation and staining.

Using a Cytosensor, which enabled us to study changes in cellular response to stimuli in real time; we investigated the effect of IGDS on fibroblasts cells plated on a membrane. This study utilised that fact that cells stimulated by an agonist release hydrogen ions thereby changing the pH. This experiment indicated that IGDS decreased the pH. However, this change could be blocked by an antibody raised against αVβ3 but was enhanced by an antibody raised against α5β1 (Schor et al, 1999).

This data indicated which receptors were involved in the binding of the IGD to the cell, as well as those involved in migration. Pre-incubation of fibroblasts with IGDS and then plating in the assay produces an increased level of migration. In our assay, the integrins that appear to be responsible for migration of fibroblasts into 3D collagen gels are α2, αV, β1, β3 and/or αVβ3 (Figure 11B). However, α5β1 does not appear important in IGDS stimulated fibroblast motility. The next step was to investigate the integrins important in binding the activating ligand. Pre-incubation of the IGDS and the anti integrin antibody prior to plating the cells on the collagen gel allowed us to determine which integrins were involved in binding the IGD motif; our data suggested, that the αVβ3 as well as the α4β1 integrin are important for cell activation leading to cell motility (Figure 11C).

Integrins are tyrosine kinase receptors (Johannsson et al, 1997). Thus once a ligand attaches to the receptor it is phosphorylated at these amino acids. The phosphorylation of these residues leads to activation of a variety of pathways due to further downstream phosphorylation and changes in protein conformation. One of the first proteins to be activated is FAK (Johannsonn et al, 1997). This protein facilitates cell motility by directly affecting various downstream events (Morgan et al, 2009). In order to determine whether our cells responded by phosphorylation of FAK, fibroblasts were incubated on plastic tissue culture dishes and exposed to IGD, Hep I, GBD and MSF The data shows an increase in FAK phosphorylation in response to these ligands (Figure 12A). An initial screen of signal transduction inhibitors was investigated in the migration assay using GBD (Figure 12B) similar results were seen with IGDS and MSF). Genistein is a naturally occurring isoflavone and a tyrosine kinase inhibitor (Akiyama et al, 1987); in the presence of genistein, GBD-stimulated cell motility of fibroblasts was inhibited, indicating the involvement of a tyrosine kinase linked receptor. A protein kinase C (PKC) inhibitor, Bis-indolylmaleimide I hydrochloride, (Wilkinson, Parker and Nixon, 1993) also inhibited the migration of fibroblasts in response to GBD (Figure 12B). This pathway was also implicated in the migration of fetal fibroblasts and fibroblasts plated at low density (data not shown).

We also investigated the MEK inhibitor, PD98059 (Cross et al, 1995). MEK is a pathway that has been implicated in migration (Anand-Apte et al, 1997). This was a very active inhibitor in the collagen gel assay; it also blocked background migration (Figure 12B). The p70 S6 kinase pathway was also important in migration (Anand-Apte et al, 1997). We found that GBD-stimulated migration was also blocked by the inhibitor Rapamycin which is a p70

S6 kinase pathway. The protein kinase A pathway was also found to be important for cell motility using a small molecule inhibitor peptide we found that cell motility was blocked (Figure 12B).

Other pathways were not found to be important for increasing fibroblast migration in the collagen gel assay. Blocking of the p38 MAPK pathway with SB203580 (Cuenda et al, 1995) had no effect on cell migration (Figure 12B). Similar results were seen for the PLC-γ pathway using the U73122 inhibitor (Chen et al, 1994). The Calcium ion-calmodulin dependant kinase pathway which is prone to inhibition by the W7 inhibitor (Cuenda et al, 1995) also had no effect on migration in response to GBD.

One of the main interests of our group, more recently, is the effect of PI3kinase/Akt inhibitors on migration (Ellis et al, 2010B). An initial screen with GBD-induced migration and the inhibitors Wortmannin and LY294002, showed a reduction in fibroblast motility (Figure 12B). Knowing that the migration was blocked by these inhibitors we investigated their effect on HA synthesis. HA synthesis was stimulated by GBD addition; however, this increase was blocked by the addition of LY294002 and wortmannin (Figure 12C). Further inhibitors of this pathway have also been investigated and blocking of this pathway is important in terms of receptor binding and activation, as well as cell migration. When MSF was incubated with inhibitors PI103, wortmannin and LY294002 in assay there was no increase in cell migration (Figure 12D). In experiments where MSF was pre-incubated with the cells and the inhibitors subsequently added in assay, the stimulated motility was blocked (Figure 12E). However, pre-incubating the cells with both MSF and inhibitor together on plastic tissue culture dishes and then plating the cells on collagen produced no stimulation of migration (Figure 12F). This suggests that this pathway is important for both MSF initial signalling and fibroblast motility in the collagen gel assay.

The main surprise of blocking the PI3/Akt pathway came from further analysis of what happened to the phosphorylation of the pathway. Most growth factors that stimulate migration via this pathway increase the phosphorylation of Akt. However, our data shows that MSF decreases phosphorylation of Akt at serine 473 whereas EGF and TGFα, both stimulators of migration, increase phosphorylation at this residue (Figure 12 G; Ellis 2010B). The Akti1/2 inhibitor used alone was shown to increase cell migration into 3D collagen gels by fibroblasts (Figure 12H). However, addition of MSF and Akti1/2 together does not stimulate migration. Other groups have shown that Akt is a family of proteins consisting of a three members, each of which have differing effects when investigated in terms of migration of cells (Zhou et al, 2006). This may explain how both MSF and the inhibitor stimulate migration by themselves but together inhibit cell migration.

The data presented so far refers mainly to fibroblasts and the collagen gel assay. However, the importance of the PI3 kinase pathway to migration has also been exemplified by using an oral squamous carcinoma cell line in the micro-chemotaxis assay, where we could inhibit the MSF stimulated migration by addition of PI3 kinase inhibitor PI3103 (Figure 12I). This inhibitor also blocked background migration. One interesting observation was that at the lowest concentration of inhibitor migration could be stimulated, by the addition of MSF, back to control levels; this suggests a competitive inhibition.

Work in Progress

This review sets out our results and interprets them in terms of the effect of N-terminal domain of Fn on the motility of cells. Whilst we have focussed on the migration assay we have not attempted to include all the data collected over the past three decades.

MSF contains a unique peptide sequence at its C-terminus. This has allowed us to raise MSF specific antibodies and to investigate the expression of MSF. Aljorani et al, (2011) investigated the expression of MSF in a small group of patients with salivary gland tumours and found that MSF was over-expressed in these tumours. Other studies are being undertaken at the present time and will be reported on in due course (Schor et al, in preparation).

MSF is also produced in response to TGFβ-1 in a matrix dependant manner (Schor et al, 2011). Initial data indicated that the addition of TGFβ-1 to MSF in the collagen gel assay resulted in no net increase in migration (Ellis et al, 1992). However, modulation of MSF by TGFβ-1 has been shown by the increase in mature MSF mRNA when TGFβ-1 was added to the cells (Kay et al, 2005). It would appear that fibroblast response to TGFβ-1 is dependent on the matrix on which gels are plated. Normal adult fibroblast cells pre-incubated with TGFβ-1 on a 'wounded' matrix such as gelatin, fibrin or plastic switch on MSF synthesis, whereas on a normal matrix of collagen, TGFβ-1 has no effect (Schor et al, 2011). However, cells that have been activated or that produce MSF normally (fetal or cancer patient fibroblasts) can have MSF expression switched off by treatment with TGFβ-1 on a collagen matrix. This has led us to speculate that MSF and perhaps breakdown products of full-length Fn have a permissive role to play in tumour cell motility and its control by a stroma which produces a matrix that allows motility.

One of the fascinating aspects of some of the recent work is the interaction between the N-terminal of Fn and proteins of bacterial origin. We have previously reported that these proteins can activate non-active Fn modules and also can inhibit stimulation of migration by smaller active peptides (Ellis et al, 2010A and Marjenberg et al, 2011). This leads to a number of possibilities how bacteria can cause pathogenesis such as cancer and delayed wound healing.

Inhibitors of MSF have previously been investigated. Jones et al, (2007) reported that N-Gal/lipocalin2 isolated from epithelial cell lines was able to inhibit MSF stimulated migration. The data suggested that this was via an indirect mechanism. Another inhibitor of MSF activity, is IGFBP-7, (unpublished data) has been isolated from endothelial cell conditioned medium.

MSF increase cell migration by modulation of the PI3 kinase/Akt pathway. This pathway is over expressed in a large percentage of oral cancers (Pontes et al, 2009; Lim et al, 2005). MSF is up-regulated in salivary gland carcinoma (Aljorani et al, 2011) and also oral squamous cell carcinoma (Aljorani et al, in preparation). We are currently investigating the role of Akt in oral cancer cell migration (Islam et al, in preparation).

The IGD sequence has been reported to be the active motif in fibroblast migration. In the past few years a number of mimetic peptides of this sequence have been made and tested for their ability to stimulate cell migration (Shpiro et al, 2005). How these sequences interact with the cell and which signal transduction pathways they utlise, remains to be ascertained.

Rather than being a inert part of the Fn protein we would suggest that the N-terminal region plays an active role in a normal development and homeostasis, as well as more pathological events such as cancer. The modulation of this region could have important consequences in the control of wound healing and tumour progression.

Conclusion

The ECM and its constituents provide the context in which cells of the stroma and epithelium are able to migrate in both a controlled and uncontrolled manner. Full length Fn has only a limited affect on cell motility but does affect adhesion. The N-terminal region of Fn whether it is MSF with its unique 10 amino acids or proteolytic fragments of the full length molecule can stimulate migration. This stimulation of migration into both collagen gels and through collagen coated membranes is controlled by the IGD amino acid sequences for fibroblastic cells. Tumour cell lines also require the HEEGH sequence, as well as IGD. The receptor that appears to be important both for activation and manifestation of migratory activity is the αvβ3 integrin. This appears to activate a number of signalling pathways; including the PI3 kinase/akt pathway. One of the effects of MSF and its active sequences is to stimulate the synthesis of HA, specifically high molecular weight HA. The control of either Fn breakdown or MSF production could lead to development of novel therapeutics with relevance to both healing and disease.

Acknowledgments

The authors wish to thank Mrs Jacqui Cox and Ms Maggie Florence for their excellent technical work. Dr Kanj Seneviratne and Dr Richard Kay for cloning of the MSF gene. Dr Jackie Banyard for early work on signalling inhibitors. Dr Nick Leslie and his lab for his work with the akt protein. Dr Rudi Marquez and Dr David Norman for peptide mimetics to IGD. Prof Iain Campbell, Dr Chris Millard, Dr David Staunton, Dr Ioannis Vakonakis, Dr Jenn Potts, Dr Uli Schwarz-Linek and their labs for making and sharing proteins for use in the assays. Emeritus Professor Seth Schor and Dr Ana Schor for their support and gaining funding for the work and we wish them well in their retirement.

Part of the work was funded by the Tattersall Fund, University of Dundee, Anonymous Trust University of Dundee. MRI is funded by an SORSAS award and LA is funded by the Iraqi Embassy UK.

References

Affolter M, Weijer CJ. Signaling to cytoskeletal dynamics during chemotaxis. *Dev. Cell.* 2005 Jul 9(1):19-34.

Akimov, S, AM Belkin, Cell-surface transglutaminase promotes fibronectin assembly via interaction with the gelatin-binding domain of fibronectin: a role in TGFbeta-dependent matrix deposition. *J.Cell Sci.* 2001 114:2989-3000.

Akiyama SK. Integrins in cell adhesion and signaling. *Hum. Cell.* 1996 9(3):181-6.

Akiyama T, Ishida J, Nakagawa S, Ogawara H, Watanabe S, Itoh N, Shibuya M, Fukami Y. Genistein, a specific inhibitor of tyrosine-specific protein kinases. *J. Biol. Chem.* 1987 262(12):5592-5.

Aljorani LE, Bankfalvi A, Carey FA, Harada K, Ohe G, Jones SJ, Ellis IR, Schor SL, Schor AM. Migration-stimulating factor as a novel biomarker in salivary gland tumours. *J. Oral Pathol. Med.* 2011 Nov 40 (10):747-54

Anand-Apte B, Zetter BR, Viswanathan A, Qiu RG, Chen J, Ruggieri R, Symons M. Platelet-derived growth factor and fibronectin-stimulated migration are differentially regulated by the Rac and extracellular signal-regulated kinase pathways. *J. Biol. Chem.* 1997 272(49):30688-92.

Ashcroft, GS, Horan, MA, Ferguson MWJ. The effects of cutaneous wound healing in mammals. *J. Anat.* 1995 187:1-26.

Bellows CG, Melcher AH, Aubin JE. Contraction and organization of collagen gels by cells cultured from periodontal ligament, gingiva and bone suggest functional differences between cell types. *J. Cell Sci.* 1981 50:299-314.

Bourguignon LY, Iida N, Welsh CF, Zhu D, Krongrad A, Pasquale D. Involvement of CD44 and its variant isoforms in membrane-cytoskeleton interaction, cell adhesion and tumor metastasis. *J. Neurooncol.* 1995 26:201-8.

Caplan, A. The ECM is instructive. *Bioessays.* 1986 5:129-32.

Carson AE, Barker TH. Emerging concepts in engineering extracellular matrix variants for directing cell phenotype. *Regen. Med.* 2009 4(4):593-600.

Chen P, Xie H, Sekar MC, Gupta K, Wells A. Epidermal growth factor receptor-mediated cell motility: phospholipase C activity is required, but mitogen-activated protein kinase activity is not sufficient for induced cell movement. *J. Cell Biol.* 1994 127(3):847-57.

Chen WY, Grant ME, Schor AM, Schor SL. Differences between adult and foetal fibroblasts in the regulation of hyaluronate synthesis: correlation with migratory activity. *J. Cell Sci.* 1989 94 (Pt 3):577-84.

Clark RA, Wikner NE, Doherty DE, Norris DA. Cryptic chemotactic activity of fibronectin for human monocytes resides in the 120-kDa fibroblastic cell-binding fragment. *J. Biol. Chem.* 1988 263(24):12115-23.

Cross DA, Alessi DR, Vandenheede JR, McDowell HE, Hundal HS, Cohen P. The inhibition of glycogen synthase kinase-3 by insulin or insulin-like growth factor 1 in the rat skeletal muscle cell line L6 is blocked by wortmannin, but not by rapamycin: evidence that wortmannin blocks activation of the mitogen-activated protein kinase pathway in L6 cells between Ras and Raf. *Biochem J.* 1994 303 (Pt 1):21-6.

Cuenda A, Rouse J, Doza YN, Meier R, Cohen P, Gallagher TF, Young PR, Lee JC. SB 203580 is a specific inhibitor of a MAP kinase homologue which is stimulated by cellular stresses and interleukin-1. *FEBS Lett.* 1995, 364(2):229-33.

Duband, JL, Rocher, S, Chen, WT, Yamada, KM, Thierry JP.

Cell adhesion and migration in early embryo: location and possible role of putative Fn receptor complex. *J. Cell Biol.* 1986 102:160-178.

Ellis IR, Schor, AM, Schor SL. EGF and TGF-alpha motogenic activities are mediated by the EGF receptor via distinct matrix-dependent mechanisms. *Exp. Cell Res.* 2007 313:732-41.

Ellis I and Schor SL. The interdependent modulation of hyaluronan synthesis by TGFß1 and extracellular matrix: Consequences for the control of cell migration. *Growth Factors.* 1995 12: 211-222.

Ellis I, Banyard J, and Schor SL. Differential response of fetal and adult fibroblasts to cytokines: cell migration and hyaluronan synthesis. *Development.*1997 124: 1593-1600.

Ellis I, Grey AM, Schor AM and Schor SL. (1992) Antagonistic effects of transforming growth factor beta and MSF on fibroblast migration and hyaluronic acid synthesis: possible implications for wound healing. *J. Cell Sci.* 1992 102: 447-456.

Ellis IR, Jones, SJ, Lindsay, Y, Schor, AM, Schor, SL, Leslie, N. Migration Stimulating Factor (MSF) promotes fibroblast migration by inhibiting AKT. *Cellular Signalling.* 2010B 22:1655-9.

Ellis IR, Jones, SJ, Staunton, D, Vakonakis, I, Norman, DG, Potts, J, Schor, AM, Schor, SL. Multi-Factorial Modulation of IGD Motogenic Potential in MSF (Migration Stimulating Factor). *Exp. Cell Res.* 2010A 316:2465-2476.

Friedl P, Bröcker EB. The biology of cell locomotion within three-dimensional extracellular matrix. *Cell Mol. Life Sci.* 2000 57(1):41-64.

Fukai F, Ohtaki M, Fujii N, Yajima H, Ishii T, Nishizawa Y, Miyazaki K, Katayama T. Release of biological activities from quiescent fibronectin by a conformational change and limited proteolysis by matrix metalloproteinases. *Biochemistry.* 1995 34(36):11453-9.

Fukai F, Suzuki H, Suzuki K, Tsugita A, Katayama T. Rat plasma fibronectin contains two distinct chemotactic domains for fibroblastic cells. *J. Biol. Chem.* 1991 266(14):8807-13.

Giancotti FG. A structural view of integrin activation and signaling. *Dev. Cell.* 2003 4(2):149-51.

Grey AM, Schor AM, Rushton G, Ellis I, Schor SL. Purification of the migration stimulating factor produced by fetal and breast cancer patient fibroblasts. *Proc. Natl. Acad. Sci. U S A.* 1989 86(7):2438-42.

Hocking DC, Sottile J, Mckeown-Longo PJ. Activation of distinct alpha 5 beta 1 mediated signalling pathways by fibronectins cell adhesion and matrix assembly domains. *J. Cell Biol.* 1998 141:241-253.

Homandberg GA, Meyers R, Xie DL. Fibronectin fragments cause chondrolysis of bovine articular cartilage slices in culture. *J. Biol. Chem.* 1992 Feb 25;267(6):3597-604.

Houard X, Germain S, Gervais M, Michaud A, van den Brûle F, Foidart JM, Noël A, Monnot C, Corvol P. Migration-stimulating factor displays HEXXH-dependent catalytic activity important for promoting tumor cell migration. *Int. J. Cancer.* 2005 116(3):378-84.

Huang YE, Iijima M, Parent CA, Funamoto S, Firtel RA, Devreotes P. Receptor-mediated regulation of PI3Ks confines PI(3,4,5)P3 to the leading edge of chemotaxing cells. *Mol. Biol. Cell.* 2003 14(5):1913-22.

Hynes RO. *Fibronectins*, Springer-Verlag, Berlin, Germany, 1990.

Iijima M, Devreotes P. Tumor suppressor PTEN mediates sensing of chemoattractant gradients. *Cell.* 2002 109(5):599-610.

Johansson S, Svineng G, Wennerberg K, Armulik A, Lohikangas L. Fibronectin-integrin interactions. *Front Biosci.* 1997 2:d126-46.

Jones SJ, Florence MM, Ellis I, Kankova K, Schor SL, Schor AM. Co-expression by keratinocytes of migration stimulating factor (MSF) and a functional inhibitor of its bioactivity (MSFI) *Exp. Cell Res.* 2007 313 (20), pp. 4145-4157.

Kay RA, Ellis IR, Jones SJ, Perrier S, Florence M, Schor AM, Schor SL. The expression of MSF, a potent onco-fetal cytokine, is uniquely controlled by 3'UTR-dependent nuclear sequestration of its precursor mRNA, *Cancer Res.* 2005 65:10742-10749.

Kielty CM, Grant ME. The collagen family: structure, assembly and organisation in the extracellular matrix. *Connective tissue and its heritable diseases.* 2003 Wiley Liss eds p159-203.

Kleinman HK, Martin GR. Matrigel: basement membrane matrix with biological activity. *Semin. Cancer Biol.* 2005 15(5):378-86.

Kölsch V, Charest PG, Firtel RA. The regulation of cell motility and chemotaxis by phospholipid signaling. *J. Cell Sci.* 2008 121(Pt 5):551-9.

Lim J, Kim J-H, Paeng J-Y, Kim M-J, Hong S-D, Lee J-I, and Hong S-P. Prognostic value of activated Akt expression in oral squamous cell carcinoma. *J. Clin. Pathol.* 2005 58: 1199-1205.

Lovell CR, Smolenski KA, Duance VC, Light ND, Young S, Dyson M. Type I and III collagen content and fibre distribution in normal human skin during ageing. *British Journal of Dermatology.* 1987 117:419-428.

Lukashev ME, Werb, Z. ECM signalling: orchestrating cell behaviour and misbehaviour. *Trends in Cell Biol.* 1998 8:437-41.

Maaser K, Wolf K, Klein CE, Niggemann B, Zänker KS, Bröcker EB, Friedl P. Functional hierarchy of simultaneously expressed adhesion receptors: integrin alpha2beta1 but not CD44 mediates MV3 melanoma cell migration and matrix reorganization within three-dimensional hyaluronan-containing collagen matrices. *Mol. Biol. Cell.* 1999 10(10):3067-79.

Marjenberg ZR, Ellis IR, Hagan RM, Prabhakaran S, Höök M, Talay SR, Potts JR, Staunton D, Schwarz-Linek U. Cooperative binding and activation of fibronectin by a bacterial surface protein. *J. Biol. Chem.* 2011 Jan 21; 286(3):1884-94.

McKeown-Longo PJ, Mosher DF. Interaction of 70 000 mol-wt-amino-terminal fragment of fibronectin with the matrix assembly receptor of fibroblasts, *J. Cell Biol.* 1985 100:364-374.

Millard CJ, Ellis IR, Pickford AR, Schor AM, Schor SL, Campbell ID. The role of the fibronectin IGD motif in stimulating fibroblast migration, *J. Biol. Chem.* 2007 282:35530-5.

Morgan MR, Byron A, Humphries MJ, Bass MD. Giving off mixed signals--distinct functions of alpha5beta1 and alphavbeta3 integrins in regulating cell behaviour. *IUBMB Life.* 2009 Jul 61(7):731-8.

Morla, A, Zhang, Z, Ruoslahti E. Superfibronectin is a functionally distinct form of fibronectin. *Nature.* 1994 367:193–196.

Newgreen D, Thiery JP. Fibronectin in early avian embryos: synthesis and distribution along the migration pathways of neural crest cells. *Cell Tissue Res.* 1980 211:269-91.

Perris R, Paulsson M, Bronner-Fraser M. Molecular mechanisms of avian neural crest cell migration on fibronectin and laminin. *Dev. Biol.* 1989 136(1):222-38.

Pontes HAR, de Aquino Xavier FC, da Silva, TSP, Fonseca, FP, Paiva, HB, Pontes, FSC, and dos Santos Pinto Jr, D. Metallothionein and p-Akt proteins in oral dysplasia and in oral squamous cell carcinoma: an immunohistochemical study. *J. Oral Pathol. and Med.* 2009 38: 644-50.

Postlethwaite AE, Keski-Oja J, Balian G, Kang AH. Induction of fibroblast chemotaxis by fibronectin. Localization of the chemotactic region to a 140,000-molecular weight non-gelatin-binding fragment. *J. Exp. Med.* 1981 153(2):494-9.

Postlethwaite AE, Seyer JM, Kang AH. Chemotactic attraction of human fibroblasts to type I, II, and III collagens and collagen-derived peptides. *Proc. Natl. Acad. Sci. U S A.* 1978 75:871-5.

Ribatti D, Nico B, Morbidelli L, Donnini S, Ziche M, Vacca A, Roncali L, Presta M. Cell-mediated delivery of fibroblast growth factor-2 and vascular endothelial growth factor onto the chick chorioallantoic membrane: endothelial fenestration and angiogenesis. *J. Vasc. Res.* 2001 Jul-Aug 38(4):389-97.

Ruoslahti E. RGD and other recognition sequences for integrins. *Annu. Rev. Cell Dev. Biol.* 1996 12:697-715.

Schor SL. Cell proliferation and migration on collagen substrata in vitro. *J. Cell Sci.* 1980 41:159-75.

Schor SL, Ellis I, Dolman, C Banyard, J Humphries, MJ , Mosher MJ, Grey, AM, Mould AP, Sottile J, Schor AM. Substratum-dependent stimulation of fibroblast migration by the gelatin-binding domain of fibronectin. *J. Cell Sci.* 1996 109:2581-90.

Schor SL, Ellis I, Banyard J, Schor AM. Motogenic activity of IGD-containing synthetic peptides. *J. Cell Sci.* 1999 112:3879-88.

Schor SL, Ellis IR, Harada K, Motegi K, Anderson ARA, Chaplain MAJ, Keatch RP, Schor AM. A novel sandwich assay for quantifying chemo-regulated cell migration within 3-Dimensional Matrices: Wound healing cytokines exhibit distinct motogenic activities compared to the transmembrane assay. *Cell motility and the cytoskeleton.* 2006 63:287-300.

Schlaepfer DD, Mitra SK. Multiple connections link FAK to cell motility and invasion. *Current Opinions in Genetics and Development.* 2004 14:92-101.

Schmidt S, Friedl P. Interstitial cell migration: integrin-dependent and alternative adhesion mechanisms. *Cell Tissue Res.* 2010 339(1):83-92.

Schor AM, Ellis I, Schor SL. Chemotaxis and Chemokinesis in 3D Macromolecular Matrices : Relevance to Angiogenesis. *Methods Mol. Med.* 2001 46:163-83.

Schor SL, Ellis IR, Jones SJ, Woolston AM, Schor AM. Bi-stable switch in MSF expression is regulated by the concerted signalling of TGFβ1 and the extracellular matrix. *Int. J. Cancer.* 2012 May 1 130(9):2024-32.

Schor SL, Schor AM, Bazill GW. The effects of fibronectin on the migration of human foreskin fibroblasts and Syrian hamster melanoma cells into three-dimensional gels of native collagen fibres. *J. Cell Sci.* 1981 48: 301-14.

Schor SL, Schor AM, Grey AM, Rushton G. Foetal and cancer patient fibroblasts produce an autocrine migration-stimulating factor not made by normal adult cells. *J. Cell Sci.* 1988 90(Pt 3):391-9.

Schor SL, Schor AM, Rushton G, Smith L. Adult, foetal and transformed fibroblasts display different migratory phenotypes on collagen gels: evidence for an isoformic transition during foetal development. *J. Cell Sci.* 1985 73:221-34.

Schor SL, Schor AM, Winn B, Rushton G. The use of three-dimensional collagen gels for the study of tumour cell invasion in vitro: experimental parameters influencing cell migration into the gel matrix. *Int. J. Cancer.* 1982 29(1):57-62.

Schor SL, Ellis IR, Jones SJ, Baillie R, Seneviratne K, Clausen J, Motegi K, Vojtesek B, Kankova K, Furrie E, Sales MJ, Schor AM, Kay RA. Migration-stimulating factor: a genetically truncated onco-fetal fibronectin isoform expressed by carcinoma and tumor-associated stromal cells. *Cancer Res.* 2003 63:8827-36.

Schor SL, Schor AM, Grey AM, Chen J, Rushton, G Grant ME,Ellis, I. Mechanism of action of the migration stimulating factor (MSF) produced by fetal and cancer patient fibroblasts: effect on hyaluronic acid synthesis. *In Vitro* 1989 25:737-746.

Shpiro N, Ellis IR, Dines TJ, Schor AM, Schor SL, Norman DG, Marquez R. Synthesis of an IGD peptidomimetic with motogenic activity. *Mol. Biosyst.* 2005 Oct1(4):318-20.

Sottile J, Mosher DF. N-terminal type I modules required for fibronectin binding to fibroblasts and to fibronectins III-1 module. *Biochem. J.* 1997 323:51-60.

Tomokowicz B, Rybinski K, Foley B, Ebel W, Kline B, Routhier E, Sass P. Nicoliades NC, Grasso L, Zhou Y. Interaction of endosialin/TEM1 with extracellular matrix proteins mediates cell adhesion and migration. *PNAS.* 2007 104:17965-70.

Toole BP. Hyaluronan in morphogenesis. *Semin. Cell Dev. Biol.* 2001 12:79-87.

Toole BP, Wight TN, Tammi MI. Hyaluronan cell interactions in cancer and vascular disease. *J. Biol. Chem.* 2002 277:4593-6.

Turley EA. Hyaluronan and cell locomotion. *Cancer Metastasis Rev.* 1992 11:21-30.

Vakonakis I, Staunton D, Ellis IR, Sarkies P, Flanagan A, Schor AM, Schor SL, Campbell ID. Motogenic sites in human fibronectin are masked by long range interactions. *J. Biol. Chem.* 2009 284:15668-75.

Valster A, Tran NL, Nakada M, Berens ME, Chan AY, Symons M. Cell migration and invasion assays. *Methods.* 2005 37(2):208-15.

Viji RI, Kumar VB, Kiran MS, Sudhakaran PR. Angiogenic response of endothelial cells to heparin-binding domain of fibronectin. *Int. J. Biochem. Cell Biol.* 2008 40(2):215-26.

Watt FM, Hodivala KJ. Cell adhesion. Fibronectin and integrin knockouts come unstuck. *Curr. Biol.* 1994 Mar 1 4(3):270-2.

West DC, Hampson IN, Arnold F, Kumar S. Angiogenesis induced by degradation products of hyaluronic acid. *Science.* 1985 228(4705):1324-6.

West DC, Kumar SH. The effect of hyaluronate and its oligosaccharides on endothelial cell proliferation and monolayer integrity. *Exp. Cell Res.* 1989 183:179-96.

Wilkinson SE, Parker PJ, Nixon JS. Isoenzyme specificity of bisindolylmaleimides, selective inhibitors of protein kinase C. *Biochem. J.* 1993 294 (Pt 2):335-7.

Woodley DT, Sarrer Y, O'Keefe EJ. Human keratinocyte locomotion on extracellular matrix. In *Wound Healing* (ed. H.Janssen, R. Rooman and J. I. S. Robertson), 1991 pp. 91-97. Petersfield: Wrightson Biomedical.

Wysocki AB, Grinnell F. Fibronectin profiles in normal and chronic wound fluid. *Lab Invest.* 1990 63(6):825-31.

Zhao Q, Liu X, Collodi P. Identification and characterization of a novel fibronectin in Zebrafish. *Exp. Cell Res.* 2001 268:211–219.

Zhou GL, Tucker DF, Bae SS, Bhatheja K, Birnbaum MJ, Field J. Opposing roles for Akt1 and Akt2 in Rac/Pak signaling and cell migration. *J. Biol. Chem.* 2006 281(47):36443-53.

Zigmond SH, Hirsch JG. Leukocyte locomotion and chemotaxis. New methods for evaluation, and demonstration of a cell-derived chemotactic factor. *J. Exp. Med.* 1973 137(2):387-410.

In: Fibronectin
Editors: James Beattie
ISBN 978-1-62257-298-4

Chapter IV

Fibronectin Assembly and Function: A Multi-Dimensional View

Wing S. To and Kim S. Midwood*
Department of Matrix Biology, Kennedy Institute of Rheumatology Division, Nuffield Department of Orthopedic Rheumatology and Musculoskeletal Sciences, Oxford University, London, The United Kingdom

Abstract

The extra cellularmatrix (ECM) acts not only as a structural support or scaffold but also plays an essential role in regulating cellular activities. Fibronectin (FN) is locally assembled by cells into a complex three-dimensional matrix on the cell surface where it controls a wide variety of cell functions. Its assembly is essential during development and tissue repair, as well as during physiological tissue homeostasis. This chapter reviews the research that examines the mechanisms of FN matrix assembly, both using cells cultured on rigid planar surfaces, as well as more recent studies that have begun to examine this process in models that more closely mimic matrix environments *in vivo*. We also discuss what is known about the effect of the FN matrix on cellular behaviour, with a focus on data which reveal the importance of the structure, mechanical properties and complexity of the ECM in activating signalling pathways that drive cell phenotype. Finally, we review how regulation of FN matrix assembly is mediated and highlight how defects in this process lead to or characterize pathological events such as fibrosis and tumorigenesis.

Understanding how FN is assembled, regulated, impacts the cell microenvironment and the influence of differential FN isoforms is vital in order to develop tools to help target and control the FN assembly process and potentially impact on disease progression.

Keywords: Fibronectin; extracellular matrix; tissue repair; fibrogenesis; tumorigenesis; plasma; cellular; isoform; EIIIA; EIIIB

* Corresponding author: e-mail:kim.midwood@kennedy@ox.ac.uk; Tel: 0208 383 4404; Fax: 0208 383 4499.

List of Abbreviations

FN	fibronectin;
ECM	extracellular matrix;
FNI	fibronectin type I repeat;
FNII	fibronectin type II repeat;
FNIII	fibronectin type III repeat;
EIIIA	alternatively spliced fibronectin type III repeat A;
EIIIB	alternatively spliced fibronectin type III repeat B;
HSPG	heparan sulphate proteoglycan;
PDGF	platelet derived growth factor;
TGF	transforming growth factor;
VEGF	vascular endothelial growth factor;
CTGF	connective tissue growth factor;
MMP	matrix metalloproteinase;
3D	three dimensional;
2D	two dimensional;
FAK	focal adhesion kinase;
ROCK	Rho kinase;
MLC	Myosin Light Chain;
SDS-PAGE	sodium dodecyl sulphate-polyacrylamide gel electrophoresis;
HUVECs	human umbilical venous endothelial cells;
CHO	chinese hamster ovary cells;
BAECs	bovine arterial endothelial cells;
MDCK	Madin-Darby canine kidney cells;
FAK	focal adhesion kinase;
FRET	fluorescence resonance energy transfer;
kDa	kilodaltons.

1. Introduction

The extracellular matrix (ECM) is a highly structured network of secreted proteins, proteoglycans and carbohydrates. However, it does not only function as an inert structural support but also plays a vital role in regulating, mediating and transmitting external signals to the cell, acting as a conduit for transmitting internal signals to the external microenvironment to regulate a host of cellular activities including adhesion, migration, proliferation and apoptosis.

Differences in the composition and organization of the ECM lead to incredibly diverse and complex range of structures, each one adapted to the tissue and its function. One example of a specialized ECM structure is the basement membrane, which is a flat lattice network composed of ECM proteins such as, collagen type IV, laminin, nidogen, perlecan and fibronectin (FN). Its organization allows it to function as a partition between the epithelia and endothelia, providing cellular cues and signals for correct cell polarization, organization and function. The glomerular basement membrane acts as an adhesive substrate for endothelial cells on the vascular internal surface and podocytes on its outer surface and functions as a molecular sieve that filters the blood plasma [1-3]. In contrast, the mesangium of the kidney is a more three-dimensional fibrillar meshwork of FN and collagen type IV, which functions to support the capillary network of the glomerulus as well as well as regulating glomerular filtration [2].

The microenvironment of a cell has long been known to affect cell behavior and the most obvious effect is on cell morphology. Paul Weiss demonstrated in 1959 that fibroblasts cultured in plasma clots demonstrated varying morphologies depending on the dimensions of the fibers formed within the clots [4]. However, many further mechanistic studies have since used cells cultured on flat, rigid, planar substrates or directly on the surface of plastic or glass plates to elucidate ECM-cell relationships. Although these studies have provided many important and essential findings pertinent to the research of cell signaling, ECM assembly and cellular processes such as migration and proliferation, the conclusions may not always applicable to cell responses in more physiological microenvironments.

Cellular response mechanisms elucidated from cells cultured within 3D ECM microenvironments better recapitulate those observed *in vivo*. Furthermore, *in vivo* developmental models including those of mouse, xenopus and chick, have produced essential information about ECM assembly mechanisms and its effects on cellular migration, proliferation and differentiation in native 3D environments [5-8].

The ECM is one of the major components of the cellular mechanotransduction machinery: the ability of cells to sense mechanical signals and convert this into biomechanical responses. Cell mechanosensing mechanisms allow cells to generate resistance against mechanical damage upon application of tensile or sheer stress [9]. Although the mechanisms for this process are still unknown, it is thought that cells likely apply force to their surrounding material to 'test' its compliance and monitor the resultant resistance and respond accordingly [10]. It is now clear that the balance of these mechanical and structural tensions detected extracellularly or generated intracellularly influence cell behavior and is the basis for the distinctive cellular responses generated upon culture within 2D compared to 3D matrices.

Here we discuss the importance of ECM assembly into complex 3D matrices and the importance of this 3D organization on cell behavior by highlighting the assembly and functions of the essential ECM glycoprotein FN.

2. Fibronectin

FN is an abundant and ubiquitous 220-250-kDa ECM glycoprotein, which is assembled on the cell surface into a complex, three-dimensional matrix [11-13]. FN is found in two distinct forms: plasma FN is synthesized by hepatocytes and secreted into the blood plasma, where it circulates at ~300-400 μg/ml [14] in a soluble, compact inactive form. Cellular FN is synthesized, locally secreted and assembled extracellularly by many cell types including fibroblasts, endothelial cells, chondrocytes, glial cells, synovial cells and myocytes [11].

FN assembly into a 3D fibrillar matrix on the cell surface has been shown to regulate a wide variety of cellular functions including adhesion [15-18], spreading [19], proliferation [16, 19-21], migration [5, 22-24], and apoptosis [25, 26]. FN is vital in directing these cellular events as FN null mice are embryonic lethal at E10-11. Mice null for FN receptors such as α5, αv or β1 integrins were also shown to be embryonic lethal highlighting the importance of FN assembly during embryogenesis [27].

As well as regulating cellular activities, FN has also been shown to support ECM maturation as it is essential in regulating the deposition of other ECM proteins including collagen type I and type III [28-35], fibrinogen [36], fibrillin-1 and -2 [37-39], fibulin [40], laminin [33, 34, 41], tenascin-C [42, 43] and reticulin [44]. The FN matrix has also been shown to regulate cell signaling by sequestering growth factors other signaling molecules including bone morphogenetic protein (BMP)-1 [45], vascular endothelial growth factor (VEGF) [46] and latent TGF-β binding proteins (LTBP) -1, -3 and -4 [47-50].

2.1. Fibronectin Structure

The ability of FN to interact with numerous binding partners including cell surface receptors, other ECM proteins and growth factors is due to its multi-modular multi-domain structure [51]. FN is composed of an array of multiple repeated and highly conserved modules: twelve FN type I repeats (FNI), two FN type II repeats (FNII), fifteen constitutively expressed and two alternatively spliced (EIIIA and EIIIB) FN type III repeats (FNIII), and a non-homologous type III connecting segment (V; variable or IIICS region). These modules are also organized into functional domains including the N-terminal 70-kDa domain (FNI_{1-9}), the 120-kDa central binding domain (CBD; $FNIII_{1-12}$) and the Heparin binding domain HepII ($FNIII_{12-14}$) (Figure 1). The multi-modular structure allows flexibility of the FN molecule which regulates its function [52-57]. FN is secreted as a dimer maintained by two disulphide S-S bonds at its C-terminus [56, 58-60] (Figure 1).

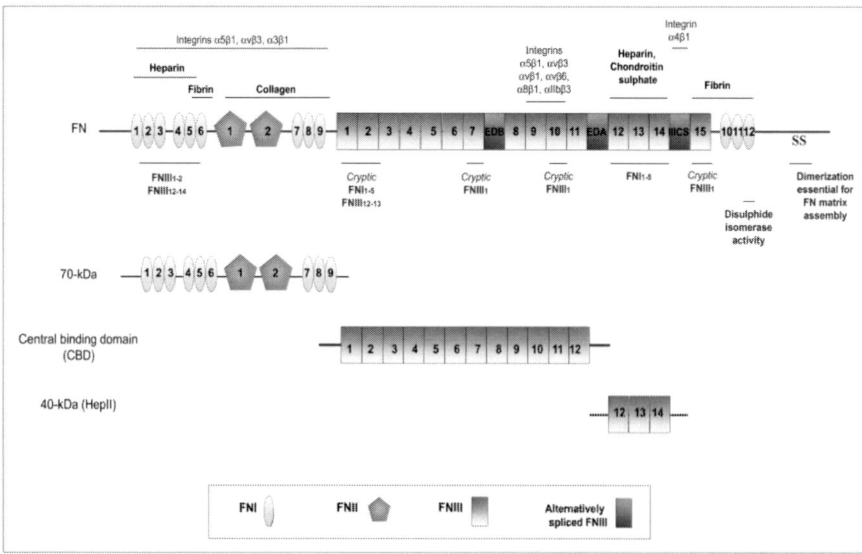

Figure 1. Modular organization of FN and its domains. FN is composed of a series of FNI repeats ( ), FNII repeats ( ), conserved FNIII repeats ( ) and alternatively spliced FNIII repeats ( ).

Although the overall tertiary fold of FNIII is highly conserved, the sequence homology between different FNIII modules $FNIII_1$-$FNIII_{15}$ within FN is actually less than 20% [61]. However, each individual FNIII module shows high cross-species homology (80-90%), suggesting each module must have a specific and important function [61-65]. For example, $FNIII_{10}$ module possesses a highly conserved RGD integrin-binding sequence, which is exposed on a β hairpin-like loop and forms a recognition site for at least 11 integrins [66-68]. FNIII modules have lower mechanical stabilities than FNI and FNII modules as they do not possess disulphide bonds. This property of FNIII is important to allow modular extension and unfolding leading to exposure of cryptic FN binding sites that are key to FN fibrillogenesis [64, 69, 70].

2.1.1. Alternatively Spliced Fibronectin

The inclusion or exclusion of alternatively spliced EIIIB and EIIIA modules (also referred to as EDB and EDA in the literature) and the more complex splicing patterns of the V/IIICS domain during transcription allows different isoforms of FN to be expressed in a distinct, tissue-dependent, temporally regulated and cell-type specific manner [3, 71-74]. This complexity in splicing means that in humans, twenty potential FN isoforms can be generated [75]. These isoforms can modulate the properties of the ECM: cellular FN shows reduced incorporation into fibrin gels compared to plasma FN [76]; EIIIA+FN and similarly, EIIIB+FN enhance cell adhesion, spreading and FN matrix assembly [15, 77-79]. The orientation and rotational interdomain flexibilities of FNIII modules are known to be affected by neighboring domains [51, 66, 68, 80-82] and inclusion or exclusion of alternatively spliced FN modules may alter the bending and rotational properties of neighboring and distant FN modules to affect cell signaling and FN-FN interactions [80, 83].

These different FN isoforms are also differentially expressed: plasma FN does not possess either the EIIIB or EIIIA alternatively spliced modules and only possess one subunit containing the V domain [74, 76, 84]. Cellular FN shows much more complex splicing patterns depending on their spatial and temporal expression patterns. Only very low levels of EIIIA and/or EIIIB+FN circulate in the blood plasma (cellular FN, 1.3 – 1.4 μg/ml) [85], but reports have shown increased levels after major traumas resulting in vascular tissue damage, inflammation and in diseases such as atherosclerosis, ischaemic heart disease and stroke [86-89]. EIIIB and EIIIA isoforms are highly upregulated in areas of physiological or pathological tissue remodeling including during development, wound healing and tissue repair. The sequences of EIIIA and EIIIB are only 29% identical, but 100% conserved between species suggesting highly distinct functions for these isoforms [90].

3. Fibronectin Matrix Assembly

3.1. Fibronectin Is Secreted in a Soluble Form

Much is still unknown about the mechanisms involved in FN fibrillogenesis and its assembly into a complex three dimensional matrix or the regulation of its deposition. FN matrix assembly is a step-wise, cell mediated process [91, 92]. FN in solution alone will not polymerize, which is important to prevent inappropriate fibrillogenesis from occurring [93].

Plasma and cellular FN are both secreted in a soluble, compact form, which is maintained by intramolecular electrostatic interactions between specific FN domains: FNI_{1-5}, $FNIII_{1-2}$, $FNIII_{2-3}$ and $FNIII_{12-14}$ [56, 58-60] (Figure 2 A). This compact quaternary structure can be observed by electron microscopy under low salt conditions [55, 57]. FRET studies also show that the N-terminal arms of the soluble FN dimer overlap in this compact structure [65].

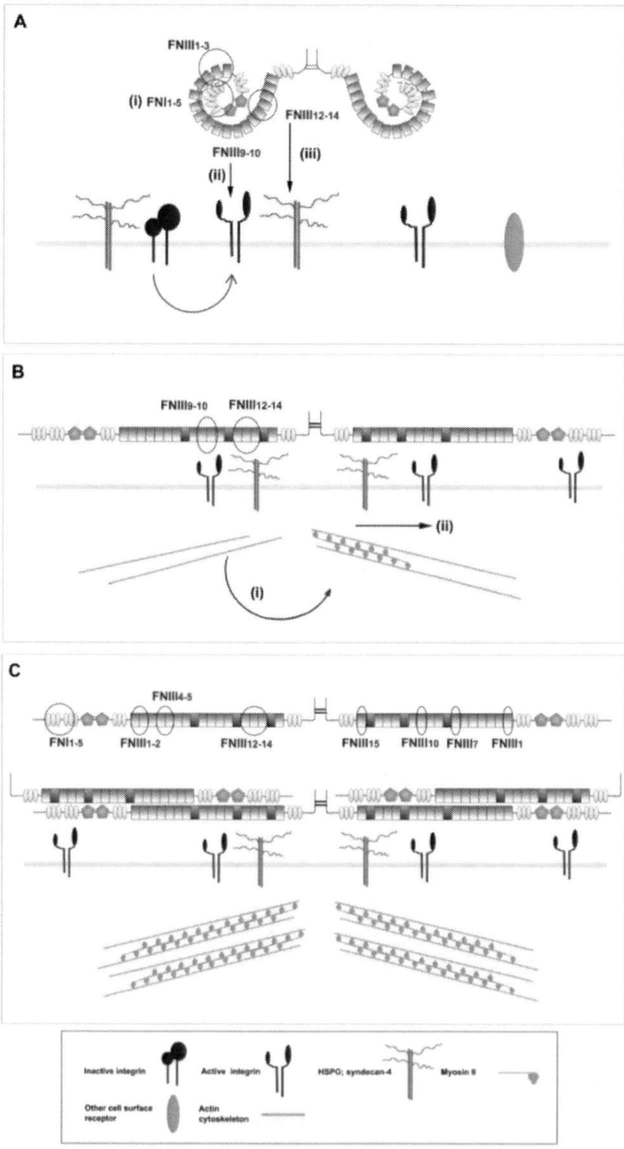

Figure 2. FN matrix assembly. (A) FN initiation requires FN to bind cell surface receptors: (i) FNI_{1-5} within 70-kDa binds to, as yet unidentified cell surface receptors, possibly including integrins, (ii) $FNIII_{9-10}$ **binds to α5β1 integrins, (iii) integrin** activation by outside-in or inside-out signaling promotes FN binding and (iv) $FNIII_{12-14}$ (HepII domain) binds to HSPGs. (B) FN unfolding: (i) FN binding of cell surface receptors induces cytoskeletal reorganization and myosin II dependent contractility (ii) with subsequent receptor clustering and translation causing FN molecules to become unfolded. (C) FN fibril formation: FN unfolding and FNIII module folding exposes FN binding sites allowing FN-FN intermolecular interactions. Domains important for each step are circled and denoted on the figure.

The importance of these intramolecular interactions have been demonstrated by FN mutants either lacking $FNIII_{12\text{-}14}$ or where this region was replaced by the alternatively spliced FN type III domains A1-A3 from tenascin-C (TN-C), as these adopted more open conformations with lower sedimentation coefficient values [56]. FN dimerisation is required for FN assembly as mutation of C-terminal cysteines results in the loss of fibrillogenesis [94].

3.2. Initiation Mechanisms of Fibronectin Assembly

Much of the mechanisms elucidated for FN matrix assembly have been focused on *de novo* initiation and FN assembly into a fibrillar matrix from various cell cultures in 2D conditions [95]. *In vitro* FN matrix assembly initiation occurs at focal contacts; initial sites of ECM contact on the cell periphery rich in paxillin, vinculin, phosphotyrosines, αv, β1 and β3 integrins [96-98]. FN initiation first involves interactions with the cell surface via the $FNI_{1\text{-}5}$ region within the 70-kDa domain: 70-kDa localizes as short, linear arrays at cell peripheries in a similar pattern as FN at early time points of assembly [99-101]; 70-kDa remains detergent soluble, which indicates its inability to assemble into a complex insoluble matrix on the cell surface [94, 99, 100]; 70-kDa demonstrates saturable binding with a similar binding affinity as full-length FN suggesting the involvement of a finite number of cell surface receptors [99, 100, 102]; 70-kDa inhibits FN matrix assembly by competing with matrix assembly sites [16, 99, 103-106], so although other regions of FN can interact with cells these are not sufficient to block fibrillogenesis [5]; mutations in the RGD-integrin binding domain still allow binding [6]; this domain and not the central binding domain is exposed in soluble FN [101, 107].

However, the receptor(s) for initiating FN matrix assembly via 70-kDa have not yet been clearly elucidated. Interaction of 70-kDa with the cell-surface requires the presence of divalent cations suggesting involvement of an integrin [108], however α5β1, the major FN binding integrin, has been shown not to be a receptor for the 70-kDa domain [106, 108, 109]. Instead αvβ3 integrin is a possible candidate: it localizes to focal contacts at the cell periphery during FN fibrillogenesis [96]; it can support FN polymerization by confluent cells lacking β1 [110]; it co-localizes to the termini of FN fibrils [110]; its interactions with 70-kDa has been demonstrated by solid phase binding [6, 111]. It has been hypothesized that αvβ3 binds to novel GNGRG motifs in $FNI_{1\text{-}5}$, which can undergo deamidation and rearrangement to *iso*DGR (a physiological ECM protein "ageing" process). These *iso*DG sequences (IGD) can interact with high affinity (Kd = 0.57 μM) to αvβ3 and other integrins [6, 111, 112]. 70-kDa possesses four IGD sequences exposed on FNI loops that could potentially act as integrin binding sites, and have been shown to mediate fibroblast migration [112]. Altogether, this data suggests that 70-kDa initiates FN matrix assembly by binding to the cell surface at focal contacts, possibly interacting with αvβ3 integrins on the cell surface (Figure 2 A).

A further essential step in the initiation and assembly of an extensive FN matrix involves α5β1 integrin binding of the RGD loop on $FNIII_{10}$ and the neighboring PHSRN sequence in $FNIII_9$ in the CBD of FN [95, 110, 113-115] (Figure 2 A). FN mutations where the RGD site was absent or mutated could not be assembled into fibrillar matrices [60] and antibodies against the RGD domain or small RGD peptides also blocked FN matrix assembly [6, 104, 105, 116, 117]. α5β1 integrin binding to CBD is also a high affinity interaction, Kd = 0.2 μM [66].

Table 1. List of FN binding integrins. (+) denotes integrins that have been shown it to be involved in FN fibrillogenesis. (–) denotes integrins that have been shown to be involved in adhesion to FN but not fibrillogenesis

FN receptors	Supports fibrillogenesis	Domain	References
α3β1	+	70-kDa	[115, 377]
α4β1	+	IIICS (V) CS1 region and EIIIA	[115, 377]
α5β1	+	$FNIII_{9-10}$ (RGD)	[11, 13, 18, 110, 378]
α9β1	-/+	EIIIA	[41, 300]
α8β1	-	$FNIII_{10}$ (RGD)	[377, 379]
αvβ1	-	$FNIII_{10}$ (RGD)	[377, 379]
αvβ3	+	$FNIII_{10}$ (RGD) or possibly 70-kDa	[101, 110, 378]
αvβ6	+	$FNIII_{10}$ (RGD)	[377]
αIIbβ3	+	$FNIII_{9-10}$ (RGD)	[114, 286, 377, 380]

Huveneers et al., (2008) demonstrated mutation of β1, or fibroblastic cells null for β1 or α5 and not αvβ3 inhibited FN matrix assembly suggesting α5β1 is required. However, it is interesting to note that FN-RGE fibroblasts expressing FN with a mutated RGD region assembled endogenous FN-RGE into short, thick, fibrils, suggesting $FNIII_{10}$ interactions with α5β1 are not essential for fibril formation [6]. The FN-RGE fibroblasts had to be plated onto laminin to allow matrix assembly to occur; when plated on vitronectin, no assembly was observed suggesting αv integrins were involved in assembling FN-RGE [6]. Huveneer's et al., (2008) plated their fibroblasts on FN, so αv integrins (in the absence of α5β1) may have been engaged in mediating cell binding to the coated FN and hence unavailable for mediating fibrillogenesis of exogenous FN. Other integrins may also be involved in mediating specific FN fibrillogenesis. For example, the α9β1 integrin was shown to be required for the matrix assembly of FNIIIA+FN by lymphatic endothelial cells [41]. Other FN binding integrins are summarized in Table 1 although not all have been shown to be active in FN matrix assembly.

Binding of α5β1 with its FN ligand induces "outside-in" signaling: binding of FN to α5β1 results in integrin activation, subsequent phosphorylation of integrin cytoplasmic domains and generation of intracellular cell signals [118]. This cell signaling induces the reorganization of the actin cytoskeleton, integrin-ligand clustering, recruitment of signaling and structural proteins and results in the maturation of focal contacts into large focal adhesion complexes, which anchor the integrin cytoskeleton to the the ECM [18, 98]. Ligation of integrins induce intracellular signaling that result in the activation of integrins to higher affinity binding states by "inside-out" signaling, enhancing FN binding and matrix assembly [118]. This integrin activation feedback loop for transmitting "outside-in" signals and translating them into "inside-out" signals regulates adhesion complex formation, integrin affinity and FN matrix assembly [95, 115, 118-120]. Integrin activation and signaling requires non-muscle myosin II ATPase, which drives the dynamic reorganization of actin and myosin filaments into contractile bundles with resultant cell-generated tension [18, 115, 120-122] (Figure 2 B).

As well as integrins, focal complexes also contain heparan sulphate proteoglycans (HSPGs), which also play an important role in FN matrix assembly (Figure 2 A). Chinese Hamster Ovary (CHO) cells defective for heparan sulphate glycosaminoglycan synthesis or expressing reduced levels of HSPGs showed loss of or reduced FN matrix assembly respectively [123]. Heparin and heparitinase treatment also inhibited exogenous FN matrix assembly by FN null fibroblasts [124]. Syndecans are type I transmembrane HSPGs implicated in regulating interactions and assembly of the ECM [125]. A heparin binding site has been mapped to the $FNIII_{13}$ module within the $FNIII_{12-14}$ HepII domain of FN [126-129] and this HepII domain is essential for the formation of focal adhesions [127, 130, 131]. The HepII domain of FN has been shown to bind the heparan sulphate side chains of syndecan-4 [132, 133] and the activation of syndecan-4 was shown to be required for cell spreading [134], cell contractility [135] and inducing fibroblasts to contract fibrin-FN gels [131]. However, although a major component of focal adhesions, Huveneers et al., (2008) showed that syndecan-4 is not essential for FN matrix assembly as syndecan-4 null mouse embryonic fibroblasts still assembled a FN matrix. Instead, syndecan-2, which is the major syndecan of fibroblasts and mesenchymal cells, is thought to be pivotal for FN matrix assembly as CHO cells expressing a truncated syndecan-2 cannot assemble FN into a matrix [136] and siRNA knockdown of syndecan-2 expression in HT1080 cells reduced FN matrix assembly [137]. Other cell surface receptors in adhesion complexes may also compensate; recently, a proteomics-based analysis determined 905 proteins were associated with focal adhesion complexes isolated from adherent human foreskin fibroblasts [138]. the complexity and multiple components that can be involved in these signaling complexes [98] and the level of compensation that occurs on the cell surface or the differences in FN assembly mechanisms between different cell types highlights the complexity and importance of this process.

3.3. FN Unfolding, Elongation and Fibril Formation

FN bound to integrins and other cell surface receptors has to be unfolded from its compact structure (Figure 2) allowing the exposure of binding domains to promote interactions between FN molecules as well as interactions with other ECM components. This unfolding process can be monitored by fluorescence resonance energy transfer (FRET) [52, 53, 65] and is dependent on the translocation of FN ligated to β1 integrins from peripheral focal contacts to central tensin-rich fibrillar adhesion complexes [96] (Figure 2 B). This translocation of FN causes it to assume extended states, exposing FN binding domains allowing FN-FN interactions resulting in FN fibril formation [96, 117, 139].

FNI_{1-5} [16, 94, 108, 140], $FNIII_{1-2}$ [140, 141], $FNIII_{4-5}$ [142] and $FNIII_{12-14}$ [103] are all important domains, which possess FN binding sites to promote FN fibrillogenesis [60, 94] (Figure 2 C). Many of these FN domains can interact with multiple FN binding sites to allow for fibril growth. For example, the 29-kDa (FNI_{1-5}) region of the 70-kDa domain, $FNIII_{1-2}$, and $FNIII_{10}$ bound to α5β1 integrins have been shown to form a ternary complex [16, 70], demonstrating the complexity of these interactions and the ability of multiple FN molecules to interact with each other. Other FN domains have been shown to be important for regulating the rate of fibrillogenesis; $FNIII_{1-7}$ deletion mutants ($FN\Delta III_{1-7}$) resulted in more rapid incorporation into detergent insoluble material, but a slower overall rate in fibrillogenesis [60]. The presence of the EIIIA or EIIIB domain have been shown to promote FN matrix assembly, which is likely to occur by exposing the integrin binding sites present in in neighboring domains: α5β1 binding site in $FNIII_{10}$ (neighboring module of EIIIB) and α4β1 binding site in the IIICS region (neighboring module of EIIIA) [78, 143]. The orientation and rotational interdomain flexibilities of FNIII modules are known to be affected by neighboring domains [51, 66, 68, 80-82] and inclusion of the EIIIB domain alters the bending and rotational angles at the EIIIB-FNIII$_8$ interface compared to FNIII$_{7\text{-}8}$ [80] exposing a cryptic loop connecting strands B and C of FNIII$_8$ [83]. This change in interdomain orientation may also impact the global conformation of FN, affecting the presentation of distant FNIII loop structures and binding sequences to also modulate FN-FN interactions during matrix assembly.

FN fibrillogenesis progresses by FNIII module unfolding and exposure of cryptic FN-FN binding sites to further promote FN fibrillogenesis [65]. Cryptic binding sites present within the folded FNIII module are shown to be exposed upon heat or chemical denaturation or by force induced unfolding [70, 144-146]. The FN linker domains and FNIII modules demonstrate an inherent elasticity; FNIII protein folds are stabilized by a hydrophobic core and do not possess any stabilizing intra-strand disulphide bonds [62, 63, 69]. Different FNIII modules are shown to have very different mechanical stabilities with FNIII$_{10}$ being the weakest and FNIII$_1$ and FNIII$_2$ the most mechanically stable [147], suggesting different roles and rates of unfolding for these different modules during the assembly process. Cryptic FN binding sites have been shown to exist in FNIII$_1$ [70, 144, 146], FNIII$_{10}$ [70], FNIII$_7$ and FNIII$_{15}$ [145] (Figure 2 C). A peptide sequence derived from FNIII$_1$ (FNIII$_1$-C) at low concentrations can induce the formation of superfibronectin from soluble FN, a high molecular weight precipitate mass of highly cross-linked aggregates of FN, which resembles cell-assembled FN fibrils by light microscopy [146, 148] and can enhance and alter FN matrix assembly including masking of the EIIIA module in matrices containing this FN isoform [148, 149]. This is thought to be due to the peptide recapitulating an intermediate state of FNIII$_1$ during its unfolding, exposing a potent cryptic FN binding site that affects FN conformation, assembly and cell signaling, although the exact mechanism is not yet well understood [150].

Modular unfolding is a mechanism that may also be used to regulate FN assembly by altering FN interactions with cell surface receptors and subsequently modifying cell signaling events. Indeed steered molecular dynamics [151] and biophysical studies of FNIII$_{9\text{-}10}$ [152] have shown the importance of correct distancing between the RGD-integrin binding site on FNIII$_{10}$ and PHRSN synergy site on FNIII$_9$ to facilitate optimal α5β1 binding, signaling and mediating cell adhesion. The unfolding of FNIII$_{10}$ may alter the exposure of the integrin binding site, reduce the synergistic interactions of the PHRSN site on FNIII$_9$ [65, 153] and it can be speculated that this loss of integrin binding affinity can then deactivate integrins, releasing the FN once it has been incorporated into fibrillar structures.

Cell generated tension is essential for FN matrix assembly as this force drives integrin translocation, induces the conformational changes of cell-surface bound FN and these forces are transmitted to the ECM resulting in its remodeling. Application of tension directly to FN in the absence of cells can expose FN-FN binding sites to promote FN-FN intermolecular interactions [154] (Figure 2 C). This cell generated tension is generated by non-muscle myosin II and requires an intact actin cytoskeleton: cytochalasin D or inhibitors of contractility (e.g. ROCK inhibitor Y-27632) can interrupt FN unfolding, FN fibrillar extension and α5β1 integrin cross-linking and clustering [18, 53, 65, 139, 155]; myosin II inhibitors inhibit cell contractility and FN matrix assembly [154]. Myosin II driven cell contractility is regulated by the myosin light chain kinase (MLCK), which phosphorylates the myosin light chain (MLC). MLCK can be activated via RhoA-ROCK II dependent [154, 156,

157] or RhoA independent signaling pathways [156, 158, 159]. Activation of myosin II promotes reorganization of myosin II and actin filaments into contractile bundles and force is generated by the "walking" of myosin II along these actin filaments [122, 160].

Fibrils are formed from the overlapping and staggered organization of extended FN dimers [29]. Initially, FN molecules are organized into thin fibrils of 5 nm diameter, which then become laterally associated into thicker fibrils of 6-22 nm diameters [29]. Further FN fibril interactions allow the formation of high molecular weight, branched, complex, fibrillar FN matrices, which are detergent insoluble [91, 92, 99].

This insoluble matrix is thought to be maintained by the cross-linking of FN molecules to form stable multimers, and studies including RNase refolding experiments have demonstrated an endogenous, partially cryptic protein disulphide isomerase activity within the FNI_{12} module of FN in [29, 161].

There is some question as to whether this covalent disulfide cross-linking is a true phenomenon that occurs in native FN matrices; there are suggestions FN associates purely via non-covalent protein-protein interactions [162, 163]. There are suggestions the high molecular weight multimers observed in non-reduced SDS-PAGE gels are a mixture of other ECM proteins such as fibrillin, which "block" FN dimers from migrating through the gel [163]. This is not altogether surprising as FN is known to interact with many different ECM components and these protein-protein interactions may well attenuate its migration through SDS-PAGE gels.

3.4. Regulation of Fibronectin Assembly

Once assembled, FN fibrils are continuously polymerized, remodeled and turned over within the fibrillar matrix on the cell surface, which is dependent on a β1 and caveolin-1 mediated endocytic mechanism [33, 48, 164]. Endocytosed FN is targeted to the lysosomes and degraded intracellularly [33, 165]. Homeostatic cell surface levels of FN are maintained by continual FN assembly and loss from the pericellular matrix [33]. Cells contract and remodel cell surface matrices including FN in response to internal and external signaling cues and this contractile response was clearly observed when cells were plated onto sheets of silicone rubber or onto isolated islands of FN [166-168]. This remodeling and maturation of the ECM has been reported to assemble networks that become less dynamically remodeled over time [48].

FN expression and assembly is stimulated by a multitude of molecules as highlighted in Table 2. Stimulation of FN expression, secretion and assembly by these agents emphasizes the complexity and requirement for tight regulation of this process as many of these cytokines are implicated in up-regulating FN expression and deposition during pathological conditions [169-171]. FN can also be degraded by multiple proteases including plasmin and MMP-9 [172, 173].

Furthermore, β1 integrin clustering can induce polarized expression of MT1-MMP to invasive structures to cause localized ECM degradation [174]. FN fragments and modules can also inhibit FN matrix assembly by competing for FN assembly sites [148], which could act as a feedback system to regulate FN levels on the cell surface. Indeed endocytosis of fibrillar FN from pre-assembled matrices was shown to be much slower than endocytosis of soluble FN [164].

Table 2. Regulators of FN mRNA expression and assembly

Positive regulators of FN mRNA	References
TGF-β1 and the TGF-β family	[171, 381-383]
PDGF-BB	[384]
IGF-1	[384]
HGF	[385]
Glucose	[386, 387]
Glucocorticoids	[388]
Negative regulators of FN mRNA	References
Cell contractility inhibitors	[52, 65]
RhoA inhibitors	[154, 156, 157]
Positive regulators of FN assembly	
Sphinogsine-1-phosphate	[389]
Estrogen	[390]
Plasminogen activator inhibitor type I (PAI-1)	[391, 392]
Urokinase plasminogen receptor (uPAR)	[393]
CTGF	[394]
LPA	[154, 157]
Low concentrations of vitronectin	[96, 395]
Anastellin *	[146, 396]
Negative regulators of FN assembly	
High concentrations of vitronectin	[328, 395, 397]
TN-C domains	**[326, 327]**
FN fragments/domains	[16, 99, 103-106, 148, 326]
FUD domain of the F1 adhesin protein from *Streptococcus pyogenes* (pUR4)	[398]

*Anastellin can induce polymerization of soluble plasma FN into "superfibronectin."

3.5. Fibronectin Assembly in Model Systems

Much has been elucidated about the FN matrix assembly process from cell monolayer studies, the culture of cells on rigid planar surfaces including substrate coated tissue culture dishes forces cells to behave very differently from those cultured in 3D [175]. Whilst cells plated onto glass or on coated substrate layers are presented with ligands organized in a single flat plane in restricted conformations, distributions and orientations, cells cultured in 3D environments are presented as irregular clusters of ECM ligands surrounding the cell [175, 176]. As well as the compositional properties of the microenvironment, the mechanical properties also play important roles in regulating cellular behavior including FN matrix assembly. In response to stiff inflexible environments, cells exert higher traction forces by forming stress fibres and strong adhesive complexes (focal adhesions). Cells align their cytoskeleton along the lines of force and it is thought this response is similar to that observed in muscle cell resistance to load [177]. This directional organization of the cell's contractile machinery also induces FN matrix assembly to occur along these lines of force [10]. In contrast, *in vivo* or when cells are cultured within ECM matrices, FN assembles around cell organizations as meshwork-like structures [178-181].

In cell cultures, FN matrix assembly is a rapid process as *de novo* FN deposition is detectable at the cell surface within 10 min of plating and within the conditioned medium by 30 min [91]. A complex insoluble FN ECM is detectable within 8 h of cell plating [95]. However, pre-existing 3D matrices act as scaffolds for further FN deposition; new fibrils co-localize with pre-existing matrix [95, 167, 182]. Cells cultured in native ECM scaffolds deposit more soluble FN onto pre-existing FN fibers, requiring lower concentrations of FN for initiation [95, 167]; pre-existing 3D FN matrices stimulate αvβ3 activation, which induces more rapid FN matrix assembly [95, 167]. Increasing the rigidity of the microenvironment by cross-linking of cell-derived ECM matrices, has also been shown to alter the kinetics of FN matrix assembly: cross-linking of cell-derived ECM enhances initial FN unfolding and changes in FN conformation but show reduced deposition and incorporation of soluble FN reduced assembly of new FN matrix by fibroblasts [167, 182]. This suggests ECM maturation is a mechanism that regulates the rate of FN deposition and hence ECM assembly.

There is also a differential pattern of deposition in matrices of different stiffness (defined as resistance to deformation); newly deposited FN fibrils on native cell-derived ECM matrices scaffolds were shown to closely co-localize with pre-existing fibrils, whereas newly deposited FN fibrils on cross-linked ECM matrices did not. On more compliant substrates i.e. the native ECM matrix, FN is assembled and remodeled with the native FN fibrils over time, whereas on stiffer substrates, the FN is already pre-stretched and assembled more rapidly without requiring the pre-existing ECM scaffold [167, 182]. However, more compliant or "soft" substrates do not support FN matrix assembly, as cells do not generate strong enough tractional forces in response, which are essential for FN matrix assembly. Similarly, FN matrix assembly by epithelial cells cultured in polyacrylamide gels with varying stiffness was shown to increase with increasing substrate stiffness [183]. This tensional dependency on FN matrix assembly is also observed *in vivo*: both development processes and pathological progression including fibrosis and tumorigenesis occurs in tissues of increased stiffness with high levels of FN assembly [9, 183-185]. For example, normal mammary interstitial tissue is soft and pliable and devoid of FN, but FN deposition increases within the stroma of many mammary tumors [186]. These processes are intractably linked: the ability of cells to detect the stiffness of the surrounding substratum and match it with the internal stiffness of the cytoskeleton requires correct FN matrix assembly and myosin ATPase activity [178].

The overall fibrillar structure of pre-existing 3D ECM can also influence FN matrix assembly: fibril scaffolds with finer fibrils showed more rapid and complex FN matrix assembly compared to thicker fibrils [187].

FN remodeling has also been shown to be a dynamic process, both *in vitro* and *in vivo* [48, 50, 139, 155, 188, 189]. Cells cultured on FN substratum can deform and remodel underlying FN ECM [168, 190, 191]. *In vivo,* the fibrillar ECM structure can also deformed by local cell migratory and protrusive activities [37, 48, 167, 178, 189, 192], as well as large-scale tissue motion, particularly during embryogenesis [37, 48]. The elasticity of FNIII modules allows mechanical unfolding and refolding of FNIII modules which facilitates this highly extensible behavior of FN [193]. This elasticity and deformation of FN fibrils has been observed by live imaging fluorescence microscopy: studies of cultured cells demonstrated the high tensional strain of FN fibrils assembled by fibroblasts, which rapidly retract when broken or detached from cells [189]. *In vivo* live imaging of FN remodeling during gastrulation events showed highly dynamic FN fibrillar movements with continuous detachment, contraction, bending, stretching, extension, retraction and annealing between neighboring FN fibrils [155, 194]. The constant remodeling of FN fibrils is a further process of ECM maturation to regulate FN deposition as well as the deposition of other ECM proteins.

The assembly and incorporation of FN into complex three-dimensional matrices surrounding the cell would present the FN ligand in a very different manner to cells cultured on coatings of FN protein or on matrix gels containing FN. Here we highlight the effects of these different cellular microenvironments on cell behavior and some of the differences on the mechanisms involved.

4. The Roles of Fibronectin in Regulating Cellular Behavior

4.1. Cell Adhesion and Morphology

Adhesive complexes are multi-functional transmembrane mass protein organizations vital to allow cells to monitor and communicate with their extracellular environment by mediating bi-directional cell signaling. These protein complexes are not static however, but are dynamic structures that assemble, mature and disassemble as required [195]. The protein network that connects the ECM and the actin cytoskeleton have been termed the 'adhesome' and over 905 proteins have recently been identified assembled at focal adhesions [138] with over 180 "interaction nodes" highlighting the complexity and interconnectivity of these adhesive structures [196].

Cell adhesion complexes and morphologies differ strikingly depending on their microenvironment and activation status. The culture of cells on rigid, planar FN substrates can force cells to adopt artificial morphologies; for example fibroblasts become ventrally and laterally polarized on FN coated surfaces [175] and flatten, spread, elongate and organize their cytoskeleton into well-defined actin stress fibers and microfilaments [187]. This has been strikingly demonstrated by studies where culture of endothelial cells on microfabricated square or round FN islands spread and constrained their shape to fit the underlying matrix [190]. The culture of neonatal rat ventricular myocytes on uniform 2D FN substrate adopt a pleomorphic shape, but when stimulated to grow on islands of FN, spread to assume the same shape of the FN islands (square, triangular etc) [168]. On 2D FN matrices, FN-integrin complexes aggregate and mature into focal adhesions: large, elongated (3-10 μm), dash-shaped, flat adhesive complexes that anchor the cells to underlying FN matrices [98, 175]. Focal adhesions are composed of: ECM receptors e.g. integrins including FN binding receptors $\alpha5\beta1$ and $\alpha v\beta3$; structural and linker proteins including vinculin, talin, paxillin, α-actinin; signaling molecules including focal adhesion kinase (FAK) and other tyrosine-phosphorylated proteins [98, 122, 134, 138, 197-199].

However, in 3D microenvironments, the adhesive complexes formed are much more variable in shape and structure. Mouse embryonic adhesive foci were formed from co-localizing α5 integrin and paxillin, which closely co-localized parallel with FN fibrils whereas these complexes were not observed by cells plated onto 2D FN substrate cultures [175]. Similarly, fibroblasts cultured in 3D tissue- or cell-derived matrices showed long and thin (19 μm long) adhesions, rich in paxillin, vinculin, FAK, phosphotyrosine, α–actinin, activated β1 and α5; β3 was absent from these 3D adhesions [175]. However, fibroblasts cultured in 3D FN matrices were reported to assemble smaller focal adhesions and fibrillar adhesions [175]. The complexity of cell adhesions in 3D cultures and *in vivo* are summarized by Harunaga and Yamada (2011) and highlight the complexity and dynamism observed in cell-ECM interactions [200].

This complexity is also reflected in the different reported cell morphologies adopted by cells in 3D cultures and *in vivo*: fibroblasts in 3D cultures can adopt irregular shapes ranging from elongated bipolar morphologies to spindly, multipolar, stellate morphologies [4, 175, 201, 202]. Endothelial cells cultured in 3D are also shown to adopt elongated morphologies and when stimulated could organize into tubular structures, which was dependent on FN matrix assembly [178]. Canine epithelial cells cultured in 3D matrices could also undergo tubulogenesis, a process also dependent on FN matrix assembly [20, 203, 204]. When fibroblasts were sandwiched between two layers of polyacrylamide coated with either FN or collagen type I, the fibroblasts adopted elongated morphologies and long, thin extensions with a loss of lamellipodia. However, more highly branched cell extensions were observed with collagen compared to FN suggesting different ECM substrates activate different integrin receptors that result in differential cell shapes [175, 205, 206].

Factors that affect fibroblast morphology include the fibril width [4, 187]. The fibrillar structure and density is an important factor in regulating cell morphology: fibroblasts in low density fibrillar matrices became elongated, bipolar spindle shaped cells, extending along the length of the ECM fibrils whereas those in high density fibrillar matrices become multi-polar, stellate shaped with multiple long protrusions [202, 207].

Differences in cell morphology are dictated by the response of the cell to its microenvironment: the mechanical properties of the ECM, its organization and composition, and hence the strength and organization of the adhesive complexes that connect the intracellular cytoskeleton with the extracellular environment.

4.2. Cell Migration

The differences observed in cell morphologies and adhesion complexes are also responsible for and a consequence of the different mechanisms involved in cell migration on 2D compared to 3D matrices. Cell migration on ECM substrates is a coordinated regulation of tractional and contractile forces generated by the cell cytoskeleton coupled and balanced with cell adhesion to the underlying ECM. Cell migration is a concerted series of spatially and temporally regulated events: cell polarization; actin polymerization; cell contraction and cellular protrusion; formation and maturation of cell adhesions at the leading edge of the cell and within cellular protrusions; retraction of the cell at the rear by disassembly of adhesive complexes [195, 208]. It is now also thought that rear cell retraction can also initiate cell migration [209]. It requires a constant remodeling of the actin cytoskeleton as actin filaments are assembled and push at the leading edge of the cells at lamellipodial structures, and are pulled from the trailing edge of the cell by contraction of the actin filaments by myosin II

[195]. Hence cell migration is highly dependent on the structural and biophysical properties of the matrix microenvironment.

FN is essential for promoting cell migration; both exogenous FN and a certain level of endogenous FN are essential for cell migration but excessive FN matrix assembly impedes this process [5, 23, 24]. Cell migration is biphasic; low concentrations of FN do not stimulate migration whereas increasingly high concentrations of FN were shown to be inhibitory for migration [23, 24, 210]. High concentrations of FN promote strong cell adhesion inducing cells to exit G1 phase and proliferate rather than migrate [23, 24]. Cell migration requires the cell-dependent assembly of FN into a matrix [24]and is dependent on the CBD, HepII and IIICS domains of FN [22]. Airway epithelial cell migration was shown to be dependent on the heparin binding C-terminal sequence within $FNIII_1$ ($FNIII_1H$) and $FNIII_{8-10}$ regions of FN [24]. Furthermore, the ability of FN to interact with other ECM components was shown to be important for promoting cell migration: FN deletion mutants of the collagen binding domain FNI_{1-6} ($FN\Delta I_{1-6}$) were shown to be defective in supporting collagen type I fibril assembly and were also unable to support cell migration [34]. This suggests that the ability of FN to support the polymerization of other ECM proteins may be important in regulating cell migration. These studies also highlight the requirement for tight regulation of FN matrix assembly to control and direct correct cell migration. It is also interesting that FN null and EIIIB-FN null bovine aortic endothelial cell motility on glass was actually enhanced in the total absence of FN or EIIIB+FN. Cells showed reduced spreading, an increase in β3 adhesions and in the case of FN null endothelial cells, a loss of cell-cell interactions and contact inhibition [179].

Matrix stiffness also plays a role in regulating cell migration; increasing matrix stiffness has been shown to induce more rapid cell motility. This directional migration induced by variable matrix stiffness has been termed durotaxis or mechanotaxis [211, 212]. Human glioma cells cultured on FN coated glass and on stiff FN coated polyacrylamide ECMs (>100 kPa) showed rapid gliding migration, whereas glioma cells cultured on matrices with an intermediate level of stiffness (10-100 kPa) showed a "stick-slip" forms of migration where cells extend towards the leading edge whilst the trailing edge at the rear of the cell would suddenly unattach and snap back towards the leading edge. In contrast, glioma cells cultured on very compliant matrices (<1 kPa) remained rounded, adhered to the matrix and were unable to migrate [211]. Incorporation of FN into 3D matrigel showed an increase in matrix stiffness confirming FN assembly into the ECM can modulate tissue stiffness that then also alters the migratory behavior of cells [185]. Migration of human prostate carcinomas through matrigel was shown to be dependent on the FN concentration, levels of FN receptors and matrigel density (stiffness) [213]. Hence, excess FN impedes epithelial cell migration [210], possibly by affecting the matrix stiffness and enhancing adhesive complex formation.

These studies in cell migration are good models for the migration of cells over stiff matrices *in vivo* for example epithelialization by keratinocytes during development or re-epitheliazation during wound healing [214]. The presence of FN as a substratum was required to initiate epidermal outgrowth even in the absence of growth factor [215] and keratinocyte migration on FN was dependent on the CBD of FN and specifically the RGD α5β1 integrin binding site [216].

Culture on 3D matrices have also been shown to induce the elongation and self-organization of endothelial cells into branched network-like structures in the absence of growth factor or angiogenic stimuli, confirming the importance of the 3D microenvironment in providing signals to self-organize into tissue like structures [159, 217]. The ability of endothelial cells to form network-like structures was dependent on matrix stiffness: cell network assembly was reduced as matrix stiffness increased [181, 218, 219]. Furthermore, FN polymerization was required for the formation of these endothelial networks as inhibition of FN assembly inhibited cell-cell contacts, enhanced cell motility and disrupted network formation [179, 181]. Also, fibroblasts cultured on 3D matrices were shown to proliferate and organize into microtissue-like structures and was also dependent on FN polymerization [220]. On stiff substrates, cells were shown to migrate away from one another whereas on soft substrates, cells aggregated and organized into tissue-like structures and networks.

In contrast to epidermal cell migration, many cell types including immune cells and tumor cells have to invade into and migrate through complex fibrillar ECMs including FN-rich interstitial tissues. Two major modes of migration through 3D matrices have been described: amoeboid mode where cells do not adhere strongly and squeeze through gaps in matrix fibrils and mesenchymal mode where cells adhere strongly to matrix fibrils and secrete proteolytic enzymes to degrade the matrix to allow their progression through the matrix [221]. It is possible for cells such as macrophages to adapt to different migratory modes depending on the fibrillar structure and composition of the matrix [221] including along the length of ECM fibrils, in a one-dimensional, rapid uniaxial migratory mode independent of ligand density [222, 223].

Within 3D matrices, neovascularization or angiogenesis also depends on FN and its polymerization, which enhances cell migration [178, 224-226]. The presence of a surrounding 3D matrix is also important to induce tubulogenesis even in the absence of angiogenic stimuli confirming the three-dimensional matrix can induce signals for the self-organization of cells into tissue like structures [159, 217]. EIIIB+FN isoforms are also highly expressed by endothelial cells during development, which is an FN isoform known to enhance cell motility and survival [72, 179]. Furthermore, endothelial tubule formation specifically requires the expression of this isoform as inhibition of EIIIB+FN expression could not be rescued by addition of exogenous plasma EIIIB-FN [72]. This is an intuitive regulatory system to ensure restricted tubule formation.

Epithelial cell migration during tubulogenesis also requires the expression and assembly of FN: Madin-Darby canine kidney (MDCK) epithelial cells express the EIIIA+ FN [204], which is stimulatory for epithelial cell migration [75, 203]. Furthermore, migrating MDCK cells express enhaced levels of EIIIA+FN [204], which assembles into a FN matrix more readily promoting cell adhesion and spreading [15, 78].

This suggests the ability to regulate cell migration by regulating FN levels within the ECM and altering the structure and stiffness of the matrix is an essential mechanism to modify a multitude of cellular processes including cell adhesion and cell signaling. Therefore misregulation of FN matrix assembly may promote many pathological processes such as tumor metastasis. Using speckle fluorescence microscopy, it has been demonstrated that the rate of actin filament debranching and capping turnover at the leading edge of the cells during their migration are higher *in vivo* migration compared to *in vitro* suggesting alternative mechanisms are utilized in 3D environments in contrast with those observed in 2D [208].

The importance of 3D environments in mimicking *in vivo* cell behavior is particularly highlighted with human umbilical venous endothelial cells (HUVECs) which cultured on the top surface of 3D fibrin matrices (effectively 2D culture of cells onto a layer of ECM substratum as cells are not stimulated to invade) cells spread to confluence but when seeded within 3D matrices organize into tubular structures [178].

Similarly 2D cultures of bovine arterial endothelial cells (BAECs) on glass resulted in polarized monolayers but culture of BAECs on top of or below an overlay of matrigel, resulted in cells organizing into capillary like structures, which invaded into the matrix [159, 179].

3.3. Cell Survival and Proliferation

FN has been shown to promote cell survival and cell growth. Multiple signaling pathways have been shown to be induced in cells upon ligation to FN substrate (Table 3 and 4). As well as the α5β1 integrins, which initiate cell signals promoting cell proliferation upon ligation to FN [227-231], αvβ3 [232] and α8β1 integrins [233] and the ERRB2 member of the epidermal growth factor tyrosine kinase receptors [234] have also been shown to activate various cell proliferation signaling pathways. Furthermore, cell growth has been shown to occur on recombinant FN lacking the RGD integrin binding sequence (FNΔRGD), which is promoted by heparin sulphate proteoglycans [124].

Many of these signaling pathways have been shown to induce the expression of cell-cycle regulatory signaling molecules including related proto-oncogenic transcription factors including c-Myc, p53 and HIF-1α which induce cell proliferation, apoptosis and promote cell cycle progression by up-regulating G1 phase proteins such as cyclin D1 [228, 229, 234]. FN also promotes cell growth by inhibiting the expression of negative regulators of cell signaling including cyclin-dependent kinase inhibitor p21 and phosphatidylinositol phosphate phosphatase (PTEN) [228, 229].

The architecture and assembly of a complex 3D FN matrix has also been shown to differentially regulate cell proliferation: CHOα5 and SV40-transformed 3T3 cells grown in a FN matrix composed of mutant FN molecules missing the $FNIII_{1-7}$ modules, which have a different rate of assembly [60] suppressed cell proliferation by arresting cells in the G_0/G_1 phase [21]. FAK phosphorylation was shown to be attenuated in cells cultured with the FN mutant suggesting that native FN matrix assembly induces FAK activation that signals to induce cell growth [21].

FN also promotes cell survival by up-regulating the expression of anti-apoptotic proteins such as B-cell leukaemia 2 (Bcl-2) [235-237] and survivin (also known as baculoviral inhibitor of apoptosis repeat-containing 5) [238] resulting in the inhibition of caspase activity [229, 237, 239] and the assembly of the apoptosome [237]. These anti-apoptotic mechanisms were again shown to be induced by signaling via the FN-binding α5β1 integrins [235, 237-240] and subsequent activation of protein kinase A (PKA), phosphatidylinositol 3-kinase (PI3-K) or Akt/Protein kinase B (PKB) serine/threonine kinase [229, 237, 238, 241]. Natural killer cells were also shown to signal via CD11b (αMβ2) to activate the Src-β catenin-ERK/JNK signaling pathway [236].

Similar survival mechanisms were shown to be induced upon culture within ECM matrices; malignant epithelial cells up-regulated FN, EIIIA+FN and α5 expression upon culture on three-dimensional laminin-rich ECMs via activation of Akt/PKB and inhibition of α5 activity resulted in the induction of cellular apoptosis [242].

Table 3. FN effects on proliferation and survival of cell cultures

Cell type	FN effect	Signaling pathway	Induced expression	Inhibits expression or activity	Ref
Non small cell lung carcinoma	Proliferation	α5β1, PKC, ERK	COX-2, prostaglandin E2		[227]
Non small cell lung carcinoma	Proliferation	ERK, Rho kinase	cyclin D1	p21	[228]
Normal human thyroid cells	Proliferation	αvβ3, Ras/Raf/MAPK kinase (Mek)/Erk and calcium (Ca2)/calcium calmodulin-dependent kinase II (CaMKII)			[232]
Immortilized human thyroid cells (TAD-2 and Hep3B)	Proliferation	Ras/Raf/MAPK kinase (Mek)/Erk and calcium (Ca2)/calcium calmodulin-dependent kinase II (CaMKII)			[241]
AtT20 neuroendocrine cell line transfected with α8	Proliferation	α8β1, PI3-K			[233]
Human retinal endothelial cells	Specific domains promote cell proliferation	α5β1, PI3-K			[230]
Schwann cells	Proliferation and migration				[399]
MCF-7 mammary adenocarcinoma cells	Proliferation and adhesion	ERBB2, HIF-1α	α5		[234]
Immortilized human thyroid cell line and normal human thyroid cells	Proliferation and survival	p21Ras/MAPK pathway (proliferation); phosphatidylinositol 3-kinase (PI3K) signal pathway (survival)			[400]
Human bronchial epithelial cells	Proliferation and survival	α5β1, NFkB, PI3-K/Akt	c-Myc, cyclin D1	p21, PTEN, caspase 3/7	[229]
Brain capillary endothelial cells	Proliferation and survival	α5β1 and αvβ3, MAP kinase			[231]
Human prostate cell lines	Promotes cell survival against TNF-α induced apoptosis	β1 integrins, Akt	Survivin		[238]
HUVECs	Survival	α5β1, PKA		Caspase 8 and 3 activity	[239]
CHO cells expressing α5β1	Survival	α5β1	Bcl2		[235]
Human bronchial epithelial cells	Survival	α5β1			[240]
Mouse natural killer cells	Survival	CD11b, Src, β-catenin, ERK, JNK	Bcl2		[236]
Canine pancreatic islets	Survival				[401]
Differentiated mature osteoblasts	Survival				[402]
Hepatic stellate cells	Survival	α5β1, FAK, PI3-K, Akt	α5β1, Bcl2	Formation of apoptosome, caspase 9 and 3 activity	[237]

Table 4. FN effects on proliferation and survival of cells cultured in 3D matrix environments

Cell type	FN effect	3D ECM culture conditions	Signaling pathway	Induced expression or activity	Ref
D2A1 mouse mammary cancer cells	Proliferation	Cultrex basement membrane extracted from a transplantable murine tumor with FN	β1, MLCK	Phospho-MLC, actin stress fibre formation	[403]
HUVEC	Morphology, migration, proliferation	3D fibrin matrices, 3D collagen matrices	myosin II dependent: MLC-2, β-actin, α-actinin	Tubule formation, stress fibre assembly; FN matrix assembly	[178]
Murine pancreatic β-cell line MIN6	Survival	FN entrapment within polyethylene glycol hydrogels			[404]
Malignant HMT-3522 T4-2 mammary epithelial cells	Survival	Three-dimensional laminin-rich ECM	α5β1, Akt	α5	[242]
Human umbilical vein endothelial cells	Survival	HUVEC covered FN coated collagen modules subcutaneously implanted *in vivo*			[244]
Human umbilical vein endothelial cells	Survival	Photocross-linked FN to hyaluronan hydrogel network			[267]

Many studies have explored the pro-adhesive, pro-migratory, proliferative and anti-apoptotic properties of FN to promote cell survival on synthetic polymers for *in vivo* use in tissue engineering. FN coating was shown to be essential for the adhesion and proliferation of HUVECs onto polymer scaffolds being developed for use as corneal prostheses [243]. HUVEC covered collagen modules that had been coated with FN were shown to enhance HUVEC survival *in vivo* after subcutaneous implantation compared to collagen modules alone.

Furthermore, this enhanced survival resulted in an increased formation of HUVEC-lined microvessels [244]. Neural stem cells within FN/collagen type I scaffolds transplanted into mouse brains that were traumatically injured showed improved survival and wider dispersion within the injured brain. However, in this case FN was not the best scaffold to improve cognitive behavior after transplantation: laminin was shown to be more efficacious [245-246] . In other cases, plasma FN was found to be essential for protecting neuronal and non-neuronal cells from apoptosis after transient focal cerebral tissue ischaemia [247] and after traumatic brain injury [248]. This suggests FN promotes the survival and differentiation of specific cell types and may not be optimal for all cell types.

These studies illustrate how intricately linked cellular behavior is to the external microenvironment and the complexity of these extracellular-intracellular signals that have to be tightly regulated. Here we highlight the function of FN during developmental and physiological ECM remodeling events and illustrate how mis-regulation of FN expression and assembly occurs in multiple pathological processes that may contribute to disease progression.

5. Roles of FN in Physiological and Pathological Remodeling

5.1. FN in Development

During development, cellular processes have to be tightly coordinated, requiring signaling events from cell-ECM, cell-cell and growth factor-cell interactions. FN is first expressed at the blastocyst stage within the trophoectoderm before blastocyst implantation but only appears to be vital for function after the onset of gastrulation [249, 250]. This suggests FN is not essential for cell proliferation or organization before this stage of development.

FN null mouse embryos showed increasing abnormalities in mesodermal development including that of the heart and extraembryonic vasculature, neural crest, trunk and limb buds suggesting FN is vital for these developmental processes [90, 251, 252]. FN and its polymerization were shown to be essential for generating polarized tension to correctly orient cell division during development [5]. Chick embryos cultured with the 70-kDa domain of FN, which inhibits FN fibrillogenesis only formed one or two somites with irregular and uneven deposits of endogenous FN covering the dorsal surface of the presomitic mesoderm rather than the uniform deposition observed in untreated embryos [7]. Loss of FN by injection of short antisense oligonucleotides into zebrafish embryos disrupted convergence and extension cell movements resulting in defective anterior somitogenesis [8]. During chick somitogenesis, FN was found to be essential not for migration or pseudopodia formation by presomitic mesodermal cells, but instead for cell elongation, orientation and organization into aster like cell clusters. FN was also essential for maintaining somites once formed and promoting the recruitment of cells to egress and incorporate into the somatic epithelium [7].

The complex co-ordination of tubule/vessel formation is regulated by signaling events from cell-ECM, cell-cell and soluble growth factor interactions [253]. Similar to the requirement for FN matrix assembly for kidney epithelial cell morphogenesis *in vitro* [20, 203] FN also plays an essential role in tubule formation and branching morphogenesis *in vivo* during the development of salivary glands, lung and kidney; FN mRNA and protein is strongly expressed in the cleft regions of budding salivary gland explant epithelium. Inhibition of FN assembly and protein function with blocking antibodies against FN, α5 or β1 or FN mRNA expression with FN siRNA inhibited branching by these epithelia explants [254].

Tubule formation requires a complex coordination of epithelial cell processes [255-257], and FN deposition in the basal laminae is important in mediating and regulating these cell processes [20, 203, 257, 258]. FN signaling also promotes epithelial cell survival during tubule formation [20] via β1 integrin signaling [259] and FAK activation [260, 261]. FN induces tubule formation by ureteric bud cells when grown in 3D collagen type I gels, a model of processes that are important during renal development [262].

FN protein expression is also highly up-regulated in regions of active neovasculogenesis and angiogenesis during development [263-266]. The assembly of a FN matrix is vital for neovessel formation by HUVECs cultured in 3D gels and angiogenesis in chick chorioallantoic membrane models [178]. This regulation of neovasculogenesis by FN matrix assembly is only observed under 3D conditions as inhibitors of FN matrix assembly had no effect on endothelial cell proliferation, morphology or migration on 2D substrates but were effective at inhibiting tubulogenesis in 3D cultures [178]. Furthermore, HUVECs cultured in FN-rich fibroblast derived matrices adopted elongated cell morphologies, remodeled the surrounding ECM and formed adhesive complexes that allowed the cells to closely co-localize and wrap around FN fibrils [217]. HUVECs cultured in 3D hydrogels with photo-crosslinked FN to the hyaluronan scaffold also enhanced tubulogenesis compared to the hyaluronan scaffold alone [267].

FN has been shown to be essential for determining cell polarity in cardiac precursors during heart tube formation; zebrafish with a null mutation in the FN1 gene showed defective adherens junction formation between myocardial precursors and poor organization of the myocardial epithelia, which lead to defective cell migration and defective heart tube formation [268]. FN-null mice developed lethal vascular defects, including impaired heart development and collapsed blood vessel lumens [90] again highlighting the importance of FN during vasculogenesis and cardiogenesis.

The EIIIA+FN and EIIIB+FN isoforms are highly expressed during development [269]. Although EIIIA or EIIIB null mice are viable [269, 270] the EIIIA/EIIIB double knockout mouse is embryonic lethal at E9.5-10.5 with severe cardiovascular defects [271]. These isoforms express in spatially and temporally distinct regions of the developing embryo suggesting specific and distinct roles for the two isoforms. FNIIIA+FN was highly expressed at sites of the developing valve and at the edges of the elongating valve leaflet whereas FNIIIA-FNIIIB-FN was shown to be highly expressed throughout all the basement membranes of developing lymphatic vessels [41]. FNIIIA+FN was shown to be vital for lymphatic valve development, and required interactions with $\alpha 9\beta 1$ integrins for the assembly by lymphatic endothelial cells [41].

This suggests that FN plays many different and distinct functional roles over the course of the developmental process. An inability to assemble FN is embryonic lethal highlighting the importance of this assembly process.

5.2. FN in Wound Healing

During wound healing, plasma FN is incorporated into the fibrin clot whilst the expression of cellular FN including the EIIIA+FN and EIIIB+FN isoforms are highly up-regulated and deposited within the granulation tissue during later wound healing events [76, 272-276]. *In vitro* experiments show that the different isoforms of FN if supplied exogenously can be assembled into newly assembling or pre-existing cellular FN matrices even if the exogenous source of FN is isolated from a different species [95, 277, 278]. However, the temporally and spatially distinct expression patterns of the different isoforms during wound healing would suggest differential and discrete roles.

Plasma FN is a major component of the fibrin clot and interacts with the fibrin matrix either by non-covalent interactions via the FNI_{1-5} and FNI_{10-12} domains of FN [12] or by covalent cross-linking, which is mediated by activated Factor XIIIa (plasma transglutaminase or coagulation factor XIII) [76, 84, 279-282]. Although cellular FN isoforms have been shown not to be efficiently incorporated into the fibrin clots *in vitro,* it is thought that they may be sufficient to allow normal wound healing events *in vivo* in the absence of plasma FN in conditional knock-out or mutant mice [85, 247, 283, 284]. The FNI_{10-12} domain has also been shown to bind connective tissue growth factor (CTGF) and enhances the affinity of FN for fibrin [285]. Furthermore, plasma FN can also bind αIIbβ3 integrins expressed on the surface of platelets and assemble into a high molecular weight, multimeric matrices on the surface of platelets by thrombin activation [286-292].

FN incorporation into the fibrin matrix is important for platelet function including platelet adhesion, migration and aggregation [282, 286, 293-295]. The fibrin-FN provisional matrix allows FN to adopt extended conformations leading to the exposure of cryptic cell binding domains [282], which function to induce cell migration and repopulation of the wound. For example, PDGF dependent fibroblast activation requires specific sequences within the heparin domain and IIICS region [22]. Fibroblast migration and invasion from the surrounding collagen-rich connective tissue ECM into the fibrin-rich provisional matrix has also been shown to require FN expression and deposition in both tissues, which acts as a conduit for migration [296-298]. The α4β1, α5β1 and αvβ3 integrins and dermatan sulphated CD44 cell surface receptors are necessary for fibroblast migration into the fibrin clot [296, 297]. α4β1 has been shown to mediate cell adhesion and contraction via interactions with the alternatively spliced V region of FN but only after Mn^{2+} induced integrin activation or in the presence of proteolysed FN cross-linked to fibrin [299]. The wound environment is highly proteolytic, with fibrin-FN matrix degradation mediated by plasmin and serine proteases and the proteolysis of FN may act as stimuli for promoting cell-mediated fibrin-FN contraction during the clearance of the provisional matrix as wound healing events progress [299].

In cutaneous wounds, keratinocytes are shown to migrate into the clot during re-epitheliazation after wounding by up-regulating the expression of FN binding integrins: α5β1, αvβ6 and αvβ1 integrins. Keratinocytes cultures stimulated by TGF-β have been shown to deposit EIIIA+FN as they migrate in a scratch wound model [172]. This isoform of FN has been shown to be essential for re-epitheliazation as EIIIA null mice were abnormal for this process [270]. Keratinocytes have also been shown to up-regulate their expression of α9β1 integrins, which is also a receptor specific for the EIIIA module of FN; α9 null mice showed reduced keratinocyte proliferation resulting in poorer re-epitheliazation after wounding [300, 301].

Cellular FN is an important and abundant component of granulation tissue, synthesized and deposited by fibroblasts and endothelial cells as they re-populate and re-vascularize the damaged tissue [44, 172, 280, 302, 303]. Within the stroma of granulation tissue fibrillar FN organizes into a dense network around fibroblasts, which orient and polarize along the FN fibrils, parallel to the epidermis [44, 304]. FN assembly into a 3D fibrillar network on the cell surface is vital for reconstituting tissue architecture, directing the deposition of other ECM components and regulating cellular processes.

Neovascularisation of granulation tissue during the resolution of tissue injury is also dependent on the presence of FN. The exposure of connective tissue ECM proteins and changes in presentation and conformations of proteins in the vascular basement membrane during wounding, acts as an important cue to regulate angiogenesis [224, 305]. The expression of FN, especially the EIIIA+FN and EIIIB+FN isoforms are highly up-regulated and deposited around neovessels and capillary sprouts within the highly vascularized granulation tissue [159, 304, 306]. The assembly of FN into a fibrillar network has been shown to be required for HUVEC proliferation and migration in three-dimensional culture conditions promoting endothelial cell survival [72, 178, 307].

Differentiation of fibroblasts into proto-myofibroblasts and then into α-smooth muscle actin (SMA) positive myofibroblasts is dependent on TGF-β, the expression and assembly of

EIIIA+FN and high extracellular mechanical tension of the granulation tissue [308-311]. High mechanical tension is generated from the tractional forces of migrating cells over the collagen- and FN-rich granulation tissue and the induction of cell contraction as the matrix becomes remodeled [299, 308]. The polymerization of FN within collagen biogels by FN-null myofibroblasts have also demonstrated that FN enhances the tensile strength and toughness of these biogels, impacting the mechanical properties of the ECM [312]. Myofibroblasts detect the mechanical properties of the ECM via specialized actin associated fibronexus adhesion complexes, which transmit intracellular actin-generated contractile forces whilst also sampling extracellular tension [308, 309, 313, 314]. EIIIA-FN null mice were shown to have impaired abnormal skin wound healing responses with reduced cell compaction and edematous-like areas within the granulation tissue and delayed re-epitheliazation confirming the importance of this isoform during late wound healing processes [270].

The revelations of FN matrix assembly *in vivo* during development and wound healing provide vital information about the functions and mechanisms involved in regulating FN fibrillogenesis and matrix assembly in physiological environments. These studies also highlight the diverse effects of *in vivo* FN assemblies on cell behavior.

5.2.1. FN in Aberrant Wound Responses

In chronic non-healing wounds, FN expression persists in the dermis around the ulcer bed [272, 275] suggesting excess and uncontrolled FN expression may be detrimental for wound healing. Similarly, in pathological skin conditions such as contact dermatitis, FN was abundant in the dermis early after hapten induction, and when applied to the ear, was up-regulated in the connective tissue [173, 315]. The continual presence of high levels of FN may enhance cell migration and infiltration into inflamed tissue, promote cell adhesion and synthesis and deposition of increased ECM proteins leading to a delayed resolution of the allergy.

FN can also be degraded by multiple proteases including plasmin and MMP-9 [172, 173]. *In vitro* FN fragments can inhibit FN matrix assembly by competing for FN assembly sites [148], which could act as a feedback system to regulate the level of FN assembled on the cell surface. The increased levels of proteases such as neutrophil elastase in chronic wound exudate, which degrade FN [316-320] contribute to this condition. Aged clots in chronic wounds contain highly cross-linked fibrin, which are stripped of other functional proteins by the highly proteolytic environment illustrating the importance of FN in regulating wound healing processes [321]. FN fragments or peptides derived from FN have also been shown to possess biologically active cryptic properties distinct from the full-length protein, and these fragments have been termed "matricryptins" [172, 322]. FN fragments have been shown to stimulate apoptosis [323, 324]. Adhesion of fibroblasts to FN fragments possessing the RGD central sequence but not full-length native FN have been shown to induce collagenase (MMP-1) and stromelysin (MMP-3) expression [325].

Other ECM components also possess matricryptic activities that can influence FN matrix assembly. Tenascin-C (TN-C) expression is highly up-regulated and deposited during wound healing and tissue remodeling with pro-longed expression in chronic non-healing wound conditions [272, 275]. Specific domains of TN-C have been shown to inhibit FN matrix assembly [326, 327]. The HepII domain of vitronectin has been shown to interact with $\alpha v\beta 3$ and $\alpha v\beta 5$ integrins, which prevents actin microfilament reorganization and loss of FN matrix assembly sites [328].

Loss of collagen type VI impairs complex FN matrix assembly, with FN fibrils oriented parallel to the long axis of the cell [329]. The degradation of ECM proteins with matricryptic activities including FN itself may act as a temporal regulatory control mechanism that functions as an indicator of the phase and progress of the wounding and remodelling process. Excess FN deposition and ECM degradation may therefore be contributing factors to conditions such as chronic non-healing wounds.

However, the matricryptic activity of specific FN sequences may be exploited for therapeutic purposes: the PHSRN synergy sequence from FNIII9 has been shown to promote keratinocyte and fibroblast migration and invasion into sea urchin embryonic matrices (which possess a natural basement membrane surrounding the matrix) and an acetylated and amidated PHSRN derivative promoted wound healing in obese, diabetic mice [330]. This sequence has also been shown to up-regulate cell-surface α5β1 expression on microvascular endothelial cells to promote cell invasion into matrigel [331].

5.3. Fibronectin in Fibrotic Conditions

Excessive deposition of connective tissue is a hallmark of fibrosis resulting in the impairment of organ structure or function. This has been compared to chronic inflammatory tissue non-resolving repair processes as the fibrotic tissue deposited has a similar structure and composition to granulation tissue [332, 333]. FN plays an important role in the development of fibrotic disease even though the predominant ECM protein within fibrotic lesions is collagen [75]. FN deposition is also observed and precedes the collagen deposition [334-341].

In the fibrosis of the kidney glomeruli and interstitial tissues, there is increased expression of FN, EIIIA+FN, EIIIB+FN and oncofetal (IIICS+) FN isoforms deposited in areas of fibrosis [3, 335-337]. Again, EIIIA+FN regulates the differentiation of fibroblasts into activated myofibroblasts which are key mediators of the chronic fibrotic response exhibiting enhanced synthesis of ECM proteins and profibrotic growth factors and cytokines [311].

EIIIA+FN assembly is also required for latent TGF-β1 activation inducing fibroblast responses to TGF-β1 and promoting the expression of profibrotic genes [171, 334, 342]. Highlighting its involvement in fibrosis, EIIIA null mice were shown to be protected from bleomycin induced pulmonary fibrosis [334] and ovalbumin induced airway fibrosis and hyper-responsiveness [343]. Fibroblasts isolated from EIIIA null mice treated with ovalbumin showed reduced cell proliferation, migration, myofibroblast differentiation (α-SMA expression), collagen deposition and TGF- β1 expression confirming the importance of EIIIA in enhancing these cellular responses [343].

FN has also been shown to play an essential role in intima-media thickening *in vivo* in mouse models of atherosclerosis [28]. Intimal, medial and adventitial thickening, collagen deposition, cell proliferation and inflammatory cell infiltration were significantly reduced after the inhibition of FN matrix assembly *in vivo* [28]. These *in vivo* studies confirm the importance of the requirement of tight regulation of FN expression and assembly to prevent aberrant cellular responses.

5.4. FN in Tumorigenesis and Metastasis

FN has been shown to be highly expressed and deposited in the stroma of many malignant tumors including squamous cell carcinomas [180, 344-346], and colorectal cancer [347]. In malignant melanomas, FN was shown to assemble and organize into tubular meshwork structures within the dermis that ensheathed the melanoma cells suggesting melanoma invasion may occur via assembly of these channel-like structures [348]. In myofibroblastic lung tumors, a linear and retircular fibrillar network was assembled and accumulated around myofibroblasts and in areas of inflammation [349].

Both alternatively spliced isoforms of FN have been described: EIIIB+FN and EIIIA+FN has been reported to be strongly expressed in tumor stroma [347, 350]. In breast carcinomas, total FN and EIIIA+FN expression were strongly increased and detectable between and around tumor cells and diffuse within the tumor stroma [242, 351, 352]. Increased FN levels have been shown to stimulate proliferation of quiescent mammary epithelial cells in differentiated acini cultured within matrigel, increasing acinar cell number with a subsequnt loss of the lumenal structure [186]. Malignant breast carcinoma cells were also shown to strongly up-regulate endogenous FN and EIIIA+FN expression as well as cell surface $\alpha5\beta1$ levels upon culture in 3D laminin-rich ECM, which allow self-assembly of cells into acinar structures under 3D culture [242]. The FN-rich stromal ECM itself has been shown to be essential for driving the tumorigenic process; the 3D matrix deposited by harvested stromal fibroblasts was sufficient to induce the differentiation of normal fibroblasts [180].

Tumorigenesis is also associated with an increasing stiffening of ECM due to enhanced levels of FN and ECM deposition and cross-linking, higher interstitial fluid pressure and compressive stresses from the tumour mass [185, 353-356]. Stiffer environments enhance cell adhesion, contractility and proliferation and also drives differentiation of quiescent cells into malignant tumour phenotypes: hepatocellular carcinomas showed increased proliferation on substrates with increasing stiffness [357]; mammary epithelial transformation was driven by ECM cross-linking with subsequent tissue stiffening [355]; glioma cells cultured within soft matrices mimicking normal brain tissue were rounded and non-migratory, whereas culture within increasingly rigid ECMs induced the tumour cells to spread, form prominent stress fibres, focal adhesions and induced migration [211]; tumour cell secretion of TGF-β could stimulate adipose progenitor cells to express and assemble FN, increasing matrix stiffness and enhancing FN deposition within mouse mammary tumours *in vivo* [185]. Fibroblasts cultured on stiff cross-linked FN gels were shown to up-regulate FAK phosphorylation, increase actin stress fibre formation, Rho dependent cytoskeletal contractility, cell spreading and focal adhesion formation [184].

Furthermore, enhanced substrate rigidity also stimulates FN deposition and assembly. FN expression has been shown to be up-regulated by mammary epithelial cells cultured on increasingly rigid FN substrates compared to softer matrigel and normal mammary tissue environments [186]. Increased matrix rigidity stimulates similar proliferation and loss of luminal structure in cultured mammary acini similar to that observed when cells were cultured in the presence of high concentrations of FN [186]. This suggests the rigidity of the substrate may also affect FN expression, which may be then a further contributory factor in altered cellular behavior on more rigid substrates such as in the tumor microenvironment.

Angiogenesis is key to the survival of large tumours and essential for tumour growth; beyond a certain critical cell mass, nutrients and oxygen cannot be supplied by simple passive diffusion [358, 359]. The promotion of tumorigenic angiogenesis also facilitates cancer cell metastasis from the primary tumor site [360]. However, cell migration, neovascularization, and self-assembly of endothelial cells into cord like structures on ECM substrates *in vitro* have been shown to be enhanced in softer matrix environments as discussed previously [181, 218, 219].

Tumor derived capillary endothelial cells showed abnormal responses to FN coated polyacrylamide matrices: they re-oriented their stress fibres perpendicular rather than parallel to the direction of the applied stress; cell spreading and shape was enhanced and stabilized on higher ECM stiffness (compared to normal cells) and more readily formed capillaries at lower cell densities [353]. These tumour endothelial cells were shown to have higher contractile phenotypes mediated by elevated Rho and ROCK baseline levels and suggest that tumor endothelial cells may have misregulated mechanosensing mechanisms that promote neovascularization even in stiffer tumor environments [353].

FN, in particular the EIIIB+ isoform is shown to be highly expressed and deposited around newly synthesized tumor blood vessels and capillary sprouts during pathological angiogenesis [269, 304, 306, 307, 346, 361-365]. For example EIIIB+FN has been demonstrated to be up-regulated in the neovasculature of invasive breast carcinomas [307], malignant human head and neck squamous cell carcinomas [346] and pancreatic tumours [269].

Both tumor and endothelial cell lines secrete and assemble EIIIB+FN and this isoform is a strong marker of rapid growing solid tumours [344, 366, 367]. EIIIB-null fibroblasts were shown to grow slower and deposited less pericellular ECM *in vitro* [79]. HUVEC cultures grown in high glucose and fibrogenic factors TGF-β1 and endothelin-1 (ET-1) showed increased EIIIB+FN expression and assembly, which induced cell proliferation and VEGF expression [72]. Furthermore, HUVECs cultured in matrigel underwent tubulogenesis in the presence of an EIIIB containing peptide [72].

Despite these studies, EIIIB null mice were not protected from tumorigenesis: tumour angiogenesis and growth were not significantly affected in the absence of EIIIB+FN [269]. In such cases, EIIIA+FN may compensate for that function in its absence. EIIIB+FN is however a primary marker for tumors due to its prominence within solid tumors and has been chosen as targets for many anti-cancer therapies including the engineering of antibodies as specific carriers for anti-angiogenic drugs [366, 368].

Matricryptic FN fragments also possess activities that may be useful therapeutically. Soluble fragments of FN that possess the heparin binding sequence ($FNI_{10\text{-}12}$) and the IIICS domain (V+ domain) were shown to enhance the invasion and spreading of oral squamous cell carcinomas into collagen-matrigel 3D matrices [369] and mutation of the heparin binding domain in these soluble FN fragments has been shown to be pro-apoptotic [370]. Short FN peptides possessing the $FNIII_{10}$ RGD sequence, FN, the CS-1 region from the IIICS alternatively spliced segment and FN-C/H-V peptide from the $FNI_{10\text{-}12}$ heparin binding domain have also been shown to promote cellular apoptosis [324]. Whether such peptides are useful as *in vivo* targets was not determined.

However, anastellin, a 76 amino acid peptide derived from the C-terminus of $FNIII_1$ have been shown to possess anti-angiogenic and anti-tumor properties leading to reduced tumor growth when subcutaneously injected into mice [371]. This appeared to only be effective at treating smaller tumors, but may be promising in targeting tumor associated angiogenesis as a useful anti-tumor therapy.

Conclusion

FN and its assembly into a 3D matrix is an essential part of the ECM, mediating and regulating essential cellular activities including adhesion, spreading, proliferation, migration and apoptosis during tissue remodeling events. The importance and role of FN modifying the composition of the extracellular environment and its impact on the tensional homeostasis existing between cells and their external environment are still being understood. Many pathological conditions such as fibrosis and tumorigenesis are considered a misregulated re-activation of developmental and wound healing processes, with re-expression of oncofetal proteins such as the EIIIA+ and EIIIB+ isoforms of FN.

So it is vital to understand how FN is assembled and regulated and the effect of these alternatively spliced FN isoforms in order to develop tools to help target and control this process and impact on disease progression.

Many of the mechanisms investigating cell-matrix interactions and fibrillar matrix assembly are elucidated from *in vitro* cell cultures on immobilized ECM substrates. However, these artificial culture environments do not faithfully recapitulate those observed *in vivo*. For example, it is particularly clear that the differences observed from culturing endothelial cells in 3D gels stimulates tubule formation contrasting with endothelial cell culture on ECM substrates that result in monolayer confluence. This tubule organization is also only regulated by FN matrix assembly when cells are cultured within 3D gels; inhibitors of FN matrix assembly had no effect on endothelial cell proliferation, morphology or migration cultured on ECM substrates but were effective at inhibiting 3D tubulogenesis [178]. Furthermore, when stromal fibroblasts are isolated and cultured on plates or on ECM substrates, in the absence of their complex stromal ECM lose their stromal phenotype [180, 239]. The importance of the ECM on cell phenotype is also highlighted by the maintenance of myofibroblasts, which rapidly lose their α-SMA stress fibres and fibronexus complexes upon culture in soft, compliant matrices [308].

Studies of FN matrix assembly by cells in planar, monolayer cultures have revealed many important mechanisms into the assembly process and the role of this FN matrix on cellular behavior. However, whether they accurately depict events *in vivo* is still unclear. Hence, many novel methods of 3D culture have been developed to try and recapitulate the behavior of cells *in vivo* [372-374].

Furthermore, the ability to examine and modify cell-FN matrix interactions and assembly during embryogenesis have been revelatory in clarifying the mechanisms involved in 3D environments [5-8, 155].

By understanding FN assembly *in vivo* and in 3D cultures will also be important towards the development of ECMs that can mimic the *in vivo* environment. This is vital in order to understand ECM dependent biological processes such as stem cell development. Also, the presence of FN linked to biodegradable 3D synthetic scaffolds has been shown to enhance cell adhesion, invasion, migration and organization [267, 375]. There is much potential for these bio-engineered materials in development of matrices that mimic *in vivo* environments allowing for better *in vitro* tools to test and develop therapeutical medical applications including improving the predictive accuracy of the drug development process as well as designing biodegradable scaffolds for use in wound healing and regenerative medical practice [376].

Acknowledgments

Our research was funded by Arthritis Research UK (ARC).

References

[1] Abrahamson DR, Leardkamolkarn V: Development of kidney tubular basement membranes. *Kidney Int.* 1991, 39:382-393.

[2] Farquhar MG: The glomerular basement membrane: not gone, just forgotten. *J. Clin. Invest.* 2006, 116:2090-2093.

[3] Van Vliet A, Baelde HJ, Vleming LJ, de Heer E, Bruijn JA: Distribution of fibronectin isoforms in human renal disease. *J. Pathol.* 2001, 193:256-262.

[4] Weiss P: Cellular Dynamics. *Reviews of Modern Physics* 1959,31:11-20.

[5] Rozario T, Dzamba B, Weber GF, Davidson LA, DeSimone DW: The physical state of fibronectin matrix differentially regulates morphogenetic movements in vivo. *Dev. Biol.* 2009, 327:386-398.

[6] Takahashi S, Leiss M, Moser M, Ohashi T, Kitao T, Heckmann D, Pfeifer A, Kessler H, Takagi J, Erickson HP, Fassler R: The RGD motif in fibronectin is essential for development but dispensable for fibril assembly. *J. Cell Biol.* 2007, 178:167-178.

[7] Martins GG, Rifes P, Amandio R, Rodrigues G, Palmeirim I, Thorsteinsdottir S: Dynamic 3D cell rearrangements guided by a fibronectin matrix underlie somitogenesis. *PLoS One* 2009, 4:e7429.

[8] Latimer A, Jessen JR: Extracellular matrix assembly and organization during zebrafish gastrulation. *Matrix Biol.* 2009, 29:89-96.

[9] Ingber DE: Mechanical control of tissue morphogenesis during embryological development. *Int. J. Dev. Biol.* 2006, 50:255-266.

[10] Schwartz MA: Integrins and extracellular matrix in mechanotransduction. *Cold,* 2:a005066. Epub 002010 Nov 005017.

[11] Mao Y, Schwarzbauer JE: Fibronectin fibrillogenesis, a cell-mediated matrix assembly process. *Matrix Biol.* 2005, 24:389-399.

[12] Pankov R, Yamada KM: Fibronectin at a glance. *J. Cell Sci.* 2002, 115:3861-3863.

[13] Wierzbicka-Patynowski I, Schwarzbauer JE: The ins and outs of fibronectin matrix assembly. *J. Cell Sci.* 2003, 116:3269-3276.

[14] Zardi L, Cecconi C, Barbieri O, Carnemolla B, Picca M, Santi L: Concentration of fibronectin in plasma of tumor-bearing mice and synthesis by Ehrlich ascites tumor cells. *Cancer Res.* 1979, 39:3774-3779.

[15] Manabe R, Ohe N, Maeda T, Fukuda T, Sekiguchi K: Modulation of cell-adhesive activity of fibronectin by the alternatively spliced EDA segment. *J. Cell Biol.* 1997, 139:295-307.

[16] Sottile J, Mosher DF: N-terminal type I modules required for fibronectin binding to fibroblasts and to fibronectin's III1 module. *Biochem. J.* 1997, 323 (Pt 1):51-60.

[17] Wang R, Clark RA, Mosher DF, Ren XD: Fibronectin's central cell-binding domain supports focal adhesion formation and Rho signal transduction. *J. Biol. Chem.* 2005, 280:28803-28810.

[18] Friedland JC, Lee MH, Boettiger D: Mechanically activated integrin switch controls alpha5beta1 function. *Science* 2009, 323:642-644.

[19] Gui L, Wojciechowski K, Gildner CD, Nedelkovska H, Hocking DC: Identification of the heparin-binding determinants within fibronectin repeat III1: role in cell spreading and growth. *J. Biol. Chem.* 2006, 281: 34816-34825.

[20] Jiang ST, Chiang HC, Cheng MH, Yang TP, Chuang WJ, Tang MJ: Role of fibronectin deposition in cystogenesis of Madin-Darby canine kidney cells. *Kidney Int.* 1999, 56:92-103.

[21] Sechler JL, Schwarzbauer JE: Control of cell cycle progression by fibronectin matrix architecture. *J. Biol. Chem.* 1998, 273:25533-25536.

[22] Clark RA, An JQ, Greiling D, Khan A, Schwarzbauer JE: Fibroblast migration on fibronectin requires three distinct functional domains. *J. Invest. Dermatol.* 2003, 121:695-705.

[23] Couchman JR, Rees DA, Green MR, Smith CG: Fibronectin has a dual role in locomotion and anchorage of primary chick fibroblasts and can promote entry into the division cycle. *J. Cell Biol.* 1982, 93:402-410.

[24] Hocking DC, Chang CH: Fibronectin matrix polymerization regulates small airway epithelial cell migration. *Am. J. Physiol. Lung Cell Mol. Physiol.* 2003, 285:L169-179.

[25] Natal C, Oses-Prieto JA, Pelacho B, Iraburu MJ, Lopez-Zabalza MJ: Regulation of apoptosis by peptides of fibronectin in human monocytes. *Apoptosis* 2006, 11: 209-219.

[26] Tafolla E, Wang S, Wong B, Leong J, Kapila YL: JNK1 and JNK2 oppositely regulate p53 in signaling linked to apoptosis triggered by an altered fibronectin matrix: JNK links FAK and p53. *J. Biol. Chem.* 2005, 280:19992-19999.

[27] Yang JT, Bader BL, Kreidberg JA, Ullman-Cullere M, Trevithick JE, Hynes RO: Overlapping and independent functions of fibronectin receptor integrins in early mesodermal development. *Dev. Biol.* 1999, 215:264-277.

[28] Chiang HY, Korshunov VA, Serour A, Shi F, Sottile J: Fibronectin is an important regulator of flow-induced vascular remodeling. *Arterioscler. Thromb. Vasc. Biol.* 2009, 29:1074-1079. Epub 2009 Apr 1030.

[29] Dzamba BJ, Peters DM: Arrangement of cellular fibronectin in noncollagenous fibrils in human fibroblast cultures. *J. Cell Sci.* 1991, 100 (Pt 3):605-612.

[30] Dzamba BJ, Wu H, Jaenisch R, Peters DM: Fibronectin binding site in type I collagen regulates fibronectin fibril formation. *J. Cell Biol.* 1993, 121: 1165-1172.

[31] Li S, Van Den Diepstraten C, D'Souza SJ, Chan BM, Pickering JG: Vascular smooth muscle cells orchestrate the assembly of type I collagen via alpha2beta1 integrin, RhoA, and fibronectin polymerization. *Am. J. Pathol.* 2003, 163:1045-1056.

[32] McDonald JA, Kelley DG, Broekelmann TJ: Role of fibronectin in collagen deposition: Fab' to the gelatin-binding domain of fibronectin inhibits both fibronectin and collagen organization in fibroblast extracellular matrix. *J. Cell Biol.* 1982, 92:485-492.

[33] Sottile J, Hocking DC: Fibronectin polymerization regulates the composition and stability of extracellular matrix fibrils and cell-matrix adhesions. *Mol. Biol. Cell* 2002, 13:3546-3559.

[34] Sottile J, Shi F, Rublyevska I, Chiang HY, Lust J, Chandler J: Fibronectin-dependent collagen I deposition modulates the cell response to fibronectin. *Am. J. Physiol. Cell Physiol.* 2007, 293:C1934-1946.

[35] Velling T, Risteli J, Wennerberg K, Mosher DF, Johansson S: Polymerization of type I and III collagens is dependent on fibronectin and enhanced by integrins alpha 11beta 1 and alpha 2beta 1. *J. Biol. Chem.* 2002, 277:37377-37381.
[36] Pereira M, Rybarczyk BJ, Odrljin TM, Hocking DC, Sottile J, Simpson-Haidaris PJ: The incorporation of fibrinogen into extracellular matrix is dependent on active assembly of a fibronectin matrix. *J. Cell Sci.* 2002, 115:609-617.
[37] Czirok A, Zamir EA, Filla MB, Little CD, Rongish BJ: Extracellular matrix macroassembly dynamics in early vertebrate embryos. *Curr. Top. Dev. Biol.* 2006, 73:237-258.
[38] Kinsey R, Williamson MR, Chaudhry S, Mellody KT, McGovern A, Takahashi S, Shuttleworth CA, Kielty CM: Fibrillin-1 microfibril deposition is dependent on fibronectin assembly. *J. Cell Sci.* 2008, 121: 2696-2704.
[39] Sabatier L, Chen D, Fagotto-Kaufmann C, Hubmacher D, McKee MD, Annis DS, Mosher DF, Reinhardt DP: Fibrillin assembly requires fibronectin. *Mol. Biol. Cell* 2009, 20:846-858.
[40] Godyna S, Mann DM, Argraves WS: A quantitative analysis of the incorporation of fibulin-1 into extracellular matrix indicates that fibronectin assembly is required. *Matrix Biol.* 1995, 14:467-477.
[41] Bazigou E, Xie S, Chen C, Weston A, Miura N, Sorokin L, Adams R, Muro AF, Sheppard D, Makinen T: Integrin-alpha9 is required for fibronectin matrix assembly during lymphatic valve morphogenesis. *Dev. Cell* 2009, 17:175-186.
[42] Chung CY, Zardi L, Erickson HP: Binding of tenascin-C to soluble fibronectin and matrix fibrils. *J. Biol. Chem.* 1995, 270:29012-29017.
[43] Ramos DM, Chen B, Regezi J, Zardi L, Pytela R: Tenascin-C matrix assembly in oral squamous cell carcinoma. *Int. J. Cancer* 1998, 75:680-687.
[44] Repesh LA, Fitzgerald TJ, Furcht LT: Fibronectin involvement in granulation tissue and wound healing in rabbits. *J. Histochem. Cytochem.* 1982, 30:351-358.
[45] Huang G, Zhang Y, Kim B, Ge G, Annis DS, Mosher DF, Greenspan DS: Fibronectin binds and enhances the activity of bone morphogenetic protein 1. *J. Biol. Chem.* 2009, 284:25879-25888.
[46] Wijelath ES, Rahman S, Namekata M, Murray J, Nishimura T, Mostafavi-Pour Z, Patel Y, Suda Y, Humphries MJ, Sobel M: Heparin-II domain of fibronectin is a vascular endothelial growth factor-binding domain: enhancement of VEGF biological activity by a singular growth factor/matrix protein synergism. *Circ. Res.* 2006, 99:853-860.
[47] Dallas SL, Sivakumar P, Jones CJ, Chen Q, Peters DM, Mosher DF, Humphries MJ, Kielty CM: Fibronectin regulates latent transforming growth factor-beta (TGF beta) by controlling matrix assembly of latent TGF beta-binding protein-1. *J. Biol. Chem.* 2005, 280:18871-18880.
[48] Dallas SL, Chen Q, Sivakumar P: Dynamics of assembly and reorganization of extracellular matrix proteins. *Curr. Top. Dev. Biol.* 2006, 75:1-24.
[49] Koli K, Hyytiainen M, Ryynanen MJ, Keski-Oja J: Sequential deposition of latent TGF-beta binding proteins (LTBPs) during formation of the extracellular matrix in human lung fibroblasts. *Exp. Cell Res.* 2005, 310:370-382.
[50] Sivakumar P, Czirok A, Rongish BJ, Divakara VP, Wang YP, Dallas SL: New insights into extracellular matrix assembly and reorganization from dynamic imaging of extracellular matrix proteins in living osteoblasts. *J. Cell Sci.* 2006, 119:1350-1360.

[51] Pickford AR, Campbell ID: NMR studies of modular protein structures and their interactions. *Chem. Rev.* 2004, 104:3557-3566.

[52] Baneyx G, Baugh L, Vogel V: Coexisting conformations of fibronectin in cell culture imaged using fluorescence resonance energy transfer. *Proc. Natl. Acad. Sci. U S A* 2001, 98:14464-14468.

[53] Baneyx G, Baugh L, Vogel V: Fibronectin extension and unfolding within cell matrix fibrils controlled by cytoskeletal tension. *Proc. Natl. Acad. Sci. U S A* 2002, 99:5139-5143.

[54] Erickson HP, Carrell N, McDonagh J: Fibronectin molecule visualized in electron microscopy: a long, thin, flexible strand. *J. Cell Biol.* 1981, 91:673-678.

[55] Erickson HP, Carrell NA: Fibronectin in extended and compact conformations. Electron microscopy and sedimentation analysis. *J. Biol. Chem.* 1983, 258:14539-14544.

[56] Johnson KJ, Sage H, Briscoe G, Erickson HP: The compact conformation of fibronectin is determined by intramolecular ionic interactions. *J. Biol. Chem.* 1999, 274:15473-15479.

[57] Rocco M, Infusini E, Daga MG, Gogioso L, Cuniberti C: Models of fibronectin. *EMBO J.* 1987, 6:2343-2349.

[58] Karuri NW, Lin Z, Rye HS, Schwarzbauer JE: Probing the conformation of the fibronectin III1-2 domain by fluorescence resonance energy transfer. *J. Biol. Chem.* 2009, 284:3445-3452.

[59] Rocco M, Carson M, Hantgan R, McDonagh J, Hermans J: Dependence of the shape of the plasma fibronectin molecule on solvent composition. Ionic strength and glycerol content. *J. Biol. Chem.* 1983, 258:14545-14549.

[60] Sechler JL, Takada Y, Schwarzbauer JE: Altered rate of fibronectin matrix assembly by deletion of the first type III repeats. *J. Cell Biol.* 1996, 134:573-583.

[61] Craig D, Gao M, Schulten K, Vogel V: Tuning the mechanical stability of fibronectin type III modules through sequence variations. *Structure* 2004, 12:21-30.

[62] Chen Y, Wu Y, Cai J: Atomic force microscopic investigation on the potential early intermediate stages of fibrillogenesis of fibronectin within fibrils. *Biochem. Biophys. Res. Commun.* 2007, 361:391-397.

[63] Gao M, Craig D, Lequin O, Campbell ID, Vogel V, Schulten K: Structure and functional significance of mechanically unfolded fibronectin type III1 intermediates. *Proc. Natl. Acad. Sci. U S A* 2003, 100:14784-14789.

[64] Krammer A, Lu H, Isralewitz B, Schulten K, Vogel V: Forced unfolding of the fibronectin type III module reveals a tensile molecular recognition switch. *Proc. Natl. Acad. Sci. U S A* 1999, 96:1351-1356.

[65] Smith ML, Gourdon D, Little WC, Kubow KE, Eguiluz RA, Luna-Morris S, Vogel V: Force-induced unfolding of fibronectin in the extracellular matrix of living cells. *PLoS Biol.* 2007, 5:e268.

[66] Altroff H, Schlinkert R, van der Walle CF, Bernini A, Campbell ID, Werner JM, Mardon HJ: Interdomain tilt angle determines integrin-dependent function of the ninth and tenth FIII domains of human fibronectin. *J. Biol. Chem.* 2004, 279:55995-56003.

[67] Carr PA, Erickson HP, Palmer AG, 3rd: Backbone dynamics of homologous fibronectin type III cell adhesion domains from fibronectin and tenascin. *Structure* 1997, 5: 949-959.

[68] Leahy DJ, Aukhil I, Erickson HP: 2.0 A crystal structure of a four-domain segment of human fibronectin encompassing the RGD loop and synergy region. *Cell* 1996, 84: 155-164.

[69] Craig D, Krammer A, Schulten K, Vogel V: Comparison of the early stages of forced unfolding for fibronectin type III modules. *Proc. Natl. Acad. Sci. U S A* 2001, 98:5590-5595.

[70] Hocking DC, Smith RK, McKeown-Longo PJ: A novel role for the integrin-binding III-10 module in fibronectin matrix assembly. *J. Cell Biol.* 1996, 133:431-444.

[71] Hershberger RP, Culp LA: Cell-type-specific expression of alternatively spliced human fibronectin IIICS mRNAs. *Mol. Cell Biol.* 1990, 10:662-671.

[72] Khan ZA, Chan BM, Uniyal S, Barbin YP, Farhangkhoee H, Chen S, Chakrabarti S: EDB fibronectin and angiogenesis -- a novel mechanistic pathway. *Angiogenesis* 2005, 8:183-196.

[73] Rybak JN, Roesli C, Kaspar M, Villa A, Neri D: The extra-domain A of fibronectin is a vascular marker of solid tumors and metastases. *Cancer Res.* 2007, 67:10948-10957.

[74] Tressel T, McCarthy JB, Calaycay J, Lee TD, Legesse K, Shively JE, Pande H: Human plasma fibronectin. Demonstration of structural differences between the A- and B-chains in the III CS region. *Biochem. J.* 1991, 274 (Pt 3):731-738.

[75] ffrench-Constant C: Alternative splicing of fibronectin--many different proteins but few different functions. *Exp. Cell Res.* 1995, 221:261-271.

[76] Wilson CL, Schwarzbauer JE: The alternatively spliced V region contributes to the differential incorporation of plasma and cellular fibronectins into fibrin clots. *J. Cell Biol.* 1992, 119:923-933.

[77] Chen W, Culp LA: Adhesion mediated by fibronectin's alternatively spliced EDb (EIIIB) and its neighboring type III repeats. *Exp. Cell Res.* 1996, 223:9-19.

[78] Guan JL, Trevithick JE, Hynes RO: Retroviral expression of alternatively spliced forms of rat fibronectin. *J. Cell Biol.* 1990, 110: 833-847.

[79] Fukuda T, Yoshida N, Kataoka Y, Manabe R, Mizuno-Horikawa Y, Sato M, Kuriyama K, Yasui N, Sekiguchi K: Mice lacking the EDB segment of fibronectin develop normally but exhibit reduced cell growth and fibronectin matrix assembly in vitro. *Cancer Res.* 2002, 62:5603-5610.

[80] Bencharit S, Cui CB, Siddiqui A, Howard-Williams EL, Sondek J, Zuobi-Hasona K, Aukhil I: Structural insights into fibronectin type III domain-mediated signaling. *J. Mol. Biol.* 2007, 367:303-309.

[81] Pickford AR, Smith SP, Staunton D, Boyd J, Campbell ID: The hairpin structure of the (6)F1(1)F2(2)F2 fragment from human fibronectin enhances gelatin binding. *EMBO J.* 2001, 20:1519-1529.

[82] Spitzfaden C, Grant RP, Mardon HJ, Campbell ID: Module-module interactions in the cell binding region of fibronectin: stability, flexibility and specificity. *J. Mol. Biol.* 1997, 265:565-579.

[83] Ventura E, Sassi F, Parodi A, Balza E, Borsi L, Castellani P, Carnemolla B, Zardi L: Alternative splicing of the angiogenesis associated extra-domain B of fibronectin regulates the accessibility of the B-C loop of the type III repeat 8. *PLoS One*, 5:e9145.

[84] Magnusson MK, Mosher DF: Fibronectin: structure, assembly, and cardiovascular implications. *Arterioscler. Thromb. Vasc. Biol.* 1998, 18: 1363-1370.

[85] Chauhan AK, Kisucka J, Cozzi MR, Walsh MT, Moretti FA, Battiston M, Mazzucato M, De Marco L, Baralle FE, Wagner DD, Muro AF: Prothrombotic effects of fibronectin isoforms containing the EDA domain. *Arterioscler. Thromb. Vasc. Biol.* 2008, 28:296-301.

[86] Peters JH, Loredo GA, Chen G, Maunder R, Hahn TJ, Willits NH, Hynes RO: Plasma levels of fibronectin bearing the alternatively spliced EIIIB segment are increased after major trauma. *J. Lab. Clin. Med.* 2003, 141:401-410.

[87] Castellanos M, Leira R, Serena J, Blanco M, Pedraza S, Castillo J, Davalos A: Plasma cellular-fibronectin concentration predicts hemorrhagic transformation after thrombolytic therapy in acute ischemic stroke. *Stroke* 2004, 35:1671-1676.

[88] Song KS, Kim HK, Shim W, Jee SH: Plasma fibronectin levels in ischemic heart disease. *Atherosclerosis* 2001, 154:449-453.

[89] Claudepierre P, Allanore Y, Belec L, Larget-Piet B, Zardi L, Chevalier X: Increased Ed-B fibronectin plasma levels in spondyloarthropathies: comparison with rheumatoid arthritis patients and a healthy population. *Rheumatology* (Oxford) 1999, 38:1099-1103.

[90] Astrof S, Hynes RO: Fibronectins in vascular morphogenesis. *Angiogenesis* 2009, 12:165-175.

[91] Choi MG, Hynes RO: Biosynthesis and processing of fibronectin in NIL.8 hamster cells. *J. Biol. Chem.* 1979, 254:12050-12055.

[92] McKeown-Longo PJ, Mosher DF: Binding of plasma fibronectin to cell layers of human skin fibroblasts. *J. Cell Biol.* 1983, 97:466-472.

[93] Mosher DF, Johnson RB: Specificity of fibronectin--fibrin cross-linking. *Ann. N Y Acad. Sci.* 1983, 408:583-594.

[94] Schwarzbauer JE: Identification of the fibronectin sequences required for assembly of a fibrillar matrix. *J. Cell Biol.* 1991, 113:1463-1473.

[95] Mao Y, Schwarzbauer JE: Stimulatory effects of a three-dimensional microenvironment on cell-mediated fibronectin fibrillogenesis. *J. Cell Sci.* 2005, 118:4427-4436.

[96] Pankov R, Cukierman E, Katz BZ, Matsumoto K, Lin DC, Lin S, Hahn C, Yamada KM: Integrin dynamics and matrix assembly: tensin-dependent translocation of alpha(5)beta(1) integrins promotes early fibronectin fibrillogenesis. *J. Cell Biol.* 2000, 148:1075-1090.

[97] Zamir E, Katz BZ, Aota S, Yamada KM, Geiger B, Kam Z: Molecular diversity of cell-matrix adhesions. *J. Cell Sci.* 1999, 112 (Pt 11):1655-1669.

[98] Geiger B, Bershadsky A, Pankov R, Yamada KM: Transmembrane crosstalk between the extracellular matrix--cytoskeleton crosstalk. *Nat. Rev. Mol. Cell Biol.* 2001, 2: 793-805.

[99] McKeown-Longo PJ, Mosher DF: Interaction of the 70,000-mol-wt amino-terminal fragment of fibronectin with the matrix-assembly receptor of fibroblasts. *J. Cell Biol.* 1985, 100:364-374.

[100] Tomasini-Johansson BR, Annis DS, Mosher DF: The N-terminal 70-kDa fragment of fibronectin binds to cell surface fibronectin assembly sites in the absence of intact fibronectin. *Matrix Biol.* 2006, 25:282-293.

[101] Xu J, Maurer LM, Hoffmann BR, Annis DS, Mosher DF: iso-DGR sequences do not mediate binding of fibronectin N-terminal modules to adherent fibronectin-null fibroblasts. *J. Biol. Chem.*, 285:8563-8571.

[102] Allio AE, McKeown-Longo PJ: Extracellular matrix assembly of cell-derived and plasma-derived fibronectins by substrate-attached fibroblasts. *J. Cell Physiol.* 1988, 135:459-466.
[103] Bultmann H, Santas AJ, Peters DM: Fibronectin fibrillogenesis involves the heparin II binding domain of fibronectin. *J. Biol. Chem.* 1998, 273: 2601-2609.
[104] McDonald JA, Quade BJ, Broekelmann TJ, LaChance R, Forsman K, Hasegawa E, Akiyama S: Fibronectin's cell-adhesive domain and an amino-terminal matrix assembly domain participate in its assembly into fibroblast pericellular matrix. *J. Biol. Chem.* 1987, 262:2957-2967.
[105] Morla A, Ruoslahti E: A fibronectin self-assembly site involved in fibronectin matrix assembly: reconstruction in a synthetic peptide. *J. Cell Biol.* 1992, 118:421-429.
[106] Zhang Q, Mosher DF: Cross-linking of the NH2-terminal region of fibronectin to molecules of large apparent molecular mass. Characterization of fibronectin assembly sites induced by the treatment of fibroblasts with lysophosphatidic acid. *J. Biol. Chem.* 1996, 271: 33284-33292.
[107] Ugarova TP, Zamarron C, Veklich Y, Bowditch RD, Ginsberg MH, Weisel JW, Plow EF: Conformational transitions in the cell binding domain of fibronectin. *Biochemistry* 1995, 34:4457-4466.
[108] Limper AH, Quade BJ, LaChance RM, Birkenmeier TM, Rangwala TS, McDonald JA: Cell surface molecules that bind fibronectin's matrix assembly domain. *J. Biol. Chem.* 1991, 266:9697-9702.
[109] Fogerty FJ, Akiyama SK, Yamada KM, Mosher DF: Inhibition of binding of fibronectin to matrix assembly sites by anti-integrin (alpha 5 beta 1) antibodies. *J. Cell Biol.* 1990, 111:699-708.
[110] Wennerberg K, Lohikangas L, Gullberg D, Pfaff M, Johansson S, Fassler R: Beta 1 integrin-dependent and -independent polymerization of fibronectin. *J. Cell Biol.* 1996, 132:227-238.
[111] Curnis F, Longhi R, Crippa L, Cattaneo A, Dondossola E, Bachi A, Corti A: Spontaneous formation of L-isoaspartate and gain of function in fibronectin. *J. Biol. Chem.* 2006, 281:36466-36476.
[112] Schor SL, Ellis I, Banyard J, Schor AM: Motogenic activity of IGD-containing synthetic peptides. *J. Cell Sci.* 1999, 112:3879-3888.
[113] Huveneers S, Truong H, Fassler R, Sonnenberg A, Danen EH: Binding of soluble fibronectin to integrin alpha5 beta1 - link to focal adhesion redistribution and contractile shape. *J. Cell Sci.* 2008, 121:2452-2462.
[114] Wu C, Bauer JS, Juliano RL, McDonald JA: The alpha 5 beta 1 integrin fibronectin receptor, but not the alpha 5 cytoplasmic domain, functions in an early and essential step in fibronectin matrix assembly. *J. Biol. Chem.* 1993, 268:21883-21888.
[115] Wu C, Keivens VM, O'Toole TE, McDonald JA, Ginsberg MH: Integrin activation and cytoskeletal interaction are essential for the assembly of a fibronectin matrix. *Cell* 1995, 83:715-724.
[116] Darribere T, Guida K, Larjava H, Johnson KE, Yamada KM, Thiery JP, Boucaut JC: In vivo analyses of integrin beta 1 subunit function in fibronectin matrix assembly. *J. Cell Biol.* 1990, 110:1813-1823.
[117] Winklbauer R, Stoltz C: Fibronectin fibril growth in the extracellular matrix of the Xenopus embryo. *J. Cell Sci.* 1995, 108 (Pt 4):1575-1586.

[118] Disatnik MH, Rando TA: Integrin-mediated muscle cell spreading. The role of protein kinase c in outside-in and inside-out signaling and evidence of integrin cross-talk. *J. Biol. Chem.* 1999, 274:32486-32492.

[119] Green JA, Berrier AL, Pankov R, Yamada KM: beta1 integrin cytoplasmic domain residues selectively modulate fibronectin matrix assembly and cell spreading through talin and Akt-1. *J. Biol. Chem.* 2009, 284:8148-8159.

[120] Somanath PR, Kandel ES, Hay N, Byzova TV: Akt1 signaling regulates integrin activation, matrix recognition, and fibronectin assembly. *J. Biol. Chem.* 2007, 282:22964-22976.

[121] Lee SE, Kamm RD, Mofrad MR: Force-induced activation of talin and its possible role in focal adhesion mechanotransduction. *J. Biomech.* 2007, 40:2096-2106.

[122] Dumbauld DW, Shin H, Gallant ND, Michael KE, Radhakrishna H, Garcia AJ: Contractility modulates cell adhesion strengthening through focal adhesion kinase and assembly of vinculin-containing focal adhesions. *J.*, 223:746-756.

[123] Chung CY, Erickson HP: Glycosaminoglycans modulate fibronectin matrix assembly and are essential for matrix incorporation of tenascin-C. *J. Cell Sci.* 1997, 110 (Pt 12):1413-1419.

[124] Sottile J, Hocking DC, Langenbach KJ: Fibronectin polymerization stimulates cell growth by RGD-dependent and -independent mechanisms. *J. Cell Sci.* 2000, 113: 4287-4299.

[125] Xian X, Gopal S, Couchman JR: Syndecans as receptors and organizers of the extracellular matrix. *Cell Tissue Res.* 2009, 339:31-46.

[126] Ingham KC, Brew SA, Atha DH: Interaction of heparin with fibronectin and isolated fibronectin domains. *Biochem. J.* 1990, 272:605-611.

[127] Bloom L, Ingham KC, Hynes RO: Fibronectin regulates assembly of actin filaments and focal contacts in cultured cells via the heparin-binding site in repeat III13. *Mol. Biol. Cell* 1999, 10:1521-1536.

[128] Ingham KC, Brew SA, Migliorini MM, Busby TF: Binding of heparin by type III domains and peptides from the carboxy terminal hep-2 region of fibronectin. *Biochemistry* 1993, 32:12548-12553.

[129] Busby TF, Argraves WS, Brew SA, Pechik I, Gilliland GL, Ingham KC: Heparin binding by fibronectin module III-13 involves six discontinuous basic residues brought together to form a cationic cradle. *J. Biol. Chem.* 1995, 270:18558-18562.

[130] Wenk MB, Midwood KS, Schwarzbauer JE: Tenascin-C suppresses Rho activation. *J. Cell Biol.* 2000, 150:913-920.

[131] Midwood KS, Valenick LV, Hsia HC, Schwarzbauer JE: Coregulation of fibronectin signaling and matrix contraction by tenascin-C and syndecan-4. *Mol. Biol. Cell* 2004, 15:5670-5677.

[132] Huang W, Chiquet-Ehrismann R, Moyano JV, Garcia-Pardo A, Orend G: Interference of tenascin-C with syndecan-4 binding to fibronectin blocks cell adhesion and stimulates tumor cell proliferation. *Cancer Res.* 2001, 61:8586-8594.

[133] Woods A, Longley RL, Tumova S, Couchman JR: Syndecan-4 binding to the high affinity heparin-binding domain of fibronectin drives focal adhesion formation in fibroblasts. *Arch. Biochem. Biophys.* 2000, 374: 66-72.

[134] Saoncella S, Echtermeyer F, Denhez F, Nowlen JK, Mosher DF, Robinson SD, Hynes RO, Goetinck PF: Syndecan-4 signals cooperatively with integrins in a Rho-dependent

manner in the assembly of focal adhesions and actin stress fibers. *Proc. Natl. Acad. Sci. U S A* 1999, 96:2805-2810.

[135] Midwood KS, Schwarzbauer JE: Tenascin-C modulates matrix contraction via focal adhesion kinase- and Rho-mediated signaling pathways. *Mol. Biol. Cell* 2002, 13:3601-3613.

[136] Klass CM, Couchman JR, Woods A: Control of extracellular matrix assembly by syndecan-2 proteoglycan. *J. Cell Sci.* 2000, 113 (Pt 3): 493-506.

[137] Galante LL, Schwarzbauer JE: Requirements for sulfate transport and the diastrophic dysplasia sulfate transporter in fibronectin matrix assembly. *J. Cell Biol.* 2007, 179:999-1009.

[138] Kuo JC, Han X, Hsiao CT, Yates JR, 3rd, Waterman CM: Analysis of the myosin-II-responsive focal adhesion proteome reveals a role for beta-Pix in negative regulation of focal adhesion maturation. *Nat.,* 13: 383-393. Epub 2011 Mar 2020.

[139] Ohashi T, Kiehart DP, Erickson HP: Dual labeling of the fibronectin matrix and actin cytoskeleton with green fluorescent protein variants. *J. Cell Sci.* 2002, 115: 1221-1229.

[140] Aguirre KM, McCormick RJ, Schwarzbauer JE: Fibronectin self-association is mediated by complementary sites within the amino-terminal one-third of the molecule. *J. Biol. Chem.* 1994, 269: 27863-27868.

[141] Sechler JL, Rao H, Cumiskey AM, Vega-Colon I, Smith MS, Murata T, Schwarzbauer JE: A novel fibronectin binding site required for fibronectin fibril growth during matrix assembly. *J. Cell Biol.* 2001, 154: 1081-1088.

[142] Maqueda A, Moyano JV, Hernandez Del Cerro M, Peters DM, Garcia-Pardo A: The heparin III-binding domain of fibronectin (III4-5 repeats) binds to fibronectin and inhibits fibronectin matrix assembly. *Matrix Biol.* 2007, 26:642-651.

[143] Abe Y, Bui-Thanh NA, Ballantyne CM, Burns AR: Extra domain A and type III connecting segment of fibronectin in assembly and cleavage. *Biochem. Biophys. Res. Commun.* 2005, 338:1640-1647.

[144] Hocking DC, Sottile J, McKeown-Longo PJ: Fibronectin's III-1 module contains a conformation-dependent binding site for the amino-terminal region of fibronectin. *J. Biol. Chem.* 1994, 269:19183-19187.

[145] Ingham KC, Brew SA, Huff S, Litvinovich SV: Cryptic self-association sites in type III modules of fibronectin. *J. Biol. Chem.* 1997, 272: 1718-1724.

[146] Morla A, Zhang Z, Ruoslahti E: Superfibronectin is a functionally distinct form of fibronectin. *Nature* 1994, 367:193-196.

[147] Oberhauser AF, Badilla-Fernandez C, Carrion-Vazquez M, Fernandez JM: The mechanical hierarchies of fibronectin observed with single-molecule AFM. *J. Mol. Biol.* 2002, 319:433-447.

[148] Bourdoulous S, Orend G, MacKenna DA, Pasqualini R, Ruoslahti E: Fibronectin matrix regulates activation of RHO and CDC42 GTPases and cell cycle progression. *J. Cell Biol.* 1998, 143:267-276.

[149] Klein RM, Zheng M, Ambesi A, Van De Water L, McKeown-Longo PJ: Stimulation of extracellular matrix remodeling by the first type III repeat in fibronectin. *J. Cell Sci.* 2003, 116:4663-4674.

[150] Neskey DM, Ambesi A, Pumiglia KM, McKeown-Longo PJ: Endostatin and anastellin inhibit distinct aspects of the angiogenic process. *J. Exp. Clin. Cancer Res.* 2008, 27:61.

[151] Krammer A, Craig D, Thomas WE, Schulten K, Vogel V: A structural model for force regulated integrin binding to fibronectin's RGD-synergy site. *Matrix Biol.* 2002, 21:139-147.

[152] Grant RP, Spitzfaden C, Altroff H, Campbell ID, Mardon HJ: Structural requirements for biological activity of the ninth and tenth FIII domains of human fibronectin. *J. Biol. Chem.* 1997, 272:6159-6166.

[153] Gee EP, Ingber DE, Stultz CM: Fibronectin unfolding revisited: modeling cell traction-mediated unfolding of the tenth type-III repeat. *PLoS One* 2008, 3:e2373. Epub 2008 Jun 2311.

[154] Zhong C, Chrzanowska-Wodnicka M, Brown J, Shaub A, Belkin AM, Burridge K: Rho-mediated contractility exposes a cryptic site in fibronectin and induces fibronectin matrix assembly. *J. Cell Biol.* 1998, 141: 539-551.

[155] Davidson LA, Dzamba BD, Keller R, Desimone DW: Live imaging of cell protrusive activity, and extracellular matrix assembly and remodeling during morphogenesis in the frog, Xenopus laevis. *Dev. Dyn.* 2008, 237:2684-2692.

[156] Yoneda A, Ushakov D, Multhaupt HA, Couchman JR: Fibronectin matrix assembly requires distinct contributions from Rho kinases I and -II. *Mol. Biol. Cell* 2007, 18: 66-75.

[157] Zhang Q, Magnusson MK, Mosher DF: Lysophosphatidic acid and microtubule-destabilizing agents stimulate fibronectin matrix assembly through Rho-dependent actin stress fiber formation and cell contraction. *Mol. Biol. Cell* 1997, 8:1415-1425.

[158] Brenner KA, Corbett SA, Schwarzbauer JE: Regulation of fibronectin matrix assembly by activated Ras in transformed cells. *Oncogene* 2000, 19:3156-3163.

[159] Fernandez-Sauze S, Grall D, Cseh B, Van Obberghen-Schilling E: Regulation of fibronectin matrix assembly and capillary morphogenesis in endothelial cells by Rho family GTPases. *Exp. Cell Res.* 2009, 315: 2092-2104.

[160] Colombelli J, Besser A, Kress H, Reynaud EG, Girard P, Caussinus E, Haselmann U, Small JV, Schwarz US, Stelzer EH: Mechanosensing in actin stress fibers revealed by a close correlation between force and protein localization. *J. Cell Sci.* 2009, 122:1665-1679. Epub 2009 Apr 1628.

[161] Langenbach KJ, Sottile J: Identification of protein-disulfide isomerase activity in fibronectin. *J. Biol. Chem.* 1999, 274:7032-7038.

[162] Chen H, Mosher DF: Formation of sodium dodecyl sulfate-stable fibronectin multimers. Failure to detect products of thiol-disulfide exchange in cyanogen bromide or limited acid digests of stabilized matrix fibronectin. *J. Biol. Chem.* 1996, 271:9084-9089.

[163] Ohashi T, Erickson HP: Revisiting the mystery of fibronectin multimers: the fibronectin matrix is composed of fibronectin dimers cross-linked by non-covalent bonds. *Matrix Biol.* 2009, 28:170-175. Epub 2009 Mar 2012.

[164] Shi F, Sottile J: Caveolin-1-dependent beta1 integrin endocytosis is a critical regulator of fibronectin turnover. *J. Cell Sci.* 2008, 121:2360-2371.

[165] Sottile J, Chandler J: Fibronectin matrix turnover occurs through a caveolin-1-dependent process. *Mol. Biol. Cell* 2005, 16:757-768.

[166] Harris AK, Stopak D, Wild P: Fibroblast traction as a mechanism for collagen morphogenesis. *Nature* 1981, 290:249-251.

[167] Kubow KE, Klotzsch E, Smith ML, Gourdon D, Little WC, Vogel V: Crosslinking of cell-derived 3D scaffolds up-regulates the stretching and unfolding of new extracellular matrix assembled by reseeded cells. *Integr. Biol.* (Camb) 2009, 1:635-648.

[168] Parker KK, Tan J, Chen CS, Tung L: Myofibrillar architecture in engineered cardiac myocytes. *Circ. Res.* 2008, 103:340-342. Epub 2008 Jul 2017.

[169] Gabrielli A, Avvedimento EV, Krieg T: Scleroderma. *N. Engl. J. Med.* 2009, 360:1989-2003.

[170] Ihn H: Scleroderma, fibroblasts, signaling, and excessive extracellular matrix. *Curr. Rheumatol. Rep.* 2005, 7:156-162.

[171] Leask A, Abraham DJ: TGF-beta signaling and the fibrotic response. *FASEB J.* 2004, 18:816-827.

[172] Larjava H, Koivisto L, Häkkinen L: Keratinocyte Interactions with Fibronectin during Wound Healing. In Madame Curie Bioscience Database: Cell Invasion. Edited by Heino J, Kähäri V-M. Austin (TX): *Landes Bioscience* 2002.

[173] Kusubata M, Hirota A, Ebihara T, Kuwaba K, Matsubara Y, Sasaki T, Kusakabe M, Tsukada T, Irie S, Koyama Y: Spatiotemporal changes of fibronectin, tenascin-C, fibulin-1, and fibulin-2 in the skin during the development of chronic contact dermatitis. *J. Invest. Dermatol.* 1999, 113: 906-912.

[174] Bravo-Cordero JJ, Marrero-Diaz R, Megias D, Genis L, Garcia-Grande A, Garcia MA, Arroyo AG, Montoya MC: MT1-MMP proinvasive activity is regulated by a novel Rab8-dependent exocytic pathway. *EMBO J.* 2007, 26:1499-1510.

[175] Cukierman E, Pankov R, Stevens DR, Yamada KM: Taking cell-matrix adhesions to the third dimension. *Science* 2001, 294:1708-1712.

[176] Wheeldon I, Farhadi A, Bick AG, Jabbari E, Khademhosseini A: Nanoscale tissue engineering: spatial control over cell-materials interactions. *Nanotechnology* 2011, 22:212001. *Epub.* 212011 Mar 212031.

[177] Walcott S, Sun SX: A mechanical model of actin stress fiber formation and substrate elasticity sensing in adherent cells. *Proc.* 2010, 107:7757-7762. *Epub.* 2010 Apr 7712.

[178] Zhou X, Rowe RG, Hiraoka N, George JP, Wirtz D, Mosher DF, Virtanen I, Chernousov MA, Weiss SJ: Fibronectin fibrillogenesis regulates three-dimensional neovessel formation. *Genes Dev.* 2008, 22: 1231-1243.

[179] Cseh B, Fernandez-Sauze S, Grall D, Schaub S, Doma E, Van Obberghen-Schilling E: Autocrine fibronectin directs matrix assembly and crosstalk between cell-matrix and cell-cell adhesion in vascular endothelial cells. *J.*, 123:3989-3999. *Epub.* 2010 Oct 3927.

[180] Amatangelo MD, Bassi DE, Klein-Szanto AJ, Cukierman E: Stroma-derived three-dimensional matrices are necessary and sufficient to promote desmoplastic differentiation of normal fibroblasts. *Am. J. Pathol.* 2005, 167:475-488.

[181] Califano JP, Reinhart-King CA: A Balance of Substrate Mechanics and Matrix Chemistry Regulates Endothelial Cell Network Assembly. *Cellular and Molecular Bioengineering* 2008, 1:122-132.

[182] Antia M, Baneyx G, Kubow KE, Vogel V: Fibronectin in aging extracellular matrix fibrils is progressively unfolded by cells and elicits an enhanced rigidity response. *Faraday Discuss* 2008, 139:229-249; *discussion* 309-225, 419-220.

[183] Eisenberg JL, Safi A, Wi X, Espinosa HD, Budinger GRS, Takawira D, Hopkinson SB, Jones JCR: Substrate stiffness regulates extracellular matrix deposition by alveolar epithelial cells. *Research and Reports in Biology* 2011, 2:1-12.

[184] Paszek MJ, Zahir N, Johnson KR, Lakins JN, Rozenberg GI, Gefen A, Reinhart-King CA, Margulies SS, Dembo M, Boettiger D, et al: Tensional homeostasis and the malignant phenotype. *Cancer Cell* 2005, 8: 241-254.

[185] Chandler EM, Saunders MP, Yoon CJ, Gourdon D, Fischbach C: Adipose progenitor cells increase fibronectin matrix strain and unfolding in breast tumors. *Phys.*, 8:015008. *Epub.* 012011 Feb 015007.

[186] Williams CM, Engler AJ, Slone RD, Galante LL, Schwarzbauer JE: Fibronectin expression modulates mammary epithelial cell proliferation during acinar differentiation. *Cancer Res.* 2008, 68:3185-3192.

[187] Hsia HC, Nair MR, Mintz RC, Corbett SA: The fiber diameter of synthetic bioresorbable extracellular matrix influences human fibroblast morphology and fibronectin matrix assembly. *Plast.,* 127:2312-2320.

[188] De Jong KL, MacLeod HC, Norton PR, Petersen NO: Fibronectin organization under and near cells. *Eur. Biophys. J.* 2006, 35:695-708.

[189] Ohashi T, Kiehart DP, Erickson HP: Dynamics and elasticity of the fibronectin matrix in living cell culture visualized by fibronectin-green fluorescent protein. *Proc. Natl. Acad. Sci. U S A* 1999, 96:2153-2158.

[190] Parker KK, Brock AL, Brangwynne C, Mannix RJ, Wang N, Ostuni E, Geisse NA, Adams JC, Whitesides GM, Ingber DE: Directional control of lamellipodia extension by constraining cell shape and orienting cell tractional forces. *Faseb. J.* 2002, 16:1195-1204.

[191] Gaultier A, Hollister M, Reynolds I, Hsieh EH, Gonias SL: LRP1 regulates remodeling of the extracellular matrix by fibroblasts. *Matrix Biol.* 2009, 29:22-30.

[192] Panorchan P, Lee JS, Kole TP, Tseng Y, Wirtz D: Microrheology and ROCK signaling of human endothelial cells embedded in a 3D matrix. *Biophys. J.* 2006, 91:3499-3507.

[193] Klotzsch E, Smith ML, Kubow KE, Muntwyler S, Little WC, Beyeler F, Gourdon D, Nelson BJ, Vogel V: Fibronectin forms the most extensible biological fibers displaying switchable force-exposed cryptic binding sites. *Proc. Natl. Acad. Sci. U S A* 2009, 106:18267-18272.

[194] Davidson LA, Keller R, DeSimone DW: Assembly and remodeling of the fibrillar fibronectin extracellular matrix during gastrulation and neurulation in Xenopus laevis. *Dev. Dyn.* 2004, 231:888-895.

[195] Lock JG, Wehrle-Haller B, Stromblad S: Cell-matrix adhesion complexes: master control machinery of cell migration. *Semin. Cancer Biol.* 2008, 18:65-76. Epub 2007 Oct 2006.

[196] Parsons JT, Horwitz AR, Schwartz MA: Cell adhesion: integrating cytoskeletal dynamics and cellular tension. *Nat.*, 11:633-643.

[197] Denhez F, Wilcox-Adelman SA, Baciu PC, Saoncella S, Lee S, French B, Neveu W, Goetinck PF: Syndesmos, a syndecan-4 cytoplasmic domain interactor, binds to the focal adhesion adaptor proteins paxillin and Hic-5. *J. Biol. Chem.* 2002, 277:12270-12274.

[198] Schaller MD: Paxillin: a focal adhesion-associated adaptor protein. *Oncogene* 2001, 20:6459-6472.

[199] Kanchanawong P, Shtengel G, Pasapera AM, Ramko EB, Davidson MW, Hess HF, Waterman CM: Nanoscale architecture of integrin-based cell adhesions. *Nature,* 468:580-584.

[200] Harunaga JS, Yamada KM: Cell-matrix adhesions in 3D. *Matrix Biol.,* 2011:23.

[201] Grinnell F: Fibroblast biology in three-dimensional collagen matrices. *Trends Cell Biol.* 2003, 13:264-269.

[202] Pizzo AM, Kokini K, Vaughn LC, Waisner BZ, Voytik-Harbin SL: Extracellular matrix (ECM) microstructural composition regulates local cell-ECM biomechanics and fundamental fibroblast behavior: a multidimensional perspective. *J. Appl. Physiol.* 2005, 98:1909-1921. *Epub.* 2004 Dec 1923.

[203] Jiang ST, Chuang WJ, Tang MJ: Role of fibronectin deposition in branching morphogenesis of Madin-Darby canine kidney cells. *Kidney Int.* 2000, 57:1860-1867.

[204] Inoue T, Nabeshima K, Shimao Y, Meng JY, Koono M: Regulation of fibronectin expression and splicing in migrating epithelial cells: migrating MDCK cells produce a lesser amount of, but more active, fibronectin. Biochem. *Biophys. Res. Commun.* 2001, 280:1262-1268.

[205] Beningo KA, Dembo M, Wang YL: Responses of fibroblasts to anchorage of dorsal extracellular matrix receptors. *Proc. Natl. Acad. Sci. U S A* 2004, 101:18024-18029. Epub 12004 Dec 18015.

[206] Shemesh T, Geiger B, Bershadsky AD, Kozlov MM: Focal adhesions as mechanosensors: a physical mechanism. *Proc. Natl. Acad. Sci. U S A* 2005, 102:12383-12388. Epub 12005 Aug 12319.

[207] Tamariz E, Grinnell F: Modulation of fibroblast morphology and adhesion during collagen matrix remodeling. *Mol. Biol. Cell* 2002, 13: 3915-3929.

[208] Le Clainche C, Carlier MF: Regulation of actin assembly associated with protrusion and adhesion in cell migration. *Physiol. Rev.* 2008, 88: 489-513.

[209] Cramer LP: Forming the cell rear first: breaking cell symmetry to trigger directed cell migration. *Nat.,* 12:628-632.

[210] Hocking DC, Kowalski K: A cryptic fragment from fibronectin's III1 module localizes to lipid rafts and stimulates cell growth and contractility. *J. Cell Biol.* 2002, 158: 175-184.

[211] Ulrich TA, de Juan Pardo EM, Kumar S: The mechanical rigidity of the extracellular matrix regulates the structure, motility, and proliferation of glioma cells. *Cancer Res.* 2009, 69:4167-4174. Epub 2009 May 4112.

[212] Lo CM, Wang HB, Dembo M, Wang YL: Cell movement is guided by the rigidity of the substrate. *Biophys. J.* 2000, 79:144-152.

[213] Zaman MH, Trapani LM, Sieminski AL, Mackellar D, Gong H, Kamm RD, Wells A, Lauffenburger DA, Matsudaira P: Migration of tumor cells in 3D matrices is governed by matrix stiffness along with cell-matrix adhesion and proteolysis. *Proc. Natl. Acad. Sci. U S A* 2006, 103: 10889-10894. Epub 12006 Jul 10810.

[214] Friedl P, Zanker KS, Brocker EB: Cell migration strategies in 3-D extracellular matrix: differences in morphology, cell matrix interactions, and integrin function. *Microsc. Res. Tech.* 1998, 43:369-378.

[215] Matsumoto R, Sugimoto M: Dermal matrix proteins initiate re-epithelialization but are not sufficient for coordinated epidermal outgrowth in a new fish skin culture model. *Cell Tissue Res.* 2007, 327:249-265. Epub 2006 Oct 2017.

[216] Kim JP, Zhang K, Chen JD, Wynn KC, Kramer RH, Woodley DT: Mechanism of human keratinocyte migration on fibronectin: unique roles of RGD site and integrins. *J. Cell Physiol.* 1992, 151:443-450.

[217] Soucy PA, Romer LH: Endothelial cell adhesion, signaling, and morphogenesis in fibroblast-derived matrix. *Matrix Biol.* 2009, 28:273-283. Epub 2009 Apr 2016.

[218] Reinhart-King CA: How matrix properties control the self-assembly and maintenance of tissues. *Ann. Biomed. Eng.* 2011, 39:1849-1856.

[219] Guo WH, Frey MT, Burnham NA, Wang YL: Substrate rigidity regulates the formation and maintenance of tissues. *Biophys. J.* 2006, 90:2213-2220. Epub 2005 Dec 2230.

[220] Sevilla CA, Dalecki D, Hocking DC: Extracellular matrix fibronectin stimulates the self-assembly of microtissues on native collagen gels. *Tissue* 2010, 16:3805-3819. Epub 2010 Sep 3806.

[221] Van Goethem E, Poincloux R, Gauffre F, Maridonneau-Parini I, Le Cabec V: Matrix architecture dictates three-dimensional migration modes of human macrophages: differential involvement of proteases and podosome-like structures. *J. Immunol.,* 184:1049-1061.

[222] Doyle AD, Wang FW, Matsumoto K, Yamada KM: One-dimensional topography underlies three-dimensional fibrillar cell migration. *J. Cell Biol.* 2009, 184:481-490. Epub 2009 Feb 2016.

[223] Friedl P, Wolf K: Plasticity of cell migration: a multiscale tuning model. *J. Cell Biol.* 2009, 188:11-19.

[224] Tonnesen MG, Feng X, Clark RA: Angiogenesis in wound healing. *J. Investig. Dermatol. Symp. Proc.* 2000, 5:40-46.

[225] Feng X, Clark RA, Galanakis D, Tonnesen MG: Fibrin and collagen differentially regulate human dermal microvascular endothelial cell integrins: stabilization of alphav/beta3 mRNA by fibrin1. *J. Invest. Dermatol.* 1999, 113:913-919.

[226] Nicosia RF, Bonanno E, Smith M: Fibronectin promotes the elongation of microvessels during angiogenesis in vitro. *J. Cell Physiol.* 1993, 154:654-661.

[227] Han S, Sidell N, Roser-Page S, Roman J: Fibronectin stimulates human lung carcinoma cell growth by inducing cyclooxygenase-2 (COX-2) expression. *Int. J. Cancer* 2004, 111:322-331.

[228] Han S, Sidell N, Roman J: Fibronectin stimulates human lung carcinoma cell proliferation by suppressing p21 gene expression via signals involving Erk and Rho kinase. *Cancer Lett.* 2005, 219:71-81.

[229] Han SW, Roman J: Fibronectin induces cell proliferation and inhibits apoptosis in human bronchial epithelial cells: pro-oncogenic effects mediated by PI3-kinase and NF-kappa B. *Oncogene* 2006, 25:4341-4349. Epub 2006 Mar 4346.

[230] Wilson SH, Ljubimov AV, Morla AO, Caballero S, Shaw LC, Spoerri PE, Tarnuzzer RW, Grant MB: Fibronectin fragments promote human retinal endothelial cell adhesion and proliferation and ERK activation through alpha5beta1 integrin and PI 3-kinase. *Invest. Ophthalmol. Vis. Sci.* 2003, 44:1704-1715.

[231] Wang J, Milner R: Fibronectin promotes brain capillary endothelial cell survival and proliferation through alpha5beta1 and alphavbeta3 integrins via MAP kinase signalling. *J. Neurochem.* 2006, 96:148-159. Epub 2005 Nov 2003.

[232] Illario M, Cavallo AL, Monaco S, Di Vito E, Mueller F, Marzano LA, Troncone G, Fenzi G, Rossi G, Vitale M: Fibronectin-induced proliferation in thyroid cells is

mediated by alphavbeta3 integrin through Ras/Raf-1/MEK/ERK and calcium/CaMKII signals. *J. Clin. Endocrinol. Metab.* 2005, 90:2865-2873. Epub 2005 Feb 2861.

[233] Farias E, Lu M, Li X, Schnapp LM: Integrin alpha8beta1-fibronectin interactions promote cell survival via PI3 kinase pathway. *Biochem. Biophys. Res. Commun.* 2005, 329:305-311.

[234] Spangenberg C, Lausch EU, Trost TM, Prawitt D, May A, Keppler R, Fees SA, Reutzel D, Bell C, Schmitt S, et al: ERBB2-mediated transcriptional up-regulation of the alpha5beta1 integrin fibronectin receptor promotes tumor cell survival under adverse conditions. *Cancer Res.* 2006, 66:3715-3725.

[235] Zhang Z, Vuori K, Reed JC, Ruoslahti E: The alpha 5 beta 1 integrin supports survival of cells on fibronectin and up-regulates Bcl-2 expression. *Proc. Natl. Acad. Sci. U S A* 1995, 92:6161-6165.

[236] Zhang T, Liu S, Yang P, Han C, Wang J, Liu J, Han Y, Yu Y, Cao X: Fibronectin maintains survival of mouse natural killer (NK) cells via CD11b/Src/beta-catenin pathway. *Blood* 2009, 114:4081-4088. Epub 2009 Sep 4088.

[237] Rodriguez-Juan C, de la Torre P, Garcia-Ruiz I, Diaz-Sanjuan T, Munoz-Yague T, Gomez-Izquierdo E, Solis-Munoz P, Solis-Herruzo JA: Fibronectin increases survival of rat hepatic stellate cells--a novel profibrogenic mechanism of fibronectin. *Cell Physiol. Biochem.* 2009, 24:271-282. Epub 2009 Aug 2003.

[238] Fornaro M, Plescia J, Chheang S, Tallini G, Zhu YM, King M, Altieri DC, Languino LR: Fibronectin protects prostate cancer cells from tumor necrosis factor-alpha-induced apoptosis via the AKT/survivin pathway. *J. Biol. Chem.* 2003, 278:50402-50411. Epub 52003 Sep 50430.

[239] Kim S, Bakre M, Yin H, Varner JA: Inhibition of endothelial cell survival and angiogenesis by protein kinase A. *J. Clin. Invest.* 2002, 110:933-941.

[240] Aoshiba K, Rennard SI, Spurzem JR: Fibronectin supports bronchial epithelial cell adhesion and survival in the absence of growth factors. *Am. J. Physiol.* 1997, 273: L684-693.

[241] Illario M, Cavallo AL, Bayer KU, Di Matola T, Fenzi G, Rossi G, Vitale M: Calcium/calmodulin-dependent protein kinase II binds to Raf-1 and modulates integrin-stimulated ERK activation. *J. Biol. Chem.* 2003, 278: 45101-45108. Epub 42003 Sep 45103.

[242] Nam JM, Onodera Y, Bissell MJ, Park CC: Breast cancer cells in three-dimensional culture display an enhanced radioresponse after coordinate targeting of integrin alpha5beta1 and fibronectin. *Cancer,* 70:5238-5248. Epub 2010 Jun 5231.

[243] Campillo-Fernandez AJ, Unger RE, Peters K, Halstenberg S, Santos M, Salmeron Sanchez M, Meseguer Duenas JM, Monleon Pradas M, Gomez Ribelles JL, Kirkpatrick CJ: Analysis of the biological response of endothelial and fibroblast cells cultured on synthetic scaffolds with various hydrophilic/hydrophobic ratios: influence of fibronectin adsorption and conformation. *Tissue Eng. Part A* 2009, 15:1331-1341.

[244] Cooper TP, Sefton MV: Fibronectin coating of collagen modules increases in vivo HUVEC survival and vessel formation in SCID mice. *Acta Biomater.* 2011, 7:1072-1083.

[245] Tate CC, Shear DA, Tate MC, Archer DR, Stein DG, LaPlaca MC: Laminin and fibronectin scaffolds enhance neural stem cell transplantation into the injured brain. *J. Tissue Eng. Regen. Med.* 2009, 3: 208-217.

[246] Stabenfeldt SE, Munglani G, Garcia AJ, LaPlaca MC: Biomimetic microenvironment modulates neural stem cell survival, migration, and differentiation. *Tissue* 2010, 16:3747-3758. Epub 2010 Oct 3748.

[247] Sakai T, Johnson KJ, Murozono M, Sakai K, Magnuson MA, Wieloch T, Cronberg T, Isshiki A, Erickson HP, Fassler R: Plasma fibronectin supports neuronal survival and reduces brain injury following transient focal cerebral ischemia but is not essential for skin-wound healing and hemostasis. *Nat. Med.* 2001, 7:324-330.

[248] Tate CC, Garcia AJ, LaPlaca MC: Plasma fibronectin is neuroprotective following traumatic brain injury. *Exp. Neurol.* 2007, 207:13-22. Epub 2007 May 2018.

[249] Morin N, Sullivan R: Expression of fibronectin and a fibronectin-binding molecule during preimplantation development in the mouse. *Hum. Reprod.* 1994, 9:894-901.

[250] Watt FM, Hodivala KJ: Cell adhesion. Fibronectin and integrin knockouts come unstuck. *Curr. Biol.* 1994, 4:270-272.

[251] George EL, Georges-Labouesse EN, Patel-King RS, Rayburn H, Hynes RO: Defects in mesoderm, neural tube and vascular development in mouse embryos lacking fibronectin. *Development* 1993, 119:1079-1091.

[252] George EL, Baldwin HS, Hynes RO: Fibronectins are essential for heart and blood vessel morphogenesis but are dispensable for initial specification of precursor cells. *Blood* 1997, 90:3073-3081.

[253] Wei WC, Kopec AK, Tang MJ: Requirement of focal adhesion kinase in branching tubulogenesis. *J. Biomed. Sci.* 2009, 16:5.

[254] Sakai T, Larsen M, Yamada KM: Fibronectin requirement in branching morphogenesis. *Nature* 2003, 423:876-881.

[255] Borenstein E, Cline. E: Cellular automata model of cystogenesis and tubulogenesis. *Proceedings of the Santa-Fe Institute Complex Systems Summer School* 2004.

[256] Montesano R, Matsumoto K, Nakamura T, Orci L: Identification of a fibroblast-derived epithelial morphogen as hepatocyte growth factor. *Cell* 1991, 67:901-908.

[257] Schoenenberger CA, Zuk A, Zinkl GM, Kendall D, Matlin KS: Integrin expression and localization in normal MDCK cells and transformed MDCK cells lacking apical polarity. *J. Cell Sci.* 1994, 107 (Pt 2): 527-541.

[258] Saelman EU, Keely PJ, Santoro SA: Loss of MDCK cell alpha 2 beta 1 integrin expression results in reduced cyst formation, failure of hepatocyte growth factor/scatter factor-induced branching morphogenesis, and increased apoptosis. *J. Cell Sci.* 1995, 108 (Pt 11): 3531-3540.

[259] Praetorius HA, Praetorius J, Nielsen S, Frokiaer J, Spring KR: Beta1-integrins in the primary cilium of MDCK cells potentiate fibronectin-induced Ca2+ signaling. *Am. J. Physiol. Renal Physiol.* 2004, 287:F969-978.

[260] Ilic D, Furuta Y, Kanazawa S, Takeda N, Sobue K, Nakatsuji N, Nomura S, Fujimoto J, Okada M, Yamamoto T: Reduced cell motility and enhanced focal adhesion contact formation in cells from FAK-deficient mice. *Nature* 1995, 377:539-544.

[261] Ilic D, Kanazawa S, Furuta Y, Yamamoto T, Aizawa S: Impairment of mobility in endodermal cells by FAK deficiency. *Exp. Cell Res.* 1996, 222:298-303.

[262] Ye P, Habib SL, Ricono JM, Kim NH, Choudhury GG, Barnes JL, Abboud HE, Arar MY: Fibronectin induces ureteric bud cells branching and cellular cord and tubule formation. *Kidney Int.* 2004, 66:1356-1364.

[263] Akhurst RJ, Lehnert SA, Faissner A, Duffie E: TGF beta in murine morphogenetic processes: the early embryo and cardiogenesis. *Development* 1990, 108:645-656.

[264] Ballard VL, Sharma A, Duignan I, Holm JM, Chin A, Choi R, Hajjar KA, Wong SC, Edelberg JM: Vascular tenascin-C regulates cardiac endothelial phenotype and neovascularization. *FASEB J.* 2006, 20:717-719.

[265] Chiquet M, Fambrough DM: Chick myotendinous antigen. I. A monoclonal antibody as a marker for tendon and muscle morphogenesis. *J. Cell Biol.* 1984, 98:1926-1936.

[266] Crossin KL, Hoffman S, Grumet M, Thiery JP, Edelman GM: Site-restricted expression of cytotactin during development of the chicken embryo. *J. Cell Biol.* 1986, 102:1917-1930.

[267] Seidlits SK, Drinnan CT, Petersen RR, Shear JB, Suggs LJ, Schmidt CE: Fibronectin-hyaluronic acid composite hydrogels for three-dimensional endothelial cell culture. *Acta,* 7:2401-2409. Epub 2011 Mar 2423.

[268] Trinh LA, Stainier DY: Fibronectin regulates epithelial organization during myocardial migration in zebrafish. *Dev. Cell* 2004, 6:371-382.

[269] Astrof S, Crowley D, George EL, Fukuda T, Sekiguchi K, Hanahan D, Hynes RO: Direct test of potential roles of EIIIA and EIIIB alternatively spliced segments of fibronectin in physiological and tumor angiogenesis. *Mol. Cell Biol.* 2004, 24:8662-8670.

[270] Muro AF, Chauhan AK, Gajovic S, Iaconcig A, Porro F, Stanta G, Baralle FE: Regulated splicing of the fibronectin EDA exon is essential for proper skin wound healing and normal lifespan. *J. Cell Biol.* 2003, 162: 149-160.

[271] Astrof S, Crowley D, Hynes RO: Multiple cardiovascular defects caused by the absence of alternatively spliced segments of fibronectin. *Dev. Biol.* 2007, 311:11-24. Epub 2007 Jul 2012.

[272] Loots MA, Lamme EN, Zeegelaar J, Mekkes JR, Bos JD, Middelkoop E: Differences in cellular infiltrate and extracellular matrix of chronic diabetic and venous ulcers versus acute wounds. *J. Invest. Dermatol.* 1998, 111:850-857.

[273] Whitby DJ, Ferguson MW: The extracellular matrix of lip wounds in fetal, neonatal and adult mice. *Development* 1991, 112:651-668.

[274] Betz P, Nerlich A, Tubel J, Penning R, Eisenmenger W: Localization of tenascin in human skin wounds--an immunohistochemical study. *Int. J. Legal Med.* 1993, 105: 325-328.

[275] Latijnhouwers MA, Bergers M, Van Bergen BH, Spruijt KI, Andriessen MP, Schalkwijk J: Tenascin expression during wound healing in human skin. *J. Pathol.* 1996, 178:30-35.

[276] Mackie EJ, Halfter W, Liverani D: Induction of tenascin in healing wounds. *J. Cell Biol.* 1988, 107:2757-2767.

[277] Hayman EG, Ruoslahti E: Distribution of fetal bovine serum fibronectin and endogenous rat cell fibronectin in extracellular matrix. *J. Cell Biol.* 1979, 83:255-259.

[278] Peters DM, Portz LM, Fullenwider J, Mosher DF: Co-assembly of plasma and cellular fibronectins into fibrils in human fibroblast cultures. *J. Cell Biol.* 1990, 111:249-256.

[279] Armstrong PB, Armstrong MT: Intercellular invasion and the organizational stability of tissues: a role for fibronectin. *Biochim. Biophys. Acta* 2000, 1470:O9-20.

[280] Midwood KS, Williams LV, Schwarzbauer JE: Tissue repair and the dynamics of the extracellular matrix. *Int. J. Biochem. Cell Biol.* 2004, 36: 1031-1037.

[281] Matsuka YV, Migliorini MM, Ingham KC: Cross-linking of fibronectin to C-terminal fragments of the fibrinogen alpha-chain by factor XIIIa. *J. Protein Chem.* 1997, 16: 739-745.

[282] Corbett SA, Lee L, Wilson CL, Schwarzbauer JE: Covalent cross-linking of fibronectin to fibrin is required for maximal cell adhesion to a fibronectin-fibrin matrix. *J. Biol. Chem.* 1997, 272:24999-25005.

[283] Schick PK, Wojensk CM, Bennett V, Denisova L: Fibronectin isoforms in megakaryocytes. *Stem Cells* 1996, 14 Suppl 1:212-219.

[284] Matuskova J, Chauhan AK, Cambien B, Astrof S, Dole VS, Piffath CL, Hynes RO, Wagner DD: Decreased plasma fibronectin leads to delayed thrombus growth in injured arterioles. *Arterioscler. Thromb. Vasc. Biol.* 2006, 26:1391-1396.

[285] Yoshida K, Munakata H: Connective tissue growth factor binds to fibronectin through the type I repeat modules and enhances the affinity of fibronectin to fibrin. *Biochim. Biophys. Acta* 2007, 1770:672-680. Epub 2006 Nov 2030.

[286] Cho J, Degen JL, Coller BS, Mosher DF: Fibrin but not adsorbed fibrinogen supports fibronectin assembly by spread platelets. Effects of the interaction of alphaIIb beta3 with the C terminus of the fibrinogen gamma-chain. *J. Biol. Chem.* 2005, 280:35490-35498.

[287] Plow EF, Birdwell C, Ginsberg MH: Identification and quantitation of platelet-associated fibronectin antigen. *J. Clin. Invest.* 1979, 63:540-543.

[288] Olorundare OE, Peyruchaud O, Albrecht RM, Mosher DF: Assembly of a fibronectin matrix by adherent platelets stimulated by lysophosphatidic acid and other agonists. *Blood* 2001, 98:117-124.

[289] Niewiarowska J, Cierniewski CS, Tuszynski GP: Association of fibronectin with the platelet cytoskeleton. *J. Biol. Chem.* 1984, 259: 6181-6187.

[290] Thurlow PJ, Kenneally DA, Connellan JM: The role of fibronectin in platelet aggregation. *Br. J. Haematol.* 1990, 75:549-556.

[291] Gardner JM, Hynes RO: Interaction of fibronectin with its receptor on platelets. *Cell* 1985, 42:439-448.

[292] Plow EF, Marguerie GA, Ginsberg MH: Fibronectin binding to thrombin-stimulated platelets: evidence for fibrin(ogen) independent and dependent pathways. *Blood* 1985, 66:26-32.

[293] Cho J, Mosher DF: Enhancement of thrombogenesis by plasma fibronectin cross-linked to fibrin and assembled in platelet thrombi. *Blood* 2006, 107:3555-3563.

[294] Ni H, Papalia JM, Degen JL, Wagner DD: Control of thrombus embolization and fibronectin internalization by integrin alpha IIb beta 3 engagement of the fibrinogen gamma chain. *Blood* 2003, 102:3609-3614.

[295] Ni H, Yuen PS, Papalia JM, Trevithick JE, Sakai T, Fassler R, Hynes RO, Wagner DD: Plasma fibronectin promotes thrombus growth and stability in injured arterioles. *Proc. Natl. Acad. Sci. U S A* 2003, 100: 2415-2419.

[296] Greiling D, Clark RA: Fibronectin provides a conduit for fibroblast transmigration from collagenous stroma into fibrin clot provisional matrix. *J. Cell Sci.* 1997, 110 (Pt 7): 861-870.

[297] Clark RA, Lin F, Greiling D, An J, Couchman JR: Fibroblast invasive migration into fibronectin/fibrin gels requires a previously uncharacterized dermatan sulfate-CD44 proteoglycan. *J. Invest. Dermatol.* 2004, 122:266-277.

[298] Knox P, Crooks S, Rimmer CS: Role of fibronectin in the migration of fibroblasts into plasma clots. *J. Cell Biol.* 1986, 102:2318-2323.

[299] Valenick LV, Hsia HC, Schwarzbauer JE: Fibronectin fragmentation promotes alpha4beta1 integrin-mediated contraction of a fibrin-fibronectin provisional matrix. *Exp. Cell Res.* 2005, 309:48-55.

[300] Liao YF, Gotwals PJ, Koteliansky VE, Sheppard D, Van De Water L: The EIIIA segment of fibronectin is a ligand for integrins alpha 9beta 1 and alpha 4beta 1 providing a novel mechanism for regulating cell adhesion by alternative splicing. *J. Biol. Chem.* 2002, 277:14467-14474.

[301] Singh P, Chen C, Pal-Ghosh S, Stepp MA, Sheppard D, Van De Water L: Loss of integrin alpha9beta1 results in defects in proliferation, causing poor re-epithelialization during cutaneous wound healing. *J. Invest. Dermatol.* 2009, 129:217-228. Epub 2008 Jul 2017.

[302] Lorenz HP, Longaker MT: Wounds: Biology, Pathology and Management. In *Surgery, Second Edition Basic Science and Clinical Evidence* Second edition: Springer New York; 2003: 77-88.

[303] Wakui S, Furusato M, Nikaido T, Yokota K, Sekiguchi J, Ohmori K, Kano Y, Ushigome S: Ultrastructural localization of fibronectin and laminin in human granulation tissue in relation to capillary development. *Cell Struct. Funct.* 1990, 15:201-210.

[304] Singh P, Reimer CL, Peters JH, Stepp MA, Hynes RO, Van De Water L: The spatial and temporal expression patterns of integrin alpha9beta1 and one of its ligands, the EIIIA segment of fibronectin, in cutaneous wound healing. *J. Invest. Dermatol.* 2004, 123:1176-1181.

[305] Kalluri R: Basement membranes: structure, assembly and role in tumour angiogenesis. *Nat. Rev. Cancer* 2003, 3:422-433.

[306] Jarnagin WR, Rockey DC, Koteliansky VE, Wang SS, Bissell DM: Expression of variant fibronectins in wound healing: cellular source and biological activity of the EIIIA segment in rat hepatic fibrogenesis. *J. Cell Biol.* 1994, 127:2037-2048.

[307] Kaczmarek J, Castellani P, Nicolo G, Spina B, Allemanni G, Zardi L: Distribution of oncofetal fibronectin isoforms in normal, hyperplastic and neoplastic human breast tissues. *Int. J. Cancer* 1994, 59:11-16.

[308] Tomasek JJ, Gabbiani G, Hinz B, Chaponnier C, Brown RA: Myofibroblasts and mechano-regulation of connective tissue remodelling. *Nat. Rev. Mol. Cell Biol.* 2002, 3:349-363.

[309] Gabbiani G: The Evolution of the Myofibroblast Concept: a Key Cell for Wound Healing and fibrotic diseases. I processi di repiarazione e fibrosi. *Giornale di Gerontologia* 2004, 52:280-282.

[310] Serini G, Bochaton-Piallat ML, Ropraz P, Geinoz A, Borsi L, Zardi L, Gabbiani G: The fibronectin domain ED-A is crucial for myofibroblastic phenotype induction by transforming growth factor-beta1. *J. Cell Biol.* 1998, 142:873-881.

[311] White ES, Lazar MH, Thannickal VJ: Pathogenetic mechanisms in usual interstitial pneumonia/idiopathic pulmonary fibrosis. *J. Pathol.* 2003, 201: 343-354.

[312] Gildner CD, Lerner AL, Hocking DC: Fibronectin matrix polymerization increases tensile strength of model tissue. *Am. J. Physiol. Heart Circ. Physiol.* 2004, 287:H46-53. Epub 2004 Mar 2004.

[313] Hinz B: Masters and servants of the force: the role of matrix adhesions in myofibroblast force perception and transmission. *Eur. J. Cell Biol.* 2006, 85:175-181.

[314] Hinz B, Phan SH, Thannickal VJ, Galli A, Bochaton-Piallat ML, Gabbiani G: The myofibroblast: one function, multiple origins. *Am. J. Pathol.* 2007, 170:1807-1816.

[315] Koyama Y, Norose K, Kusubata M, Irie S, Kusakabe M: Differential expression of tenascin in the skin during hapten-induced dermatitis. *Histochem. Cell Biol.* 1996, 106:263-273.

[316] Grinnell F, Zhu M: Fibronectin degradation in chronic wounds depends on the relative levels of elastase, alpha1-proteinase inhibitor, and alpha2-macroglobulin. *J. Invest. Dermatol.* 1996, 106:335-341.

[317] Latijnhouwers MA, Bergers M, Veenhuis RT, Beekman B, Ankersmit-Ter Horst MF, Schalkwijk J: Tenascin-C degradation in chronic wounds is dependent on serine proteinase activity. *Arch. Dermatol. Res.* 1998, 290: 490-496.

[318] Palolahti M, Lauharanta J, Stephens RW, Kuusela P, Vaheri A: Proteolytic activity in leg ulcer exudate. *Exp. Dermatol.* 1993, 2:29-37.

[319] Rao CN, Ladin DA, Liu YY, Chilukuri K, Hou ZZ, Woodley DT: Alpha 1-antitrypsin is degraded and non-functional in chronic wounds but intact and functional in acute wounds: the inhibitor protects fibronectin from degradation by chronic wound fluid enzymes. *J. Invest. Dermatol.* 1995, 105:572-578.

[320] Trengove NJ, Stacey MC, MacAuley S, Bennett N, Gibson J, Burslem F, Murphy G, Schultz G: Analysis of the acute and chronic wound environments: the role of proteases and their inhibitors. *Wound Repair Regen.* 1999, 7:442-452.

[321] Clark RAF: *The Molecular and Cellular Biology of Wound Repair.* 2nd edn. New York, N. Y.: Plenum Pres; 1996.

[322] Tran KT, Lamb P, Deng JS: Matrikines and matricryptins: Implications for cutaneous cancers and skin repair. *J. Dermatol. Sci.* 2005, 40:11-20.

[323] McGill G, Shimamura A, Bates RC, Savage RE, Fisher DE: Loss of matrix adhesion triggers rapid transformation-selective apoptosis in fibroblasts. *J. Cell Biol.* 1997, 138:901-911.

[324] Hadden HL, Henke CA: Induction of lung fibroblast apoptosis by soluble fibronectin peptides. *Am. J. Respir. Crit. Care Med.* 2000, 162: 1553-1560.

[325] Werb Z, Tremble PM, Behrendtsen O, Crowley E, Damsky CH: Signal transduction through the fibronectin receptor induces collagenase and stromelysin gene expression. *J. Cell Biol.* 1989, 109:877-889.

[326] To WS, Midwood KS: Cryptic domains of tenascin-C differentially control fibronectin fibrillogenesis. *Matrix Biol.* 2010.

[327] To WS, Midwood KS: Identification of novel and distinct binding sites within tenascin-C for soluble and fibrillar fibronectin. *J. Biol. Chem.*

[328] Hocking DC, Sottile J, Reho T, Fassler R, McKeown-Longo PJ: Inhibition of fibronectin matrix assembly by the heparin-binding domain of vitronectin. *J. Biol. Chem.* 1999, 274:27257-27264.

[329] Sabatelli P, Bonaldo P, Lattanzi G, Braghetta P, Bergamin N, Capanni C, Mattioli E, Columbaro M, Ognibene A, Pepe G, et al: Collagen VI deficiency affects the organization of fibronectin in the extracellular matrix of cultured fibroblasts. *Matrix Biol.* 2001, 20:475-486.

[330] Livant DL, Brabec RK, Kurachi K, Allen DL, Wu Y, Haaseth R, Andrews P, Ethier SP, Markwart S: The PHSRN sequence induces extracellular matrix invasion and accelerates wound healing in obese diabetic mice. *J. Clin. Invest.* 2000, 105:1537-1545.

[331] Zeng ZZ, Yao H, Staszewski ED, Rockwood KF, Markwart SM, Fay KS, Spalding AC, Livant DL: alpha(5)beta(1) Integrin Ligand PHSRN Induces Invasion and alpha(5) mRNA in Endothelial Cells to Stimulate Angiogenesis. *Transl. Oncol.* 2009, 2:8-20.

[332] Diegelmann RF: Cellular and biochemical aspects of normal and abnormal wound healing: an overview. *J. Urol.* 1997, 157:298-302.

[333] van der Straaten HM, Canninga-van Dijk MR, Verdonck LF, Castigliego D, Borst HP, Aten J, Fijnheer R: Extra-domain-A fibronectin: a new marker of fibrosis in cutaneous graft-versus-host disease. *J. Invest. Dermatol.* 2004, 123:1057-1062.

[334] Muro AF, Moretti FA, Moore BB, Yan M, Atrasz RG, Wilke CA, Flaherty KR, Martinez FJ, Tsui JL, Sheppard D, et al: An essential role for fibronectin extra type III domain A in pulmonary fibrosis. *Am. J. Respir. Crit. Care Med.* 2008, 177:638-645.

[335] Barnes VL, Musa J, Mitchell RJ, Barnes JL: Expression of embryonic fibronectin isoform EIIIA parallels alpha-smooth muscle actin in maturing and diseased kidney. *J. Histochem. Cytochem.* 1999, 47: 787-798.

[336] Oomura A, Nakamura T, Arakawa M, Ooshima A, Isemura M: Alterations in the extracellular matrix components in human glomerular diseases. *Virchows Arch. A Pathol. Anat. Histopathol.* 1989, 415: 151-159.

[337] Buyukbabani N, Droz D: Distribution of the extracellular matrix components in human glomerular lesions. *J. Pathol.* 1994, 172:199-207.

[338] Babu M, Diegelmann R, Oliver N: Fibronectin is overproduced by keloid fibroblasts during abnormal wound healing. *Mol. Cell Biol.* 1989, 9: 1642-1650.

[339] Chin GS, Liu W, Steinbrech D, Hsu M, Levinson H, Longaker MT: Cellular signaling by tyrosine phosphorylation in keloid and normal human dermal fibroblasts. *Plast. Reconstr. Surg.* 2000, 106:1532-1540.

[340] McCarty SM, Syed F, Bayat A: Influence of the human leukocyte antigen complex on the development of cutaneous fibrosis: an immunogenetic perspective. *Acta Derm. Venereol.*, 90:563-574.

[341] Kischer CW, Hendrix MJ: Fibronectin (FN) in hypertrophic scars and keloids. *Cell Tissue Res.* 1983, 231:29-37.

[342] White ES, Baralle FE, Muro AF: New insights into form and function of fibronectin splice variants. *J. Pathol.* 2008, 216:1-14.

[343] Kohan M, Muro AF, Bader R, Berkman N: The extra domain A of fibronectin is essential for allergen-induced airway fibrosis and hyperresponsiveness in mice. *J.*, 127:439-446.e431-435. Epub 2010 Dec 2017.

[344] Kosmehl H, Berndt A, Strassburger S, Borsi L, Rousselle P, Mandel U, Hyckel P, Zardi L, Katenkamp D: Distribution of laminin and fibronectin isoforms in oral mucosa and oral squamous cell carcinoma. *Br. J. Cancer* 1999, 81:1071-1079.

[345] Mhawech P, Dulguerov P, Assaly M, Ares C, Allal AS: EB-D fibronectin expression in squamous cell carcinoma of the head and neck. *Oral. Oncol.* 2005, 41:82-88.

[346] Birchler MT, Milisavlijevic D, Pfaltz M, Neri D, Odermatt B, Schmid S, Stoeckli SJ: Expression of the extra domain B of fibronectin, a marker of angiogenesis, in head and neck tumors. *Laryngoscope* 2003, 113: 1231-1237.

[347] Pujuguet P, Hammann A, Moutet M, Samuel JL, Martin F, Martin M: Expression of fibronectin ED-A+ and ED-B+ isoforms by human and experimental colorectal cancer. Contribution of cancer cells and tumor-associated myofibroblasts. *Am. J. Pathol.* 1996, 148:579-592.

[348] Kaariainen E, Nummela P, Soikkeli J, Yin M, Lukk M, Jahkola T, Virolainen S, Ora A, Ukkonen E, Saksela O, Holtta E: Switch to an invasive growth phase in melanoma is associated with tenascin-C, fibronectin, and procollagen-I forming specific channel structures for invasion. *J. Pathol.* 2006, 210:181-191.

[349] Kaarteenaho-Wiik R, Lakari E, Soini Y, Pollanen R, Kinnula VL, Paakko P: Tenascin expression and distribution in pleural inflammatory and fibrotic diseases. *J. Histochem. Cytochem.* 2000, 48:1257-1268.

[350] Midulla M, Verma R, Pignatelli M, Ritter MA, Courtenay-Luck NS, George AJ: Source of oncofetal ED-B-containing fibronectin: implications of production by both tumor and endothelial cells. *Cancer Res.* 2000, 60:164-169.

[351] Ioachim E, Kamina S, Kontostolis M, Agnantis NJ: Immunohistochemical expression of cathepsin D in correlation with extracellular matrix component, steroid receptor status and proliferative indices in breast cancer. *Virchows Arch.* 1997, 431:311-316.

[352] Ioachim E, Charchanti A, Briasoulis E, Karavasilis V, Tsanou H, Arvanitis DL, Agnantis NJ, Pavlidis N: Immunohistochemical expression of extracellular matrix components tenascin, fibronectin, collagen type IV and laminin in breast cancer: their prognostic value and role in tumour invasion and progression. *Eur. J. Cancer* 2002, 38:2362-2370.

[353] Ghosh K, Thodeti CK, Dudley AC, Mammoto A, Klagsbrun M, Ingber DE: Tumor-derived endothelial cells exhibit aberrant Rho-mediated mechanosensing and abnormal angiogenesis in vitro. *Proc. Natl. Acad. Sci. U S A* 2008, 105:11305-11310. Epub 12008 Aug 11306.

[354] Padera TP, Stoll BR, Tooredman JB, Capen D, di Tomaso E, Jain RK: Pathology: cancer cells compress intratumour vessels. *Nature* 2004, 427: 695.

[355] Levental KR, Yu H, Kass L, Lakins JN, Egeblad M, Erler JT, Fong SF, Csiszar K, Giaccia A, Weninger W, et al: Matrix crosslinking forces tumor progression by enhancing integrin signaling. *Cell* 2009, 139:891-906.

[356] Butcher DT, Alliston T, Weaver VM: A tense situation: forcing tumour progression. *Nat. Rev. Cancer* 2009, 9:108-122.

[357] Schrader J, Gordon-Walker TT, Aucott RL, van Deemter M, Quaas A, Walsh S, Benten D, Forbes SJ, Wells RG, Iredale JP: Matrix stiffness modulates proliferation, chemotherapeutic response, and dormancy in hepatocellular carcinoma cells. *Hepatology* 2011, 53:1192-1205.

[358] Kerbel RS: Tumor angiogenesis: past, present and the near future. *Carcinogenesis* 2000, 21:505-515.

[359] Folkman J: Tumor angiogenesis: therapeutic implications. *N. Engl. J. Med.* 1971, 285:1182-1186.

[360] Zetter BR: Angiogenesis and tumor metastasis. *Annu. Rev. Med.* 1998, 49: 407-424.

[361] Hanamura N, Yoshida T, Matsumoto E, Kawarada Y, Sakakura T: Expression of fibronectin and tenascin-C mRNA by myofibroblasts, vascular cells and epithelial cells in human colon adenomas and carcinomas. *Int. J. Cancer* 1997, 73:10-15.

[362] Schnyder B, Semadeni RO, Fischer RW, Vaughan L, Car BD, Heitz PU, Winterhalter KH, Odermatt BF: Distribution pattern of tenascin-C in normal and neoplastic mesenchymal tissues. *Int. J. Cancer* 1997, 72: 217-224.

[363] Talts JF, Wirl G, Dictor M, Muller WJ, Fassler R: Tenascin-C modulates tumor stroma and monocyte/macrophage recruitment but not tumor growth or metastasis in a mouse strain with spontaneous mammary cancer. *J. Cell Sci.* 1999, 112 (Pt 12):1855-1864.

[364] D'Ovidio MC, Mastracchio A, Marzullo A, Ciabatta M, Pini B, Uccini S, Zardi L, Ruco LP: Intratumoral microvessel density and expression of ED-A/ED-B sequences of fibronectin in breast carcinoma. *Eur. J. Cancer* 1998, 34:1081-1085.

[365] Tominaga K, Higuchi K, Watanabe T, Fujiwara Y, Kim S, Arakawa T, Iwao H, Kuroki T: Expression of gene for EIIIA- and EIIIB- fibronectin, fetal types of fibronectin, during gastric ulcer healing in rats. *Dig. Dis. Sci.* 2001, 46:311-317.

[366] Kaspar M, Zardi L, Neri D: Fibronectin as target for tumor therapy. *Int. J. Cancer* 2006, 118:1331-1339.

[367] Zijlstra A, McCabe NR, Schelling ME: Expression and assembly of the angiogenic marker B-fibronectin by endothelial cells in vitro: regulation by confluency. *Angiogenesis* 1999, 3:77-87.

[368] Berndt A, Kollner R, Richter P, Franz M, Voigt A, Borsi L, Giavazzi R, Neri D, Kosmehl H: A comparative analysis of oncofetal fibronectin and tenascin-C incorporation in tumour vessels using human recombinant SIP format antibodies. *Histochem.,* 133:467-475. Epub 2010 Mar 2017.

[369] Kapila YL, Niu J, Johnson PW: The high affinity heparin-binding domain and the V region of fibronectin mediate invasion of human oral squamous cell carcinoma cells in vitro. *J. Biol. Chem.* 1997, 272:18932-18938.

[370] Ghosh A, Joo NE, Chen TC, Kapila YL: Proapoptotic fibronectin fragment induces the degradation of ubiquitinated p53 via proteasomes in periodontal ligament cells. *J.,* 45:481-487. Epub 2010 Mar 2019.

[371] Yi M, Ruoslahti E: A fibronectin fragment inhibits tumor growth, angiogenesis, and metastasis. *Proc. Natl. Acad. Sci. U S A* 2001, 98:620-624.

[372] Souza GR, Molina JR, Raphael RM, Ozawa MG, Stark DJ, Levin CS, Bronk LF, Ananta JS, Mandelin J, Georgescu MM, et al: Three-dimensional tissue culture based on magnetic cell levitation. *Nat.,* 5:291-296. Epub 2010 Mar 2014.

[373] Kim JB: Three-dimensional tissue culture models in cancer biology. *Semin. Cancer Biol.* 2005, 15:365-377.

[374] Yoshii Y, Waki A, Yoshida K, Kakezuka A, Kobayashi M, Namiki H, Kuroda Y, Kiyono Y, Yoshii H, Furukawa T, et al: The use of nanoimprinted scaffolds as 3D culture models to facilitate spontaneous tumor cell migration and well-regulated spheroid formation. *Biomaterials,* 32:6052-6058. Epub 2011 Jun 6052.

[375] Dubey G, Mequanint K: Conjugation of fibronectin onto three-dimensional porous scaffolds for vascular tissue engineering applications. *Acta Biomater.* 1114, 7:1114-1125.

[376] Sun T, Jackson S, Haycock JW, MacNeil S: Culture of skin cells in 3D rather than 2D improves their ability to survive exposure to cytotoxic agents. *J. Biotechnol.* 2006, 122:372-381. Epub 2006 Jan 2030.

[377] Johansson S, Svineng G, Wennerberg K, Armulik A, Lohikangas L: Fibronectin-integrin interactions. *Front Biosci.* 1997, 2:d126-146.

[378] Takagi J, Strokovich K, Springer TA, Walz T: Structure of integrin alpha5beta1 in complex with fibronectin. *EMBO J.* 2003, 22:4607-4615.

[379] Schnapp LM, Hatch N, Ramos DM, Klimanskaya IV, Sheppard D, Pytela R: The human integrin alpha 8 beta 1 functions as a receptor for tenascin, fibronectin, and vitronectin. *J. Biol. Chem.* 1995, 270:23196-23202.

[380] Kauf AC, Hough SM, Bowditch RD: Recognition of fibronectin by the platelet integrin alpha IIb beta 3 involves an extended interface with multiple electrostatic interactions. *Biochemistry* 2001, 40:9159-9166.

[381] Hocevar BA, Brown TL, Howe PH: TGF-beta induces fibronectin synthesis through a c-Jun N-terminal kinase-dependent, Smad4-independent pathway. *EMBO J.* 1999, 18:1345-1356.

[382] Qi W, Chen X, Holian J, Mreich E, Twigg S, Gilbert RE, Pollock CA: Transforming growth factor-beta1 differentially mediates fibronectin and inflammatory cytokine expression in kidney tubular cells. *Am. J. Physiol. Renal Physiol.* 2006, 291: F1070-1077.

[383] Roberts CJ, Birkenmeier TM, McQuillan JJ, Akiyama SK, Yamada SS, Chen WT, Yamada KM, McDonald JA: Transforming growth factor beta stimulates the expression of fibronectin and of both subunits of the human fibronectin receptor by cultured human lung fibroblasts. *J. Biol. Chem.* 1988, 263:4586-4592.

[384] Zhang J, Lo C: Regulation of fibronectin expression by PDGF-BB and IGF-I in cultured rat thoracic aortic adventitial fibroblasts. *Cell Biol. Int.* 1995, 19:517-525.

[385] Liu Y, Centracchio JN, Lin L, Sun AM, Dworkin LD: Constitutive expression of HGF modulates renal epithelial cell phenotype and induces c-met and fibronectin expression. *Exp. Cell Res.* 1998, 242:174-185.

[386] Lam S, Verhagen NA, Strutz F, van der Pijl JW, Daha MR, van Kooten C: Glucose-induced fibronectin and collagen type III expression in renal fibroblasts can occur independent of TGF-beta1. *Kidney Int.* 2003, 63: 878-888.

[387] Oh JH, Ha H, Yu MR, Lee HB: Sequential effects of high glucose on mesangial cell transforming growth factor-beta 1 and fibronectin synthesis. *Kidney Int.* 1998, 54:1872-1878.

[388] Oliver N, Babu M, Diegelmann R: Fibronectin gene transcription is enhanced in abnormal wound healing. *J. Invest. Dermatol.* 1992, 99: 579-586.

[389] Zhang Q, Peyruchaud O, French KJ, Magnusson MK, Mosher DF: Sphingosine 1-phosphate stimulates fibronectin matrix assembly through a Rho-dependent signal pathway. *Blood* 1999, 93:2984-2990.

[390] Quinn JA, Graeber CT, Frackelton AR, Jr., Kim M, Schwarzbauer JE, Filardo EJ: Coordinate regulation of estrogen-mediated fibronectin matrix assembly and epidermal growth factor receptor transactivation by the G protein-coupled receptor, GPR30. *Mol. Endocrinol.* 2009, 23: 1052-1064.

[391] Vial D, Monaghan-Benson E, McKeown-Longo PJ: Coordinate regulation of fibronectin matrix assembly by the plasminogen activator system and vitronectin in human osteosarcoma cells. *Cancer Cell Int.* 2006, 6:8.

[392] Vial D, McKeown-Longo PJ: PAI1 stimulates assembly of the fibronectin matrix in osteosarcoma cells through crosstalk between the alphavbeta5 and alpha5beta1 integrins. *J. Cell Sci.* 2008,121:1661-1670.

[393] Monaghan-Benson E, Mastick CC, McKeown-Longo PJ: A dual role for caveolin-1 in the regulation of fibronectin matrix assembly by uPAR. *J. Cell Sci.* 2008, 121:3693-3703.

[394] Weston BS, Wahab NA, Mason RM: CTGF mediates TGF-beta-induced fibronectin matrix deposition by upregulating active alpha5beta1 integrin in human mesangial cells. *J. Am. Soc. Nephrol.* 2003, 14: 601-610.

[395] Zheng M, Ambesi A, Yu L, McKeown-Longo PJ: Quantification of fibronectin matrix assembly sites using a novel ELISA assay. *Matrix Biol.* 2007, 26:330-333.

[396] You R, Klein RM, Zheng M, McKeown-Longo PJ: Regulation of p38 MAP kinase by anastellin is independent of anastellin's effect on matrix fibronectin. *Matrix Biol.* 2009, 28:101-109. Epub 2009 Feb 2004.

[397] Rivera-Torres A: Fibronectin Matrix Assembly: Localization of ß1 Integrin and Effects of Laminin on Formation of 70-kDa Fragment Linear Arrays or Fibronectin Fibrils. In *Integrated Biological Sciences Summer Research Program University of Wisconsin;* 2001.

[398] Tomasini-Johansson BR, Kaufman NR, Ensenberger MG, Ozeri V, Hanski E, Mosher DF: A 49-residue peptide from adhesin F1 of Streptococcus pyogenes inhibits fibronectin matrix assembly. *J. Biol. Chem.* 2001, 276:23430-23439. Epub 22001 Apr 23425.

[399] Baron-Van Evercooren A, Kleinman HK, Seppa HE, Rentier B, Dubois-Dalcq M: Fibronectin promotes rat Schwann cell growth and motility. *J. Cell Biol.* 1982, 93: 211-216.

[400] Illario M, Amideo V, Casamassima A, Andreucci M, di Matola T, Miele C, Rossi G, Fenzi G, Vitale M: Integrin-dependent cell growth and survival are mediated by different signals in thyroid cells. *J. Clin. Endocrinol. Metab.* 2003, 88:260-269.

[401] Wang RN, Rosenberg L: Maintenance of beta-cell function and survival following islet isolation requires re-establishment of the islet-matrix relationship. *J. Endocrinol.* 1999, 163:181-190.

[402] Globus RK, Doty SB, Lull JC, Holmuhamedov E, Humphries MJ, Damsky CH: Fibronectin is a survival factor for differentiated osteoblasts. *J. Cell Sci.* 1998, 111:1385-1393.

[403] Barkan D, Kleinman H, Simmons JL, Asmussen H, Kamaraju AK, Hoenorhoff MJ, Liu ZY, Costes SV, Cho EH, Lockett S, et al: Inhibition of metastatic outgrowth from single dormant tumor cells by targeting the cytoskeleton. *Cancer Res.* 2008, 68:6241-6250.

[404] Weber LM, Hayda KN, Anseth KS: Cell-matrix interactions improve beta-cell survival and insulin secretion in three-dimensional culture. *Tissue Eng. Part A* 2008, 14:1959-1968.

In: Fibronectin
Editors: James Beattie

ISBN 978-1-62257-298-4

Chapter V

Vesicular Trafficking of the Fibronectin Receptor α5β1 Integrin

Chuan Hu*[*], *Nazarul Hasan and Krista A. Riggs
Department of Biochemistry and Molecular Biology,
University of Louisville School of Medicine, Louisville, Kentucky, US

Abstract

α5β1 integrin, the prototypic receptor for fibronectin, plays important roles in extracellular matrix remodeling, angiogenesis and embryonic development. α5β1 is overexpressed in human cancers and is a potential cancer therapeutic target. The α5 and β1 integrin subunits are synthesized and paired in the endoplasmic reticulum, and delivered to the plasma membrane by vesicular trafficking. Cell surface α5β1 molecules are constantly endocytosed, transported into endosomes, and then recycled back to the plasma membrane. Vesicular trafficking controls the density and distribution of α5β1 at the cell surface. In this chapter, we review recent progress in the molecular mechanism of α5β1 integrin trafficking. We discuss the roles of α5β1 trafficking in cell migration and cancer metastasis.

1. Introduction

Integrins are primary receptors for cell adhesion to the extracellular matrix (ECM) proteins such as fibronectin, laminin, collagen and vitronectin. The receptors play key roles in various cellular processes including adhesion, migration, proliferation, survival and differentiation (Hynes, 2002), and in many pathological conditions including cancer progression and metastasis (Guo and Giancotti, 2004; Hood and Cheresh, 2002). Integrins are heterodimers of α and β subunits. While there are at least 18 different α subunits and 8

[*] Email: chuan.hu@louisville.edu.

different β subunits, only 24 α/β combinations have been described (Clark and Brugge, 1995; Ruegg and Mariotti, 2003). Each α/β heterodimer has its own ligand subset and specificity. With the exception of the α4 subunit, both the α and β subunits have large ligand-binding extracellular domains and short cytoplasmic domains. Integrins link extracellular matrix proteins to the cytoskeletal network in the cytoplasm.

The α5 integrin (ITGA5) only assembles with β1 integrin (ITGB1) to form a heterodimeric receptor. α5β1 recognizes an Arg-Gly-Asp (RGD) motif in fibronectin, fibrinogen, vitronectin and other large glycoproteins (Pierschbacher and Ruoslahti, 1984), and is a major fibronectin receptor in cells. Here, we review the vesicular trafficking pathways that deliver α5β1 integrin to and retrieve it from the plasma membrane, focusing on the proteins that are involved in this process.

2. The Fibronectin Receptor α5β1 Integrin

2.1. Structure of α5β1 Integrin

In humans, the ITGA5 gene encodes a protein of 1049 amino acids (a. a.). After cleavage of the signal peptide, delivery and expression at the cell surface, α5 integrin has 28 a. a. located in the cytoplasm. The ITGB1 gene encodes a 798 a. a. protein which contains 47 a. a. in the cytoplasmic domain. In addition, the α5 subunit undergoes post-translational cleavage in its extracellular domain to yield disulfide-linked light and heavy chains that join with β1 to form a heterodimer. Both α5 and β1 integrins have a single transmembrane domain and are heavily glycosylated. α5β1 integrin binds to fibronectin by recognizing an RGD motif in the 10th type III module of fibronectin (Pankov and Yamada, 2002). Synthetic peptides that contain the RGD motif block the interaction of α5β1 and fibronectin. The interaction of α5β1 and fibronectin is required for fibronectin fibril formation and dictates ECM assembly (Cukierman et al., 2001). α5β1 knockout mouse embryos die around day 10-11 of gestation, and exhibit pronounced defects in mesoderm, neural tube and the vascular system (Yang et al., 1993). The defects of α5β1 knockout embryos mirror the defects in fibronectin knockout embryos (George et al., 1993), further indicating that the interaction between α5β1 and fibronectin is important in embryonic development.

Although the crystal structure of α5β1 integrin has not been solved, the structures of other integrins shed light on how α5β1 interacts with fibronectin at the atomic level. For α2β1 integrin with an I domain in the α subunit, the collagen ligand binds to the top surface of the I domain causing conformational changes in the heterodimer (Emsley et al., 2000). For αvβ3 integrin that does not have an I domain in the α subunit, the ligand binding pocket is located between a seven-bladed β-propeller domain of the αv subunit and an I-like domain from β3 (Xiong et al., 2001; Xiong et al., 2002). Molecular electron microscopy analysis suggests that fibronectin binding to α5β1 is similar to αvβ3 integrin. Unliganded α5β1 extracellular domain exhibits a 'closed' conformation, whereas the binding of fibronectin induces an 'open' conformation with a dramatic change in the β subunit (Takagi et al., 2003). A fibronectin fragment that contains the RGD motif binds to the interface between the β-propeller domain in α5 and the I-like domain in β1 (Takagi et al., 2003).

2.2. α5β1 Integrin in Signaling

Integrins exist in at least two functional states: an 'inactive' state or an 'active' state. The association of the α and β cytoplasmic domains retain integrins in the 'inactive state' (O'Toole et al., 1994; O'Toole et al., 1991). Signaling events, such as binding of the cytosolic protein talin to the β cytoplasmic domain separates the association of the α and β cytoplasmic domains and activates integrins, resulting in 'inside-out' signaling (Calderwood et al., 2002; Calderwood et al., 1999). Reciprocally, the binding of ECM ligands to integrins stimulates the separation of the cytoplasmic domains, promoting their interactions with cytoskeleton and signaling molecules, resulting in 'outside-in' signaling. Although their cytoplasmic domains have no intrinsic kinase activity, integrins are important mediators of cell signaling (ffrench-Constant and Colognato, 2004) (Figure 1). The binding of fibronectin leads to α5β1 integrin clustering and the recruitment of additional structural (*e.g.,* vinculin, α-actinin), adaptor (*e.g.,* p130Cas, paxillin) and signaling proteins to form large protein complexes called focal adhesion complexes (Wozniak et al., 2004). Two of the major tyrosine kinases activated in focal adhesion complexes are focal adhesion kinase (FAK) and Src. Upon integrin activation, FAK is recruited to sites of integrin clustering through interactions with integrin-binding proteins, *e.g.*, talin and paxillin (Parsons, 2003). Upon recruitment, FAK is activated via autophosphorylation at tyrosine 397 (Y397) (Wozniak et al., 2004; Zhao and Guan, 2009), creating the binding site for Src, which is activated by phosphorylation at tyrosine 416 (Y416) (Mitra and Schlaepfer, 2006).

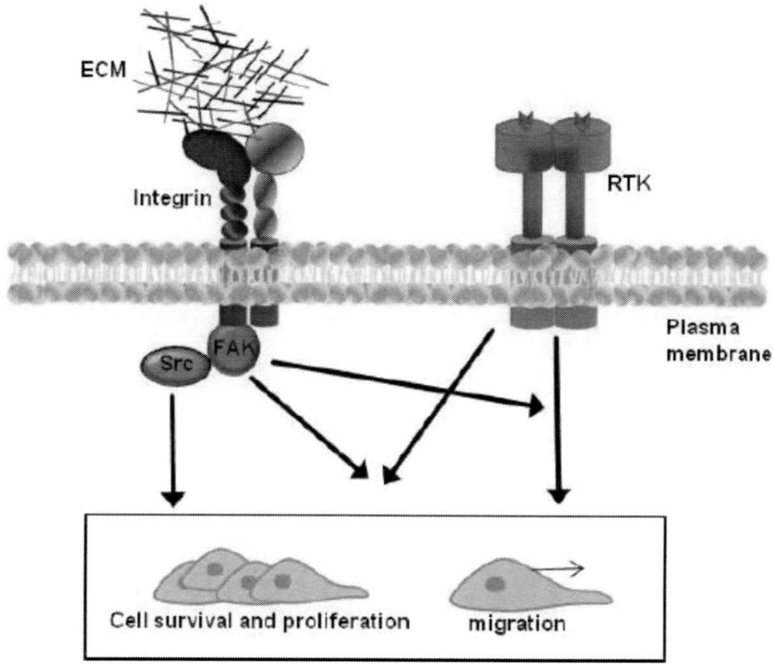

Figure 1. Integrin signaling. The binding of ECM ligands to the extracellular domains of integrins leads to integrin clustering and activation. Integrins recruit and activate focal adhesion kinase (FAK) and Src kinase. FAK and Src further activate downstream signaling cascades, which promote cell survival, proliferation and migration. In addition, the integrin signaling pathway cross-talks with receptor tyrosine kinases(RTK) such as EGFR.

FAK-Src activation regulates a variety of downstream signaling pathways that are important in cell proliferation, survival and migration (Guo and Giancotti, 2004) (Figure 1). Furthermore, α5β1 cross-talks with and activates receptor tyrosine kinases such as the epidermal growth factor receptor (EGFR), leading to the activation of mitogen-activated protein kinase (MAPK) (Kuwada and Li, 2000).

2.3. α5β1 Integrin in Cancer

Integrins play key roles in cancer progression, metastasis and drug resistance (Damiano, 2002; Guo and Giancotti, 2004). Cancer cells tend to increase the expression of integrins that favor their proliferation, survival and migration, whereas they tend to lose the expression of integrins that exert the opposite effect (Guo and Giancotti, 2004).

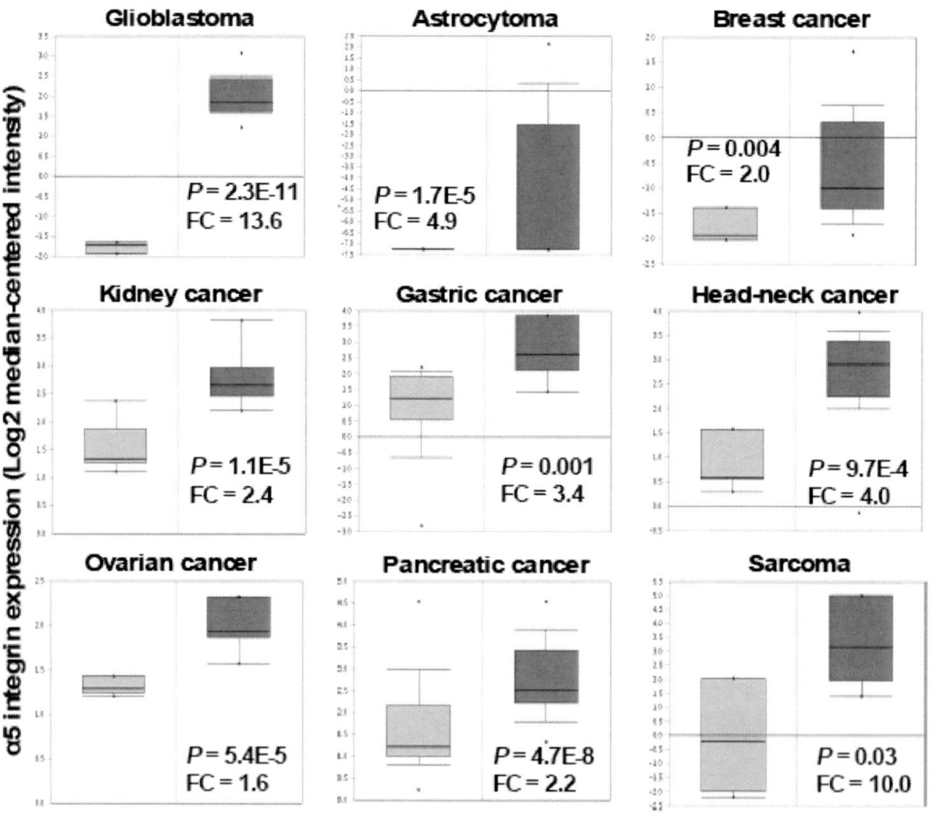

Figure 2. Expression of α5 integrin is elevated in human cancers. α5 integrin mRNA expression in 3 normal neural stem cell, 22 glioblastoma samples (Lee et al., 2006); 6 normal temporal lobe, 45 astrocytoma samples (Rickman et al., 2001); 3 normal breast, 19 lobular breast carcinoma samples (Zhao et al., 2004); 9 normal kidney, 9 renal carcinoma samples (Lenburg et al., 2003); 31 normal gastric mucosa, 7 gastric adenocarcinoma samples (D'Errico et al., 2009); 4 normal uvula, 34 head-neck carcinoma samples (Cromer et al., 2004); 5 normal ovarian surface epithelium, 45 ovarian adenocarcinoma samples (Lu et al., 2004); 39 normal pancreas, 39 pancreatic ductal adenocarcinoma samples (Badea et al., 2008); and 4 normal myometrium, 20 smooth muscle sarcoma samples (Quade et al., 2004) is analyzed using microarrays. Dark boxes represent tumor tissues, whereas light boxes represent normal tissues. Box plots were generated at www.oncomine.org. *P*, *P*-value; FC, fold change.

Elevated expression of α5β1 integrin has been reported in cervical cancer (Contag et al., 2004; Lee et al., 2008), ovarian cancer (Sawada et al., 2008), prostate cancer (Knudsen and Miranti, 2006), and non-small cell lung cancer (Adachi et al., 2000). By analyzing the ONCOMINE cancer gene expression microarray database (Rhodes et al., 2004), we find that α5 integrin is overexpressed in many types of human cancers (Figure 2), including glioblastoma (Lee et al., 2006), astrocytoma (Rickman et al., 2001), breast cancer (Zhao et al., 2004), kidney cancer (Lenburg et al., 2003), gastric cancer (D'Errico et al., 2009), head-neck cancer (Cromer et al., 2004), ovarian cancer (Lu et al., 2004), pancreatic cancer (Badea et al., 2008), and sarcoma (Quade et al., 2004). Since α5 integrin only assembles with the β1 subunit, the expression level of α5 in cancers represents the level of α5β1. Accordingly, fibronectin, the ligand of α5β1, is overexpressed in cancers including non-small cell lung carcinoma (Han et al., 2006).

Antibody, peptide and nonpeptide small molecule antagonists of α5β1 have been shown to inhibit tumor angiogenesis and cause regression of human tumors in animal models (Kim et al., 2000; Martinkova et al., 2010), suggesting that α5β1 promotes tumor growth. Importantly, myeloma cells, which express α5β1 integrin, are more resistant to cytotoxic drugs when adhere to fibronectin compared with cells grown in suspension, indicating that the interaction of α5β1 and fibronectin confers drug resistance to cancer cells (Damiano et al., 1999).

3. Vesicular Trafficking of α5β1 Integrin

Eukaryotic cells are compartmentalized into membrane-bound organelles. Transport vesicles deliver transmembrane proteins such as integrins and secretory factors from one compartment to another and in and out of cells. The density of α5β1 integrin at the cell surface is regulated by two vesicular trafficking pathways: the biosynthetic-secretory pathway and the endocytic recycling pathway (Figure 3).

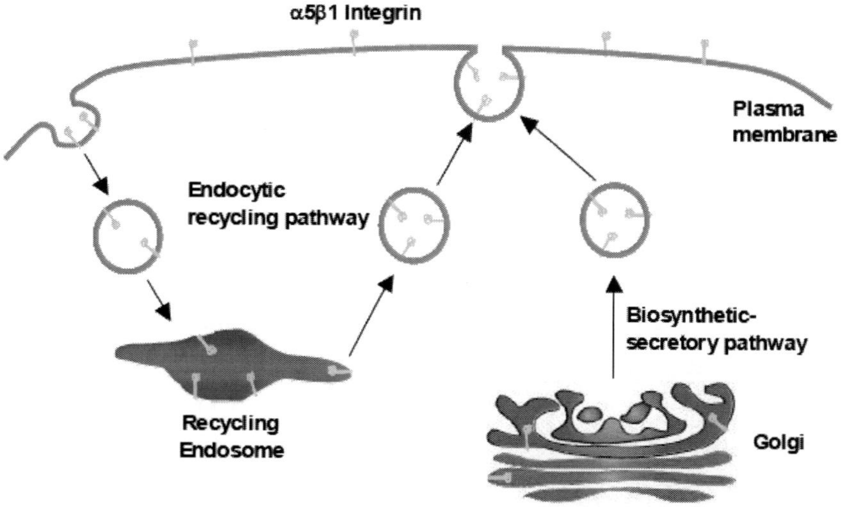

Figure 3. Vesicular trafficking pathways of α5β1 integrin.

In the biosynthetic-secretory pathway, α5β1 is synthesized in the endoplasmic reticulum (ER), transported through the Golgi apparatus, and delivered to the plasma membrane by transport vesicles. In the endocytic recycling pathway, endocytosis is the internalization of α5β1 by invagination of the plasma membrane and formation of membrane-bound vesicles. Recycling is the returning process of endocytosed α5β1 to the cell surface.

3.1. Delivery of Newly Synthesized α5β1 Integrin

After translation, the α5 and β1 integrins are paired in the ER before being delivered to the cell surface. The correct folding and association of α5 and β1 require Ca^{2+} in the ER (Tiwari et al., 2011). The initial association of α5 and β1 is facilitated by a chaperone proteins named calnexin (Hotchin et al., 1995; Isaji et al., 2006). Additionally, N-glycosylation of the β-propeller domain of α5 is necessary for α5β1 heterodimerization and cell surface delivery (Isaji et al., 2006). In talin-deficient cells, α5β1 is accumulated in the ER to Golgi intermediate compartment (ERGIC), indicating that talin is involved in the export of newly synthesized α5β1 (Martel et al., 2000).

3.2. Endocytic Recycling of α5β1 Integrin

The pioneering works of Mark Bretscher showed that cell surface α5β1 is endocytosed constantly at a rate of ~ 1%/min, and cells maintain an intracellular α5β1 pool of 14% of total α5β1 (Bretscher, 1992). Endocytosis of α5β1 can occur by multiple mechanisms. The cytoplasmic domain of β1 integrin contains NXXY (in which X denotes any amino acids) motifs that recruit transmembrane proteins to clathrin-coated pits at the plasma membrane by interactions with adaptor proteins such as AP2 (Pellinen et al., 2008). Silencing of AP2 or clathrin heavy chain (CHC) increases cell surface expression of α5 integrin (Teckchandani et al., 2009), indicating that α5β1 integrin undergoes clathrin-dependent endocytosis. In addition, α5β1 can be endocytosed in a caveolin-dependent manner (Shi and Sottile, 2008), and caveolin-dependent α5β1 endocytosis is essential for the endocytosis and turnover of fibronectin in the ECM (Shi and Sottile, 2008). In endothelial cells, neuropilin 1 (NRP1), a co-receptor for vascular endothelial growth factor A, interacts with α5β1 integrin and stimulates the endocytosis of active α5β1 (Valdembri et al., 2009). At present, it is not clear how the different α5β1 endocytic pathways are called to action in cells. A genome-wide screen reveals that depletion of α5 and β1 integrins leads to endosome clustering at the cell center, suggesting that in addition to being transported by vesicular trafficking, α5β1 integrin is an important regulator of endocytosis (Collinet et al., 2010).

A fraction of endocytosed α5β1 molecules bound with fibronectin are sorted into multivesicular endosomes and degraded in lysosomes (Lobert et al., 2010). This sorting involves fibronectin-triggered ubiquitination of the α5 cytoplasmic domain, and is required in cell migration (Lobert et al., 2010). However, most endocytosed α5β1 molecules are recycled to the cell surface, rather than being targeted to lysosomes for degradation (Bretscher, 1989). We next discuss the proteins that mediate and regulate α5β1 recycling.

3.3. Rab Proteins Involved in α5β1 Trafficking

The vesicular delivery process involves the successive steps of vesicle budding, transport, tethering, docking and fusion. Rab proteins regulate vesicle budding, transport, tethering and fusion (Novick and Zerial, 1997). Rabs are small monomeric GTPases that belong to the Ras superfamily. Rabs are anchored to cellular membranes by prenyl lipid anchors, and switch between an inactive form that binds to GDP and an active form that binds to GTP. There are more than 60 Rabs in human, and each Rab is distributed to distinct intracellular compartments. Jim Norman and colleagues identified several Rab proteins that regulate integrin trafficking. In a "long-loop" pathway, α5β1 integrin is delivered to Rab4-containing early endosomes and then to the perinuclear recycling endosomes that contain Rab11 (Powelka et al., 2004; Roberts et al., 2004) before being recycled to the plasma membrane in a Rab11-dependent manner (Roberts et al., 2001). Protein kinase B/Akt promotes α5β1 recycling by phosphorylating and inactivating glycogen synthase kinase 3 (GSK-3) (Roberts et al., 2004). When cells are treated with platelet-derived growth factor (PDGF), αvβ3 but not α5β1 integrin, is recycled directly from early endosomes to the plasma membrane (Roberts et al., 2001). This "short-loop" pathway depends on Rab4 (Roberts et al., 2001). By association with the cytoplasmic domain of β3 integrin, the serine/threonine kinase protein kinase D1 (PKD1) promotes αvβ3 recycling (Woods et al., 2004).

Rab25, another member of the Rab11 subfamily, has been reported to stimulate the aggressiveness of breast and ovarian cancers (Cheng et al., 2004). By interacting with the β1 integrin cytoplasmic domain, Rab25 retains a pool of recycling α5β1 vesicles at the cell front and promotes cancer cell invasion in three-dimensional extracellular matrix (Caswell et al., 2007). During cytokinesis, α5β1 integrin is transported in Rab21-containing vesicles to the cleavage furrow, which is required for successful cytokinesis (Pellinen et al., 2008). After cytokinesis, Rab21-containing vesicles transport α5β1 away from the cleavage furrow towards the leading edge so that the daughter cells can spread and separate from each other (Pellinen et al., 2008).

3.4. SNARE Proteins Involved in α5β1 Trafficking

The final step in integrin trafficking is the fusion of integrin-transporting vesicles with target compartments. A large body of work has demonstrated that intracellular vesicle fusion is catalyzed by the interactions of SNARE (soluble N-ethylmaleimide-sensitive factor attachment protein receptor) proteins on transport vesicles (v-SNAREs) and on target membranes (t-SNAREs) (Bonifacino and Glick, 2004; Jahn et al., 2003; Rothman, 1994). SNAREs are cytoplasmic oriented type I membrane proteins. The cytoplasmic domains of v- and t-SNAREs form an extremely stable complex (Hayashi et al., 1994; Yang et al., 1999) to drive membrane fusion (Hu et al., 2003; Weber et al., 1998). After membrane fusion, the adapter protein SNAP (soluble NSF attachment protein) and the ATPase NSF (N-ethylmaleimide-sensitive factor) dissociate the v-/t-SNARE complexes at the expense of ATP (Mayer et al., 1996; Sollner et al., 1993) to free SNAREs for the next round of fusion. Individual members of the SNARE family localize to distinct subcellular compartments (Chen and Scheller, 2001), suggesting that each SNARE has a selective role in vesicular trafficking events. Since α5β1 integrin travels through various membrane-bound compartments (Figure 3), multiple v-/t-SNARE complexes are expected to mediate the various vesicle fusion events in α5β1 trafficking.

Significant progress has been made to understand integrin exocytosis, *i.e.*, the fusion of integrin-containing vesicles with the plasma membrane. Coppolino and colleagues showed that the expression of a dominant-negative form of NSF (E329Q-NSF) reduces cell surface α5β1 without altering total cellular levels of α5β1, and inhibits fibronectin-dependent integrin signaling, suggesting that SNAREs are involved in α5β1 trafficking to the cell surface (Skalski et al., 2011; Tayeb et al., 2005). VAMP3 (cellubrevin) is a v-SNARE present in recycling endosomes and endosome-derived vesicles (Galli et al., 1994; McMahon et al., 1993). Galli and colleagues showed that cleavage of VAMP3 by the catalytic light chain of tetanus toxin (TeTx-LC) impairs migration of MDCK cells (Proux-Gillardeaux et al., 2005). In CHO cells, TeTx-LC reduces cell surface α5β1 integrin expression, inhibits cell spreading on fibronectin, and delays trafficking of α5β1 from Rab11-containing recycling endosomes (Skalski and Coppolino, 2005). Together, these data suggest that VAMP3 is involved in α5β1 trafficking to the plasma membrane. The v-SNARE VAMP2 (synaptobrevin 2) mediates regulated exocytosis in neurons and endocrine cells (Hanson et al., 1997; Morgenthaler et al., 2003). Using siRNA-induced gene silencing, we showed that depletion of VAMP2 in HeLa human cervical cancer cells markedly reduces cell surface α5β1 and inhibits chemotactic migration to fibronectin, without altering cell surface expression of α2β1 or α3β1 integrins, indicating that in HeLa cells VAMP2 mediates α5β1 trafficking to the plasma membrane (Hasan and Hu, 2010). Further studies are needed to investigate whether VAMP2 and VAMP3 have redundant roles in α5β1 trafficking.

The fusion of transporting vesicles with the plasma membrane depends on the interactions of v-SNAREs on transport vesicles and t-SNAREs at the plasma membrane. Several plasma membrane t-SNAREs have been implicated in integrin exocytosis. In macrophages, VAMP3 interacts with the plasma membrane t-SNAREs syntaxin 4 and SNAP-23 to drive endocytic recycling of α5β1 integrin (Veale et al., 2010). We reported that in HeLa cells depletion of the t-SNAREs syntaxin 3 and syntaxin 4, but not syntaxin 2, reduces cell surface expression of α5β1 and accumulated α5β1 in cytoplasmic vesicles. These data indicate that syntaxin 3 and syntaxin 4 mediate the delivery of α5β1 integrin to the cell surface (Day et al., 2011). Compared with knockdown of syntaxins 3 or 4, double knockdown of syntaxins 3 and 4 has no additive effect on α5β1 cell surface expression, suggesting that syntaxins 3 and 4 function in the same pathway of α5β1 trafficking (Day et al., 2011). Collectively, our results suggest that in HeLa cells VAMP2 interacts with syntaxins 3 and 4 to mediate the fusion of α5β1-containing vesicles with the plasma membrane.

3.5. Roles of α5β1 Trafficking in Cell Migration and Cancer Metastasis

The endocytosis and recycling of integrins is particularly important during cell migration. Cell migration begins when a cell responds to an external signal by polarizing and extending a protrusion in the direction of movement. Integrins mediate cell adhesion to the ECM at the leading edge to stabilize the protrusion and serve as traction points so that contraction moves the cell body forward (Ridley et al., 2003; Webb et al., 2002). To provide a fresh source of integrins at the cell front, integrins are endocytosed and recycled to the leading edge (Kiosses et al., 2001; Laukaitis et al., 2001; Lawson and Maxfield, 1995; Palecek et al., 1996; Pierini et al., 2000; Regen and Horwitz, 1992; Woods et al., 2004). Consistent with this view, in motile neutrophils on fibronectin, endocytosed α5 integrin is present in endocytic recycling compartments that are oriented toward the direction of cell migration (Pierini et al., 2000). When fibroblasts migrate on fibronectin, endocytosed α5 integrin is localized in recycling endosomes, and is delivered to the base of lamellipodia (Laukaitis et al., 2001).

Both α5β1 and αvβ3 integrins bind fibronectin. α5β1 promotes random cell migration on fibronectin, whereas αvβ3 promotes persistent migration, because α5β1 activates the small GTPase RhoA, which phosphorylates and inactivates the actin-severing protein cofilin (Danen et al., 2005). Interestingly, modulations in the relative rates of αvβ3 and α5β1 trafficking can influence the selection of persistent or random mode of cell migration (White et al., 2007).

hen αvβ3 is undergoing active endocytic recycling, endocytic recycling of α5β1 is suppressed (White et al., 2007). Although the exact mechanism is still not clear, the Rab-coupling protein (RCP) is involved in this process. When cancer cells are treated with αvβ3 antagonists, RCP dissociates from αvβ3 and binds to β1 integrin to link α5β1 to Rab11, resulting in increased α5β1 recycling and random cell migration (Caswell et al., 2008; Caswell et al., 2009). Furthermore, by binding to β1 integrin and the kinase domain of EGFR, RCP couples α5β1 integrin and EGFR in a complex and coordinates their recycling to the plasma membrane (Caswell et al., 2008).

Metastasis, the spreading of cancer from a primary site to distant organs, is responsible for 90 percent of cancer deaths. Metastasis consists of a series of sequential steps, including local tumor growth, migration of cancer cells through basement membrane and invasion into surrounding tissues, intravasation into lymphatic and blood vessels, spread and survival in the circulation, and extravasation and establishing colonies at distant sites (Huber et al., 2005). Integrins are involved in each step of the metastatic cascade, especially in migration and invasion of cancer cells (Goel et al., 2008; Guo and Giancotti, 2004; Hood and Cheresh, 2002).

An elegant study shows that mutants of the tumor suppressor protein p53 stimulates cancer invasion and metastasis by promoting endocytic recycling of α5β1 integrin. p53 is a transcription factor that regulates the expression of genes important in DNA repair, cell-cycle arrest, apoptosis and autophage (Crighton et al., 2006; Levine, 1997). p53 is mutated in 50% of human cancers.

Most of p53 mutations are missense point mutations, and about a third occur at six residues, often referred to as 'hot spots', in the DNA binding domain. By gain of function, the 'hot spot' p53 mutants R273H and R175H promote cancer cell invasion and metastasis (Muller et al., 2009). Strikingly, R273H and R175H stimulate endocytic recycling of α5β1 integrin and EGFR, but not the transferrin receptor. The p53 mutants stimulate α5β1 recycling by inhibiting the function of another member of the p53 family, p63 (Muller et al., 2009). The increased expression (Figure 2) and endocytic recycling of α5β1 integrin may be a major propelling force in cancer metastasis.

Conclusion

The temporal and spatial distribution of the fibronectin receptor α5β1 integrin at the cell surface is regulated by vesicular trafficking. α5β1 at the plasma membrane is constantly endocytosed, transported into endosomes and then recycled back to the cell surface. Disruption of α5β1 trafficking inhibits cell migration, cytokinesis, ECM remodeling and cancer metastasis. Recent studies have identified some of the SNARE and Rab proteins that mediate and regulate α5β1 trafficking. Further work is required to determine how parallel α5β1 trafficking pathways are selected and how α5β1 trafficking is regulated. Further work is needed to elucidate how integrin trafficking is integrated with signal transduction cascades, particularly in metastatic cancers.

Acknowledgments

This work was supported by startup funds from the University of Louisville and CA135123 from the National Institutes of Health (to C.H.).

References

Adachi, M., Taki, T., Higashiyama, M., Kohno, N., Inufusa, H., and Miyake, M. (2000). Significance of integrin alpha5 gene expression as a prognostic factor in node-negative non-small cell lung cancer. *Clin. Cancer Res.*, Vol.6, pp. 96-101.

Badea, L., Herlea, V., Dima, S.O., Dumitrascu, T., and Popescu, I. (2008). Combined gene expression analysis of whole-tissue and microdissected pancreatic ductal adenocarcinoma identifies genes specifically overexpressed in tumor epithelia. *Hepatogastroenterology*, Vol.55, pp. 2016-27.

Bonifacino, J.S., and Glick, B.S. (2004). The mechanisms of vesicle budding and fusion. *Cell*, Vol.116, pp. 153-66.

Bretscher, M.S. (1989). Endocytosis and recycling of the fibronectin receptor in CHO cells. *EMBO J.*, Vol.8, pp. 1341-8.

Bretscher, M.S. (1992). Circulating integrins: alpha 5 beta 1, alpha 6 beta 4 and Mac-1, but not alpha 3 beta 1, alpha 4 beta 1 or LFA-1. *EMBO J.*, Vol.11, pp. 405-10.

Calderwood, D.A., Yan, B., de Pereda, J.M., Alvarez, B.G., Fujioka, Y., Liddington, R.C., and Ginsberg, M.H. (2002). The phosphotyrosine binding-like domain of talin activates integrins. *J. Biol. Chem.*, Vol.277, pp. 21749-58.

Calderwood, D.A., Zent, R., Grant, R., Rees, D.J., Hynes, R.O., and Ginsberg, M.H. (1999). The Talin head domain binds to integrin beta subunit cytoplasmic tails and regulates integrin activation. *J. Biol. Chem.*, Vol.274, pp. 28071-4.

Caswell, P.T., Chan, M., Lindsay, A.J., McCaffrey, M.W., Boettiger, D., and Norman, J.C. (2008). Rab-coupling protein coordinates recycling of alpha5beta1 integrin and EGFR1 to promote cell migration in 3D microenvironments. *J. Cell Biol.*, Vol.183, pp. 143-55.

Caswell, P.T., Spence, H.J., Parsons, M., White, D.P., Clark, K., Cheng, K.W., Mills, G.B., Humphries, M.J., Messent, A.J., Anderson, K.I., McCaffrey, M.W., Ozanne, B.W., and Norman, J.C. (2007). Rab25 associates with alpha5beta1 integrin to promote invasive migration in 3D microenvironments. *Dev. Cell*, Vol.13, pp. 496-510.

Caswell, P.T., Vadrevu, S., and Norman, J.C. (2009). Integrins: masters and slaves of endocytic transport. *Nat. Rev. Mol. Cell Biol.*, Vol.10, pp. 843-53.

Chen, Y.A., and Scheller, R.H. (2001). SNARE-mediated membrane fusion. *Nat. Rev. Mol. Cell Biol.*, Vol.2, pp. 98-106.

Cheng, K.W., Lahad, J.P., Kuo, W.L., Lapuk, A., Yamada, K., Auersperg, N., Liu, J., Smith-McCune, K., Lu, K.H., Fishman, D., Gray, J.W., and Mills, G.B. (2004). The RAB25 small GTPase determines aggressiveness of ovarian and breast cancers. *Nat. Med.*, Vol.10, pp. 1251-6.

Clark, E.A., and Brugge, J.S. (1995). Integrins and signal transduction pathways: the road taken. *Science*, Vol.268, pp. 233-9.

Collinet, C., Stoter, M., Bradshaw, C.R., Samusik, N., Rink, J.C., Kenski, D., Habermann, B., Buchholz, F., Henschel, R., Mueller, M.S., Nagel, W.E., Fava, E., Kalaidzidis, Y., and Zerial, M. (2010). Systems survey of endocytosis by multiparametric image analysis. *Nature*, Vol.464, pp. 243-9.

Contag, S.A., Gostout, B.S., Clayton, A.C., Dixon, M.H., McGovern, R.M., and Calhoun, E.S. (2004). Comparison of gene expression in squamous cell carcinoma and adenocarcinoma of the uterine cervix. *Gynecol. Oncol.*, Vol.95, pp. 610-7.

Crighton, D., Wilkinson, S., O'Prey, J., Syed, N., Smith, P., Harrison, P.R., Gasco, M., Garrone, O., Crook, T., and Ryan, K.M. (2006). DRAM, a p53-induced modulator of autophagy, is critical for apoptosis. *Cell*, Vol.126, pp. 121-34.

Cromer, A., Carles, A., Millon, R., Ganguli, G., Chalmel, F., Lemaire, F., Young, J., Dembele, D., Thibault, C., Muller, D., Poch, O., Abecassis, J., and Wasylyk, B. (2004). Identification of genes associated with tumorigenesis and metastatic potential of hypopharyngeal cancer by microarray analysis. *Oncogene*, Vol.23, pp. 2484-98.

Cukierman, E., Pankov, R., Stevens, D.R., and Yamada, K.M. (2001). Taking cell-matrix adhesions to the third dimension. *Science*, Vol.294, pp. 1708-12.

D'Errico, M., de Rinaldis, E., Blasi, M.F., Viti, V., Falchetti, M., Calcagnile, A., Sera, F., Saieva, C., Ottini, L., Palli, D., Palombo, F., Giuliani, A., and Dogliotti, E. (2009). Genome-wide expression profile of sporadic gastric cancers with microsatellite instability. *Eur. J. Cancer*, Vol.45, pp. 461-9.

Damiano, J.S. (2002). Integrins as novel drug targets for overcoming innate drug resistance. *Curr. Cancer Drug Targets*, Vol.2, pp. 37-43.

Damiano, J.S., Cress, A.E., Hazlehurst, L.A., Shtil, A.A., and Dalton, W.S. (1999). Cell adhesion mediated drug resistance (CAM-DR): role of integrins and resistance to apoptosis in human myeloma cell lines. *Blood*, Vol.93, pp. 1658-67.

Danen, E.H., van Rheenen, J., Franken, W., Huveneers, S., Sonneveld, P., Jalink, K., and Sonnenberg, A. (2005). Integrins control motile strategy through a Rho-cofilin pathway. *J. Cell Biol.*, Vol.169, pp. 515-26.

Day, P., Riggs, K.A., Hasan, N., Corbin, D., Humphrey, D., and Hu, C. (2011). Syntaxins 3 and 4 mediate vesicular trafficking of alpha5beta1 and alpha3beta1 integrins and cancer cell migration. *Int. J. Oncol.*, Vol.39, pp. 863-71.

Emsley, J., Knight, C.G., Farndale, R.W., Barnes, M.J., and Liddington, R.C. (2000). Structural basis of collagen recognition by integrin alpha2beta1. *Cell*, Vol.101, pp. 47-56.

ffrench-Constant, C., and Colognato, H. (2004). Integrins: versatile integrators of extracellular signals. *Trends Cell Biol*, Vol.14, pp. 678-86.

Galli, T., Chilcote, T., Mundigl, O., Binz, T., Niemann, H., and De Camilli, P. (1994). Tetanus toxin-mediated cleavage of cellubrevin impairs exocytosis of transferrin receptor-containing vesicles in CHO cells. *J. Cell Biol.*, Vol.125, pp. 1015-24.

George, E.L., Georges-Labouesse, E.N., Patel-King, R.S., Rayburn, H., and Hynes, R.O. (1993). Defects in mesoderm, neural tube and vascular development in mouse embryos lacking fibronectin. *Development*, Vol.119, pp. 1079-91.

Goel, H.L., Li, J., Kogan, S., and Languino, L.R. (2008). Integrins in prostate cancer progression. *Endocr. Relat. Cancer*, Vol.15, pp. 657-64.

Guo, W., and Giancotti, F.G. (2004). Integrin signalling during tumour progression. *Nat. Rev. Mol. Cell Biol.*, Vol.5, pp. 816-26.

Han, S., Khuri, F.R., and Roman, J. (2006). Fibronectin stimulates non-small cell lung carcinoma cell growth through activation of Akt/mammalian target of rapamycin/S6 kinase and inactivation of LKB1/AMP-activated protein kinase signal pathways. *Cancer Res.*, Vol.66, pp. 315-23.

Hanson, P.I., Heuser, J.E., and Jahn, R. (1997). Neurotransmitter release - four years of SNARE complexes. *Curr. Opin. Neurobiol.*, Vol.7, pp. 310-5.

Hasan, N., and Hu, C. (2010). Vesicle-associated membrane protein 2 mediates trafficking of alpha5beta1 integrin to the plasma membrane. *Exp. Cell Res.*, Vol.316, pp. 12-23.

Hayashi, T., McMahon, H., Yamasaki, S., Binz, T., Hata, Y., Sudhof, T.C., and Niemann, H. (1994). Synaptic vesicle membrane fusion complex: action of clostridial neurotoxins on assembly. *EMBO J.*, Vol.13, pp. 5051-61.

Hood, J.D., and Cheresh, D.A. (2002). Role of integrins in cell invasion and migration. *Nat. Rev. Cancer*, Vol.2, pp. 91-100.

Hotchin, N.A., Gandarillas, A., and Watt, F.M. (1995). Regulation of cell surface beta 1 integrin levels during keratinocyte terminal differentiation. *J. Cell Biol.*, Vol.128, pp. 1209-19.

Hu, C., Ahmed, M., Melia, T.J., Sollner, T.H., Mayer, T., and Rothman, J.E. (2003). Fusion of cells by flipped SNAREs. *Science*, Vol.300, pp. 1745-49.

Huber, M.A., Kraut, N., and Beug, H. (2005). Molecular requirements for epithelial-mesenchymal transition during tumor progression. *Curr. Opin. Cell Biol.*, Vol.17, pp. 548-58.

Hynes, R.O. (2002). Integrins: bidirectional, allosteric signaling machines. *Cell*, Vol.110, pp. 673-87.

Isaji, T., Sato, Y., Zhao, Y., Miyoshi, E., Wada, Y., Taniguchi, N., and Gu, J. (2006). N-glycosylation of the beta-propeller domain of the integrin alpha5 subunit is essential for alpha5beta1 heterodimerization, expression on the cell surface, and its biological function. *J. Biol. Chem.*, Vol.281, pp. 33258-67.

Jahn, R., Lang, T., and Sudhof, T.C. (2003). Membrane fusion. *Cell*, Vol.112, pp. 519-33.

Kim, S., Bell, K., Mousa, S.A., and Varner, J.A. (2000). Regulation of angiogenesis in vivo by ligation of integrin alpha5beta1 with the central cell-binding domain of fibronectin. *Am. J. Pathol.*, Vol.156, pp. 1345-62.

Kiosses, W.B., Shattil, S.J., Pampori, N., and Schwartz, M.A. (2001). Rac recruits high-affinity integrin alphavbeta3 to lamellipodia in endothelial cell migration. *Nat. Cell Biol.*, Vol.3, pp. 316-20.

Knudsen, B.S., and Miranti, C.K. (2006). The impact of cell adhesion changes on proliferation and survival during prostate cancer development and progression. *J. Cell Biochem.*, Vol.99, pp. 345-61.

Kuwada, S.K., and Li, X. (2000). Integrin alpha5/beta1 mediates fibronectin-dependent epithelial cell proliferation through epidermal growth factor receptor activation. *Mol. Biol. Cell*, Vol.11, pp. 2485-96.

Laukaitis, C.M., Webb, D.J., Donais, K., and Horwitz, A.F. (2001). Differential dynamics of alpha 5 integrin, paxillin, and alpha-actinin during formation and disassembly of adhesions in migrating cells. *J. Cell Biol.*, Vol.153, pp. 1427-40.

Lawson, M.A., and Maxfield, F.R. (1995). Ca(2+)- and calcineurin-dependent recycling of an integrin to the front of migrating neutrophils. *Nature*, Vol.377, pp. 75-9.

Lee, J., Kotliarova, S., Kotliarov, Y., Li, A., Su, Q., Donin, N.M., Pastorino, S., Purow, B.W., Christopher, N., Zhang, W., Park, J.K., and Fine, H.A. (2006). Tumor stem cells derived from glioblastomas cultured in bFGF and EGF more closely mirror the phenotype and genotype of primary tumors than do serum-cultured cell lines. *Cancer Cell*, Vol.9, pp. 391-403.

Lee, M.Y., Chou, C.Y., Tang, M.J., and Shen, M.R. (2008). Epithelial-mesenchymal transition in cervical cancer: correlation with tumor progression, epidermal growth factor receptor overexpression, and snail up-regulation. *Clin. Cancer Res.*, Vol.14, pp. 4743-50.

Lenburg, M.E., Liou, L.S., Gerry, N.P., Frampton, G.M., Cohen, H.T., and Christman, M.F. (2003). Previously unidentified changes in renal cell carcinoma gene expression identified by parametric analysis of microarray data. *BMC Cancer*, Vol.3, pp. 31.

Levine, A.J. (1997). p53, the cellular gatekeeper for growth and division. *Cell*, Vol.88, pp. 323-31.

Lobert, V.H., Brech, A., Pedersen, N.M., Wesche, J., Oppelt, A., Malerod, L., and Stenmark, H. (2010). Ubiquitination of alpha 5 beta 1 integrin controls fibroblast migration through lysosomal degradation of fibronectin-integrin complexes. *Dev. Cell*, Vol.19, pp. 148-59.

Lu, K.H., Patterson, A.P., Wang, L., Marquez, R.T., Atkinson, E.N., Baggerly, K.A., Ramoth, L.R., Rosen, D.G., Liu, J., Hellstrom, I., Smith, D., Hartmann, L., Fishman, D., Berchuck, A., Schmandt, R., Whitaker, R., Gershenson, D.M., Mills, G.B., and Bast, R.C., Jr. (2004). Selection of potential markers for epithelial ovarian cancer with gene expression arrays and recursive descent partition analysis. *Clin. Cancer Res.*, Vol.10, pp. 3291-300.

Martel, V., Vignoud, L., Dupe, S., Frachet, P., Block, M.R., and Albiges-Rizo, C. (2000). Talin controls the exit of the integrin alpha 5 beta 1 from an early compartment of the secretory pathway. *J. Cell Sci.*, Vol.113 (Pt 11), pp. 1951-61.

Martinkova, E., Maglott, A., Leger, D.Y., Bonnet, D., Stiborova, M., Takeda, K., Martin, S., and Dontenwill, M. (2010). alpha5beta1 integrin antagonists reduce chemotherapy-induced premature senescence and facilitate apoptosis in human glioblastoma cells. *Int. J. Cancer*, Vol.127, pp. 1240-8.

Mayer, A., Wickner, W., and Haas, A. (1996). Sec18p (NSF)-driven release of Sec17p (alpha-SNAP) can precede docking and fusion of yeast vacuoles. *Cell*, Vol.85, pp. 83-94.

McMahon, H.T., Ushkaryov, Y.A., Edelmann, L., Link, E., Binz, T., Niemann, H., Jahn, R., and Sudhof, T.C. (1993). Cellubrevin is a ubiquitous tetanus-toxin substrate homologous to a putative synaptic vesicle fusion protein. *Nature*, Vol.364, pp. 346-9.

Mitra, S.K., and Schlaepfer, D.D. (2006). Integrin-regulated FAK-Src signaling in normal and cancer cells. *Curr. Opin. Cell Biol.*, Vol.18, pp. 516-23.

Morgenthaler, F.D., Knott, G.W., Floyd Sarria, J.C., Wang, X., Staple, J.K., Catsicas, S., and Hirling, H. (2003). Morphological and molecular heterogeneity in release sites of single neurons. *Eur. J. Neurosci.*, Vol.17, pp. 1365-74.

Muller, P.A., Caswell, P.T., Doyle, B., Iwanicki, M.P., Tan, E.H., Karim, S., Lukashchuk, N., Gillespie, D.A., Ludwig, R.L., Gosselin, P., Cromer, A., Brugge, J.S., Sansom, O.J., Norman, J.C., and Vousden, K.H. (2009). Mutant p53 drives invasion by promoting integrin recycling. *Cell*, Vol.139, pp. 1327-41.

Novick, P., and Zerial, M. (1997). The diversity of Rab proteins in vesicle transport. *Curr. Opin. Cell Biol.*, Vol.9, pp. 496-504.

O'Toole, T.E., Katagiri, Y., Faull, R.J., Peter, K., Tamura, R., Quaranta, V., Loftus, J.C., Shattil, S.J., and Ginsberg, M.H. (1994). Integrin cytoplasmic domains mediate inside-out signal transduction. *J. Cell Biol.*, Vol.124, pp. 1047-59.

O'Toole, T.E., Mandelman, D., Forsyth, J., Shattil, S.J., Plow, E.F., and Ginsberg, M.H. (1991). Modulation of the affinity of integrin alpha IIb beta 3 (GPIIb-IIIa) by the cytoplasmic domain of alpha IIb. *Science*, Vol.254, pp. 845-7.

Palecek, S.P., Schmidt, C.E., Lauffenburger, D.A., and Horwitz, A.F. (1996). Integrin dynamics on the tail region of migrating fibroblasts. *J Cell Sci*, Vol.109 (Pt 5),pp.941-52.

Pankov, R., and Yamada, K.M. (2002). Fibronectin at a glance. *J. Cell Sci.*, Vol.115, pp. 3861-3.

Parsons, J.T. (2003). Focal adhesion kinase: the first ten years. *J. Cell Sci.*, Vol.116, pp. 1409-16.

Pellinen, T., Tuomi, S., Arjonen, A., Wolf, M., Edgren, H., Meyer, H., Grosse, R., Kitzing, T., Rantala, J.K., Kallioniemi, O., Fassler, R., Kallio, M., and Ivaska, J. (2008). Integrin trafficking regulated by Rab21 is necessary for cytokinesis. *Dev. Cell*, Vol.15, pp.371-85.

Pierini, L.M., Lawson, M.A., Eddy, R.J., Hendey, B., and Maxfield, F.R. (2000). Oriented endocytic recycling of alpha5beta1 in motile neutrophils. *Blood*, Vol.95, pp. 2471-80.

Pierschbacher, M.D., and Ruoslahti, E. (1984). Cell attachment activity of fibronectin can be duplicated by small synthetic fragments of the molecule. *Nature*, Vol.309, pp. 30-3.

Powelka, A.M., Sun, J., Li, J., Gao, M., Shaw, L.M., Sonnenberg, A., and Hsu, V.W. (2004). Stimulation-dependent recycling of integrin beta1 regulated by ARF6 and Rab11. *Traffic*, Vol.5, pp. 20-36.

Proux-Gillardeaux, V., Gavard, J., Irinopoulou, T., Mege, R.M., and Galli, T. (2005). Tetanus neurotoxin-mediated cleavage of cellubrevin impairs epithelial cell migration and integrin-dependent cell adhesion. *Proc. Natl. Acad. Sci. U S A*, Vol.102, pp. 6362-7.

Quade, B.J., Wang, T.Y., Sornberger, K., Dal Cin, P., Mutter, G.L., and Morton, C.C. (2004). Molecular pathogenesis of uterine smooth muscle tumors from transcriptional profiling. *Genes Chromosomes Cancer*, Vol.40, pp. 97-108.

Regen, C.M., and Horwitz, A.F. (1992). Dynamics of beta 1 integrin-mediated adhesive contacts in motile fibroblasts. *J. Cell Biol.*, Vol.119, pp. 1347-59.

Rhodes, D.R., Yu, J., Shanker, K., Deshpande, N., Varambally, R., Ghosh, D., Barrette, T., Pandey, A., and Chinnaiyan, A.M. (2004). ONCOMINE: a cancer microarray database and integrated data-mining platform. *Neoplasia*, Vol.6, pp. 1-6.

Rickman, D.S., Bobek, M.P., Misek, D.E., Kuick, R., Blaivas, M., Kurnit, D.M., Taylor, J., and Hanash, S.M. (2001). Distinctive molecular profiles of high-grade and low-grade gliomas based on oligonucleotide microarray analysis. *Cancer Res.*, Vol.61, pp. 6885-91.

Ridley, A.J., Schwartz, M.A., Burridge, K., Firtel, R.A., Ginsberg, M.H., Borisy, G., Parsons, J.T., and Horwitz, A.R. (2003). Cell migration: integrating signals from front to back. *Science*, Vol.302, pp. 1704-9.

Roberts, M., Barry, S., Woods, A., van der Sluijs, P., and Norman, J. (2001). PDGF-regulated rab4-dependent recycling of alphavbeta3 integrin from early endosomes is necessary for cell adhesion and spreading. *Curr. Biol.*, Vol.11, pp. 1392-402.

Roberts, M.S., Woods, A.J., Dale, T.C., Van Der Sluijs, P., and Norman, J.C. (2004). Protein kinase B/Akt acts via glycogen synthase kinase 3 to regulate recycling of alpha v beta 3 and alpha 5 beta 1 integrins. *Mol. Cell Biol.*, Vol.24, pp. 1505-15.

Rothman, J.E. (1994). Mechanisms of intracellular protein transport. *Nature*, Vol.372, pp. 55-63.

Ruegg, C., and Mariotti, A. (2003). Vascular integrins: pleiotropic adhesion and signaling molecules in vascular homeostasis and angiogenesis. *Cell Mol. Life Sci.*, Vol.60, pp. 1135-57.

Sawada, K., Mitra, A.K., Radjabi, A.R., Bhaskar, V., Kistner, E.O., Tretiakova, M., Jagadeeswaran, S., Montag, A., Becker, A., Kenny, H.A., Peter, M.E., Ramakrishnan, V., Yamada, S.D., and Lengyel, E. (2008). Loss of E-cadherin promotes ovarian cancer metastasis via alpha 5-integrin, which is a therapeutic target. *Cancer Res.*, Vol.68, pp. 2329-39.

Shi, F., and Sottile, J. (2008). Caveolin-1-dependent beta1 integrin endocytosis is a critical regulator of fibronectin turnover. *J. Cell Sci.*, Vol.121, pp. 2360-71.

Skalski, M., and Coppolino, M.G. (2005). SNARE-mediated trafficking of alpha5beta1 integrin is required for spreading in CHO cells. *Biochem. Biophys. Res. Commun.*, Vol.335, pp. 1199-210.

Skalski, M., Sharma, N., Williams, K., Kruspe, A., and Coppolino, M.G. (2011). SNARE-mediated membrane traffic is required for focal adhesion kinase signaling and Src-regulated focal adhesion turnover. *Biochim. Biophys. Acta*, Vol.1813, pp. 148-58.

Sollner, T., Bennett, M.K., Whiteheart, S.W., Scheller, R.H., and Rothman, J.E. (1993). A protein assembly-disassembly pathway in vitro that may correspond to sequential steps of synaptic vesicle docking, activation, and fusion. *Cell*, Vol.75, pp. 409-18.

Takagi, J., Strokovich, K., Springer, T.A., and Walz, T. (2003). Structure of integrin alpha5beta1 in complex with fibronectin. *EMBO J.*, Vol.22, pp. 4607-15.

Tayeb, M.A., Skalski, M., Cha, M.C., Kean, M.J., Scaife, M., and Coppolino, M.G. (2005). Inhibition of SNARE-mediated membrane traffic impairs cell migration. *Exp. Cell Res.*, Vol.305, pp. 63-73.

Teckchandani, A., Toida, N., Goodchild, J., Henderson, C., Watts, J., Wollscheid, B., and Cooper, J.A. (2009). Quantitative proteomics identifies a Dab2/integrin module regulating cell migration. *J. Cell Biol.*, Vol.186, pp. 99-111.

Tiwari, S., Askari, J.A., Humphries, M.J., and Bulleid, N.J. (2011). Divalent cations regulate the folding and activation status of integrins during their intracellular trafficking. *J. Cell Sci.*, Vol.124, pp. 1672-80.

Valdembri, D., Caswell, P.T., Anderson, K.I., Schwarz, J.P., Konig, I., Astanina, E., Caccavari, F., Norman, J.C., Humphries, M.J., Bussolino, F., and Serini, G. (2009). Neuropilin-1/GIPC1 signaling regulates alpha5beta1 integrin traffic and function in endothelial cells. *PLoS Biol.*, Vol.7, pp. e25.

Veale, K.J., Offenhauser, C., Whittaker, S.P., Estrella, R.P., and Murray, R.Z. (2010). Recycling endosome membrane incorporation into the leading edge regulates lamellipodia formation and macrophage migration. *Traffic*, Vol.11, pp. 1370-79.

Webb, D.J., Parsons, J.T., and Horwitz, A.F. (2002). Adhesion assembly, disassembly and turnover in migrating cells -- over and over and over again. *Nat. Cell Biol.*, Vol.4, pp. E97-100.

Weber, T., Zemelman, B.V., McNew, J.A., Westermann, B., Gmachl, M., Parlati, F., Sollner, T.H., and Rothman, J.E. (1998). SNAREpins: minimal machinery for membrane fusion. *Cell*, Vol.92, pp. 759-72.

White, D.P., Caswell, P.T., and Norman, J.C. (2007). alpha v beta3 and alpha5beta1 integrin recycling pathways dictate downstream Rho kinase signaling to regulate persistent cell migration. *J. Cell Biol.*, Vol.177, pp. 515-25.

Woods, A.J., White, D.P., Caswell, P.T., and Norman, J.C. (2004). PKD1/PKCmu promotes alphavbeta3 integrin recycling and delivery to nascent focal adhesions. *EMBO J.*, Vol.23, pp. 2531-43.

Wozniak, M.A., Modzelewska, K., Kwong, L., and Keely, P.J. (2004). Focal adhesion regulation of cell behavior. *Biochim. Biophys. Acta*, Vol.1692, pp. 103-19.

Xiong, J.P., Stehle, T., Diefenbach, B., Zhang, R., Dunker, R., Scott, D.L., Joachimiak, A., Goodman, S.L., and Arnaout, M.A. (2001). Crystal structure of the extracellular segment of integrin alpha Vbeta3. *Science*, Vol.294, pp. 339-45.

Xiong, J.P., Stehle, T., Zhang, R., Joachimiak, A., Frech, M., Goodman, S.L., and Arnaout, M.A. (2002). Crystal structure of the extracellular segment of integrin alpha Vbeta3 in complex with an Arg-Gly-Asp ligand. *Science*, Vol.296, pp. 151-5.

Yang, B., Gonzalez, L., Jr., Prekeris, R., Steegmaier, M., Advani, R.J., and Scheller, R.H. (1999). SNARE interactions are not selective. Implications for membrane fusion specificity. *J. Biol. Chem.*, Vol.274, pp. 5649-53.

Yang, J.T., Rayburn, H., and Hynes, R.O. (1993). Embryonic mesodermal defects in alpha 5 integrin-deficient mice. *Development*, Vol.119, pp. 1093-105.

Zhao, H., Langerod, A., Ji, Y., Nowels, K.W., Nesland, J.M., Tibshirani, R., Bukholm, I.K., Karesen, R., Botstein, D., Borresen-Dale, A.L., and Jeffrey, S.S. (2004). Different gene expression patterns in invasive lobular and ductal carcinomas of the breast. *Mol. Biol. Cell*, Vol.15, pp. 2523-36.

Zhao, J., and Guan, J.L. (2009). Signal transduction by focal adhesion kinase in cancer. *Cancer Metastasis Rev.*, Vol.28, pp. 35-49.

In: Fibronectin
Editors: James Beattie

ISBN 978-1-62257-298-4

Chapter VI

Methods to Study Fibronectin Secretion and Matrix Assembly by Fibroblasts *In Vitro* and *In Vivo*

Anna Goc[1,2], Maha Abdalla[1,2] and Payaningal R. Somanath[1,2,3,*]

[1]Clinical and Experimental Therapeutics, College of Pharmacy, University of Georgia,
[2]Charlie Norwood VA Medical Center, Augusta, Georgia, US
[3]Department of Medicine, Georgia Health Sciences University, Augusta, GA

Abstract

Fibronectin (FN) is a high molecular weight glycoprotein composed of two structurally similar monomers forming a dimer linked by a pair of disulfide bonds. It is the product of a single *FN* gene, undergoing alternative splicing leading to 20 potential isoforms. Fibronectin exists in two forms: soluble (major component of blood plasma) and insoluble (major component of the extracellular matrix). It is secreted by many different cell types including endothelial cells, keratinocytes and monocytes. However, fibroblasts are the major source of fibronectin synthesis in a variety of tissues, which not only secrete fibronectin, but also assemble the soluble dimers into an insoluble matrix enabling a remodeling process in many tissues including such aspects as cell-cell communication, cell-to-basement-membrane attachment, and regulation of inflammation as well as maintaining tissue integrity. Through regulation of cell adhesion, growth, migration, and differentiation, fibronectin plays a crucial role in various essential processes such as embryogenesis, nerve regeneration, wound healing, angiogenesis, and cancer metastasis. Altered fibronectin synthesis, secretion, degradation, and organization have been associated with a number of pathological conditions, including fibrosis and cancer. Thus, it is important to better understand the significance of this multifunctional

* Corresponding author: Payaningal R. Somanath, PhD, Assistant Professor. Tel: 706-721-4250; Fax: 706-721-3994. Email: sshenoy@georgiahealth.edu.

protein in both physiological and pathological events. In this chapter, we summarize the methods used to study the synthesis and assembly of fibronectin *in vitro* and *in vivo*, and may also provide useful tools to investigate various aspects of matrix biology.

Abbreviations

APS	ammonium persulfate;
βME	β-mercapthoethanol;
ECM	extracellular matrix proteins;
FN	fibronectin;
GF	growth factor;
HRP	horseradish peroxide;
KO	knock out;
SDS	sodium dodecyl sulfate;
PAGE	polyacrylamide gel electrophoresis;
PBS	phosphate buffered saline;
PVDF	membrane, polyvinylidene fluoride membrane;
TBM	3,3',5,5'-tetramethybenzidine;
TBS	Tris-buffered saline;
TBST	Tris-buffered saline with Twen-20;
TCA	tri-chloride acetate, TEMED, N, N, N', N'-tetramethylethylenediamine, Tris, (tri-hydroxy methyl) aminomethane.

Introduction

The extracellular matrix (ECM) is the multifunctional, dynamic component of tissue that serves as a scaffold for the cells to maintain their integrity through enabling cell-cell communication and preserving correct cellular morphology [1-2].

It consists of the interstitial matrix and the basement membrane, whose central function is a regulation of cell growth, proliferation, migration and tissue formation and remodeling [3-5]. Interstitial matrix is a mesh of fibrous proteins (e.g. fibronectin, collagens, and elastin) associated with glycosaminoglycans (e. g. heparan sulfate, chondroitin sulfate and keratan sulfate).

ECM is present between the cells (i.e. in the intercellular spaces) and acts as a shock absorber against the strain and pressure generated in tissues and in the ECM itself. Basement membrane, on the other hand, is a sheet-like deposition of fibers formed from the ECM proteins and mostly deposited by the endothelial cells.

There are many cell types that bind to the extracellular matrix but fibroblasts are the most abundant. Present in ECM, they synthesize the precursor components of the ECM (mainly the soluble form of fibronectin) and secrete them into the extracellular environment. Once secreted, fibronectins are aggregated and assembled by the fibroblasts in the web through a multi-step cascade of events which sustain the existing structural framework [6-8].

The synthesis is initiated by stimuli such as: growth factors (bFGF, TGF-β, PDGF) and drugs (bleomycin, cyclophosphamide), which activate a number of intracellular signaling cascades including the PI3 Kinase-Akt signaling pathway. This is followed by the activation of the fibronectin (*FN)* gene and synthesis of fibronectin dimers [7, 9-10], which are deposited outside the cells to be assembled. Fibronectin matrix assembly begins when the secreted soluble fibronectin dimers are attached to the existing extracellular matrix and associated with the cellular cytoskeleton via many adaptor molecules and cell surface integrin receptors.

The assembly of soluble fibronectin into insoluble matrix is a complex cell-mediated process that requires the engagement of fibronectin with integrins, elongation of fibronectin, formation of fibrils and subsequent cross-linking with adjacent fibronectin [11-12]. These soluble fibronectin dimers bind to $\alpha_5\beta_1$ integrin receptors that are present on the fibroblast surface, and cause clustering of these cell surface integrins. As matrix assembly proceeds, fibronectin-fibronectin interactions enable the soluble, cell-associated fibrils, to branch and stabilize into an insoluble fibronectin matrix [13-14].

Fibronectin assembly is strictly regulated *in vivo*. This is because up- or down-regulation of fibronectin secretion and assembly is linked with many physiological and pathological events [15-17].

For example, inhibition of fibronectin fibril formation is associated with vascular leakage as well as augmented tumor growth by disturbing the normal development and vascularization of tissue/organs. On the other hand, an excessive deposition of matrix fibronectin in response to growth factors such as TGFβ is typical for fibrosis and, in consequence, various ECM related clinical conditions such as, scleroderma or pulmonary fibrosis.

To be able to understand ECM formation and comprehend the complex dynamics of that structure, a number of methods were developed in the last several years to study fibronectin secretion and assembly [18-20].

Below we provide easy to follow methods supported with clear descriptions of the principle behind the experiment including materials required and detailed procedures to perform the experiments:

1. Study of Fibronectin Secretion and Assembly by Fibroblasts *In Vitro*

1.1. Detection of Fibronectin Secretion In Vitro

There are two methods to quantitatively and/or qualitatively determine secretion of fibronectin by cells: by Western Blot analysis and by ELISA method. These analyses are usually performed to detect the secreted soluble fibronectin from the cell conditioned medium for a given period of time in the presence and absence of growth factor stimuli.

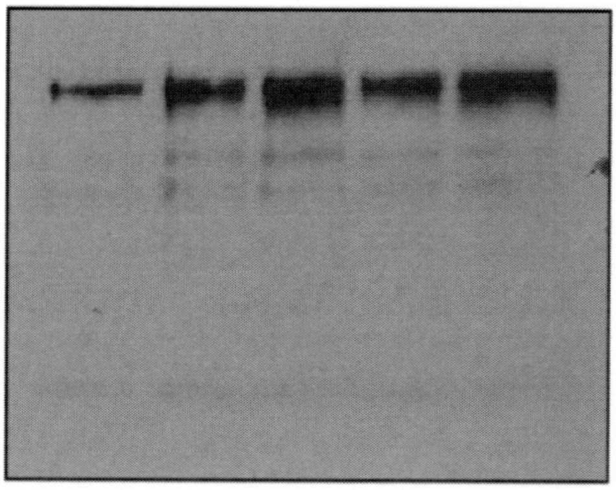

Figure 1. Western blot showing enhanced secretion of fibronectin by the fibroblasts in the conditioned media in response to different doses of TGF-β. NIH 3T3 cells were treated with 0, 25, 50, 100, and 250 pM of TGF-β for 16 h. Conditioned media was collected, subjected for TCA precipitation and analyzed by Western blotting fusing antibodies against fibronectin. Multiple bands of various multimer sizes and products of proteolytic degradation are seen on the blot.

1.1.1. Determination of Fibronectin Secretion by Western Analysis [7, 21-25]

Materials

1. *Fibronectin precipitation:* 100 % trichloroacetic acid TCA (Fisher Scientific), ice cold acetone (Sigma), 1X PBS. 100% TCA is prepared by dissolving 500g of TCA in 350 ml of dH_20.
2. *Sodium dodecyl sulfate-Poly acrylamide gel electrophoresis (SDS-PAGE):* 6 % polyacrylamide gel, 1X running buffer (0.25 M Tris, 1.92 M Glycine, 0.1 % SDS, pH=8.3)
3. *Electro-transfer:* PVDF (Polyvinylidene fluoride) membrane 10 cm X 10 cm (BioRad), 1X transfer buffer (0.25 M Tris, 1.92 M Glycine, pH=8.3)
4. *Immunodetection of fibronectin:* Primary antibody against fibronectin (Sigma; 1:2500 dilution), HRP-conjugated secondary antibody against primary antibody (BioRad; 1:5000 dilution), blocking solution (BioRad; 5 % non-fat milk in 1X TBST), washing solution (1X TBST: 1X TBS, 0.1 % Tween-20), and developing solution (peroxidase solution and luminol substrate e.g. commercially available chemiluminescent substrate for detection HRP such as: SuperSignal® West Pico from Thermo Fisher or Immobilon® Western from Millipore)

Method

1. *Fibronectin precipitation:*

   Plate the fibroblasts in the amounts of 2-5x10^6/well on 6-well plates, 24-48 h before the collection of conditioned medium. Replace serum-containing medium with 1 ml/well of serum-free medium and treat fibroblasts (or other cell types of study such as endothelial cells) with necessary growth factors or a drugs of interest.

Collect conditioned serum-free media from cells and centrifuge for 10 min at 13,000 rpm in an Eppendorf refrigerated micro-centrifuge (12,000 g) to pellet and remove the cell debris.

Carefully remove the supernatant by pipetting, and transfer it to a new tube; discard the pellet. Add 1 volume of 100 % TCA to 4 volumes of collected conditioned medium (1 to 5 dilution; i.e. add 250 μl of 100 % TCA to 1 ml of conditioned serum-free medium) and incubate overnight at 4°C. Centrifuge the sample for 10 min at 13,000 rpm (12,000 g) and discard the supernatant, leaving the protein pellet intact. Care should be taken to prevent the loss of any precipitated proteins, which are largely adhered to the sides of the tube and form a white layer. Next, wash the pellet twice with 250 μl of ice-cold acetone, following centrifugation for 20 min at 13,000 rpm (12,000 g). Dry the pellet at room temperature (RT) (approx. 2-5 min) and add 10 μl of 10X PBS. Finally, add 50 μl of Laemmli sample buffer with β-mercaptoethanol and boil the samples for 5 min at 95 °C. Load 30 μl of sample on resolving 6 % polyacrylamide gel.

2. *SDS-electrophoresis:*

   Note: This SDS-PAGE method is designed for a vertical slab gel with a maximum size of 1.5 mm X 7.5 cm X 9.0 cm. For thicker gels or mini-gels, the volumes of loaded sample and the operating current must be adjusted.

   Prepare 6 % resolving gel and overlay with 4 % stacking gel. Leave the gel for 20-30 minutes to allow the polymerization and solidification. Load 30 μl of sample (as described above) in two sets and run having set up the power supply on 100 volts for 1.5 h or until the blue dye reaches the bottom of the gel. Do not allow the dye front to go out of the gel.

3. *Wet electro-transfer:*

   Note: While the gel is running, prepare the following:

   a) Put filter papers and transfer pads in 1X transfer buffer for 30 min
   b) Next, soak the PVDF membrane in methanol for 20 second
   c) Then immerse it in 1X transfer buffer for about 1 min.

   After completion of the run, take out the gel and stain one part of run sample set in Coomassie staining solution (40 % methanol, 10 % glacial acetic acid, 0.1 % Coomassie R250) for 1h and insert the second inside the 1X transfer buffer for electrode-transfer.

   Pre-wet the PVDF membrane with methanol for 20 seconds, and then wash the PVDF in 1X transfer buffer. Take out the transfer cassette, pre-soaked transfer pad, filter papers, and gel and construct the sandwich carefully, avoiding air bubble formation. Next, place the aforementioned sandwich inside the transfer tank filled with 1X transfer buffer. Connect the cathode and anode end of the lid of the tank to the power pack to their respective points. Run the transfer at 210 milliamps for 1.5 h. Meanwhile, de-stain the Coomassie stained gel in de-staining solution (30 % methanol, 15 % glacial acetic acid and water) until you observe clearly visible bands.

4. *Immuno-detection:*

   After transfer, block the PVDF membrane by incubating it for 30 min in blocking buffer (approx. 10 ml) to saturate nonspecific protein binding sites. Then, discard the blocking solution and wash the PVDF membrane with 1X of washing buffer (approx. 10 ml) with 2 changes.

Meanwhile prepare the primary antibody solution by adding 5 µl of primary antibody in 10 ml of blocking buffer. Incubate the PVDF membrane at 4°C overnight, if possible, with gentle shaking. Discard the primary antibody solution and wash the PVDF membrane 4 times for 1 h with 4 changes (at an interval of 15 minutes each) in 1X washing buffer (approx. 20-30 ml each wash). In the meantime, prepare the HRP-conjugated secondary antibody solution by adding 2.5 µl of secondary antibody in 10 ml of blocking buffer. Incubate the strips with secondary antibody at room temperature for 60 min, with gentle shaking. Discard the secondary antibody solution and wash the PVDF membrane for 1 h with 4 changes at an interval of 15 min in 1X of wash buffer (with approx. 20-30 ml each wash). During the last wash, prepare substrate solution.

5. *Developing:*

   Take 1ml of each of substrate solutions (stable peroxidase solution and luminol substrate) and pour all 2 ml on PVDF membrane. Make sure that the entire PVDF membrane is covered with the chemiluminescent solution. Leave for 5 min and develop using autoradiography film (e.g. Kodak or HyBlot CL). Protein bands can be normalized either to the Coomassie-stained membrane or densitometry can be performed.

1.1.2. Determination of Fibronectin Secretion by ELISA Method [7, 23-24]

Materials

Primary antibody against fibronectin (1:10000 dilution), HRP-conjugated secondary antibody against primary antibody (1:10000 dilution), blocking solution (1 % BSA in 1X TBST), washing solution (1X TBST: 1X TBS, 0.1 % Tween-20), developing solution (TMB substrate commercially available from Sigma), stopping solution. (2 % phosphoric acid)

Method

1. *96-well plate preparation:*

   Coat the 96-well plate with the conditioned medium collected from the cells (as performed in the previous protocol) for 1 h. Next, block the wells with blocking solution (1 % BSA in 1X TBST) for 1 h at RT.

2. *Immuno-detection:*

   Aspirate the blocking solution and add primary antibody (i.e. 1 µl of primary antibody in 10 ml of blocking buffer; 1:10000 dilution) incubate overnight at 4°C. Wash the well three times (10 min intervals) at RT with 1X TBST and block once again with 1 % BSA in 1X TBST for 10 min. Next, incubate the well with HRP-conjugated secondary antibody (i.e. 1 µl of primary antibody in 10 ml of blocking buffer; 1:10000 dilution) for 1 h at RT and wash 5X (10 min intervals) using 1X TBST.

3. *Developing:*

   Add 50 µl of 1X TBS to each well followed by 50 µl of TBM substrate reagent. Next, incubate the wells for *exactly* 30 min at RT and end the reaction by adding 50 µl of stopping solution (2 % phosphoric acid). Read the wells at 480 nm. Perform each experiment in triplicate. Normalization can be performed by comparing either to the control or to the standard curve with known concentration of fibronectin used to coat the well.

1.2. Study of Fibronectin Assembly by Fibroblasts *in Vitro* [24-27]

Materials

1. *Cells:* (e.g. fibroblasts, endothelial cells, cancer cells etc.)
2. *Immuno-staining:* fixing solution (4 % parafomaldehyde in 1X TBS), primary antibody against fibronectin (1:1000 dilution), blocking solution (5 % BSA in 1X TBS), washing solution (1X TBS)
3. *Detection:* Rhodamine- or FITC- conjugated secondary antibody (i.e. Alexa Flour-labeled antibody against primary antibody, 1:500 dilution, both available from Invitrogen), Fluorescent mounting medium (4,6-Diamidino-2-phenylindole-containing Vectashield; Vector Laboratories, Burlingame, CA)
4. *Sterile glass cover-slips:* 18 mm coverslips from Fisher (Cat # 12-545-84) cleaned with 70% Alcohol

Method

1. *Plating cells:*

   Plate the fibroblasts at 0.5-1x10^7/well on sterile glass cover-slips without matrix coating in 12-well plates, 24-48h before immuno-detection. Replace serum-containing medium with 0.5 ml/well of serum-free medium and treat fibroblasts (or other cells types) with growth factors or the drugs of interest. After incubation, remove the conditioned medium, gently wash the confluent cell monolayer 3 times with 0.5 ml of 1X TBS and fix with 0.5 ml of fixing solution for 30-60 min. Wash once the fixed cell monolayer carefully with 1 ml of 1X TBS.
2. *Immuno-staining:*

   Block the fixed cells with 1 ml of blocking buffer for 1h at RT and wash the cover slip 3 times with 1 ml (each wash) of washing solution. Carefully remove the entire washing buffer and add 0.5 ml anti-fibronectin primary antibody solution (i.e. 5 µl of primary antibody in 5 ml of blocking buffer). Incubate the cover slip overnight at 4°C, and then wash 3 times with 1 ml (each wash) of washing solution. Next, gently aspirate the entire washing solution and incubate with 0.5 ml of Alexa Fluor- labeled secondary antibody for 1h at RT either in darkness or under dim light. Repeat washing under the same conditions as described above. Remove the cover slip and dry by touching corners with 'Kim-wipes'. Wipe very carefully (*DO NOT* touch the glass surface). Mount the cover slip with fluorescent mounting medium containing DAPI (Vectashield from Vector laboratories) and leave to dry at RT in the dark. If microscopic analysis is not performed immediately, store slides at 4°C in the dark by covering the slide-box with aluminum foils.

3. *Immuno-detection of assembled fibronectin matrix:*

   Photograph the slides using a fluorescence microscope. Fibronectin mesh-work will be visible as green (FITC) or Red (Alexa Red), depending upon the nature of fluorescent probe (Figure 2). Representative areas of the field can be quantified by many commercially available software such as Image-Pro Plus, Kodak imaging software or Image J program from NIH. The assembled matrix density can be quantified at 20X objective and expressed as area per field. Four fields should be quantified for each section of control slide and treated slide. The value of quantified fibronectin matrix may be normalized to the number of cells per field as visible by the DAPI stained blue nuclei in the Figure 2.

2. Analysis of Fibronectin Secretion and Assembly by the Fibroblasts *In Vivo*

2.1. Detection of Fibronectin Secretion by Fibroblasts in Vivo [21-24]

Detection of secreted fibronectin can be performed in tissue biopsies using either Western blot or ELISA method as described above (see section 1.1.1 and 1.1.2 above) with minor modifications provided below. These modifications also apply if the detection of fibronectin is to be carried out in cell lysates *in vitro.*

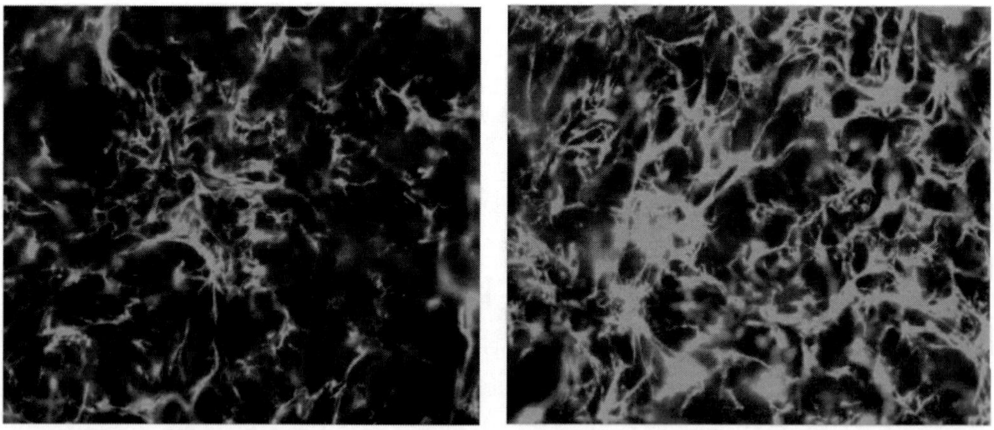

Figure 2. Fluorescent images of fibronectin meshwork created by the NIH 3T3 fibroblasts visualized using fluorescence microscopy. Freshly prepared NIH 3T3 cell monolayers were incubated for 4 hours in the presence and absence of 20ng/ml fibroblast growth factor and 10μg/ml extracellular soluble fibronectin. Monolayers were then fixed in 2% paraformaldehyde and stained with anti-fibronectin antibody to study matrix assembly. Assembled fibronectin fibrils are shown in green fluorescence with Alexa Fluor. Nuclei are stained in blue with DAPI.

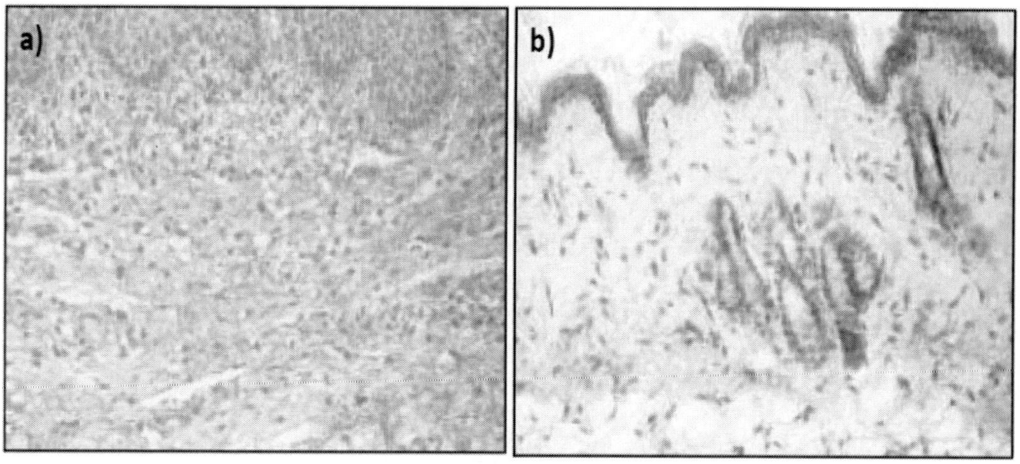

Figure 3. Immunohistochemical images of extracellular matrix skin tissue. a) Human skin section with fibronectin stained using DXPMAHT4C antibodies (adapted with permission from www.drixmed.com); b) Mouse skin section with fibronectin stained using DXPRAMF antibodies (adapted with permission from www.drixmed.com). Stained fibronectin is seen in brown staining as indicated by the arrows.

Materials

1. *Tissue lysis:* 0.5 cm X 0.5 cm biopsy of tissue of interest, RIPA buffer (20mM Tris-HCl, pH7.4; 1 %Triton X-100, 5mM EDTA, phosphatase inhibitors [10mM sodium pyrophosphate, 5mM sodium orthovanadate, 5mM sodium fluoride], protease inhibitor cocktail from Roche Diagnostics, Basel, Switzerland)
2. *SDS-PAGE:* 6 % polyacrylamide gel, 1X running buffer (0.25 M Tris, 1.92 M Glycine, 0.1 % SDS, pH=8.3)
3. *Electro-transfer:* PVDF membrane 10 cm X 10cm, 1X transfer buffer (0.25 M Tris, 1.92 M Glycine, pH=8.3)
4. *Immuno-detection:* Primary antibody against fibronectin (1:2500 dilution), HRP-conjugated secondary antibody against primary antibody (1:5000 dilution), blocking solution (5 % non-fat milk in 1X TBST), washing solution (1X TBST: 1X TBS, 0.1 % Tween-20), developing solution (peroxidase solution and luminol substrate e.g. commercially available chemiluminescent substrate for detection HRP such as: SuperSignal® West Pico from Thermo Fisher or Immobilon® Western from Millipore)

Method

1. Tissue lysis: Collect the 0.5 cm X 0.5 cm biopsy of tissue (lower sample size may be used, but the amount of RIPA buffer has to be adjusted) and freeze either in -80 °C for 30 min or liquid nitrogen for 30-60 sec. Take the frozen tissue sample and place it in ceramic grader and add 250 μl of RIPA buffer followed by pulverization of the sample with pestle until the tissue is completely homogenized (smooth, gravy-like texture). Pipette the entire sample into Eppendorf- tube and add additional 100 μl of RIPA buffer to the tissue lysate to collect the remaining homogenized sample.

Next, centrifuge sample for 15 min at 13,000 rpm (12,000 *g*) and save the supernatant at 4°C (should be clear and transparent; if not repeat centrifugation under the same conditions). A layer of while effervescence may be present on the top of the solution, which is a lipid rich fraction. In order to collect lipid free supernatant, introduce a 200 μl pipette into the solution piercing through the lipid layer by holding the dispenser down up to the threshold and release the dispenser to collect and transfer the supernatant to a fresh Eppendorf tube. Discard the pellet.

Finally, add 50 μl of Laemmli sample buffer with βME and boil the sample for 5 min in 95°C. Load 20 μl of sample on resolving 6 % polyacrylamide gel. Next, proceed according to steps 1 to 5 described in paragraph 1.1.1. to achieve immuno-detection.

2.2. Detection of Fibronectin Assembly by Fibroblasts in Vivo [28]

Materials

1. Frozen or paraffin sections of tissue of interest (e.g. skin, heart, lung, tumor etc).
2. Staining: fixing solution (10 % formalin), primary antibodies against fibronectin (1:1000 dilution), blocking solution (10 % normal serum in 1X TBST), and washing solution (1X TBST: 1X TBS, 0.1 % Triton X-100)
3. Detection: HRP-streptavidin secondary antibodies against fibronectin (1:500 dilution), peroxidase substrate solution (3, 3'-Diamino- benzidine - DAB, commercially available e.g. from Abcam)

Method

1. Preparation of sections:

   Paraffin sections

   Fix tissue with fixing solution for 24-48 hours at RT (fixative volume should be 5-10 times of tissue volume). Close the tissue in embedding cassette and process for paraffin embedding as follows (total 16h):

   - 70% ethanol, 2 changes, 1h each
   - 80% ethanol, 1 change, 1h
   - 95% ethanol, 1 change, 1h
   - 100% ethanol, 3 changes, 1.5h each
   - xylene, 3 changes, 1.5h each
   - paraffin wax (58-60 °C), 2 changes, 2h each

   Next, embed the tissue biopsy into paraffin, cut at 5 μm, and affix into slides. Allow section to air dry for 30 min and then place in 45-50 °C oven overnight. After that, deparaffinize sections as follow:

   - xylene, 3 changes, 10 min each
   - 100% ethanol, 2 changes, 3 min each
   - 95% ethanol, 2 changes, 1 min each

Then rinse the section in distilled water and keep the slide in the tap water until ready to perform staining.

Frozen Sections

Snap freeze fresh tissue in liquid nitrogen, embed in OCT compound in cryomolds and cut 5 μm thick cryostat sections followed by mounting on super frost plus slides or gelatin coated slides. If sections were kept frozen warm slides at RT for 30-60 minutes and fix in ice-cold acetone or 100-proof acetone for 2-5 minutes. Next, air-dry the slide for 30-60 minutes and wash once in 1X TBS.

Note: If staining is not be performed immediately, store slides at - 80 °C until needed.

2. Staining

   Wash the slides twice with washing solution for 5 min each. Next, incubate the section in blocking solution for 2h at RT in a humid chamber. Remove blocking solution by gentle tapping and wash 3 times with washing solution for 5 min each. Remove the washing solution and wipe around the sections with a paper wipe. Add primary antibody diluted in 1X TBST with 1 % BSA (approximately 200μl/ section). Incubate the section overnight at RT in humid chamber. Aspirate the entire antibody solution very gently, repeat the washing procedure and incubate slides in 0.3 % H_2O_2 in 1X TBS for 15 min at RT to block endogenous peroxidase activity. Repeat washing and apply biotinylated secondary antibody diluted in1X TBST with 1 % BSA onto the slide, and incubate for 1 h at RT. Wash the section 3 times with washing solution for 5 minutes each. Next, very carefully dry the section by touching its corners with paper wipe (DO NOT touch glass surface) and tapping. Incubate sections in HRP-Streptavidin solution for 30 min at RT. Next, rinse in washing solution 3 times for 2 min each and incubate section in peroxidase substrate solution. Rinse slide in 1X TBST 3 times for 1 min each. Then, dehydrate 1X through 95 % ethanol for 1 min and 2X in 100 % ethanol for 3 min. Finally, clear section in xylene 2X for 5min and cover slip with mounting medium.

3. Detection

   Representative sections can be photographed by Image-Pro Plus software. The matrix density can be quantified at 20X objective and expressed as area per field. Four fields should be quantified for each section of control section and treated section.

Acknowledgments

Research in the laboratory is supported by funding from the National Institutes of Health (R01HL103952), American Heart Association (0830326N) and University of Georgia Research Foundation to PRS.

References

[1] Alberts B, B.D., Lewis J, Raff M, Roberts K, Watson JD The Extracellular Matrix of Animals. *The Molecular Biology of the Cell,* 1996. 3rd edition(Chapter 19).

[2] Alberts B, B.D., Hopin K, Johnson A, Lewis J, Raff M, Roberts K, Walter P Tissues and Cancer. *Essential cell biology*. New York and London: Garland Science, 2004.

[3] Liotta, L.A., et al., Metastatic potential correlates with enzymatic degradation of basement membrane collagen. *Nature,* 1980. 284(5751): p. 67-8.

[4] Vlodavsky, I., et al., Extracellular matrix-resident growth factors and enzymes: possible involvement in tumor metastasis and angiogenesis. *Cancer Metastasis Rev.,* 1990. 9(3): p. 203-26.

[5] Ingber, D., Extracellular matrix and cell shape: potential control points for inhibition of angiogenesis. *J. Cell Biochem.,* 1991. 47(3): p. 236-41.

[6] Clark, R.A., Fibronectin matrix deposition and fibronectin receptor expression in healing and normal skin. *J. Invest. Dermatol.*, 1990. 94(6 Suppl): p. 128S-134S.

[7] Goc, A., et al., TGFbeta- and bleomycin-induced extracellular matrix synthesis is mediated through Akt and mammalian target of rapamycin (mTOR). *J. Cell Physiol.*, 2011. 226(11): p. 3004-13.

[8] Couchman, J.R., M.R. Austria, and A. Woods, Fibronectin-cell interactions. *J. Invest. Dermatol.,* 1990. 94(6 Suppl): p. 7S-14S.

[9] Adamson, I.Y., Drug-induced pulmonary fibrosis. *Environ. Health Perspect.,* 1984. 55: p. 25-36.

[10] Adamson, I.Y. and D.H. Bowden, The pathogenesis of bloemycin-induced pulmonary fibrosis in mice. *Am. J. Pathol.,*1974.77(2):p.185-97.

[11] Mosher, D.F., et al., Assembly of extracellular matrix. *Curr. Opin. Cell Biol.,* 1992. 4(5): p. 810-8.

[12] Schwarzbauer, J.E. and J.L. Sechler, Fibronectin fibrillogenesis: a paradigm for extracellular matrix assembly. *Curr. Opin. Cell Biol.,* 1999. 11(5): p. 622-7.

[13] Hynes, R., *Fibronectins.* Berlin: Springer-Verlag NY 1990. ISBN 0-387-97050-9.

[14] Hynes, R.O., The dynamic dialogue between cells and matrices: implications of fibronectin's elasticity. *Proc. Natl. Acad. Sci. U S A,* 1999. 96(6): p. 2588-90.

[15] Larsen, M., et al., The matrix reorganized: extracellular matrix remodeling and integrin signaling. *Curr. Opin. Cell Biol.,* 2006. 18(5): p. 463-71.

[16] Morrissey, C. and R.L. Vessella, The role of tumor microenvironment in prostate cancer bone metastasis. *J. Cell Biochem.*, 2007.101(4):p.873-86.

[17] Pasqualini, R., et al., A study of the structure, function and distribution of beta 5 integrins using novel anti-beta 5 monoclonal antibodies. *J. Cell Sci.,* 1993. 105 (Pt 1): p. 101-11.

[18] Bourdoulous, S., et al., Fibronectin matrix regulates activation of RHO and CDC42 GTPases and cell cycle progression. *J. Cell Biol.*, 1998. 143(1): p. 267-76.

[19] Sechler, J.L., et al., Modulation of cell-extracellular matrix interactions. *Ann. N Y Acad. Sci.,* 1998. 857: p. 143-54.

[20] Sottile, J., D.C. Hocking, and P.J. Swiatek, Fibronectin matrix assembly enhances adhesion-dependent cell growth. *J. Cell Sci.,* 1998. 111 (Pt 19): p. 2933-43.

[21] Laemmli, U.K., Cleavage of structural proteins during the assembly of the head of bacteriophage T4. *Nature*, 1970. 227(5259): p. 680-5.

[22] Burnette, W.N., "Western blotting": electrophoretic transfer of proteins from sodium dodecyl sulfate--polyacrylamide gels to unmodified nitrocellulose and radiographic detection with antibody and radioiodinated protein A. *Anal. Biochem.,* 1981. 112(2): p. 195-203.

[23] Somanath, P.R., J. Chen, and T.V. Byzova, Akt1 is necessary for the vascular maturation and angiogenesis during cutaneous wound healing. *Angiogenesis,* 2008. 11(3): p. 277-88.

[24] Somanath, P.R., et al., Akt1 signaling regulates integrin activation, matrix recognition, and fibronectin assembly. *J. Biol. Chem.,* 2007. 282(31): p. 22964-76.

[25] Chen, J., et al., Akt1 regulates pathological angiogenesis, vascular maturation and permeability in vivo. *Nat. Med.,* 2005. 11(11):p.1188-96.

[26] Kowalczyk, A.P., R.H. Tulloh, and P.J. McKeown-Longo, Polarized fibronectin secretion and localized matrix assembly sites correlate with subendothelial matrix formation. *Blood,* 1990. 75(12): p. 2335-42.

[27] Somanath, P.R. and T.V. Byzova, 14-3-3beta-Rac1-p21 activated kinase signaling regulates Akt1-mediated cytoskeletal organization, lamellipodia formation and fibronectin matrix assembly. *J. Cell Physiol.,* 2009. 218(2): p. 394-404.

[28] Ali, S., et al., Immunohistochemical detection of fibronectin in human thymus-comparison between groups of different ages. *Pak. J. Physiol.*, 2009. 5(1): p. 45-50.

In: Fibronectin
Editors: James Beattie

ISBN 978-1-62257-298-4

Chapter VII

Pathways Regulating the Secretion and Assembly of Fibronectin Matrix: An Update Methods to Study Fibronectin Secretion and Matrix Assembly by Fibroblasts *In Vitro* and *In Vivo*

Maha Abdalla[1,2], *Anna Goc*[1,2]
and Payaningal R. Somanath[1,2,3,*]
[1]Clinical and Experimental Therapeutics, College of Pharmacy, University of Georgia, Georgia, US
[2]Charlie Norwood VA Medical Center, Augusta, Georgia, US
[3]Department of Medicine, Georgia Health Sciences University, Augusta, Georgia, US

Abstract

Fibronectin (FN) is an essential extracellular matrix (ECM) protein that plays an important role in tissue repair both physiologically and pathophysiologically. The polymerization of FN into the matrix is a highly regulated process to ensure ECM accurately regulates various cellular functions including adhesion, differentiation, proliferation, migration and apoptosis and that it relays accurate signals intracellularly. FN, which is synthesized and secreted by different cell types, is organized into a fibrillar network through a highly regulated process termed fibrillogenesis. This process requires

* Address for correspondence: Payaningal R. Somanath, PhD - Clinical and Experimental Therapeutics, College of Pharmacy, University of Georgia, HM1200, Georgia Health Sciences University, Augusta, GA 30912. Tele: 706-721-4250; Fax: 706-721-3994. Email: sshenoy@georgiahealth.edu.

direct interaction with cell-surface adhesion receptors such as integrins and exertion of mechanical force to induce FN conformational change from the soluble, compact and non-functional state to the insoluble, extended and highly adhesive state. This process is mediated through the binding of the activated cell surface integrins, primarily α5β1 integrin, to the RGD (Arg-Gly-Asp) cell-binding domain on FN.

In integrating the extracellular environment with the interior of the cell, integrins have a unique ability to transduce signals via "outside-in" and "inside-out" signaling. Growth factors, particularly transforming growth factor beta (TGF-βTGFβ), modulate integrin activation in fibroblasts. TGF-βTGFβ controls the transcription of genes encoding integrins via binding to its receptor, leading to the phosphorylation of intracellular effector proteins Smad2 and 3 and activation of downstream signaling and translocation to the nucleus. The crosstalk between integrin and TGF-βTGFβ is essential for FN matrix assembly; however, the molecular mechanisms underlying FN secretion and assembly are not well known. Studies have shown that modulation of focal adhesion kinase (FAK), Rho-GTPases, protein kinase Bα (Akt1), mammalian target of rapamycin (mTOR), and MAP kinase pathways are among the regulatory mechanisms that play an integral role in FN secretion and assembly.

Abnormalities in the crosstalk between integrins, growth factors and FN is characterized by excessive differentiation and ECM remodeling resulting in aberrant response to tissue injury and contributing to a variety of pathological processes. The aim of this review is to describe the mechanisms and consequences of FN secretion and matrix assembly, and provide a brief overview of FN on its clinical impact. First, we describe FN structure and the important features responsible for interaction with integrins and subsequent cell functioning. Next, we focus on the matrix assembly and essential mechanisms underlying this process. Finally, we discuss how FN can drive pathophysiological processes in fibrosis and cancer.

Abbreviations

Akt	protein kinase B
ARF	ADP-ribosylation factor
Cav-1	caveolin-1
Cdc42	cell division cycle 42
EC	endothelial cell
ECM	extracellular matrix
EGF	epidermal growth factor
EMT	epithelial mesenchymal transitions
ERK	extracellular signal regulated kinase
FAT	focal adhesion targeting,
FGF	fibroblast growth factor
FN	fibronectin
FAK	focal adhesion kinase
GAP	GTPase-activating protein
GEF	GDP–GTP exchange factor
GSK3β	glycogen synthetase kinase 3 type β,
JNK	jun N-terminal kinase
MAPK	mitogen activated protein kinase
MEF	mice embryonic fibroblast

MLC	myosin light chain
MLCP	myosin light chain phosphatase
mTOR	mammalian target of rapamycin
MMP	matrix metalloproteinase
Src	steroid receptor co-activator
PAK	p21-activated kinase
PDGF	platelet derived growth factor
PI3K	phosphatidylinositol-3 kinase
PP2A	protein phosphatase 2A
PTEN	phosphatase and tensin homologue deleted on chromosome ten
PTKs	protein tyrosine kinases
PYK2	Proline-rich tyrosine Kinase 2
RGD	Arg-Gly-Asp cell-binding sequence
ROCK	Rho-associated kinase
RTK	tyrosine kinase receptor
Smurf	Smad –ubiquitination regulatory factor
TGFβ	Transforming growth factor β
VEGF	vascular endothelial growth factor

Introduction

The fundamental importance of the extracellular matrix (ECM) protein fibronectin (FN) can be inferred by its ubiquitous presence in tissues and by the non-viability of FN-deficient embryos. Under normal conditions, the ECM, specifically FN, is an important determinant of biological functions as illustrated by its high efficiency in promoting anchorage dependent cell cycle progression [1] and cell survival and proliferation [2]. Of the ECM proteins, adhesion to FN specifically stimulates cell cycle progression by the activation of RhoA in addition to growth factor mediated activation of Ras/MAPK signaling [1]. More interestingly, adhesion to FN decreases growth factor requirement for DNA synthesis by up to 1000-fold [3, 4] and selectively desensitizes specific growth factor stimulation in fibroblasts [5].

A principal component of the ECM, FN is especially abundant in tissues undergoing development or wound healing [6]. FN is critical for embryogenesis and is instrumental in various physiologic and pathologic processes. FN-rich matrices serve as regulators of cellular processes including proliferation, survival and differentiation, and provide structural support for cell adhesion and migration in ECM remodeling and tissue organization [6, 7]. Given the complexity and variability of its roles in basic cellular processes and binding partners, FN is involved in a large number of critical events including embryogenesis [8] where it has been shown to be pivotal for development as demonstrated by early embryonic lethality in mice with null mutation in FN gene.

Furthermore, FN is critical in wound healing processes, such as tissue morphogenesis [8, 9] thrombosis [10] and inflammation [11], and maintenance of normal tissue architecture [12, 13]. Abnormalities in FN, such as increased matrix deposition and assembly, have also been associated with several pathologies, including tumorigenesis and metastasis [14], fibrosis [15], and atherosclerosis [11]. Chronically elevated TGFβ1 levels are associated with excessive fibrosis which is characterized by increased FN deposition [16-19]. The hallmark of fibrotic disease is increased ECM protein synthesis by fibroblasts persistently stimulated by TGFβ1.

FN is a multi-domain adhesive glycoprotein that is multifunctional and is expressed and secreted by different cell types, but most particularly fibroblasts. FN regulates cell-ECM contact and cell-cell interactions by binding to different components of the ECM and to the membrane bound cell surface receptors, particularly integrins. FN that is secreted from a variety of cells is assembled into matrix by integrins, mainly α5β1, which is the primary integrin for FN matrix assembly [20]. Furthermore, cross talk between integrins and TGF-βTGFβ is an important modulator of FN transcription and assembly through a complex network of signaling pathways such as MAP Kinases and PI3K-Akt pathways [21].

Dynamic alterations in the structure and composition of the ECM, termed ECM remodeling, occur at different stages of development, during tissue repair and remodeling, and as a result of many pathologies and malignancies [22, 23]. Cells are continually integrating a plethora of signals to determine cell fate, survival and proliferation. It is in this light that the PI3K-Akt pathway can be considered as a central integrator of a tangled web of signaling networks with direct and indirect effects on each other. Thus, better understanding of FN regulatory mechanisms is crucial in order to develop novel therapeutics.

Although extensive research has established the role of MAP kinases in regulating FN mediated cell processes, we are only just beginning to understand the pivotal role of Akt [24-26]. Numerous studies have implicated the essential role of Akt, however the mechanisms by which Akt mediates and modulates FN induced signaling are just beginning to be explored. . Thus, the purpose of this review is to explore the molecular mechanisms mediating FN synthesis and matrix assembly, focusing on Akt, a major intracellular signaling protein.

Structure of FN

FN is considered a novel, highly conserved, multi-domain glycoprotein. While encoded by a single FN gene, *FN1* on chromosome 2q35, alternative splicing can generate as many as 20 isoforms in humans [17, 27]. FN is found in two different forms: "plasma FN or pFN", a soluble dimer in the circulation which is secreted by hepatocytes directly, and "cellular FN or cFN" which is synthesized locally in tissues by many cell types including fibroblasts and endothelial cells [19, 28]. A key distinctive feature of cFN is the inclusion of variable amounts of either or both extra-domain (ED)-A and ED-B, Type III modules subject to alternative splicing by exon skipping, also termed exon shuffling. Additionally, almost all cFN subunits contain the variable, V region (also termed Type III connecting segment (IIICS) element. In contrast, pFN lacks ED-A and ED-B and only one of its subunits contains the V region [17]. Embryos that lack both ED-A and ED-B show cardiovascular defects due to altered FN functions [29].

ED-A is suggested to play a central role in the differentiation of fibroblasts to myofibroblasts [30, 31]. Indeed, the importance of the EDA domain in wound healing and fibrosis were shown in EDA null mice, which demonstrated abnormal wound healing [32] and an absence of pulmonary fibrosis despite bleomycin treatment [15]. Thus, EDA cFN plays a critical role in fibroblasts activation and mediating fibrotic disease.

FN is secreted as a soluble dimeric glycoprotein linked by inter-chain disulfide bonds near the C-terminus. Each FN monomer, 220-250 kDa, is folded into five or six domains connected by a flexible polypeptide chain. Each domain is composed of a mosaic of secondary structure modular repeats: 12 type I modules (FN-I), 2 type II modules (FN-II) and 15 to 17 type III modules (FN-III), and a variable sequence (v) [16].

Within each fragment, there are several discrete domains responsible for the binding versatility of FN. C-terminal FN- I_{10-12} is a FN binding domain, also known as an assembly domain, necessary for FN – FN interaction and initiation of FN assembly.

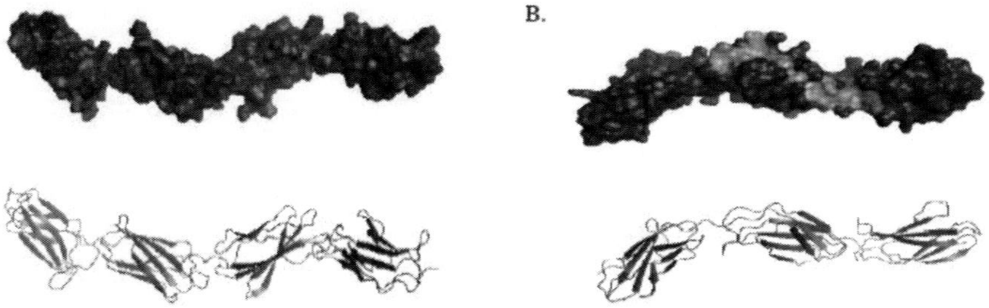

Figure 1. RGD site and integrin binding sites in human FN: A. RGD sequence (residues 1493–1496) (red). B. Integrin $\alpha_{IIb}\beta3$ (green) and α4β1 (red) binding motif, and binding motif containing Asp184 residue (yellow).

The cell binding domain of FN is composed of FN- III_{1-11}, including the alternatively spliced ED-A and ED-B modules, specifically FN- III_{10} and FN- III_9. Several sites within this domain play a role in FN matrix assembly and/or interact with integrin. FN- III_{10} contains the most critical and primary cell binding domain, Arg-Gly-Asp (RGD). This sequence is necessary for mediating the interaction between FN and α5β1 integrin; FN is the only adhesion molecule recognized by α5β1 integrin, which play an integral role in FN polymerization and matrix assembly [33-35]. FN also binds to other integrins including αvβ1, αvβ3 and αIIbβ3 integrins, but with a lower affinity [36, 37].

In addition, FN- III_9 is also important for interaction with certain integrins through the amino acid sequence 'Pro-His-Ser-Arg-Asn (PHSRN)' [37]. This sequence is known as a "synergy site", and is unique to FN [38] as it further enhances the binding affinity of α5β1 integrin for FN [39] (Figure 1). FN Type III_1 has self-association activities and a cryptic heparin binding sites [40]. The exposure of that matricryptic site is necessary for matrix FN to localize to lipid rafts, caveolae, where it regulates cytoskeletal organization and cell growth [41]. The variable composition of FN molecules facilitates a diversity of interactions with cell surface receptors that suggest a role beyond the structural considerations of the extracellular matrix.

Secretion, Polymerization and Assembly of FN

FN is synthesized and secreted by different cell types including fibroblasts, one of the most ubiquitous cell types in the body, along with endothelial cells and macrophages. Following FN protein translation, FN is folded, dimerized and glycosylated in the endoplasmic reticulum [42, 43]. FN then is secreted as a soluble compact globular dimer, a process mediated by electrostatic interaction, a regulatory mechanism likely to prevent intermolecular interactions and cell aggregation prior to assembly [44]. Emerging evidence indicates that the polymerized insoluble FN is functionally distinct from soluble protomeric FN. FN matrix assembly regulates cell growth, contractility, cell migration. FN-null mice embryonic fibroblasts (MEFs) fail to spread or proliferate on type I collagen gel, while addition of FN to these cells results in increased cell proliferation and micro-tissue formation [45]. FN matrix assembly plays an important role in preserving the structural integrity of blood vessels. FN regulates the composition and organization of the ECM. FN deposition regulates tenascin C fibrils, and the deposition of fibulin and fibrinogen in the ECM. Notably, FN secretion and FN polymerization regulate both the deposition and maintenance of collagen types I and III [46].

Cells exert mechanical force on their surroundings, and the mechanism varies according to stimuli, environment, morphology and differentiated state of cells. FN undergoes cell mediated assembly to form insoluble, elastic matrix fibrils. Cells can apply cytoskeleton-generated tension to unfold FN leading to changes in cell morphology, cell signaling and proliferation [38]. This mechanical force permit cells to change biochemical signals received from the environment, in a force- and time-dependent manner, by altering molecular recognition through unfolding and extending FN. This is achieved by exposing cryptic sites for FN-FN binding; changing the spatial positions of the RGD and synergy sites altering $\alpha5\beta1$ integrin binding selectivity and signaling pathways; a tensile molecular recognition switch achieved by unraveling of modules such as FN-III_{10}, which was shown to straighten the RGD loop, thus, reducing accessibility and affinity to cell surface integrins. FN-III_{10} is the first module to unfold in FN-III module repeats due to its low mechanical stability [38]. Cell mediated assembly of FN into fibrils is essential to fully exploit the mechano-chemical characteristics of FN to regulate cell signaling.

FN also exhibits mechanical properties that allow the coupling of cytoskeletal generated tension to modulate fiber directionality and extension during polymerization. The mechanical coupling to the cytoskeleton exposes cryptic sites on FN that affect its affinity and selectivity to different members of the integrin family. Unmasking of previously unrecognizable receptor-binding sites may result from conformational changes due to proteolytic cleavage within a domain of an ECM molecule. Different molecular effects resulting from proteolytic cleavage of an ECM protein by matrix metalloproteinase 9 (MMP9) may lead to the exposure of a previously hidden, or 'cryptic', binding site for integrin receptors, such as an RGD-containing binding site for a $\beta1$ integrin receptor [47]. FN matrix assembly regulates cell morphology and functions by regulating the interaction of integrin with the actin cytoskeleton. The translocation and formation of $\alpha5\beta1$ integrin-containing fibrillar adhesion requires FN matrix assembly [46] and occurs via a tensin-dependent mechanism [48]. Early FN fibrillogenesis requires $\alpha5\beta1$ integrin translocation from focal contacts to the ECM contacts a process dependent on tensin, a primary cytoskeletal component of ECM contacts and which is also an actin binding protein. It has been suggested that loss of cell-substrate adhesion does not result in loss of FN matrix. Several studies have observed that despite disrupting FN matrix, cells remained attached and spread, and expressed well-developed stress fiber and focal contacts [46]. Studies have shown FN matrix turnover process to involve the release of proteases into the ECM and may also include FN endocytosis [49]. Changes in cytoskeletal dynamics are suggested to be critical for FN matrix turnover. Indeed, a dynamic, reciprocal relationship exists, where disruption of actin cytoskeleton disrupts FN matrix organization, and FN polymerization regulates actin cytoskeletal assembly. FN mechanically couples the ECM of cells to the cytoskeleton via integrin; mechanical forces are among the cues that determine cell shape and ultimately gene expression and eventual fate [38]. Integrin mediated anchorage to the ECM is required for cell proliferation by initiating signaling pathways that are co-regulated by growth factors. Integrins mediate a "dynamic reciprocity" between cell-cell and cell-ECM [77].

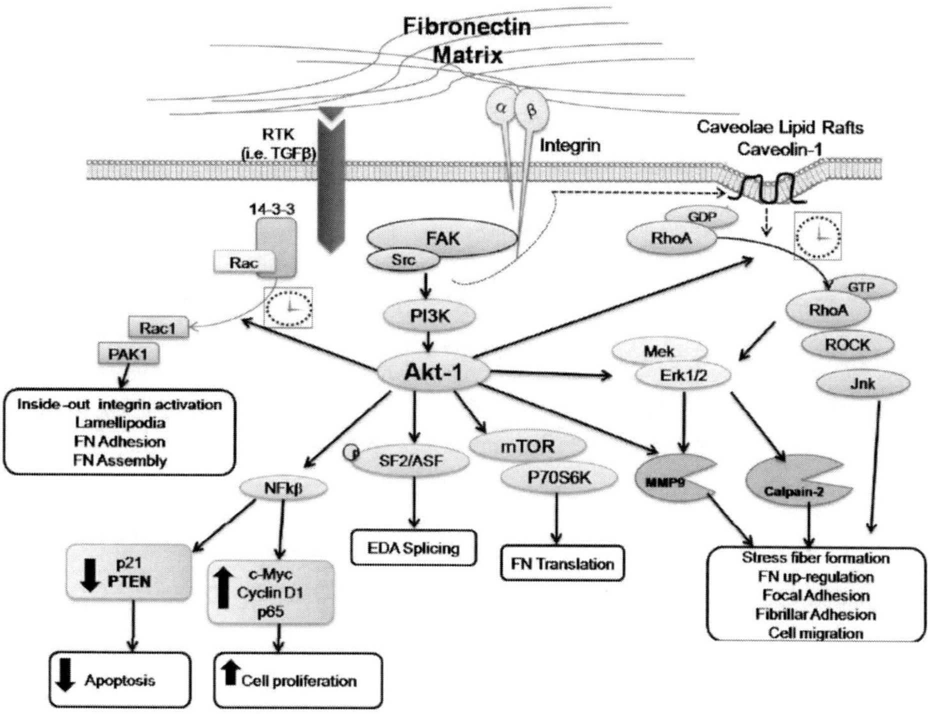

Figure 2. Schematic representation of the role of Akt1 in mediating the dynamic reciprocity between FN and integrins: Upon integrin- and/or RTK-mediated cell-FN attachment, FN induces integrin activation and clustering into focal adhesions, a process that requires the presence of FAK-Src complex. FAK/Src complexes serve as mechanosensors that converge extracellular signaling into the cytoskeleton by cross-linking signaling from integrins and RTKs, via proximal interactions. In certain cells (e.g. embryonic stem cells), activation of caveolin-1 by FN/integrin/FAK-Src is necessary for cell survival and proliferation. PI3K/Akt dynamically regulates FN-integrin interaction. FN ligation can induce the activation of Akt1, outside-in signaling, through β1 integrin cytoplasmic domain and FAK-Src complex. Conversely, Akt1 mediates FN synthesis and matrix assembly via inside-out signaling pathways. *FN stimulates its own synthesis and matrix assembly*. Akt1 enhances FN translation and ED-A alternative splicing via the activation of mTOR/P70S6k pathway and direct phosphorylation of SF2/ASF, respectively. FN is exported to ECM where it binds integrins, and Akt1 enhances inside-out activation and accumulation of integrins in focal adhesion through its cross talk with RhoA (contraction in lagging edge) and Rac (protrusion in leading edge) at different stages during cell motility and integrin clustering, simultaneously. Thus, enhanced integrin affinity for FN generates mechanical forces that reveal cryptic sites on FN necessary for FN matrix assembly. Furthermore, FN regulates cell survival and proliferation through Akt1 mediated down-regulation of PTEN and up-regulation of c-Myc and cyclin D1. *FN regulates matrix turnover*. FN binding to integrin activates Akt1 and Erk pathways that up-regulate calpain-2 and MMP9 that are necessary for de-adhesion during cell motility and matrix degradation.

Molecular Mechanisms Regulating Secretion and Assembly of FN Matrix

Although the molecular mechanisms governing FN secretion and assembly are not fully understood, a hierarchy is evident in the assembly process. FN interaction with integrins and polymerization into the ECM is influenced by several growth factors and signaling proteins [50]. Transforming growth factor β (TGFβ), an extracellular growth factor, enhances secretion of ECM proteins including FN, increases the number of matrix assembly sites and enhances assembly of fibronectin matrix [51]. Conversely, an increase in intracellular cAMP levels decreases the number of assembly sites and inhibits FN polymerization [52]. Other signaling molecules that have been implicated in the regulation of FN matrix assembly include Src kinase, phosphatidylinositol 3-kinase [50], other growth factors [51], and protein kinase C [53].

Growth factors such as TGFβ [54-56] and cytokines [57] can regulate FN synthesis at the transcriptional level and translational level, through modulation of mRNA stability or translational efficiency. Regulation of FN synthesis also occurs at the level of protein degradation. Turnover of the FN matrix *in vivo* occurs relatively quickly, with as much as half of labeled FN that was incorporated into the ECM being lost within 24 hours [58]. ECM degradation and turnover can be mediated by many proteases, including plasmin, plasminogen activators, and MMPs [59], and by caveolin-dependent endocytosis [60] and intracellular proteolysis [61].

Integrin-FN Interactions in Cellular Adhesion and Assembly of FN Matrix

Integrins, α-β heterodimeric transmembrane molecules, are the principal receptors for ECM [34]. The α5β1 integrins, "classic" FN receptor, is the major integrin receptor responsible for initiating FN matrix assembly. Other integrins can bind FN but do not induce FN assembly [48]. In the absence of RGD-α5β1 integrin interaction, some FN matrices can still be assembled, however less efficiently, by interacting with α_v integrins, such as αvβ3, or αIIbβ3 [18]. Binding of FN induces integrin clustering, further increasing their affinity to FN [18, 48]. More generally, an individual integrin will recognize several distinct proteins [34]; for example, the αvβ3 integrin has been reported to bind collagen VI, laminin, FN, vitronectin, thrombospondin, Von Willebrand factor, and fibrinogen [62]. The β cytoplasmic domain is essential for integrins recruitment to focal contact as evident by impaired recruitment due to its truncation/mutation. The β cytoplasmic domain alone, expressed as a fusion chimera with another membrane protein, has been shown to be sufficient for focal contact localization [63, 64]. The α cytoplasmic domain serves as an auto-inhibitor of certain functions of the β cytoplasmic domain such as focal contact recruitment. Ligand binding induces conformation changes that relieve this auto-inhibition [18, 48]. Furthermore, the β cytoplasmic domain is important in signal transduction, particularly integrin activation of FAK whereas truncation or mutation of the α cytoplasmic domain does not impact this process [65].

Integrin α5β1 is critically important in vertebrate development, as demonstrated by the early embryonic lethality of null mice [66]. Notably, this phenotype is even more severe and occurs at an earlier stage than that observed in FN-null mice. Interestingly, α5 knock-out mice are embryonic lethal despite their ability to develop normal FN matrix [67, 68]. This suggests that α5β1-FN interaction conveys critical signals in development that are not compensated by other integrins including αv, thus further illustrating the importance of this integrin and FN in embryogenesis . Furthermore, different cell types depend on different integrins for survival. For example, α5β1 integrin promotes survival in fibroblasts, endothelial cells (ECs) and epithelial cells [69, 70], and αvβ3 integrin is important for angiogenic ECs survival [71]. It has also become apparent that integrin mediated signaling pathways can be integrin specific. Whereas α5β1 integrin mediated adhesion to FN supports cell survival by up-regulating Bcl-2 and inhibiting Bad, caspase-9, and p53 through activation of FAK, PI3K/Akt, and CAMK IV signaling [69, 72-74], αvβ1 is inefficient in supporting FN matrix assembly and cell migration [75]. Cell adhesion is essential for embryonic development and tissue function and repair following injury [76] and is essential for maintenance of tissue integrity and polarity. Integrin mediated adhesions are known to undergo dynamic changes in structure and molecular composition from focal complexes to early focal adhesions- stress fiber associated focal contacts.

Integrins drive early FN fibrillogenesis, thus providing direct "dynamic reciprocity" [77] between the ECM and the cytoskeleton [48]. Inactive integrins, which are diffusely distributed in the plasma membrane, are recruited into focal contacts where they can be re-activated [18]. Cytoskeletal generated tension can occur through two types of integrin-containing structures, focal contact and focal/fibrillar adhesions [48, 78]. Focal contact mediates transmembrane linkage between the ECM and the cytoskeleton, and is composed of structural, cytoskeletal, and signaling proteins, where talin and α-catenin bind the β-integrin cytoplasmic tail domain. Soluble, compact FN dimer selectively binds to integrin via the RGD and synergy sites simultaneously, thus, initiating FN polymerization and fibril matrix assembly [46].

Integrins modulate the activity and localization of Rho-GTPases and their effector proteins that in turn regulate actin cytoskeletal dynamics that govern the morphological changes during cell adhesion and spreading on ECM [79]. Rac1 and Cdc42 stimulate the formation of "focal complexes", small cell matrix adhesions, associated with lamellipodia and filopodia [80]. RhoA stimulates actin stress fibers formation and the maturation of focal complexes to focal contacts and fibrillar adhesions [81, 82].

Fibrillar adhesions, also termed cell-matrix adhesions, contain tensin and α5β1 integrins, and co-localize with FN fibrils [46, 83]. Tensin is one of the major cytoskeletal proteins in fibrillar adhesions [78, 84]. Tensin has multiple actin-binding sites [85] and associates directly with the β1 integrin cytoplasmic tail via its phosphotyrosine-binding domain [86, 87]. There is evidence that suggests tensin is required for the formation of fibrillar adhesions [48]. In addition, when FN matrix assembly is blocked, fibrillar adhesions are not generated and tensin resides in focal adhesion plaques [46], suggesting that fibronectin regulates the distribution of tensin.

In addition to mediating adhesion to ECM, integrins are important in ECM assembly. Integrin connection to the actin cytoskeleton is essential for FN fibrillogenesis. FN fibrillogenesis requires integrin-mediated binding to soluble FN dimers and conformation changes in FN mediated by Rho-dependent cytoskeletal contractility to stretch integrin-bound FN molecules exposing cryptic FN binding sites [88-90]. The specific translocation of α5β1 integrin together with tensin from focal contacts to another cell-matrix adhesion "fibrillar adhesions" is of critical importance in FN matrix assembly [48, 78]. A fragment of tensin inhibits both the translocation of α5β1 integrin and FN fibrillogenesis without affecting focal contacts [48], thus demonstrating that the high efficiency with which α5β1 modulates FN is partly due to its association with tensin.

Cell-cell adhesion is mediated by an analogous process but with different players which include tight junctions and adherens junctions, which include cadherin mediated adhesion [91]. The adhesive functions of both integrins and cadherins are well established and are known to require cytoskeletal interactions [92]. However the dynamic interplay between them is not well understood. Lefort and colleagues have recently identified α5β1 integrin and tensin as novel components of N-cadherin adhesion complexes and demonstrated the regulatory role of matrix FN on cell-cell and cell-ECM dynamics. In the absence of FN, α5β1 integrin and tensin localize to cell-cell adhesion complexes Ligation of FN to α5β1 integrin triggers the dissociation of tensin from N-cadherin and promotes tensin association with α5β1 integrin. The α5β1 integrin and tensin complex then translocates to the ECM fibrillar adhesions [93]. Thus, FN matrix assembly coordinates the interaction between cell-ECM and cell–cell adhesion.

Integrin-mediated adhesion to matrix proteins such as FN and collagen results in activation of complex pro-survival signal transduction pathways. Cross linking of different integrin heterodimers initiates different signaling pathways [94]. Mutual and cooperative crosstalk of integrins with growth factor receptors such as TGF-βTGFβR regulates intracellular signaling pathways such as FAK/Src, MAPK, PI3K signaling and the activity of Rho family of small GTPases [95].

Integrin-Growth Factor Cross-Talk in the Assembly of FN Matrix and Fibrogenesis

Integrins can also regulate cell-ECM interaction through their association with growth factor receptors [96]. There is extensive cross-talk between integrins and growth factors that modulate mitogenic signaling; For example, whereas an acute and robust activation of Erk-MAPK signaling occurs in response to growth factor stimulation, FN adhesion leads to milder Erk activation with slower kinetics [79, 97].

Several integrins can regulate TGFβ signaling directly by TGFβ activation, or indirectly through regulating SMAD dependent and independent signaling. The importance of the synergy between growth factors and integrin has been demonstrated best in cell migration through FAK/Src signaling. FAK (focal adhesion kinase) a non-receptor tyrosine kinase, has been shown to act as a receptor proximal regulatory protein that directly integrates growth factor and integrin signals.

FAK binds though its C-terminus, to the ECM-ligated integrins induces autophosphorylation at tyrosine 397 (Y397) which leads to binding to another tyrosine kinase, Src, which in turn phosphorylates FAK at tyrosine 861 (Y861). In addition to the localization through the C-terminus into the integrin adhesion complex, FAK proximally regulates and directly interacts with the cytoplasmic domain of cognate growth factor receptors, thus conveying both integrin and growth factors mediated signaling [98].

TGFβ Signaling in FN Secretion and Matrix Assembly

TGFβ is crucial due to its important role as a transcriptional regulator where it stimulates the production of ECM proteins including FN and tissue inhibitors of metalloproteinases, and decreases the synthesis of FN degrading MMPs.

TGFβ superfamily ligands are secreted morphogenes, some of which are crucial regulatory cytokines that are essential in activating myriad of signaling cascades that modulate diverse cellular processes in embryonic development and adult tissue homeostasis [99].The source of the activity was identified as two distinct factors, TGF-α and –β Roberts [100-102]of which TGF- α belongs to the epidermal growth factor (EGF) family of growth factors. In mammals, TGFβ has three different isoforms TGFβ -1, -2, and -3 that are known to function as modulators of ECM proteins by induction of both gene activation and protein formation. Studies showed that mechanical strain/loading/stretch induces TGFβ expression and FN interaction.

TGFβ 1, the prototypical cytokine, acts primarily through the SMAD protein to stimulate fibroblast activation and fibronectin deposition. Mature TGFβ remains associated with latency associated peptide (LAP), thus inhibiting TGFβ activation [103]. Upon release from the latent complex, mature dimeric TGFβ ligand binds to a serine-threonine transmembrane kinase, TGFβ Receptor type II (TGFβRII), which is intrinsically active. This is followed by the recruitment and phosphorylation of TGFβR type I (TGFβRI) kinase resulting in the formation of a heterodimeric serine/threonine kinase receptor complex and propagation of signal cascade via downstream Smad proteins. Activated TGFβRI initiates the canonical signaling pathway where it directly phosphorylates and activates effector proteins Smad 2 and Smad 3, also known as receptor-regulated Smad (R-Smad) which are regulators of transcription. Smad-2 and -3 are anchored by the cytoskeletal protein filamin A. Phosphorylated Smad 2/3 undergo conformational changes through unfolding and subsequently forming a heterotrimeric complex with Co-Smad4 and translocating into the nucleus. Smad4 is not crucial for TGFβ activity [104], and nuclear import of Smads does not require Smad4. The TGFβR complex remains active for at least 3 to 4 hours after ligand binding and continuous receptor stimulation maintains Smad complex accumulation in the nucleus where they regulate gene transcription. Smads interacts with transcriptional co-activators such as CBP, p300 or ARC105 to regulate the expression of target genes. Smad proteins can mediate the activation or repression of target genes, including those that encode integrins; a process regulated by cell type and Smad protein associated partners [99, 105, 106]. Initially, studies have identified Smad target specific genes through direct binding to DNA. However, it has become apparent that Smads act as transcriptional co-modulators and require the association with other transcription factors such as ATF/CREB, AP-1, FAST and SP1 to mediate transcriptional activation or repression of different genes [107-110].

TGFβ signaling is negatively regulated by inhibitory Smads (I-Smad), Smad 6/7, which compete with R-Smads for TGFβRI or Co-Smad complex formation. I-Smads regulate the transcription of Smurf (Smad –ubiquitination regulatory factor) proteins, which promotes the degradation of TGFβRI through proteosomal pathways [106, 111].

More recently, it has been shown that ligand-activated TGFβ receptors can stimulate signaling through non-canonical (Smad-independent) signaling predominantly via PI3K/Akt/mTOR, RhoGTPases and MAPK pathways. This in turn controls a variety of cellular responses which modify cell growth and survival. Some of these pathways have been shown to also cross-talk and regulate Smad activation and signaling, including JNK and p38 MAPK pathways [112].

In wound healing, TGFβ suppresses the inflammatory response and promotes the formation of granulation tissue by inducing fibroblast proliferation and differentiation. TGFβ induces the expression of FN and different integrins including α1β1 and α2β1, which mediate fibroblast contraction, and αv integrins that activate latent TGFβ. Additionally, αvβ3 can suppress TGFβ signaling by inhibiting TGFβR expression. Over-expression of the α5β1 integrin leads to up-regulation of TGFβR expression and establishes a growth inhibitory autocrine loop for TGFβ [113]. Conversely, TGFβ treatment of NRK fibroblasts leads to upregulation of α5β1 expression and a loss of anchorage dependence of growth [114]. Additionally, latent forms of TGFβ are ligands of αvβ6 and αvβ1 integrins, and the association of TGFβ with αvβ6 leads to spatially restricted activation of TGFβ.

Interestingly, the transcriptional control exerted by TGFβ leads to regulation of integrin activation and function in a feedback loop. TGFβ regulates integrin expression and activation via 'inside-out' mechanisms, it also regulates the expression of integrin ligands, including FN and tenascin, and integrin associated proteins, such as paxilin and ILK. (Fransvea et al 2009, Pechkovsky et al, 2008).Cell adhesion and spreading, migration, survival and proliferation all depend on integrin signaling through FAK, PI3K activated downstream of growth factor stimulation and ECM attachment [50].

FAK/Src Pathway Regulating Assembly of FN Matrix

Mounting evidence lends its weight to a critical role of Focal adhesion kinase (FAK) in FN-integrin mediated signaling. Immunoprecipitation studies demonstrated that FAK undergoes enhanced tyrosine phosphorylation in response to FN [115] [116], and its tyrosine kinase activity in NIH 3T3 fibroblasts increases 2 to 3 folds and is further enhanced in v-Src transfected cells [115]. It is ubiquitously expressed in mammalian cells, and the targeting and localization of FAK to the focal adhesion has been demonstrated in a variety of cells including fibroblasts, endothelial cells (ECs), keratinocytes, and osteoclasts. Following its discovery as a focal adhesion protein that was phosphorylated in response to Src transformation [117, 118], extensive research has revealed that FAK plays an important role in normal development and has been implicated in diseases including cancer. Germ line deletion of FAK results in embryonic lethality as demonstrated in FAK deficient mice, embryos showed gastrulation abnormalities mesodermal deficiency similar to those found in the FN-deficient mouse [8, 119]. Furthermore, FAK-/- cells showed a dramatic reduction in cell motility, survival and migration due to the inability to turnover adhesion complexes [119]. Further studies showed poor vascularization and defects in the fusion of cardiac plates [120].

Due to the lack of enzymatic activity of the short integrin cytoplasmic tails, recruitment of adaptor and signaling proteins are essential for effector functions [83]. Activation of FAK and Src family protein tyrosine kinases (PTKs) and subsequent protein tyrosine phosphorylation is a prominent event in integrin outside-in signaling [121]. Hence, the activation of the non- receptor protein tyrosine kinases, FAK and Src, are the best characterized key coordinators of integrin signaling.

FAK plays a central role in integrin signaling and is essential for fibroblast assembly of FN fibrils [120]. FAK is a unique member of the FA-PTK family, which also contains PYK2. Unlike other non-receptor tyrosine kinases, FAK does not contain SH2 or SH3 domains. Instead, it contains a central catalytic domain flanked by large non- catalytic domains, the N-terminal FERM homology domain that interacts with b1-containing integrins, and a C-terminal domain containing a focal adhesion targeting (FAT) sequence [117, 118]. These contain multiple binding sites for proteins including Src and paxillin and PI3K; Tyr-397 on FAK serves as a binding site for PI3K. Tyrosine phosphorylation of FAK is regarded as not only important in integrin mediated signaling, but also in integrin-independent signaling.

The activity of FAK can be influenced by diverse input signals, from integrins to growth factors. Thereby, the multiple tyrosine residues of FAK act as a platform for many different interaction proteins and because of this, FAK plays an indispensable role in multiple signal transduction pathways. It has been clearly shown, that FAK is necessary for cells to survive and resist anoikis, a term describing apoptosis as a result of loss of adherence. Moreover, FAK is indispensable in various aspects of cell motility. Due to its very upstream position in signaling, it is still unknown in which way certain signal inputs lead to specific and well-timed arrangements of FAK-protein complexes and to very distinct downstream effects.

Cell-ECM attachment induces FAK auto-phosphorylation at Tyr397, which correlates with increased catalytic activity, either directly through integrin clustering or indirectly through Src [122, 123]. Upon integrin clustering, FAK is recruited to the focal adhesion complex. In the inactive form, FAK is in a closed conformation in which the N-terminal FERM domain, auto-inhibitor, directly binds and inhibits the catalytic domain rendering tyrosine sites inaccessible. Upon competitive binding of integrin with the FERM domain, FAK unfolds and undergoes rapid auto-phosphorylation at Tyr397, which creates high affinity binding for Src through its SH2 domain. Although integrin ligation is the main activator of FAK, FAK can also be activated by growth factors, mechanical stimuli and GPCR agonists.

FAK and paxillin localize to focal complexes - a process mediated by Rho family members Rac and Cdc42, not RhoA by itself. The interaction with actin cytoskeleton and Rho modulated actino-myosin tension regulates the maturation of focal complexes into focal adhesions. Rho also mediates FA turnover. Paxilin, which is phosphorylated following integrin activation, serves as a scaffold protein to mediate downstream integrin-signaling. Germ line deletion of paxilin showed results in a similar phenotype as the deletion of FAK and FN. In ex vivo studies, paxillin null cells showed deficient focal adhesion formation, migration and localization of FAK to focal adhesion complexes. Therefore, paxillin is considered to be a critical transducer of FN signals in development [124].

Rho Pathway in FN Matrix Assembly and Fibrogenesis

RhoA was isolated by Richard Axel's group in 1985 and was the first member of Rho/Rac family identified [125, 126]. Of the six subfamilies, we will focus on Rho, Rac, and Cdc42 owing to their essential role as molecular switches in regulating cellular responses such as cytoskeletal changes, microtubule dynamics, cell polarity and cell cycle progression [127].

Renand and colleagues [128] developed as assay to detect GTP-Rho and reported initial observations that underscored the importance of these molecular switches and laid the foundation for extensive studies aimed at understanding Rho regulation. Utilizing direct measurement of Rho activity, they reported that Rho can be activated upon ECM ligation or by soluble factors such as LPA and that the protein exhibits a biphasic regulation during cell spreading on FN. The initial phase is associated with a period of low Rho activity that corresponds with rapid spreading on FN with cells showing small focal complexes at periphery with minimal stress fiber or focal adhesions. This is followed by a transient increase in Rho activity leading to integrin clustering and formation of focal adhesions and stress fibers. Furthermore, cell adhesion to FN down-regulates growth factor induced Rho activation whereas in suspended cells its activity and MLC phosphorylation remain elevated. They suggested the existence of a negative-feedback loop following Rho-dependent focal adhesion formation that serves the dual purpose of preventing excessive formation of focal adhesion and contraction, and regulating focal adhesion disassembly necessary for cell migration.

Rho activation generates contractile force through phosphorylation of myosin light chain (MLC) on Ser19 [129] and inhibition of MLC phosphatase (MLCP) through the action of Rho-associated kinase (ROCK). Furthermore, RhoA can mediate TGFβ-induced matrix accumulation. For example, TGFβ-induced RhoA activation correlates with fibronectin upregulation in mesangial cells [110].

The initial cell attachment and spreading over a substrate is accompanied by reduced activity of the GTPase RhoA and simultaneous activation of the GTPases Rac1 and cell division cycle 42 (Cdc42), which suppresses actomyosin contractility and enhances actin-mediated protrusion. During early stages of cell spreading, α5β1 integrins inhibit RhoA activity through a Src-dependent tyrosine phosphorylation of p190RhoGAP [130]. Integrins also activate Rac1 and Cdc42 by activating FAK-Src-mediated tyrosine phosphorylation of p130CAS [130, 131]. The FAK–Src complex also phosphorylates paxillin, which recruits ADP-ribosylation factor (ARF)–GTPase-activating protein (GAP), as well as recruiting Rho guanine nucleotide exchange factor 7 (ARHGEF7; also termed β-PIX), the GDP–GTP exchange factor (GEF) for Cdc42 and Rac1 activation. Through a direct interaction within focal adhesions and membrane ruffles, ARHGEF7 recruits and activates Rac1 [132]. Thus, integrin signaling through the Src family kinases and FAK regulates the localization and activity of GEFs and GAPs to coordinate actin cytoskeleton stabilization with membrane protrusion thus mediating cell matrix adhesion and migration.

Adhesion to FN stimulates "outside-in" signaling through integrins that triggers a complex network of intracellular signaling events. For instance, fibroblasts plated on FN induce the activation of MAPK [133-135]. In converging adhesive signals between the ECM with cytoskeleton, integrins stimulates certain pathways that are critical for the initiation and maintenance of matrix assembly. FAK [120] and PI3K [50] are essential for assembly of FN fibrils. Members of the Src kinase family are essential for maintaining matrix association with the cell. FN matrix assembly was delayed in SYF cells (which lack Src family kinases Src, Yes, and Fyn) but not eliminated and this delay is overcome following increased Src expression and activity [131]. This suggests that Src plays an important supportive role in matrix assembly dynamics. After matrix accumulation, inhibition of Src activity causes rapid loss of matrix [136]. Consequently, these pathways are essential for cell adhesion, spreading, migration survival and proliferation.

PI3K/Akt Cascade in FN Secretion, Assembly and Fibrogenesis

Perhaps the most pivotal mediator and modulator of FN activity is protein kinase B (Akt), a serine/threonine kinase. Akt, a major effector downstream of PI3K, is a key survival kinase and is a multi-functional protein that regulates a plethora of cellular functions including survival, proliferation, and transcription, integrin trafficking and recycling, actin organization, angiogenesis and apoptosis. Although not fully understood, mounting evidence suggests that the versatility of Akt activity is associated with the complex network formed between Akt and its substrates such as GSK-3, mTOR, TSC, FOXO as well as cross-talk with other signaling pathways such as p38 MAPK and Rac –PAK [137-143].

The Akt family consists of three highly homologous isoforms that are transcribed from independent genes and have overlapping but distinct functions [138, 144]. Of the three Akt isoforms, we identified Akt1 as the predominant isoform that accounts for approximately 75% of total Akt activity in vascular cells [145].

It has been shown that owing to its complexity in regulating a variety of cell responses, Akt signaling is very fine-tuned. Prolonged Akt signaling is associated with embryonic lethality, edema and vascular malformation and remodeling abnormalities [146]. Loss of the Akt1 gene leads to growth retardation accompanied by a decrease in size of multiple organs and increased cell apoptosis [147]. Environmental cues including growth factors, cytokines, and ECM regulate the highly orchestrated multistep process of cell migration involving Akt [148].

Research from our laboratory has established the central role of Akt in the regulation of integrin activation in ECs and fibroblasts, ECM assembly, tissue remodeling and angiogenesis [56, 145, 149-152]. Several *in vivo* studies have shown the crucial role of Akt1 in wound healing and vascular maturation [151]. We previously reported several interesting features of an Akt1 knock-out mouse model including impaired migration, impaired secretion and assembly of ECM proteins and enhanced angiogenesis resulting in vascular leakage and skin abnormalities [145]. Akt mediates VEGF induced integrin activation in ECs [153, 154].

Specifically, VEGF-165 was shown to activate integrin αvβ3 leading to enhanced adhesion and migration of ECs a process that involves PI3K and Akt. Integrin activation is important for EC and fibroblast adhesion and migration by regulating ECM assembly. ECs isolated from Akt1-/- mice exhibited impaired integrin activation which was associated with impaired migration on matrix proteins including FN, and impaired ECM assembly in skin and vascular basement membrane of Akt1-/- mice [145]. Additionally, genetic ablation of Akt1 but not Akt2, leads to impaired wound healing characterized by a reduced vascular area and impaired vascular maturation, further demonstrating the importance of Akt1[151]. These finding suggests that Akt1 is essential for inside-out integrin signaling.

The tryptophan residue at position 775 of β1 integrin cytoplasmic domain was shown to be a specific regulator of Akt activation by repressing the activity of the protein phosphatase 2A (PP2A) in ECM assembly [143]. Mutation of Trp775 (W775A) in β1 integrin specifically suppresses Akt activation and fibroblast survival without interfering with other integrin-mediated signaling pathways. This has been shown to be due to rapid Akt de-phosphorylation through increased local activation of PP2A, which was 2.5 times higher compared to β1 WT in response to PDGF [143].A subsequent study demonstrated similar results and further illustrated the importance of two distinct amino acid residues on β1integrin cytoplasmic domain, Trp-775 and Argenine-760 (Arg-760)in modulating cell spreading and FN matrix assembly.Trp-775 and Arg-760 differentially modulate FN assembly and cell spreading through talin and Akt-1respectively. While both Trp-775 and Arg-760 exhibit overlapping roles in regulating FN assembly through recruiting talin to the β1integrin cytoplasmic complex, Trp-775 selectively regulates Akt1 signaling in cell spreading on FN. W775A mutation disrupts Akt1 and talin activity leading to impaired cell adhesion and spreading, and defects in FN matrix assembly and fibrillar adhesion. Although over-expression of Akt1 in W775A mutant cells rescues cell spreading activity, it does not reverse impaired FN fibrillogenesis. This suggests that Akt specifically regulates cell spreading not matrix assembly, although the effect of Akt knockdown on FN assembly in β1WT integrin cells was not investigated directly. Thus the role of Akt in mediating inside-out FN assembly remains unclear [155].

Originally, it was thought that Akt mediates cellular functions in response to outside-in signaling induced by ECM-integrin ligation [156]. However, recent studies have emphasized the role of Akt in regulating integrin activity and, subsequently, mediating inside-out signaling [138, 145, 154].

FN matrix assembly is impaired in Akt1-/- fibroblasts [150]. Akt1over-expression results in significant increase in integrin activation, while Akt deletion impairs integrin activation and, in turn, impairs EC and fibroblast adhesion and migration [150]. Fibroblasts are necessary for effective repair in tissue injury and are known to express α5β1 integrin, while αvβ3, which mainly binds vitronectin, is expressed to a lower extent [89]. Deletion of Akt in fibroblasts significantly impairs adhesion to FN without affecting integrin expression in the presence and absence of growth factor stimulation; implying that Akt regulates integrin functional activity and FN recognition. Although Akt -/- and WT fibroblasts adhere to vitronectin to the same extent, it was to a much lower level compared to FN for both WT and Akt-/- cells. Similarly, Akt1-/- fibroblasts show a significant decrease in migration on FN compared to vitronectin [150].

Additionally, bFGF, not VEGF, induces PI3K dependent Akt phosphorylation and activation in fibroblasts. In addition to impaired FN matrix recognition, Akt deficiency – Akt1-/-, PI3K inhibitor, or Akt inhibitor (SH-5) - results in impaired FN matrix assembly in fibroblasts. Treatment with integrin β1blocking antibody significantly blunted FN assembly in Akt1-/- and Akt1+/+ fibroblasts. Notably, exogenousβ1 integrin activation, using TS2/16, rescues the defect on FN assembly caused by SH5. Similarly, restoration of Akt activity by reconstitution of Akt in Akt-/- ECs and fibroblasts rescues the defect in FN assembly and leads to enhanced migration. Thus, demonstrating that, in the absence of Akt1, impaired FN assembly is due to impaired ability of β1 integrin to recognize FN. Collectively, we demonstrate the essential role of Akt1 in regulating inside-out activation of integrins, specifically β1 integrin, FN recognition and matrix assembly in fibroblasts and ECs [150]. Furthermore, while α5β1 and αvβ3 internalization and transport to the recycling compartment are Akt-independent, PI3K and Akt regulate the trafficking and recycling of integrins back to the plasma membrane by regulating GSK-3 [157].

Numerous studies demonstrate that FN stimulates cytoskeletal remodeling, cell growth, survival, and migration via the PI3K/Akt/mTOR pathway [7, 158-160]. However, evidence of PI3K/Akt induced FN expression is not fully understood. Our previous findings indicate that impaired adhesion and migration of Akt -/- fibroblasts results in impaired FN assembly in vitro [150] and in vivo [145, 151]. Akt1 regulates FN assembly through the activation of α5β1 integrin. We proposed that Akt1 also regulates FN synthesis and secretion [56]. Indeed, in TGFβ stimulated fibroblasts, FN synthesis and secretion in the conditioned media increases in a dose and time dependent manner; an effect that is further enhanced with over expression of Akt and impaired in Akt deficient cells. This suggests that Akt can modulate FN expression possibly through mTOR, known to be involved in protein translation and control of the overall rate of protein synthesis. Specific inhibition of mTOR results in diminished growth factor mediated FN synthesis and secretion.

Interestingly, we observed that mTOR regulates FN at translational not transcriptional level through activation of P70S6Kinase and S6Ribosomal Protein, both of which are downstream effectors of mTOR [56]. Similar results are observed in bleomycin stimulated fibroblasts; bleomycin is known to induce fibrosis potentially through TGFβ signaling pathways. Notably, we observed that mTOR is involved in transcriptional repression of several proteins including α4 and β3 integrins, matrix metalloprotease-8, -10, -11, and laminin. All together, we conclude that PI3K/Akt/mTOR regulates FN by controlling protein levels at translational level. A recent study supported this finding and also showed that Akt enhances EDA alternative splicing by directly phosphorylating and activating SF2/ASF [161]. To date, evidence for the mechanism of transcriptional regulation by mTOR has not been clearly illustrated this area requires further investigation. Collectively these data suggests a reciprocal dynamic regulation between FN and the Akt pathway.

FN-Mediated Regulation of Directional Cell Migration Through Akt Signaling Pathway

Recent studies have demonstrated caveolin-1 (cav-1) can serve as a scaffold protein that plays an important role in sequestering and modulating signaling pathways (reviewed in [162]) as evident in TGFβR internalization and degradation [163, 164] and association with β integrins to promote cell growth [165] and proliferation [166, 167]. In cav-1 deficient cells, TGFβ-mediated RhoA-activation is repressed, and cav-1 is required for TGFβ1 mediated FN expression [110]. Absence of cav-1 correlates with loss of membrane anchorage and cell cycle arrest [168]. A novel dynamic regulatory function between FN and cav-1 was recently illustrated by two groups. First, it was shown that FN stimulates cell proliferation, in part, by initiating and maintaining the proximal association between β1 integrin/FAK-Src and cav-1,and phosphorylation of cav-1is necessary to activate RhoA-PI3K/Akt-Erk1/2 pathways [169]. This study also suggested that the localization of RhoA to the caveolae microdomain is necessary for mediating cell functions. Indeed, another study demonstrated that the cav-1-Rho-GTPase interaction is crucial for α5 integrin expression and the activation of Src, Ras and Erk [170].

The Rho family of GTPases (namely, Rho, Rac and Cdc42) acting downstream of integrins, are critical regulators of the cytoskeletal dynamics involved in cell morphology, cell polarization and directed migration, and ECM assembly in vascular cells [81, 171-174]. Additionally, activation and localization of Rac and Cdc42 activate p21-activated kinase (PAK) auto-phosphorylation, a major downstream effector [175, 176] which is responsible for myosin light chain phosphorylation which regulates cytoskeletal organization and cell migration. The process of cell migration and cell guidance migratory mechanisms has been extensively investigated [148, 177]. PAK1 is the major isoform of the PAK family, which contains seven isoforms, and which was recently identified as a substrate of Akt [178].

Although most studies on Rho have illuminated its role in stress fiber and focal adhesion formation in fibroblasts, it has become apparent that Rho also plays an important role in physiological cell migration and invasion in tumors.

The process of tread milling (please explain this term) and directed migration is a highly regulated step wise process. Chemo-attractants activate RhoA and Rac/Cdc24 and localize them to the trailing edge and leading edge, respectively. Cdc42 induces actin spikes and promotes filopodia extension leading to directional sensing, and establishment of cell polarity during adhesion [179]. Rac is required for F-actin formation and promotes lamellipodia and membrane ruffles formation leading to protrusion and extension of the leading edge. Rho activation is required for focal adhesion assembly, actin stress fiber formation which inhibits protrusion while promoting attachment and contraction at the trailing edge [180]. Contraction at the lagging end can in turn push the membrane at the other side outwards forming polarized directed migration. There exists a localized incompatibility of the two actin cytoskeletal responses, actin-myosin and actin polymerization, due to the ability of Rho and Rac to inhibit signals that promotes the other. Thus, each actin assembly gradually segregates spatially to the lagging or leading edge of the cell and, simultaneously, promotes the other's activity at a distance [181]. PI3K is necessary for localization of Rac- Cdc42/PAK1 signaling at the leading edge. Simultaneously, active PAK1 regulates PTEN and F-actin localization to the trailing edge [182].

Much attention has been focused on identifying the downstream signaling proteins of PI3K involved in regulating cell morphology and motility. It is known that PI3K is required for growth factor, and integrin mediated cell migration and is involved in actin filament remodeling induced by growth factors, integrins [183] and Ras. For example, whereas insulin stimulated cells exhibit rapid actin filament reorganization which is correlated with PI3K recruitment and activation in that region, PI3K inhibition abolishes insulin induced actin remodeling and membrane ruffling. Similarly, inhibiting actin filament formation disrupts PI3K distribution and signal transduction [184-187]. Furthermore, inhibition of actin filament formation also results in attenuated PI3K- and Akt-mediated cell migration [158]. This suggests an important crosstalk between RhoGTPases such as Rac and PI3K/Akt signaling pathways.

Previous studies have revealed the importance of Akt in modulating Rac-PAK activity and subsequent actin cytoskeleton reorganization and cell migration [178, 188-190]. Through an SH3 binding domain, PAK is localized in focal adhesions via the adaptor proteins Nck and PIX. Akt directly phosphorylates serine 21 on PAK, releasing it from Nck. PAK then translocates to the cytosol, therefore facilitating PAK activity in directed cell migration [178]. The cross-talk between PI3K and Rho-GTPase family is complex, since PI3K can act both up- and down-stream of the GTPases.

'Akt–Integrin–FN' Paradigm

Akt1 regulates inside-out activation of integrins, cell adhesion and migration and FN matrix assembly through cross talk with 14-3-3β-Rac1-PAK [152]. Rac1 localization is Akt1 dependent; Akt1 activity correlates with the formation of lamellipodia, but not with filopodia or stress fibers. In Akt-/- fibroblasts and ECs, lamellipodia formation and directional migration on FN is impaired.

In both cell types over-expression of Akt results in Rac1 translocation to membrane ruffles and enhanced Rac1 activity and PAK phosphorylation, which is blunted by treatment with inhibitors of PI3K and Akt, but not GSK-3, or in cells over expressing inactive Akt. These findings emphasize the strong regulatory role of Akt on Rac1-Pak signaling [152].

The14-3-3 proteins are a family of regulator molecules expressed in all eukaryotic cells. These adaptor proteins bind a multitude of functionally diverse signaling proteins; hence 14-3-3 is implicated in a wide range of vital cellular processes. Akt substrates, such as Gsk-3, FOXO, and cRaf, are known to interact with 14-3-3 [191]. Also, 14-3-3 mediates Rac1 localization. This suggests that Akt modulates Rac1 activity through 14-3-3. Indeed, enhanced Rac1 localization and formation of lamellipodia, adhesion to FN, and FN assembly are observed in fibroblasts over-expressing 14-3-3 and inhibited with co-expression of inactive Akt (DN-Akt). Furthermore, lamellipodia formation and cell adhesion and spreading on FN is enhanced in DN-Akt fibroblasts co-expressing active- Rac1 or -PAK1, and inhibited in fibroblasts over-expressing Akt with inactive-Rac1 or -PAK1. This indicates that lamellipodia formation and cell adhesion, as discussed above is through Akt mediated inside-out activation of integrins and is regulated by 14-3-3-Rac-Pak signaling downstream of Akt1 [152].

Notably, the defect in Akt-/- fibroblast to adhere to FN is rescued by co-expression of active-Rac1 or PAK1, but not 14-3-3. Interestingly, constitutively active myristoylated Akt (myr-Akt) fibroblasts co-expressing inactive-Rac1 or –PAK1 show impaired FN assembly. In contrast, impaired FN matrix assembly in DN-Akt and Akt null fibroblasts is rescued with co-expression of active-Rac1 or -Pak1 [152]. Collectively, we identified a novel cross talk between 14-3-3-Rac-Pak and Akt in regulating inside-out integrin activation, cell-FN adhesion, FN matrix assembly.

FN in Disease

Under normal physiological conditions, signaling pathways are highly regulated resulting in dynamic and finely orchestrated intra- and inter-molecular interactions to modulate a wide range of cellular functions [169, 192-196]. However, in tumorigenesis [197, 198] and fibrogenesis [31, 199, 200], many essential pathways are de-regulated and dynamic interaction between pathways is perturbed. FN expression is up-regulated in several cancers, especially in non-small cell lung carcinoma (NSCLC) [201]. FN stimulates NSCLC cell growth and resistance to apoptosis through the activation of $\alpha5\beta1$ integrin and inhibition of cyclin-dependent kinase inhibitor p21 gene expression; hence FN is implicated in drug resistance [202]. Notably, these effects are specific to FN and are not shared by other ECM proteins such as type I collagen.

Mutations in Smad proteins, specifically Smad4, were found in a wide range of human cancers. Recently it has been shown that Smad4 C324Y mutation plays an important role in metastasis of papillary thyroid carcinoma. Such mutation leads to the acquisition of transformed and invasive phenotype through significant activation of TGFβ signaling, lengthening of nuclear localization with R-Smads survival and growth in anchorage independent conditions and decreased adhesion to FN and levels of E-cadherin, which is crucial for the EMT process [203].

FN stimulates fibroblast proliferation and differentiation into myofibroblasts in many fibrotic diseases including pulmonary fibrosis. Excess myofibroblasts in fibroblast foci represent the hallmark lesions of IPF. They are responsible for increased ECM deposition which leads to impaired gas exchange and ultimately fibrosis. Glomerular sclerosis of diverse etiologies is characterized by excess matrix accumulation mediated by up-regulation of TGFβ and FN as an important pathogenic factor [110].

FibronectinFN-mediated aberrant activation of the classic PI3K/Akt and MAP Kinase cascades has been implicated in signaling toward pro-oncogenic outcomes in many malignancies including prostate cancer [18, 204, 205] as well as in fibrotic disease [18, 206, 207]. Transformed cells are characterized by their ability to survive in the absence of matrix adhesion [208]. For example, PI3K/Akt is a key mediator of survival of both primary and Ras transformed cells in the presence or absence of matrix adhesion [209]. One important implication of this observation is that activated Ras can induce cell survival in the absence of integrin signaling, thus contributing to tumor cell survival.

Several studies have demonstrated that up-regulation of both PI3K/Akt and MAPK is required for metabolic transformation, thus, initiating oncogenic transformation in oncoproteins such as Ras, Src, Akt, Raf, and c-Myc [198, 210]. This creates a dysregulated continuous positive feedback loop leading to sustained growth, proliferation and metastasis. Additionally, loss of tumor suppressor genes encoding p53 and PTEN, and secretion of MMPs are involved in this process [198]. Furthermore, induction of epithelial-mesenchymal transition has been identified as a maker of aggressive phenotypes in certain cancers and fibrotic diseases. EMT has been shown to be PI3K-Rho dependent or MAPK-Smad dependent [99, 211-215].

The observation of increased plasma levels of MMP9 in NSCLC patients suggests that the metastatic and invasive capability of tumor cells correlates with enhanced MMP9 expression [201, 216].

Indeed, a study showed that, in lung cancer cells, FN induces the activation of the FAK-Src complex that subsequently activates PI3K/Akt and Erk1/2 pathways. Further, FN induces cell invasion through enhanced expression of MMP9 that degrades matrix components. FN also induces cell de-adhesion and migration via enhanced expression of calpain proteases, specifically calpain-2, and RhoA, respectively [49].

In fibrosis excessive synthesis and matrix assembly of FN induces cell differentiation towards a contractile phenotype as evident in fibroblast to myofibroblast differentiation [31, 217]. Also, sustained FN accumulation mediates sustained activation of myofibroblasts, resistance to apoptosis, and deposition of excessive collagen through Akt-mTOR-PTEN pathways in various fibrotic diseases including lung [218, 219] and liver fibrosis [220, 221].

Therapeutic targeting of PI3K and Akt has proven more difficult and the progress of a number of compounds has been limited by specificity and / or toxicity. This is disappointing as Akt is often activated in response to FN adhesion in the *in vitro* setting and inhibiting this pathway with the laboratory based agents' wortmannin and/or LY294002 has been shown to have a beneficial effect in combination with radiation in many cell systems [222]. mTOR, however, which is downstream of Akt , and controls the translation of proteins for cell cycle progression and cell proliferation, is perhaps a more promising target. mTOR inhibitors are currently under evaluation, particularly in endometrial cancer, glioblastoma and pulmonary fibrosis, all of which commonly possess mutations in PTEN [223]. Identifying the principal elements of the Akt-mTOR-PTEN signaling network and determining the dynamic with FN will improve our understanding of cancer and fibrosis pathogenesis and lead to the rational development of novel therapeutics.

Acknowledgment

Research in the laboratory is supported by funding from the National Institutes of Health (R01HL103952), American Heart Association (0830326N) and University of Georgia Research Foundation to PRS.

References

[1] Danen EH, Sonneveld P, Sonnenberg A, Yamada KM. Dual stimulation of Ras/mitogen-activated protein kinase and RhoA by cell adhesion to fibronectin supports growth factor-stimulated cell cycle progression. *The Journal of cell biology*. 2000 Dec 25;151(7):1413-22.

[2] Kuwada SK, Li X. Integrin alpha5/beta1 mediates fibronectin-dependent epithelial cell proliferation through epidermal growth factor receptor activation. *Molecular biology of the cell*. 2000 Jul;11(7): 2485-96.

[3] Asthagiri AR, Nelson CM, Horwitz AF, Lauffenburger DA. Quantitative relationship among integrin-ligand binding, adhesion, and signaling via focal adhesion kinase and extracellular signal-regulated kinase 2. *The Journal of biological chemistry*. 1999 Sep 17;274(38): 27119-27.

[4] Asthagiri AR, Reinhart CA, Horwitz AF, Lauffenburger DA. The role of transient ERK2 signals in fibronectin- and insulin-mediated DNA synthesis. *Journal of cell science*. 2000 Dec;113 Pt 24:4499-510.

[5] Galownia NC, Kushiro K, Gong Y, Asthagiri AR. Selective desensitization of growth factor signaling by cell adhesion to fibronectin. *The Journal of biological chemistry*. 2007 Jul 27;282(30): 21758-66.

[6] Hynes RO. The dynamic dialogue between cells and matrices: implications of fibronectin's elasticity. *Proceedings of the National Academy of Sciences of the United States of America*. 1999 Mar 16; 96(6):2588-90.

[7] Hay N. The Akt-mTOR tango and its relevance to cancer. *Cancer cell*. 2005 Sep; 8(3):179-83.

[8] George EL, Georges-Labouesse EN, Patel-King RS, Rayburn H, Hynes RO. Defects in mesoderm, neural tube and vascular development in mouse embryos lacking fibronectin. *Development* (Cambridge, England). 1993 Dec;119(4):1079-91.

[9] Sakai T, Johnson KJ, Murozono M, Sakai K, Magnuson MA, Wieloch T, et al. Plasma fibronectin supports neuronal survival and reduces brain injury following transient focal cerebral ischemia but is not essential for skin-wound healing and hemostasis. *Nature medicine*. 2001 Mar; 7(3): 324-30.

[10] Dixit VM, Haverstick DM, O'Rourke K, Hennessy SW, Broekelmann TJ, McDonald JA, et al. Inhibition of platelet aggregation by a monoclonal antibody against human fibronectin. *Proceedings of the National Academy of Sciences of the United States of America*. 1985 Jun;82(11):3844-8.

[11] Ross R. The pathogenesis of atherosclerosis: a perspective for the 1990s. *Nature*. 1993 Apr 29;362(6423):801-9.

[12] Ruoslahti E, Engvall E. Integrins and vascular extracellular matrix assembly. *The Journal of clinical investigation*. 1997 Dec 1;100(11 Suppl):S53-6.

[13] Wang JH, Thampatty BP, Lin JS, Im HJ. Mechanoregulation of gene expression in fibroblasts. *Gene*. 2007 Apr 15;391(1-2):1-15.

[14] Han S, Ritzenthaler JD, Sitaraman SV, Roman J. Fibronectin increases matrix metalloproteinase 9 expression through activation of c-Fos via extracellular-regulated kinase and phosphatidylinositol 3-kinase pathways in human lung carcinoma cells. *The Journal of biological chemistry*. 2006 Oct 6;281(40):29614-24.

[15] Muro AF, Moretti FA, Moore BB, Yan M, Atrasz RG, Wilke CA, et al. An essential role for fibronectin extra type III domain A in pulmonary fibrosis. *American journal of respiratory and critical care medicine*. 2008 Mar 15;177(6):638-45.

[16] Pankov R, Yamada KM. Fibronectin at a glance. *Journal of cell science*. 2002 Oct 15;115(Pt 20):3861-3.

[17] White ES, Baralle FE, Muro AF. New insights into form and function of fibronectin splice variants. *The Journal of pathology*. 2008 Sep; 216(1): 1-14.

[18] Singh P, Carraher C, Schwarzbauer JE. Assembly of fibronectin extracellular matrix. *Annu. Rev. Cell Dev. Biol.* Nov 10;26:397-419.

[19] To WS, Midwood KS. Plasma and cellular fibronectin: distinct and independent functions during tissue repair. *Fibrogenesis and tissue repair*. 2011;4:21.

[20] Hynes RO. Integrins: bidirectional, allosteric signaling machines. *Cell*. 2002 Sep 20; 110 (6): 673-87.

[21] Hocevar BA, Brown TL, Howe PH. TGF-beta induces fibronectin synthesis through a c-Jun N-terminal kinase-dependent, Smad4-independent pathway. *The EMBO Journal*. 1999 Mar 1; 18 (5): 1345-56.

[22] Larsen M, Artym VV, Green JA, Yamada KM. The matrix reorganized: extracellular matrix remodeling and integrin signaling. *Current opinion in cell biology*. 2006 Oct; 18(5): 463-71.

[23] Nelson CM, Bissell MJ. Of extracellular matrix, scaffolds, and signaling: tissue architecture regulates development, homeostasis, and cancer. *Annual review of cell and developmental biology*. 2006;22: 287-309.

[24] Huang C, Jacobson K, Schaller MD. MAP kinases and cell migration. *Journal of cell science*. 2004 Sep 15;117(Pt 20):4619-28.

[25] Glading A, Chang P, Lauffenburger DA, Wells A. Epidermal growth factor receptor activation of calpain is required for fibroblast motility and occurs via an ERK/MAP kinase signaling pathway. *The Journal of biological chemistry*. 2000 Jan 28; 275(4): 2390-8.

[26] Boutros T, Chevet E, Metrakos P. Mitogen-activated protein (MAP) kinase/MAP kinase phosphatase regulation: roles in cell growth, death, and cancer. *Pharmacological reviews*. 2008 Sep;60(3):261-310.

[27] ffrench-Constant C. Alternative splicing of fibronectin--many different proteins but few different functions. *Experimental cell research.* 1995 Dec; 221(2):261-71.

[28] Ohashi T, Kiehart DP, Erickson HP. Dynamics and elasticity of the fibronectin matrix in living cell culture visualized by fibronectin-green fluorescent protein. *Proceedings of the National Academy of Sciences of the United States of America.* 1999 Mar 2;96(5): 2153-8.

[29] Astrof S, Crowley D, Hynes RO. Multiple cardiovascular defects caused by the absence of alternatively spliced segments of fibronectin. *Developmental biology.* 2007 Nov 1; 311(1): 11-24.

[30] Serini G, Bochaton-Piallat ML, Ropraz P, Geinoz A, Borsi L, Zardi L, et al. The fibronectin domain ED-A is crucial for myofibroblastic phenotype induction by transforming growth factor-beta1. *The Journal of cell biology.* 1998 Aug 10; 142(3): 873-81.

[31] Tomasek JJ, Gabbiani G, Hinz B, Chaponnier C, Brown RA. Myofibroblasts and mechano-regulation of connective tissue remodelling. *Nature reviews Molecular cell biology.* 2002 May;3(5): 349-63.

[32] Muro AF, Chauhan AK, Gajovic S, Iaconcig A, Porro F, Stanta G, et al. Regulated splicing of the fibronectin EDA exon is essential for proper skin wound healing and normal lifespan. *The Journal of cell biology.* 2003 Jul 7;162(1):149-60.

[33] Pierschbacher MD, Ruoslahti E. Cell attachment activity of fibronectin can be duplicated by small synthetic fragments of the molecule. *Nature.* 1984 May 3-9; 309 (5963): 30-3.

[34] Hynes RO. Integrins: versatility, modulation, and signaling in cell adhesion. *Cell.* 1992 Apr 3; 69(1):11-25.

[35] Magnusson MK, Mosher DF. Fibronectin: structure, assembly, and cardiovascular implications. *Arteriosclerosis, thrombosis, and vascular biology.* 1998 Sep; 18 (9): 1363-70.

[36] Guan JL, Trevithick JE, Hynes RO. Retroviral expression of alternatively spliced forms of rat fibronectin. *The Journal of cell biology.* 1990 Mar;110(3):833-47.

[37] Redick SD, Settles DL, Briscoe G, Erickson HP. Defining fibronectin's cell adhesion synergy site by site-directed mutagenesis. *The Journal of cell biology.* 2000 Apr 17; 149 (2): 521-7.

[38] Vogel V, Baneyx G. The tissue engineeting puzzle: a molecular perspective. *Annual review of biomedical engineering.* 2003;5:441-63.

[39] Aota S, Nomizu M, Yamada KM. The short amino acid sequence Pro-His-Ser-Arg-Asn in human fibronectin enhances cell-adhesive function. *The Journal of biological chemistry.* 1994 Oct 7;269(40): 24756-61.

[40] Hocking DC, Sottile J, McKeown-Longo PJ. Fibronectin's III-1 module contains a conformation-dependent binding site for the amino-terminal region of fibronectin. *The Journal of biological chemistry.* 1994 Jul 22;269(29):19183-7.

[41] Hocking DC, Kowalski K. A cryptic fragment from fibronectin's III1 module localizes to lipid rafts and stimulates cell growth and contractility. *The Journal of cell biology.* 2002 Jul 8;158(1):175-84.

[42] Hynes RO, Yamada KM. Fibronectins: multifunctional modular glycoproteins. *The Journal of cell biology.* 1982 Nov; 95(2 Pt 1): 369-77.

[43] Wierzbicka-Patynowski I, Schwarzbauer JE. The ins and outs of fibronectin matrix assembly. *Journal of cell science*. 2003 Aug 15;116(Pt 16):3269-76.
[44] Johnson KJ, Sage H, Briscoe G, Erickson HP. The compact conformation of fibronectin is determined by intramolecular ionic interactions. *The Journal of biological chemistry*. 1999 May 28; 274(22): 15473-9.
[45] Sevilla CA, Dalecki D, Hocking DC. Extracellular matrix fibronectin stimulates the self-assembly of microtissues on native collagen gels. *Tissue engineering Part A*. 2010 Dec; 16(12):3805-19.
[46] Sottile J, Hocking DC. Fibronectin polymerization regulates the composition and stability of extracellular matrix fibrils and cell-matrix adhesions. *Molecular biology of the cell*. 2002 Oct; 13 (10): 3546-59.
[47] Dityatev A, Schachner M, Sonderegger P. The dual role of the extracellular matrix in synaptic plasticity and homeostasis. *Nature reviews Neuroscience*. 2010 Nov; 11(11): 735-46.
[48] Pankov R, Cukierman E, Katz BZ, Matsumoto K, Lin DC, Lin S, et al. Integrin dynamics and matrix assembly: tensin-dependent translocation of alpha(5)beta(1) integrins promotes early fibronectin fibrillogenesis. *The Journal of cell biology*. 2000 Mar 6; 148(5):1075-90.
[49] Meng XN, Jin Y, Yu Y, Bai J, Liu GY, Zhu J, et al. Characterisation of fibronectin-mediated FAK signalling pathways in lung cancer cell migration and invasion. *British journal of cancer*. 2009 Jul 21;101(2): 327-34.
[50] Wierzbicka-Patynowski I, Schwarzbauer JE. Regulatory role for SRC and phosphatidylinositol 3-kinase in initiation of fibronectin matrix assembly. *The Journal of biological chemistry*. 2002 May 31;277(22): 19703-8.
[51] Allen-Hoffmann BL, Crankshaw CL, Mosher DF. Transforming growth factor beta increases cell surface binding and assembly of exogenous (plasma) fibronectin by normal human fibroblasts. *Molecular and cellular biology*. 1988 Oct;8(10):4234-42.
[52] Allen-Hoffmann BL, Mosher DF. Matrix assembly sites for exogenous fibronectin are decreased on human fibroblasts after treatment with agents which increase intracellular cAMP. *The Journal of biological chemistry*. 1987 Oct 15;262(29):14361-5.
[53] Somers CE, Mosher DF. Protein kinase C modulation of fibronectin matrix assembly. *The Journal of biological chemistry*. 1993 Oct 25; 268(30): 22277-80.
[54] Blatti S, Foster D, Ranganathan G, Moses H, Getz M. Induction of fibronectin gene transcription and mRNA is a primary response to growth-factor stimulation of AKR-2B cells. *Proc. Natl. Acad. Sci. USA*. 1988 85(4):1119-23.
[55] Liu C, Yao J, Mercola D, Adamson E. The transcription factor EGR-1 directly transactivates the fibronectin gene and enhances attachment of human glioblastoma cell line U251. *The Journal of biological chemistry*. 2000 Jul 7;275(27):20315-23.
[56] Goc A, Choudhary M, Byzova TV, Somanath PR. TGFbeta- and bleomycin-induced extracellular matrix synthesis is mediated through Akt and mammalian target of rapamycin (mTOR). *Journal of cellular physiology*. 2011 Nov;226(11):3004-13.
[57] Diaz A, Jimenez SA. Interferon-gamma regulates collagen and fibronectin gene expression by transcriptional and post-transcriptional mechanisms. *The international journal of biochemistry and cell biology*. 1997 Jan;29(1):251-60.

[58] Rebres RA, McKeown-Longo PJ, Vincent PA, Cho E, Saba TM. Extracellular matrix incorporation of normal and NEM-alkylated fibronectin: liver and spleen deposition. *The American journal of physiology.* 1995 Dec;269(6 Pt 1):G902-12.

[59] Sternlicht MD, Werb Z. How matrix metalloproteinases regulate cell behavior. *Annual review of cell and developmental biology.* 2001; 17: 463-516.

[60] Sottile J, Chandler J. Fibronectin matrix turnover occurs through a caveolin-1-dependent process. *Molecular biology of the cell.* 2005 Feb;16(2):757-68.

[61] Rotundo RF, Vincent PA, McKeown-Longo PJ, Blumenstock FA, Saba TM. Hepatic fibronectin matrix turnover in rats: involvement of the asialoglycoprotein receptor. *The American journal of physiology.* 1999 Dec;277(6 Pt 1):G1189-99.

[62] Kuhn K, Eble J. The structural bases of integrin-ligand interactions. *Trends in cell biology.* 1994 Jul;4(7):256-61.

[63] LaFlamme SE, Akiyama SK, Yamada KM. Regulation of fibronectin receptor distribution. *The Journal of cell biology.* 1992 Apr;117(2): 437-47.

[64] Rosales C, Gresham HD, Brown EJ. Expression of the 50-kDa integrin-associated protein on myeloid cells and erythrocytes. *J. Immunol.* 1992 Oct 15;149(8):2759-64.

[65] Tahiliani PD, Singh L, Auer KL, LaFlamme SE. The role of conserved amino acid motifs within the integrin beta3 cytoplasmic domain in triggering focal adhesion kinase phosphorylation. *The Journal of biological chemistry.* 1997 Mar 21;272(12):7892-8.

[66] Yang JT, Rayburn H, Hynes RO. Embryonic mesodermal defects in alpha 5 integrin-deficient mice. *Development* (Cambridge, England). 1993 Dec;119(4):1093-105.

[67] Bouvard D, Brakebusch C, Gustafsson E, Aszodi A, Bengtsson T, Berna A, et al. Functional consequences of integrin gene mutations in mice. *Circulation research.* 2001 Aug 3;89(3):211-23.

[68] Takahashi S, Leiss M, Moser M, Ohashi T, Kitao T, Heckmann D, et al. The RGD motif in fibronectin is essential for development but dispensable for fibril assembly. *The Journal of cell biology.* 2007 Jul 2; 178(1):167-78.

[69] Zhang Z, Vuori K, Reed JC, Ruoslahti E. The alpha 5 beta 1 integrin supports survival of cells on fibronectin and up-regulates Bcl-2 expression. *Proceedings of the National Academy of Sciences of the United States of America.* 1995 Jun 20;92(13):6161-5.

[70] Lee JW, Juliano RL. alpha5beta1 integrin protects intestinal epithelial cells from apoptosis through a phosphatidylinositol 3-kinase and protein kinase B-dependent pathway. *Molecular biology of the cell.* 2000 Jun;11(6):1973-87.

[71] Brooks PC, Montgomery AM, Rosenfeld M, Reisfeld RA, Hu T, Klier G, et al. Integrin alpha v beta 3 antagonists promote tumor regression by inducing apoptosis of angiogenic blood vessels. *Cell.* 1994 Dec 30; 79(7):1157-64.

[72] Ilic D, Almeida EA, Schlaepfer DD, Dazin P, Aizawa S, Damsky CH. Extracellular matrix survival signals transduced by focal adhesion kinase suppress p53-mediated apoptosis. *The Journal of cell biology.* 1998 Oct 19;143(2):547-60.

[73] Matter ML, Ruoslahti E. A signaling pathway from the alpha5beta1 and alpha(v)beta3 integrins that elevates bcl-2 transcription. *The Journal of biological chemistry.* 2001 Jul 27; 276(30):27757-63.

[74] Lee BH, Ruoslahti E. alpha5beta1 integrin stimulates Bcl-2 expression and cell survival through Akt, focal adhesion kinase, and Ca2+/calmodulin-dependent protein kinase IV. *Journal of cellular biochemistry.* 2005 Aug 15;95(6):1214-23.

[75] Zhang Z, Morla AO, Vuori K, Bauer JS, Juliano RL, Ruoslahti E. The alpha v beta 1 integrin functions as a fibronectin receptor but does not support fibronectin matrix assembly and cell migration on fibronectin. *The Journal of cell biology*. 1993 Jul; 122(1):235-42.

[76] Hynes RO, Zhao Q. The evolution of cell adhesion. *The Journal of cell biology*. 2000 Jul 24;150(2):F89-96.

[77] Roskelley CD, Bissell MJ. Dynamic reciprocity revisited: a continuous, bidirectional flow of information between cells and the extracellular matrix regulates mammary epithelial cell function. *Biochemistry and cell biology = Biochimie et biologie cellulaire*. 1995 Jul-Aug;73(7-8):391-7.

[78] Zamir E, Katz BZ, Aota S, Yamada KM, Geiger B, Kam Z. Molecular diversity of cell-matrix adhesions. *Journal of cell science*. 1999 Jun; 112 (Pt 11):1655-69.

[79] Schwartz MA, Shattil SJ. Signaling networks linking integrins and rho family GTPases. *Trends in biochemical sciences*. 2000 Aug; 25(8): 388-91.

[80] Nobes CD, Hall A. Rho, rac, and cdc42 GTPases regulate the assembly of multimolecular focal complexes associated with actin stress fibers, lamellipodia, and filopodia. *Cell.* 1995 Apr 7;81(1):53-62.

[81] Cernuda-Morollon E, Ridley AJ. Rho GTPases and leukocyte adhesion receptor expression and function in endothelial cells. *Circulation research.* 2006 Mar 31; 98(6): 757-67.

[82] Ridley AJ. Rho proteins: linking signaling with membrane trafficking. *Traffic* (Copenhagen, Denmark). 2001 May;2(5):303-10.

[83] Geiger B, Bershadsky A. Assembly and mechanosensory function of focal contacts. *Current opinion in cell biology*. 2001 Oct;13(5):584-92.

[84] Lo SH. Tensin. *The international journal of biochemistry and cell biology*. 2004 Jan; 36(1):31-4.

[85] Lo SH, Janmey PA, Hartwig JH, Chen LB. Interactions of tensin with actin and identification of its three distinct actin-binding domains. *The Journal of cell biology.* 1994 Jun; 125(5):1067-75.

[86] Calderwood DA, Fujioka Y, de Pereda JM, Garcia-Alvarez B, Nakamoto T, Margolis B, et al. Integrin beta cytoplasmic domain interactions with phosphotyrosine-binding domains: a structural prototype for diversity in integrin signaling. *Proceedings of the National Academy of Sciences of the United States of America.* 2003 Mar 4;100(5): 2272-7.

[87] McCleverty CJ, Lin DC, Liddington RC. Structure of the PTB domain of tensin1 and a model for its recruitment to fibrillar adhesions. *Protein Sci.* 2007 Jun;16(6):1223-9.

[88] Geiger B, Bershadsky A, Pankov R, Yamada KM. Transmembrane crosstalk between the extracellular matrix--cytoskeleton crosstalk. *Nature reviews Molecular cell biology.* 2001 Nov; 2(11):793-805.

[89] Mao Y, Schwarzbauer JE. Fibronectin fibrillogenesis, a cell-mediated matrix assembly process. Matrix biology : *Journal of the International Society for Matrix Biology.* 2005 Sep;24(6):389-99.

[90] Zhong C, Chrzanowska-Wodnicka M, Brown J, Shaub A, Belkin AM, Burridge K. Rho-mediated contractility exposes a cryptic site in fibronectin and induces fibronectin matrix assembly. *The Journal of cell biology.* 1998 Apr 20;141(2):539-51.

[91] Fukata M, Kaibuchi K. Rho-family GTPases in cadherin-mediated cell-cell adhesion. *Nature reviews. 2001* Dec;2(12):887-97.

[92] Chen X, Gumbiner BM. Crosstalk between different adhesion molecules. *Current opinion in cell biology*. 2006 Oct;18(5):572-8.

[93] Lefort CT, Wojciechowski K, Hocking DC. N-cadherin cell-cell adhesion complexes are regulated by fibronectin matrix assembly. *The Journal of biological chemistry.* Jan 28; 286(4):3149-60.

[94] Jewell K, Kapron-Bras C, Jeevaratnam P, Dedhar S. Stimulation of tyrosine phosphorylation of distinct proteins in response to antibody-mediated ligation and clustering of alpha 3 and alpha 6 integrins. *Journal of cell science.* 1995 Mar;108 (Pt3): 1165-74.

[95] Miao H, Li S, Hu YL, Yuan S, Zhao Y, Chen BP, et al. Differential regulation of Rho GTPases by beta1 and beta3 integrins: the role of an extracellular domain of integrin in intracellular signaling. *Journal of cell science.* 2002 May 15;115(Pt 10):2199-206.

[96] Borges E, Jan Y, Ruoslahti E. Platelet-derived growth factor receptor beta and vascular endothelial growth factor receptor 2 bind to the beta 3 integrin through its extracellular domain. *The Journal of biological chemistry.* 2000 Dec 22;275(51):39867-73.

[97] Clark EA, Brugge JS. Integrins and signal transduction pathways: the road taken. *Science.* 1995 Apr 14;268(5208):233-9.

[98] Sieg DJ, Hauck CR, Ilic D, Klingbeil CK, Schaefer E, Damsky CH, et al. FAK integrates growth-factor and integrin signals to promote cell migration. *Nature cell biology*. 2000 May;2(5):249-56.

[99] Piek E, Moustakas A, Kurisaki A, Heldin CH, ten Dijke P. TGF-(beta) type I receptor/ALK-5 and Smad proteins mediate epithelial to mesenchymal transdifferentiation in NMuMG breast epithelial cells. *Journal of cell science.* 1999 Dec; 112 (Pt 24):4557-68.

[100] Roberts AB, Sporn MB, Assoian RK, Smith JM, Roche NS, Wakefield LM, et al. Transforming growth factor type beta: rapid induction of fibrosis and angiogenesis in vivo and stimulation of collagen formation in vitro. *Proceedings of the National Academy of Sciences of the United States of America.* 1986 Jun;83(12):4167-71.

[101] Anzano MA, Roberts AB, Sporn MB. Anchorage-independent growth of primary rat embryo cells is induced by platelet-derived growth factor and inhibited by type-beta transforming growth factor. *Journal of cellular physiology.* 1986 Feb;126(2):312-8.

[102] de Caestecker MP, Piek E, Roberts AB. Role of transforming growth factor-beta signaling in cancer. *Journal of the National Cancer Institute.* 2000 Sep 6; 92 (17): 1388-402.

[103] Gentry LE, Lioubin MN, Purchio AF, Marquardt H. Molecular events in the processing of recombinant type 1 pre-pro-transforming growth factor beta to the mature polypeptide. *Molecular and cellular biology.* 1988 Oct;8(10):4162-8.

[104] Sirard C, Kim S, Mirtsos C, Tadich P, Hoodless PA, Itie A, et al. Targeted disruption in murine cells reveals variable requirement for Smad4 in transforming growth factor beta-related signaling. *The Journal of biological chemistry.* 2000 Jan 21; 275 (3): 2063-70.

[105] Whitman M. Smads and early developmental signaling by the TGFbeta superfamily. *Genes and development.* 1998 Aug 15;12(16):2445-62.

[106] Massague J, Chen YG. Controlling TGF-beta signaling. *Genes and development.* 2000 Mar 15;14(6):627-44.

[107] Shi Y, Wang YF, Jayaraman L, Yang H, Massague J, Pavletich NP. Crystal structure of a Smad MH1 domain bound to DNA: insights on DNA binding in TGF-beta signaling. *Cell.* 1998 Sep 4;94(5):585-94.

[108] Attisano L, Wrana JL. Smads as transcriptional co-modulators. *Current opinion in cell biology.* 2000 Apr;12(2):235-43.

[109] Hassane S, Leonhard WN, van der Wal A, Hawinkels LJ, Lantinga-van Leeuwen IS, ten Dijke P, et al. Elevated TGFbeta-Smad signalling in experimental Pkd1 models and human patients with polycystic kidney disease. *The Journal of pathology.* 2010 Sep; 222(1): 21-31.

[110] Peng F, Zhang B, Wu D, Ingram AJ, Gao B, Krepinsky JC. TGFbeta-induced RhoA activation and fibronectin production in mesangial cells require caveolae. *American Journal of physiology Renal physiology.* 2008 Jul;295(1):F153-64.

[111] Moustakas A, Souchelnytskyi S, Heldin CH. Smad regulation in TGF-beta signal transduction. *Journal of cell science.* 2001 Dec;114(Pt 24): 4359-69.

[112] Moustakas A, Heldin CH. The regulation of TGFbeta signal transduction. *Development* (Cambridge, England). 2009 Nov;136(22): 3699-714.

[113] Wang D, Sun L, Zborowska E, Willson JK, Gong J, Verraraghavan J, et al. Control of type II transforming growth factor-beta receptor expression by integrin ligation. *The Journal of biological chemistry.* 1999 Apr 30;274(18):12840-7.

[114] Dalton SL, Scharf E, Davey G, Assoian RK. Transforming growth factor-beta overrides the adhesion requirement for surface expression of alpha(5)beta(1) integrin in normal rat kidney fibroblasts. A necessary effect for induction of anchorage-independent growth. *The Journal of biological chemistry.* 1999 Oct 15;274(42):30139-45.

[115] Guan JL, Shalloway D. Regulation of focal adhesion-associated protein tyrosine kinase by both cellular adhesion and oncogenic transformation. *Nature.* 1992 Aug 20; 358 (6388): 690-2.

[116] Kornberg L, Earp HS, Parsons JT, Schaller M, Juliano RL. Cell adhesion or integrin clustering increases phosphorylation of a focal adhesion-associated tyrosine kinase. *The Journal of biological chemistry.* 1992 Nov 25;267(33):23439-42.

[117] Hanks SK, Calalb MB, Harper MC, Patel SK. Focal adhesion protein-tyrosine kinase phosphorylated in response to cell attachment to fibronectin. *Proceedings of the National Academy of Sciences of the United States of America.* 1992 Sep 15; 89(18): 8487-91.

[118] Schaller MD, Borgman CA, Cobb BS, Vines RR, Reynolds AB, Parsons JT. pp125FAK a structurally distinctive protein-tyrosine kinase associated with focal adhesions. *Proceedings of the National Academy of Sciences of the United States of America.* 1992 Jun 1;89(11):5192-6.

[119] Ilic D, Furuta Y, Kanazawa S, Takeda N, Sobue K, Nakatsuji N, et al. Reduced cell motility and enhanced focal adhesion contact formation in cells from FAK-deficient mice. *Nature.* 1995 Oct 12; 377 (6549): 539-44.

[120] Ilic D, Kovacic B, Johkura K, Schlaepfer DD, Tomasevic N, Han Q, et al. FAK promotes organization of fibronectin matrix and fibrillar adhesions. *Journal of cell science.* 2004 Jan 15;117(Pt 2):177-87.

[121] Jiang G, Giannone G, Critchley DR, Fukumoto E, Sheetz MP. Two-piconewton slip bond between fibronectin and the cytoskeleton depends on talin. *Nature.* 2003 Jul 17; 424(6946):334-7.

[122] Calalb MB, Polte TR, Hanks SK. Tyrosine phosphorylation of focal adhesion kinase at sites in the catalytic domain regulates kinase activity: a role for Src family kinases. *Molecular and cellular biology.* 1995 Feb;15(2):954-63.

[123] Toutant M, Costa A, Studler JM, Kadare G, Carnaud M, Girault JA. Alternative splicing controls the mechanisms of FAK autophosphorylation. *Molecular and cellular biology.* 2002 Nov; 22(22): 7731-43.

[124] Turner CE. Paxillin and focal adhesion signalling. *Nature cell biology.* 2000 Dec; 2(12):E231-6.

[125] Malumbres M, Barbacid M. RAS oncogenes: the first 30 years. *Nature reviews Cancer.* 2003 Jun;3(6):459-65.

[126] Bustelo XR, Sauzeau V, Berenjeno IM. GTP-binding proteins of the Rho/Rac family: regulation, effectors and functions in vivo. *BioEssays: news and reviews in molecular, cellular and developmental biology.* 2007 Apr;29(4):356-70.

[127] Wennerberg K, Rossman KL, Der CJ. The Ras superfamily at a glance. *Journal of cell science.* 2005 Mar 1;118(Pt 5):843-6.

[128] Ren XD, Kiosses WB, Schwartz MA. Regulation of the small GTP-binding protein Rho by cell adhesion and the cytoskeleton. *The EMBO Journal.* 1999 Feb 1;18(3):578-85.

[129] Amano M, Ito M, Kimura K, Fukata Y, Chihara K, Nakano T, et al. Phosphorylation and activation of myosin by Rho-associated kinase (Rho-kinase). *The Journal of biological chemistry.* 1996 Aug 23; 271 (34): 20246-9.

[130] Bass MD, Morgan MR, Roach KA, Settleman J, Goryachev AB, Humphries MJ. p190RhoGAP is the convergence point of adhesion signals from alpha 5 beta 1 integrin and syndecan-4. *The Journal of cell biology.* 2008 Jun 16;181(6):1013-26.

[131] Huveneers S, Danen EH. Adhesion signaling - crosstalk between integrins, Src and Rho. *Journal of cell science.* 2009 Apr 15;122(Pt 8): 1059-69.

[132] ten Klooster JP, Jaffer ZM, Chernoff J, Hordijk PL. Targeting and activation of Rac1 are mediated by the exchange factor beta-Pix. *The Journal of cell biology.* 2006 Feb 27;172(5):759-69.

[133] McNamee HP, Ingber DE, Schwartz MA. Adhesion to fibronectin stimulates inositol lipid synthesis and enhances PDGF-induced inositol lipid breakdown. *The Journal of cell biology.* 1993 May;121(3):673-8.

[134] Zhu X, Assoian RK. Integrin-dependent activation of MAP kinase: a link to shape-dependent cell proliferation. *Molecular biology of the cell.* 1995 Mar;6(3):273-82.

[135] Chen Q, Kinch MS, Lin TH, Burridge K, Juliano RL. Integrin-mediated cell adhesion activates mitogen-activated protein kinases. *The Journal of biological chemistry.* 1994 Oct 28;269(43):26602-5.

[136] Wierzbicka-Patynowski I, Mao Y, Schwarzbauer JE. Continuous requirement for pp60-Src and phospho-paxillin during fibronectin matrix assembly by transformed cells. *Journal of cellular physiology.* 2007 Mar;210(3):750-6.

[137] Manning BD, Cantley LC. AKT/PKB signaling: navigating downstream. *Cell.* 2007 Jun 29;129(7):1261-74.

[138] Somanath PR, Razorenova OV, Chen J, Byzova TV. Akt1 in endothelial cell and angiogenesis. *Cell cycle.* 2006 Mar;5(5):512-8.

[139] Dong LQ, Liu F. PDK2: the missing piece in the receptor tyrosine kinase signaling pathway puzzle. *American Journal of physiology Endocrinology and metabolism.* 2005 Aug;289(2):E187-96.

[140] Danen EH, Yamada KM. Fibronectin, integrins, and growth control. *Journal of cellular physiology.* 2001 Oct;189(1):1-13.

[141] Accili D, Arden KC. FoxOs at the crossroads of cellular metabolism, differentiation, and transformation. *Cell.* 2004 May 14; 117 (4):421-6.

[142] Kandel ES, Hay N. The regulation and activities of the multifunctional serine/threonine kinase Akt/PKB. *Experimental cell research.* 1999 Nov 25; 253(1):210-29.

[143] Pankov R, Cukierman E, Clark K, Matsumoto K, Hahn C, Poulin B, et al. Specific beta1 integrin site selectively regulates Akt/protein kinase B signaling via local activation of protein phosphatase 2A. *The Journal of biological chemistry.* 2003 May 16;278(20):18671-81.

[144] Bellacosa A, Testa JR, Moore R, Larue L. A portrait of AKT kinases: human cancer and animal models depict a family with strong individualities. *Cancer biology and therapy.* 2004 Mar;3(3):268-75.

[145] Chen J, Somanath PR, Razorenova O, Chen WS, Hay N, Bornstein P, et al. Akt1 regulates pathological angiogenesis, vascular maturation and permeability in vivo. *Nature medicine.* 2005 Nov;11(11):1188-96.

[146] Sun JF, Phung T, Shiojima I, Felske T, Upalakalin JN, Feng D, et al. Microvascular patterning is controlled by fine-tuning the Akt signal. *Proceedings of the National Academy of Sciences of the United States of America.* 2005 Jan 4;102(1):128-33.

[147] Chen WS, Xu PZ, Gottlob K, Chen ML, Sokol K, Shiyanova T, et al. Growth retardation and increased apoptosis in mice with homozygous disruption of the Akt1 gene. *Genes and development.* 2001 Sep 1;15 (17): 2203-8.

[148] Le Clainche C, Carlier MF. Regulation of actin assembly associated with protrusion and adhesion in cell migration. *Physiological reviews.* 2008 Apr;88(2):489-513.

[149] Chen YL, Law PY, Loh HH. Inhibition of PI3K/Akt signaling: an emerging paradigm for targeted cancer therapy. *Current medicinal chemistry Anti-cancer agents.* 2005 Nov; 5(6):575-89.

[150] Somanath PR, Kandel ES, Hay N, Byzova TV. Akt1 signaling regulates integrin activation, matrix recognition, and fibronectin assembly. *The Journal of biological chemistry.* 2007 Aug 3;282(31): 22964-76.

[151] Somanath PR, Chen J, Byzova TV. Akt1 is necessary for the vascular maturation and angiogenesis during cutaneous wound healing. *Angiogenesis.* 2008;11(3):277-88.

[152] Somanath PR, Byzova TV. 14-3-3beta-Rac1-p21 activated kinase signaling regulates Akt1-mediated cytoskeletal organization, lamellipodia formation and fibronectin matrix assembly. *Journal of cellular physiology.* 2009 Feb;218(2):394-404.

[153] Byzova TV, Goldman CK, Pampori N, Thomas KA, Bett A, Shattil SJ, et al. A mechanism for modulation of cellular responses to VEGF: activation of the integrins. *Molecular cell.* 2000 Oct;6(4):851-60.

[154] Byzova TV, Goldman CK, Jankau J, Chen J, Cabrera G, Achen MG, et al. Adenovirus encoding vascular endothelial growth factor-D induces tissue-specific vascular patterns in vivo. *Blood.* 2002 Jun 15;99(12): 4434-42.

[155] Green JA, Berrier AL, Pankov R, Yamada KM. beta1 integrin cytoplasmic domain residues selectively modulate fibronectin matrix assembly and cell spreading through talin and Akt-1. *The Journal of biological chemistry.* 2009 Mar 20;284(12):8148-59.

[156] Shiojima I, Walsh K. Role of Akt signaling in vascular homeostasis and angiogenesis. *Circulation research.* 2002 Jun 28;90(12):1243-50.

[157] Roberts MS, Woods AJ, Dale TC, Van Der Sluijs P, Norman JC. Protein kinase B/Akt acts via glycogen synthase kinase 3 to regulate recycling of alpha v beta 3 and alpha 5 beta 1 integrins. *Molecular and cellular biology.* 2004 Feb;24(4):1505-15.

[158] Qian Y, Corum L, Meng Q, Blenis J, Zheng JZ, Shi X, et al. PI3K induced actin filament remodeling through Akt and p70S6K1: implication of essential role in cell migration. *American Journal of physiology Cell physiology.* 2004 Jan;286(1):C153-63.

[159] Efeyan A, Sabatini DM. mTOR and cancer: many loops in one pathway. *Current opinion in cell biology.* 2010 Apr;22(2):169-76.

[160] Dobashi Y, Watanabe Y, Miwa C, Suzuki S, Koyama S. Mammalian target of rapamycin: a central node of complex signaling cascades. *International journal of clinical and experimental pathology.* 2011 Jun 20;4(5):476-95.

[161] White ES, Sagana RL, Booth AJ, Yan M, Cornett AM, Bloomheart CA, et al. Control of fibroblast fibronectin expression and alternative splicing via the PI3K/Akt/mTOR pathway. *Experimental cell research.* 2010 Oct 1;316(16):2644-53.

[162] Cohen AW, Hnasko R, Schubert W, Lisanti MP. Role of caveolae and caveolins in health and disease. *Physiological reviews.* 2004 Oct;84(4): 1341-79.

[163] Di Guglielmo GM, Le Roy C, Goodfellow AF, Wrana JL. Distinct endocytic pathways regulate TGF-beta receptor signalling and turnover. *Nature cell biology.* 2003 May; 5(5): 410-21.

[164] Razani B, Woodman SE, Lisanti MP. Caveolae: from cell biology to animal physiology. *Pharmacological reviews.* 2002 Sep;54(3):431-67.

[165] Heiser LM, Wang NJ, Talcott CL, Laderoute KR, Knapp M, Guan Y, et al. Integrated analysis of breast cancer cell lines reveals unique signaling pathways. *Genome biology.* 2009;10(3):R31.

[166] Illario M, Cavallo AL, Monaco S, Di Vito E, Mueller F, Marzano LA, et al. Fibronectin-induced proliferation in thyroid cells is mediated by alphavbeta3 integrin through Ras/Raf-1/MEK/ERK and calcium/ CaMKII signals. *The Journal of clinical endocrinology and metabolism.* 2005 May;90(5):2865-73.

[167] Matthews LC, Taggart MJ, Westwood M. Modulation of caveolin-1 expression can affect signalling through the phosphatidylinositol 3-kinase/Akt pathway and cellular proliferation in response to insulin-like growth factor I. *Endocrinology.* 2008 Oct;149(10):5199-208.

[168] Cerezo A, Guadamillas MC, Goetz JG, Sanchez-Perales S, Klein E, Assoian RK, et al. The absence of caveolin-1 increases proliferation and anchorage- independent growth by a Rac-dependent, Erk-independent mechanism. *Molecular and cellular biology.* 2009 Sep; 29(18): 5046-59.

[169] Park SY, Shin HW, Lee KB, Lee MJ, Jang JJ. Differential expression of matrix metalloproteinases and tissue inhibitors of metalloproteinases in thioacetamide-induced chronic liver injury. *Journal of Korean medical science.* 2010 Apr;25(4):570-6.

[170] Arpaia E, Blaser H, Quintela-Fandino M, Duncan G, Leong HS, Ablack A, et al. The interaction between caveolin-1 and Rho-GTPases promotes metastasis by controlling the expression of alpha5-integrin and the activation of Src, Ras and Erk. *Oncogene.* Jul 18.

[171] Bos JL, Rehmann H, Wittinghofer A. GEFs and GAPs: critical elements in the control of small G proteins. *Cell.* 2007 Jun 1;129(5): 865-77.

[172] Linseman DA, Loucks FA. Diverse roles of Rho family GTPases in neuronal development, survival, and death. Frontiers in bioscience : *a Journal and virtual library.* 2008;13:657-76.

[173] Mammoto A, Mammoto T, Ingber DE. Rho signaling and mechanical control of vascular development. *Current opinion in hematology.* 2008 May;15(3):228-34.

[174] Nourshargh S, Hordijk PL, Sixt M. Breaching multiple barriers: leukocyte motility through venular walls and the interstitium. *Nature reviews Molecular cell biology.* 2010 May;11(5):366-78.

[175] Kichina JV, Goc A, Al-Husein B, Somanath PR, Kandel ES. PAK1 as a therapeutic target. *Expert opinion on therapeutic targets.* 2010 Jul; 14(7): 703-25.

[176] Kiosses WB, Daniels RH, Otey C, Bokoch GM, Schwartz MA. A role for p21-activated kinase in endothelial cell migration. *The Journal of cell biology.* 1999 Nov 15; 147(4): 831-44.

[177] Lamalice L, Le Boeuf F, Huot J. Endothelial cell migration during angiogenesis. *Circulation research.* 2007 Mar 30;100(6):782-94.

[178] Zhou GL, Zhuo Y, King CC, Fryer BH, Bokoch GM, Field J. Akt phosphorylation of serine 21 on Pak1 modulates Nck binding and cell migration. *Molecular and cellular biology.* 2003 Nov;23(22):8058-69.

[179] Benard V, Bohl BP, Bokoch GM. Characterization of rac and cdc42 activation in chemoattractant-stimulated human neutrophils using a novel assay for active GTPases. *The Journal of biological chemistry.* 1999 May 7;274(19):13198-204.

[180] Narumiya S, Tanji M, Ishizaki T. Rho signaling, ROCK and mDia1, in transformation, metastasis and invasion. *Cancer metastasis reviews.* 2009 Jun;28(1-2):65-76.

[181] Xu J, Wang F, Van Keymeulen A, Herzmark P, Straight A, Kelly K, et al. Divergent signals and cytoskeletal assemblies regulate self-organizing polarity in neutrophils. *Cell.* 2003 Jul 25;114(2):201-14.

[182] Li Z, Hannigan M, Mo Z, Liu B, Lu W, Wu Y, et al. Directional sensing requires G beta gamma-mediated PAK1 and PIX alpha-dependent activation of Cdc42. *Cell.* 2003 Jul 25;114(2):215-27.

[183] Mercurio AM, Rabinovitz I. Towards a mechanistic understanding of tumor invasion--lessons from the alpha6beta 4 integrin. *Seminars in cancer biology.* 2001 Apr; 11(2): 129-41.

[184] Defilippi P, Olivo C, Venturino M, Dolce L, Silengo L, Tarone G. Actin cytoskeleton organization in response to integrin-mediated adhesion. *Microscopy research and technique.* 1999 Oct 1; 47 (1): 67-78.

[185] Tsakiridis T, Tong P, Matthews B, Tsiani E, Bilan PJ, Klip A, et al. Role of the actin cytoskeleton in insulin action. *Microscopy research and technique.* 1999 Oct 15; 47(2): 79-92.

[186] Aoki M, Blazek E, Vogt PK. A role of the kinase mTOR in cellular transformation induced by the oncoproteins P3k and Akt. *Proceedings of the National Academy of Sciences of the United States of America.* 2001 Jan 2;98(1):136-41.
[187] Khayat ZA, Tong P, Yaworsky K, Bloch RJ, Klip A. Insulin-induced actin filament remodeling colocalizes actin with phosphatidylinositol 3-kinase and GLUT4 in L6 myotubes. *Journal of cell science.* 2000 Jan; 113 Pt 2:279-90.
[188] Pankov R, Endo Y, Even-Ram S, Araki M, Clark K, Cukierman E, et al. A Rac switch regulates random versus directionally persistent cell migration. *The Journal of cell biology.* 2005 Aug 29;170(5):793-802.
[189] Guo F, Debidda M, Yang L, Williams DA, Zheng Y. Genetic deletion of Rac1 GTPase reveals its critical role in actin stress fiber formation and focal adhesion complex assembly. *The Journal of biological chemistry.* 2006 Jul 7;281(27):18652-9.
[190] Vandermoere F, El Yazidi-Belkoura I, Demont Y, Slomianny C, Antol J, Lemoine J, et al. Proteomics exploration reveals that actin is a signaling target of the kinase Akt. Molecular and cellular proteomics : *MCP.* 2007 Jan;6(1):114-24.
[191] Wilker E, Yaffe MB. 14-3-3 Proteins--a focus on cancer and human disease. *Journal of molecular and cellular cardiology.* 2004 Sep;37(3): 633-42.
[192] Mattson MP. Pathways towards and away from Alzheimer's disease. *Nature.* 2004 Aug 5;430(7000):631-9.
[193] Rathmell JC, Thompson CB. Pathways of apoptosis in lymphocyte development, homeostasis, and disease. *Cell.* 2002 Apr;109 Suppl: S97-107.
[194] Patel K, Thompson AJ, Chuang WL, Lee CM, Peng CY, Shanmuganathan G, et al. Insulin resistance is independently associated with significant hepatic fibrosis in Asian chronic hepatitis C genotype 2 or 3 patients. *Journal of gastroenterology and hepatology.* 2011 Jul;26(7):1182-8.
[195] Zubilewicz A, Hecquet C, Jeanny J, Soubrane G, Courtois Y, Mascarelli F. Proliferation of CECs requires dual signaling through both MAPK/ERK and PI 3-K/Akt pathways. *Investigative ophthalmology and visual science.* 2001 Feb;42(2):488-96.
[196] Rossi DJ, Jamieson CH, Weissman IL. Stems cells and the pathways to aging and cancer. *Cell.* 2008 Feb 22;132(4):681-96.
[197] Singh S, Upadhyay AK, Ajay AK, Bhat MK. p53 regulates ERK activation in carboplatin induced apoptosis in cervical carcinoma: a novel target of p53 in apoptosis. *FEBS letters.* 2007 Jan 23;581(2): 289-95.
[198] Fritz V, Fajas L. Metabolism and proliferation share common regulatory pathways in cancer cells. *Oncogene.* 2010 Aug 5;29(31): 4369-77.
[199] Wynn TA. Integrating mechanisms of pulmonary fibrosis. *The Journal of experimental medicine.* 2011 Jul 4;208(7):1339-50.
[200] Hinz B, Gabbiani G. Fibrosis: recent advances in myofibroblast biology and new therapeutic perspectives. *F1000 biology reports.* 2010; 2:78.
[201] Yang SF, Hsieh YS, Lin CL, Hsu NY, Chiou HL, Chou FP, et al. Increased plasma levels of urokinase plasminogen activator and matrix metalloproteinase-9 in nonsmall cell lung cancer patients. Clinica chimica acta; *international Journal of clinical chemistry.* 2005 Apr; 354(1-2):91-9.
[202] Rintoul RC, Sethi T. Extracellular matrix regulation of drug resistance in small-cell lung cancer. *Clinical science.* 2002 Apr;102(4):417-24.

[203] D'Inzeo S, Nicolussi A, Donini CF, Zani M, Mancini P, Nardi F, et al. A Novel Human SMAD4 Mutation Is Involved In Papillary Thyroid Carcinoma Progression. *Endocrine-related cancer.* 2011 Nov 22.

[204] Altomare DA, Testa JR. Perturbations of the AKT signaling pathway in human cancer. *Oncogene.* 2005 Nov 14;24(50):7455-64.

[205] Baron V, Adamson ED, Calogero A, Ragona G, Mercola D. The transcription factor Egr1 is a direct regulator of multiple tumor suppressors including TGFbeta1, PTEN, p53, and fibronectin. *Cancer gene therapy.* 2006 Feb;13(2):115-24.

[206] Price PM. Diabetes: caught in the Akt? *Kidney international.* 2007 May; 71(9):839-41.

[207] Parapuram SK, Shi-wen X, Elliott C, Welch ID, Jones H, Baron M, et al. Loss of PTEN expression by dermal fibroblasts causes skin fibrosis. *The Journal of investigative dermatology.* 2011 Oct;131(10):1996-2003.

[208] Stoker M, O'Neill C, Berryman S, Waxman V. Anchorage and growth regulation in normal and virus-transformed cells. *International Journal of cancer Journal international du cancer.* 1968 Sep 15;3(5):683-93.

[209] Khwaja A, Rodriguez-Viciana P, Wennstrom S, Warne PH, Downward J. Matrix adhesion and Ras transformation both activate a phosphoinositide 3-OH kinase and protein kinase B/Akt cellular survival pathway. *The EMBO Journal.* 1997 May 15; 16(10): 2783-93.

[210] Yang YA, Han WF, Morin PJ, Chrest FJ, Pizer ES. Activation of fatty acid synthesis during neoplastic transformation: role of mitogen-activated protein kinase and phosphatidylinositol 3-kinase. *Experimental cell research.* 2002 Sep 10;279(1):80-90.

[211] Horowitz JC, Lee DY, Waghray M, Keshamouni VG, Thomas PE, Zhang H, et al. Activation of the pro-survival phosphatidylinositol 3-kinase/AKT pathway by transforming growth factor-beta1 in mesenchymal cells is mediated by p38 MAPK-dependent induction of an autocrine growth factor. *The Journal of biological chemistry.* 2004 Jan 9;279(2):1359-67.

[212] Oft M, Peli J, Rudaz C, Schwarz H, Beug H, Reichmann E. TGF-beta1 and Ha-Ras collaborate in modulating the phenotypic plasticity and invasiveness of epithelial tumor cells. *Genes and development.* 1996 Oct 1;10(19):2462-77.

[213] Oft M, Akhurst RJ, Balmain A. Metastasis is driven by sequential elevation of H-ras and Smad2 levels. *Nature cell biology.* 2002 Jul; 4(7): 487-94.

[214] Bhowmick NA, Ghiassi M, Bakin A, Aakre M, Lundquist CA, Engel ME, et al. Transforming growth factor-beta1 mediates epithelial to mesenchymal transdifferentiation through a RhoA-dependent mechanism. *Molecular biology of the cell.* 2001 Jan;12(1):27-36.

[215] Bakin AV, Tomlinson AK, Bhowmick NA, Moses HL, Arteaga CL. Phosphatidylinositol 3-kinase function is required for transforming growth factor beta-mediated epithelial to mesenchymal transition and cell migration. *The Journal of biological chemistry.* 2000 Nov 24; 275 (47): 36803-10.

[216] Bjorklund M, Koivunen E. Gelatinase-mediated migration and invasion of cancer cells. *Biochimica et biophysica acta.* 2005 May 25; 1755(1): 37-69.

[217] Garcia-Gonzalez E, Selvi E, Balistreri E, Lorenzini S, Maggio R, Natale MR, et al. Cannabinoids inhibit fibrogenesis in diffuse systemic sclerosis fibroblasts. *Rheumatology.* 2009 Sep;48(9):1050-6.

[218] Lu Y, Azad N, Wang L, Iyer AK, Castranova V, Jiang BH, et al. Phosphatidylinositol-3-kinase/akt regulates bleomycin-induced fibroblast proliferation and collagen production. *American journal of respiratory cell and molecular biology.* Apr; 42(4): 432-41.

[219] Goodwin A, Jenkins G. Role of integrin-mediated TGFbeta activation in the pathogenesis of pulmonary fibrosis. *Biochemical Society transactions.* 2009 Aug; 37 (Pt 4):849-54.

[220] Huang JF, Chuang YH, Dai CY, Yu ML, Huang CF, Hsiao PJ, et al. Hepatic Akt expression correlates with advanced fibrosis in patients with chronic hepatitis C infection. *Hepatol. Res.* May;41(5):430-6.

[221] Osna NA. An annual topic highlight: alcohol and liver, 2011. *World Journal of gastroenterology : WJG.* 2011 May 28;17(20):2455.

[222] Graham K, Olson MF. The ras signalling pathway as a target in cancer therapy. *Recent results in cancer research Fortschritte der Krebsforschung.* 2007;172:125-53.

[223] Sansal I, Sellers WR. The biology and clinical relevance of the PTEN tumor suppressor pathway. *J. Clin. Oncol.* 2004 Jul 15; 22(14): 2954-63.

In: Fibronectin
Editors: James Beattie

ISBN 978-1-62257-298-4

Chapter VIII

Modulation of Fibronectin Expression in Astrocytes Affects Neuritogenesis: Relevance in Developmental Neurotoxicity

Lucio G. Costa[1,2,*], *Marina Guizzetti*[3], *Daniella Pizzurro*[1] *and Gennaro Giordano*[1]
[1]Dept. of Environmental and Occupational Health Sciences, University of Washington, Seattle, Washington, US
[2]Dept. of Human Anatomy, Pharmacology and Forensic Science, University of Parma Medical School, Parma, Italy
[3]Jesse Brown VA Medical Center and Dept. of Psychiatry, University of Illinois at Chicago, Chicago, Illinois, US

Abstract

Neurite outgrowth in CNS neurons is primarily dependent on extracellular matrix proteins that are produced and released by glial cells. One of these proteins, fibronectin, together with laminin, plays a pivotal role in neuronal differentiation mediated by astrocytes. When co-cultured together, astrocytes interact with neurons and promote their differentiation; interference with this interaction can indirectly affect neuronal development. Cholinergic agonists, through activation of muscarinic M3 receptors in astrocytes increase their ability to promote neuritogenesis, largely by increasing the expression and release of fibronectin and laminin. This effect is mediated by activation of multiple signal transduction pathways in astrocytes. Ethanol, a known developmental neurotoxicant, inhibits muscarinic receptor signal transduction at the level of phosphor-

* Correspondence: Dr. Lucio G. Costa. Dept. of Environmental and Occupational Health Sciences, University of Washington, 4225 Roosevelt #100, Seattle, WA 98105. Tel: 206 5432831. Fax: 206 6854696. E-mail: lgcosta@u.washington.edu.

lipase D, and antagonizes the astrocyte-mediated neuritogenic effect on hippocampal neurons. Similar effects of muscarinic agonists and of ethanol have been also seen in the *in vitro* hippocampal slice. Thyroid hormones (T3) can increase fibronectin expression in astrocytes, and hence increase neuritogenesis in cerebellar neurons.

Conversely, astrocytes from hypothyroid animals, a condition know to cause profound developmental neurotoxicity, have decreased levels of fibronectin and decreased ability to promote neuritogenesis. The metal manganese, also increasingly recognized as a developmental neurotoxicant, accumulates in astrocytes, where it causes oxidative stress. Upon exposure to manganese, astrocytes have a decreased expression /release of fibronectin, and a decreased ability to foster neuritogenesis in hippocampal neurons. These effects are due to manganese-induced oxidative stress and are antagonized by antioxidants. Other oxidants, such as hydrogen peroxide or 2,3-dimethoxy-1,4-naphtoquinone, cause similar effects on astrocytic fibronectin and on neuritogenesis. Additionally, certain organophosphorus insecticides, such as diazinon and its oxygen analog diazoxon, which can also cause oxidative stress, are also capable of altering fibronectin expression and neuritogenesis. These studies indicate that interference with fibronectin expression/release in astrocytes greatly affects their ability to exert neuritogenic effects on developing neurons.

Introduction

Glial cells, and in particular astrocytes, exert a profound effect on neuronal development, as they provide trophic support for neuronal survival, and are involved in neuronal migration, in axonal and dendritic growth, and in synaptogenesis (Christopherson et al. 2005; Barres, 2008; Pfrieger, 2009; Faissner et al. 2010). Neurite outgrowth is a fundamental event in brain development, and is accomplished by signals from the extracellular space such as extracellular matrix proteins, cell adhesion molecules, and soluble factors that can promote or inhibit the growth of neurites (Kyryusko et al. 2004). Astrocyte-neuron interactions play a key role in this process, as they express and release factors that can promote neurite outgrowth, such as laminin-1, N-cadherin, thrombospondin, fibronectin (Price and Hynes, 1985; Kiryushko et al. 2004), or even inhibit neurite outgrowth (e.g. chondroitin sulfate proteoglycans such as neurocan or brevican; Friedlander et al. 1994).

The extracellular matrix, with its main proteins laminin, fibronectin, and thrombospondins, is a major contributor of molecular signals that may regulate different stages of neuronal development (Hynes, 2009). The extracellular matrix should indeed not be seen as an inert scaffold, but as a dynamic network providing cells with signals vital for their survival, differentiation, proliferation, and migration (Hynes, 2009; Singh et al. 2010). Neurite outgrowth is a fundamental event in brain development, but also in the regeneration of damaged nervous tissue. Fibronectin in particular, plays a key role in nerve regeneration (Previtali et al. 2008) by fostering neurite outgrowth (Mukhatyar et al. 2011). Fibronectin is highly expressed in the developing brain, particularly in astrocytes, and induces neuritogenesis both *in vitro* and *in vivo*, by activating receptors of the integrin family that link fibronectin to the cytoskeleton (Price and Hynes, 1985; Kiryushko et al. 2004; Mao and Schwartzbauer, 2005).

This chapter examines aspects of astrocyte-neuron interaction involving primarily fibronectin, and leading to neurite outgrowth. A neurotransmitter, acetylcholine, and a hormone, thyroid hormone, will be discussed for their ability to increase the expression and release of fibronectin in astrocytes, and to foster neuronal differentiation. Inhibition of the effects of acetylcholine (as in case of exposure to ethanol), or a diminished level of thyroid hormone (as in hypothyroidism) causes a decrease in astrocyte fibronectin and a reduced astrocyte-mediated neuritogenesis, thereby inducing developmental neurotoxic effects. Other developmental neurotoxicants, such as manganese chloride or organophosphorus insecticides, which cause oxidative stress in astrocytes, also inhibit fibronectin expression and consequent neuritogenesis.

Modulation of Astrocyte Fibronectin by Cholinergic Muscarinic Agonists and Effect of Ethanol

Astrocytes express various neurotransmitter receptors whose functions in these cells remain for the most part elusive (Porter and McCarthy, 1997). Among these, astrocytes also express cholinergic muscarinic receptors, particularly the M2 and M3 subtypes (Guizzetti et al. 1996). There is substantial evidence that acetylcholine may influence various aspects of brain development (Schmidt, 1981; Buznikov, 1984; Costa 1993; Weiss et al. 1998; Lauder and Schambra, 1999; Guizzetti et al. 1996; Abreu-Villaca et al. 2011). Components of the cholinergic system, including choline acetyltransferase and acetylcholine receptors, are present early in development, long before the appearance of synapses (Costa et al. 1993). Furthermore, acetylcholine levels are high in the neonatal rat brain (80-90% of adult values), and muscarinic receptor-activated signal transduction systems (notably phospholipases C and D) are greatly enhanced in the neonatal rat (Balduini et al. 1987; Costa et al. 1995).

Acetylcholine is believed to have non-transmitter effects during development, as it can regulate morphogenic cell movements during gastrulation, glial cell proliferation, and neuronal differentiation and survival (Lipton and Kater, 1989; Buznikov, 1984; Lauder and Schambra, 1999; Nguyen et al. 2001). Developing neurons may fire action potentials and trigger acetylcholine secretion from the axonal growth cone while the axon is still growing, and neuron-released acetylcholine may contribute to further axonal growth and formation of synaptic contacts (Yao et al. 2000). Furthermore, astrocytes may themselves release acetylcholine, which may act in an autocrine fashion (Wessler et al. 1997).

It has been recently shown that acetylcholine, by activating muscarinic M3 receptors on astrocytes, increases their ability to foster neuritogenesis in neurons, and that this effect is mediated primarily by fibronectin (Guizzetti et al. 2008). Rat cortical astrocytes were incubated for 24 h with the cholinergic agonist carbachol, an analogue of acetylcholine which is not hydrolyzed by acetylcholinesterase (AChE). After complete wash-out, astrocytes were co-cultured for an additional 24 h with rat hippocampal neurons, plated on glass coverslips (Guizzetti et al. 2008).

At the end of this co-culture, neurons were fixed and labeled with anti-βIII-tubulin antibody followed by a secondary antibody and mounted on microscope slides for morphological assessment of neurites (Guizzetti et al. 2008). Upon co-culture with carbachol-primed astrocytes, the average length of the longest neurite (identified as the axon by Tau-1 staining) was increased by 2 to 3-fold with the longest axon measuring 310 um compared to

170 um for control astrocytes. Length of minor neurites (identified by staining with MAP2), and the number of neurites per cell were also increased 2-fold by carbachol-treated astrocytes (Table 1). This effect of carbachol was due to activation of muscarinic M3 receptors, and was mediated by activation of multiple signal transduction pathways, including phospholipase D, phospholipase C and phosphoinositide 3-kinase and their respective downstream effectors, particularly Erk1,2- mitogen activated protein kinase (MAPK) and protein kinase C ζ (Guizzetti et al. 2008; 2011).

As said, neurite outgrowth in the central nervous system is primarily dependent on extracellular matrix proteins that are produced and released by glial cells. In particular, fibronectin and laminin play a pivotal role in neuronal differentiation mediated by astrocytes (Liesi et al. 1986; Matthiessen et al. 1989; Kiryushko et al. 2004). Confocal analysis of carbachol-treated astrocytes indicated that immunofluorescence for fibronectin and laminin-1 was increased; furthermore, in carbachol-treated astrocytes, fibronectin was assembled in organized fibrils, which were not present in unstimulated cells (Guizzetti et al. 2008).

Carbachol also increased fibronectin and laminin protein levels in astrocyte lysate and in astrocyte medium, as measured by Western blot. Of note is that fibronectin levels were considerably higher than those of laminin, in agreement with previous reports (Matthiessen et al. 1989). Levels of mRNA of both fibronectin and laminin were also increased by carbachol in astrocytes (Guizzetti et al. 2008).

Table 1. Effect of carbachol and ethanol on astrocyte-mediated neuritogenesis in hippocampal neurons and on fibronectin expression in astrocytes

	Control	Carbachol (1 mM)	+EtOH 25 mM	+EtOH 50 mM	+EtOH 100 mM
Average length longest neurite (um)	57	158	140 *	65 #	58#
Average length minor neurites (um)	22	42	41	27 #	23 #
Number of neurites/ cell	3.8	7.9	7.0 *	4.2 #	3.9 #
Fibronectin (%)	100	160	140 *	122 #	110 #

Rat cortical astrocytes were incubated with the indicated compounds at the indicated concentrations for 24 h. After washout, astrocytes were co-cultured with hippocampal neurons for 24 h. Neurite length was determined by morphometric analysis. Fibronectin was measured by confocal microscopy in astrocytes, 24 h after incubation with the indicated compounds. Data are from Guizzetti et al. (2008; 2010). Significantly different from carbachol alone * $p<0.05$; # $p<0.01$. Values for carbachol are all significantly different from control, $p<0.01$.

The effect of carbachol on fibronectin was due to activation of muscarinic M3 receptors, and involved activation of the same signal transduction pathways indicated above, i.e. phospholipase D, phospholipase C, and phosphoinositide 3-kinase (Guizzetti et al. 2008; 2011).

Extracellular matrix proteins can be regulated by changes in their expression and/or in their rate of degradation. An important mechanism of fibronectin and laminin proteolysis involves the plasminogen activator system, which leads to the activation of plasminogen into the proteolytic enzyme, plasmin (Vassalli et al. 1991). A proteomic analysis of astrocyte secretoma had identified a number of proteins involved in neuritogenesis, including plasminogen activator inhibitor-1 (PAI-1) (Moore et al. 2009). PAI-1, a member of the serine protease inhibitor superfamily (SERPIN), also called SERPINE-1, inhibits the formation of plasmin, and hence the degradation of fibronectin and laminin (Irigoyen et al. 1999). Carbachol increased the levels of PAI-1 in the astrocyte medium by 2-fold, and up-regulated PAI-1 mRNA levels by a maximum of 3.5-fold (Guizzetti et al. 2008). PAI-1 increased the levels of extracellular matrix proteins, particularly that of fibronectin, and was also capable to stimulate neuritogenesis by a direct action on neurons (Guizzetti et al. 2008).

Two additional experiments add evidence to the fact that fibronectin plays a primary role in astrocyte-mediated neurite outgrowth. First, while both fibronectin and laminin levels remain high in the extracellular matrix 24 h after carbachol wash-out, likely because of their slow turnover, only fibronectin levels are significantly elevated in astrocyte medium at the same time-point, while laminin levels do not differ from control values (Guizzetti et al. 2008). Second, after carbachol wash-out, incubation of astrocytes with a function-blocking fibronectin antibody potently antagonized the effect of carbachol on elongation of the longest neurite and of minor neurites; in contrast, a laminin-1 function-blocking antibody inhibited only minor neurite elongation without affecting the length of the longest process (Guizzetti et al. 2008).

Altogether, these studies identified a novel mechanism by which acetylcholine may affect brain development. By activating muscarinic M3 receptors in astrocytes and associated signal transduction pathways, carbachol induces a transcriptionally-mediated increase in fibronectin (and laminin-1, to a minor extent) expression and release. Through this mechanism, carbachol-primed astrocytes can induce neuritogenesis in neurons. PAI-1 also contributes to the latter event, by inhibiting the degradation of fibronectin and laminin-1, and by exerting a direct neuritogenic action on neurons (Guizzetti et al. 2008).

Ethanol is a known human developmental neurotoxicant. In utero exposure to ethanol can cause Fetal Alcohol Spectrum Disorders (FASD), whose most severe manifestation is represented by the Fetal Alcohol Syndrome (FAS). FAS is characterized by growth retardation, facial dysmorphology, and neurodevelopmental abnormalities, leading to long-lasting cognitive and behavioral deficits (Jones, 2011; Riley et al. 2011). The mechanisms involved in the toxic effects of ethanol on the developing nervous system remain elusive; alcohol has been shown to cause apoptotic neuronal death, to inhibit glial cell proliferation, and to alter, through multiple mechanisms, neuronal development. A possible action of ethanol is inhibition of neurite outgrowth, and recent evidence suggests that ethanol may interfere with the ability of astrocytes to induce neurite outgrowth (Pascual and Guerri, 2007; Chen and Charness, 2008). As ethanol had been reported to inhibit the signaling pathways activated by M3 muscarinic receptors in astrocytes (Costa et al. 2004), it was hypothesized that it may also inhibit carbachol-stimulated, astrocyte-induced neuritogenesis in hippocampal neurons, by inhibiting the release of fibronectin and other neuritogenic molecules. Indeed, incubation of astrocytes with ethanol, at concentrations as low as 25 mM (corresponding to 0.115 mg/dl), in the presence of carbachol, followed by wash-out, inhibited the effects of carbachol on axonal length and on minor neurite length (Table 1; Guizzetti et al. 2010). The primary target for ethanol appeared to be phospholipase D. 1-Butanol, an alcohol which, like ethanol, undergoes a transphosphatidylation reaction thereby inhibiting phospholipase D (Gustavsson, 1995), inhibited the outgrowth of the axon and of minor neurites. In contrast, its tertiary analog *tert*-butanol, which does not inhibit phospholipase D, was devoid of effect

(Guizzetti et al. 2010). Ethanol also inhibited carbachol-induced increases in fibronectin and laminin in the astrocyte lysate and in the medium, as well as the release of PAI-1. These effects were mimicked by 1-butanol, but not by *tert*-butanol, supporting the notion that inhibition of phospholipase D is the main mechanism of ethanol's action. This was further substantiated by the finding that ethanol did not affect the laminin level increase caused in astrocytes by exogenous phosphatidic acid, which lies downstream of phospholipase D (Guizzetti et al. 2010).

The same effects of carbachol and ethanol on neuritogenesis and fibronectin expression were also found in another *in vitro* system, the hippocampal slice (Giordano et al. 2011). Organotypic hippocampal slices, prepared from postnatal day-5 rats, were transfected with a gene gun, and GFP (green fluorescent protein) was expressed for 72 hr (to visualize neurons) before treatments for 24 hr with carbachol (1 mM) and/or ethanol (50 mM). Carbachol increased the length of the longest neurite (the axon), the length of the minor neurites and the number of branches per neuron, without affecting the number of neurites per cell; ethanol inhibited all effects of carbachol on neurite outgrowth. Carbachol also increased the levels of fibronectin and of laminin-1, and this also was antagonized by ethanol (Giordano et al., 2011). It had previously been shown that in addition to stimulating M3 receptors on astrocytes, leading to increased fibronectin expression and release, and increased neuritogenesis (Guizzetti et al. 2008), carbachol could also exert some neuritogenic effects directly on hippocampal neurons (VanDeMark et al. 2009). Such action consisted in an increased axonal length, without any effect on minor neurites, and was mediated by M1 muscarinic receptors (VanDeMark et al. 2009). In the hippocampal slice, a muscarinic M3 receptor antagonist almost completely abolished the effect of carbachol on neuritogenesis, while a muscarinic M1 antagonist had only a minor effect. Furthermore, an M3 antagonist also blocked carbachol-induced increase in fibronectin and laminin. Finally, function-blocking antibodies to fibronectin or to laminin-1 antagonized the effect of carbachol on neurite outgrowth (Giordano et al., 2011). These results indicate that in a more complex system such as the hippocampal slice, which conserves the regional brain architecture and astrocyte-neuron interactions, carbachol-stimulated neuritogenesis is mainly due to an astrocyte/fibronectin-mediated effect. Altogether these findings show that the interaction of ethanol with muscarinic receptor-activated signal transduction (notably phospholipase D) in astrocytes, leads to an impairment of astrocyte-neuron communication; inhibition of the expression and release of fibronectin and of other neuritogenic factors (e.g. laminin, PAI-1) leads to an impaired neuritogenesis in neurons, which may underlie some of the developmental neurotoxic effects observed upon alcohol exposure.

Modulation of Astrocyte Fibronectin by Thyroid Hormones and Effect of Hypothyroidism

Thyroid hormone (3,5,3'-triiodothyronine; T3) is essential for normal development of the nervous system, as it influences several processes of brain development, such as neuronal migration, neurite outgrowth, synaptogenesis, myelination, and glial cell differentiation (Anderson, 2001; Gomes et al. 2001). In the fetal and neonatal periods, hypothyroidism may produce severe mental retardation, deaf-mutism, and spastic diplegia (cretinism) (Trentin, 2006). Astrocytes are believed to be important mediators of T3 action in the developing nervous system (Trentin, 2006). A series of studies in the past several years has shown that T3 can act on astrocytes, which express thyroid hormone receptors (Carlson et al. 1996), and, by affecting astrocyte-neuron interactions, indirectly modulate neuronal development. Cerebellar astrocytes treated with T3, and then co-cultured with cerebellar granule neurons, significantly increased neurite length compared to neurons co-cultured with control astrocytes (Martinez and Gomes, 2002). This effect of T3 appeared to be mediated by epidermal growth factor (EGF); treatment of astrocytes with EGF stimulated neuritogenesis similarly to T3, and exposure of astrocytes to T3 in the presence of an inhibitor of the EGF receptor, inhibited the effect of T3 on neurite outgrowth (Martinez and Gomes, 2002). However, though T3 induced EGF release from astrocytes, the observed effect on neuritogenesis was not due to a direct action of EGF on neurons, but was mediated by an autocrine effect of EGF on astrocytes. Indeed, both T3 and EGF increased the expression of fibronectin and laminin in astrocytes, and, as predicted, in the presence of EGF receptor inhibitors the effect of T3 was significantly inhibited (Martinez and Gomes, 2002). By using specific pharmacological inhibitors, it was also shown that the increase of fibronectin and laminin expression in astrocytes by T3 or EGF was mediated by the MAPK and phosphoinositide 3-kinase pathways (Martinez and Gomes, 2002). Trentin et al. (2003) also found that T3 was able to induce reorganization of fibronectin and laminin distribution, and an increase in their levels, in cerebellar astrocytes. This latter study indicated that the effect of T3 was mediated by fibroblast growth factor 2 (FGF2), which was produced autocrinally by astrocytes upon stimulation with the thyroid hormone. Co-culturing T3-treated astrocytes with cerebellar granule neurons, significantly increased (by about 2-fold) neurite length (Trentin et al. 2003).

These studies clearly indicate that T3 can increase fibronectin and laminin expression in astrocytes, and that this effect, as in the case of activation of muscarinic receptors, leads to an increase in astrocyte-mediated neuritogenesis. As a consequence, a decreased availability of T3, as present in hypothyroidism, would be expected to decrease expression of extracellular matrix proteins in astrocytes, and hence decrease neuronal differentiation. Indeed, when mouse cerebral cortical astrocytes were cultured in thyroid hormone-depleted serum, the distribution of fibronectin was dramatically affected, and fibronectin levels were decreased by 50% (Dezonne et al. 2009). Of interest is the unexpected finding that in this *in vitro* model of hypothyroidism, levels and cellular distribution of laminin were not affected (Dezonne et al. 2009). *In vivo* studies provide additional support to these *in vitro* findings. Rats were treated with 0.02% methimazole in drinking water from gestational day 10 to day of sacrifice (postnatal day 3) to induce hypothyroidism (Calloni et al. 2005). Cerebellar astrocytes from hypothyroid rats had an altered distribution of fibronectin and laminin and a 26-28% decreased in their levels (Mendes-de-Aguiar et al. 2010). When normal or hypothyroid cerebellar granule neurons were co-cultured with hypothyroid astrocytes, neurite length, and the proportion of neurons with neurites was significantly decreased. In contrast, normal astrocytes were able to support equal neuritogenesis in control and hypothyroid neurons (Mendes-de-Aguilar et al. 2010). The effects of hypothyroidism on fibronectin and laminin expression, as well as on astrocyte-mediated neuritogenesis were reversed by *in vitro* treatment of astrocytes with T3 (Mendes-de-Aguiar et al. 2010).

In summary, these studies indicate that T3 can positively modulate fibronectin (and laminin) expression in astrocytes, and indirectly affect neuronal differentiation.

Hypothyroidism, a condition known to cause severe developmental abnormalities, is associated with lower levels of fibronectin and laminin, and with shorter neurite length, which can be reversed by addition of T3.

Modulation of Astrocyte Fibronectin by Agents Causing Oxidative Stress

Manganese (Mn) is an essential trace metal, necessary for normal functioning of a variety of physiological processes in several tissues and organ systems. Mn is also an important cofactor for a number of enzymes such as glutamine synthetase or superoxide dismutase. While Mn deficiency is rare, elevated exposure can occur, leading to accumulation of Mn in brain which results in significant neurotoxicity. In adults, Mn neurotoxicity is characterized by a variety of psychiatric, cognitive, and motor disturbances, the latter resembling Parkinsonism (Levy and Nassetta, 2003). In children, neurotoxicity caused primarily as a result of Mn exposure through parenteral nutrition, is characterized by various neurological and behavioral abnormalities (Erikson et al. 2007; Ljung and Vahter, 2007). Mn is transported into the brain through multiple transport systems and accumulates in certain brain areas (e.g. striatum, substantia nigra), where concentrations of 200-300 uM have been reported (Aschner et al. 2007). Astrocytes have high capacity transporters for this metal, and in these cells concentrations of Mn are 50-60-fold higher than in neurons (Aschner et al. 1999). A number of effects exerted by Mn in astrocytes have been reported, including oxidative stress, mitochondrial dysfunction, alteration of the glutamate/glutamine cycling, changes in antioxidant enzymes, and cellular swelling (Hazel, 2002; Milatovic et al. 2007).

A recent study has shown that Mn interferes with the ability of astrocytes to promote neuritogenesis in hippocampal neurons (Giordano et al. 2009). Exposure of rat astrocytes to $MnCl_2$ (100-500 uM), followed by wash-out, significantly impaired their ability to promote neurite outgrowth. Mn caused a concentration-dependent decrease in axon length, in the length of minor neurites, and in the number of neurites per cell. For example, at the concentration of 200 uM, Mn decreased average axon length by 41%, and average neurite length by 36%. Such effects were shown not to be related to cytotoxicity, or to a direct effect of Mn on neurons (Giordano et al. 2009). Mn caused a concentration-dependent decrease of fibronectin protein in the astrocyte lysate as well as in the medium (Table 2) and a decrease in the levels of fibronectin mRNA (Giordano et al. 2009).

The effect of Mn on astrocyte-mediated neurite outgrowth, and on fibronectin expression appeared to be mediated by oxidative stress, as suggested by the following findings (Giordano et al. 2009): 1) Mn caused a time-, and concentration-dependent increase of the levels of ROS in astrocytes; 2) Two antioxidants, melatonin (200 uM) and N-t-butyl-alpha-phenylnitrone (100 uM), antagonized the effects of Mn (Giordano et al. 2009); 3) increasing intracellular glutathione (GSH) levels by means of GSH ethyl ester (which caused a 58% increase in GSH) antagonized the effect of Mn; 4) Depletion of intracellular GSH with buthionine sulfoximine (which decreased GSH levels in astrocytes by 70%), potentiated the effect of Mn.

Table 2. Modulation of fibronectin protein levels in astrocyte lysate and medium by agents causing oxidative stress

Treatment	Fibronectin (% of control)	
	Lysate	Medium
Control	100	100
Mn (100 uM)	83	85
Mn (200 uM)	60 *	35 #
Mn (500 uM)	11 #	10 #
DMNQ (10 uM)	50 #	51 #
H_2O_2 (10 uM)	28 #	30 #
DZ (10 uM)	62 *	ND
DZO (10 uM)	60 *	ND

Rat cortical astrocytes were incubated with the indicated compounds at the indicated concentrations for 24 h, after which fibronectin was measured by Western blot in astrocyte lysate and in medium. Data are from Giordano et al. (2009) and from Pizzurro et al. (unpublished). ND=Not determined. Significantly different from control * $p<0.05$; # $p<0.01$.

Additional studies carried out with agents known to specifically cause oxidative stress substantiate this hypothesis. Astrocytes were incubated with hydrogen peroxide (H_2O_2; 5 or 10 uM) or 2,3-dimethoxy-1,4-naphtoquinone (DMNQ; 2.5 or 10 uM) for 24 h, before being washed-out and co-cultured with hippocampal neurons for an additional 48 h (Giordano et al. 2009). Both compounds significantly impaired the ability of astrocytes to promote neurite outgrowth in hippocampal neurons at concentrations which did not cause any cytotoxicity in astrocytes. Furthermore, both H_2O_2 and DMNQ caused a significant decrease of fibronectin levels in the lysate and in the medium (Table 2; Giordano et al. 2009).

Two additional chemicals, the organophosphorus (OP) insecticide diazinon (DZ) and its oxygen analog diazoxon (DZO) were also investigated for their ability to alter astrocyte-mediated neurite outgrowth in hippocampal neurons, and astrocytic expression of fibronectin. OP insecticides are one of the main classes of insecticides since the 1940s (Costa, 2008). Their primary target of toxicity is AChE whose inhibition by OPs causes accumulation of acetylcholine at cholinergic synapses, with overstimulation of cholinergic receptors and an ensuing cholinergic syndrome. One aspect of continuous concern in OP toxicology is related to their potential developmental neurotoxicity (Garry, 2004; Costa 2006). It has been suggested that some of the observed effects of OPs on the developing brain may be independent of their cholinergic action and may be due to either the parent compound or other metabolites, which, at difference with the oxon, do not inhibit AChE (Costa, 2006).

One of such alternative mechanisms may be oxidative stress, as several reports have indicated that various OPs and their oxons can induce oxidative stress, which is independent from AChE inhibition, both *in vitro* (Crumpton et al. 2000; Qiao et al. 2005) or *in vivo* (Milatovic et al. 2006; Dettbarn et al. 2006).

In an earlier study, DZ and DZO were found to increase reactive oxygen species (ROS) levels in mouse cerebellar granule neurons, and their cytotoxicity in these cells was related to their ability to induce oxidative stress (Giordano et al. 2007). More recently, it has been found that both compounds also cause an increase of ROS in rat cortical astrocytes, though to a minor extent than in neurons (Pizzurro et al., unpublished results). Utilizing the same experimental protocol as indicated above, it was found that exposure of astrocytes to DZ or DZO (both at 10 uM), followed by wash-out and by co-culture with hippocampal neurons for 48 h, significantly decreased the ability of astrocytes to foster neuritogenesis. In the presence of DZ- or DZO-treated astrocytes, axonal length was ~45-50% less than that observed in the presence of control astrocytes. A similar reduction (~35-40%) of minor neurite length was also observed, while number of neurites/cell was not altered (Pizzurro et al., unpublished observations). As expected, DZ and DZO decreased fibronectin expression in astrocytes (Table 2). Neither compound, at this concentration (10 uM) had any affect on astrocyte viability (Pizzurro et al., unpublished results).

Another recent study reported that exposure to rat astrocytes to cobalt chloride ($CoCl_2$) increased ROS production by 2-fold, and reduced their ability to stimulate neurite outgrowth in Neuro2A mouse neuroblastoma cells (Chen et al. 2011). An oxidative stress-induced decreased in the astrocytic expression of thrombospondin-1 was also observed, but fibronectin was not measured in this study.

Altogether, this series of experiments indicate that induction of oxidative stress in astrocytes leads to alterations in the expression of fibronectin; this, in turn, hampers the ability of astrocytes to promote neurite outgrowth in neurons.

These findings in astrocytes stand in contrast with results obtained in other cellular systems, in which ROS was reported to cause an increase in fibronectin levels. For example, high glucose has been shown to cause a ROS-mediated increase of fibronectin mRNA and protein in human peritoneal mesothelial cells, human umbilical vein endothelial cells, and mesangial cells (Lee et al. 2004; Ihnat et al. 2007; Yeh et al. 2009). It is possible that fibronectin is differentially regulated in astrocytes than in other cell types. Astrocyte-released fibronectin is known to be a 230kDa protein with an extracellular domain lacking in plasma fibronectin (Mahler et al. 1997). Of interest is that two other compounds that have been shown to cause a decrease of fibronectin in astrocytes, ethanol and kainic acid (Mahler et al. 1997; Ren et al. 2000), are also known to cause oxidative stress in these cells (Muscoli et al. 2002; Giordano et al. 2006).

It is also possible that in astrocytes oxidative stress leads to an altered expression of proteins such as PAI-1, which modulate fibronectin levels. Altered expression and release, by ROS-exposed astrocytes, of other neuritogenic (e.g. laminin, N-cadherin, meteorin) or neurite-inhibiting (e.g. neurocan, brevican) factors, which have been identified in the astrocyte secretoma (Moore et al. 2009), may also contribute to the observed effects on neuritogenesis.

Conclusion

The extracellular matrix plays a major role in astrocyte-neuron communication, leading to enhanced neurite outgrowth, and fibronectin is a main player in this regard. As neuritogenesis is a fundamental aspect of brain development, negative modulators of fibronectin expression and release would be expected to cause developmental neurotoxicity.

Indeed, a number of known developmental neurotoxicants were shown to affect fibronectin levels in astrocytes thereby impairing astrocyte-mediated neuronal differentiation.

Further studies should explore potential mechanisms and molecular events involved in the regulation of fibronectin expression in astrocytes.

Furthermore, given the role of fibronectin, and other proteins such as laminin, in fostering neuritogenesis during brain development and upon nerve injury, it is not surprising that novel strategies are directed toward the creation of novel artificial extracellular matrix proteins with potential therapeutic efficacy. In this regard, it is of interest that an artificially designed fusion protein containing a sequence from laminin and one from fibronectin was shown to increase neurite outgrowth in N1E-115 neuroblastoma cells more than laminin-1 itself (Nakamura et al. 2009).

Acknowledgments

Research by the authors was supported by grants W81XWH-05-1-0239 from the Manganese Health Research Program (DoD), and by NIH (AA08154; ES07032; ES07033; ES09601).

References

Abreu-Villaca Y, Filgueiras CC, Manhaes AC. Developmental aspects of the cholinergic system. *Behav. Brain Res.* 221: 367-378, 2011.

Anderson GW. Thyroid hormone and the brain. *Front. Neuroendocrinol.* 22: 1-17, 2001.

Aschner M, Vrana KE, Zhang W. Manganese uptake and distribution in the central nervous system (CNS). *NeuroToxicology* 20: 173-180, 1999.

Aschner M, Guilarte TR, Schneider JS, Zheng W. Manganese: recent advances in understanding its transport and neurotoxicity. *Toxicol. Appl. Pharmacol.* 221: 131-147, 2007.

Balduini W, Murphy SD, Costa LG. Developmental changes in muscarinic receptor-stimulated phosphoinositide metabolism in rat brain. *J. Pharmacol. Exp. Ther.* 241: 421-427, 1987.

Barres BA. The mistery and magic of glia: a perspective on their roles in health and disease. *Neuron* 60: 430-440, 2008.

Buznikov GA. The action of neurotransmitters and related substances on early embryogenesis. *Pharmacol. Ther.* 25: 23-59, 1984.

Calloni GW, Penno CA, Cordova FM, Trentin AG, Neto VM, Leal RB. Congenital hypothyroidism alters the phosphorylation of ERK1/2 and p38MAPK in the hippocampus of neonatal rats. *Brain Res. Dev. Brain Res.* 154: 141-145, 2005.

Carlson DJ, Strait KA, Schwartz HL, Oppenheimer JH. Thyroid hormone receptor isoform content in cultured type 1 and type 2 astrocytes. *Endocrinology* 126: 121-124, 1996.

Chen S, Charness ME. Ethanol inhibits neuronal differentiation by disrupting activity-dependent neuroprotective protein signaling. *Proc. Natl. Acad. Sci. USA* 105: 19962-19967, 2008.

Chen JK, Zhan YJ, Yang CS, Tzeng SF. Oxidative stress-induced attenuation of thrombospondin-1 expression in primary astrocytes. *J. Cell. Biochem.* 112: 59-70, 2011.

Christopherson KS, Ullian EM, Stokes CC, Mullowney CE, Hell JW, Agah A, Lawler J, Mosher DF, Bornstein P, Barres BA. Thrombospondins are astrocyte-secreted proteins that promote CNS synaptogenesis. *Cell* 120: 421-433, 2005.

Costa LG. Muscarinic receptors in the developing nervous system. In: *Receptors in the Developing Nervous System* (Zagon I, McLaughlin P, eds.), Chapman and Hall, London, 21-42, 1993.

Costa LG. Current issues in organophosphate toxicology. *Clin. Chim. Acta* 366: 1-13, 2006.

Costa LG. Toxic effects of Pesticides. In: Casarett and Doull's Toxicology. *The Basic Science of Poisons*. 7th Edition. (Klaassen CD, ed.), McGraw-Hill, New York, NY, pp. 883-930, 2008.

Costa LG, Balduini W, Renò F. Muscarinic receptor stimulation of phospholipase D activity in the developing brain. *Neurosci. Res. Comm.* 17:169-176, 1995.

Costa LG, Vitalone A, Guizzetti M. Signal transduction mechanisms involved in the antiproliferative effects of ethanol in glial cells. *Toxicol. Lett.* 149: 67-73, 2004.

Crumpton TL, Seidler FJ, Slotkin TA. Is oxidative stress involved in the developmental neurotoxicity of chlorpyrifos? *Dev. Brain Res.* 121: 189-195, 2000.

Dettbarn, WF, Milatovic D, Gupta RC. Oxidative stress in anticholinesterase-induced excitotoxicity. In: *Toxicology of Organophosphate and Carbamate Compounds* (Gupta RC, ed.), Elsevier, Amsterdam, pp. 511-532, 2006.

Dezonne RS, Stipursky J, Gomes FCA. Effect of thyroid hormone depletion on cultured murine cerebral cortex astrocytes. *Neurosci. Lett.* 467: 58-62, 2009.

Erikson KM, Thompson K, Aschner J, Aschner M. Manganese neurotoxicity: a focus on the neonate. *Pharmacol. Ther.* 113: 369-377, 2007.

Faissner A, Pyka' M, Geissler M, Sobik T, Frischknecht R, Guldenfinger ED et al. Contributions of astrocytes to synapse formation and maturation-Potential functions of the perisynaptic extracellular matrix. *Brain Res. Rev.* 63: 26-38, 2010.

Friedlander DR, Milev P, Kathikeyan L, Margolis RK, Margolis RU, Grumet M. The neuronal chondroitin sulfate proteoglycan neurocan binds to the neural cell adhesion molecules Ng-CAM/L1/NILE and N-CAM, and inhibits neuronal adhesion and neurite outgrowth. *J. Cell Biol.* 125: 669-680, 1994.

Garry VF. Pesticides and children. *Toxicol Appl. Pharmacol.* 198: 152-163, 2004.

Giordano G, White CC, McConnachie LA, Fernandez C, Kavanagh TJ, Costa LG. Neurotoxicity of domoic acid in cerebellar granule cells in a genetic model of glutathione deficiency. *Mol. Pharmacol.* 70: 2116-2126, 2006.

Giordano G, Afsharinejad Z, Guizzetti M, Vitalone A, Kavanagh TJ, Costa LG. Organophosphorus insecticides chlorpyrifos and diazinon and oxidative stress in a genetic model of glutathione deficiency. *Toxicol. Appl. Pharmacol.* 219: 181-189, 2007.

Giordano G, Pizzurro D, VanDeMark KL, Guizzetti M, Costa LG. Manganese inhibits the ability of astrocytes to promote neuronal differentiation. *Toxicol. Appl. Pharmacol.* 240: 226-235, 2009.

Giordano G, Guizzetti M, Dao K, Mattison H, Costa LG. Ethanol impairs muscarinic-receptor-induced neuritogenesis in rat hippocampal slices: role of astrocytes and extracellular matrix proteins. *Biochem. Pharmacol.* 82: 1792-1799, 2011.

Gomes FCA, Lima FRS, Trentin AG, Moura Neto V. Thyroid hormone role in nervous system morphogenesis. *Prog. Brain Res.* 132; 41-50, 2001.

Guizzetti M, Costa P, Peters J, Costa LG. Acetylcholine as a mitogen: muscarinic receptor-mediated proliferation of rat astrocytes and human astrocytoma cells. *Eur. J. Pharmacol.* 297: 265-273, 1996.

Guizzetti M, Moore NH, Giordano G. Costa LG. Modulation of neuritogenesis by astrocyte muscarinic receptors. *J. Biol. Chem.* 283: 31884-31897, 2008.

Guizzetti M, Moore NH, Giordano LG, VanDeMark KL, Costa LG. Ethanol inhibits neuritogenesis induced by astrocyte muscarinic receptors. *Glia* 58: 1395-1406, 2010.

Guizzetti M, Moore NH, VanDeMark KL, Giordano G, Costa LG. Muscarinic receptor-activated signal transduction pathways involved in the neuritogenic effect of astrocytes in hippocampal neurons. *Eur. J. Pharmacol.* 659: 102-107, 2011.

Gustavsson L. Phosphatidylethanol formation: specific effects of ethanol mediated via phospholipase D. *Alcohol Alcohol.* 30: 391-406, 1995.

Hazell AS. Astrocytes and manganese neurotoxicity. *Neurochem Int.* 41: 271-277, 2002.

Hynes RO. The extracellular matrix: not just pretty fibrils. *Science* 326: 1216-1219, 2009.

Ihnat MA, Thorpe JE, Kamat CD, Szabo C, Gree DE, Warnke LA, Lacza Z, Cselenyak A, Ross K, Shakir S, Piconi L, Kaltreider RC, Ceriello A. Reactive oxygen species mediate a cellular "memory" of high glucose stress signaling. *Diabetologia* 50: 1523-1530, 2007.

Irigoyen JP, Munoz-Canoves P, Montero L, Koziczak M, Nagamine Y. The plasminogen activator system: biology and regulation. *Cell. Mol. Life Sci.* 56: 104-132, 1999.

Jones KL. The effects of alcohol on fetal development. *Birth Def. Res.* (Pt. C) 93: 3-11, 2011.

Kiryushko D, Berezin V, Bock E. Regulators of neurite outgrowth: role of cell adhesion molecules. *Ann. N.Y. Acad. Sci.* 1014: 140154, 2004.

Lauder JM, Schambra UB. Morphogenetic roles of acetylcholine. *Environ. Health Perspect.* 107 (Suppl. 1): 65-69, 1999.

Lee HB, Yu MR, Song JS, Ha H. Reactive oxygen species amplify protein kinase C signaling in high glucose-induced fibronectin expression by human peritoneal mesothelial cells. *Kidney Int.* 65: 1170-1179, 2004.

Levy BS, Nassetta WJ. Neurologic effects of manganese in humans: a review. *Int. J. Occup. Environ. Health* 9: 153-163, 2003.

Liesi P, Kirkwood T, Vaheri A. Fibronectin is expressed by astrocytes cultured from embryonic and early postnatal rat brain. *Exp. Cell Res.* 163: 175-185, 1986.

Lipton SA, Kater SB. Neurotransmitter regulation of neuronal outgrowth, plasticity and survival. *Trends Neurosci.* 12: 265-270, 1989.

Ljung K, Vahter M. Time to re-evaluate the guideline value for manganese in drinking water? *Environ. Health Perspect.* 115: 1533-1538, 2007.

Mahler M, Be-Ari Y, Represa A. Differential expression of fibronectin, tenascin-C, and NCAMs in cultured hippocampal astrocytes activated by kainite, bacterial lipopolys accharide or basic fibroblast growth factor. *Brain Res.* 775: 63-73, 1997.

Mao Y, Schwartzbauer JE. Fibronectin fibrillogenenesis, a cell-mediated matrix assembly process. *Matrix Biol.* 24: 389-399, 2005.

Martinez R, Gomes FC. Neuritogenesis induced by thyroid hormone-treated astrocytes is mediated by epidermal growth factor/mitogen-activated protein kinase-phosphatidylinositol 3-kinase pathways and involves modulation of extracellular matrix proteins. *J. Biol. Chem.* 277: 49311-49318, 2002.

Matthiessen HP, Schmalenbach C, Muller HW. Astroglia-released neurite growth-inducing activity for embryonic hippocampal neurons is associated with laminin bound in a sulfated complex and free fibronectin. *Glia* 2: 177-188, 1989.

Mendes-de-Aguiar CBN, Alchini R, Zucco JKZ, Costa-Silva B, Decker H, Alvarez-Silva M, Tasca CI, Trentin AG. Impaired astrocytic extracellular matrix distribution under congenital hypothyroidism affects neuronal development in vitro. *J. Neurosci. Res.* 88: 3350-3360, 2010.

Milatovic D, Gupta RC, Aschner M. Anticholinesterase toxicity and oxidative stress. *Sci. World* 6: 295-310, 2006.

Milatovic D, Yin Z, Gupta RC, Sidoryk M, Albrecht J, Aschner JL, Aschner M. Manganese induces oxidative impairment in cultured rat astrocytes. *Toxicol. Sci.* 98: 198-205, 2007.

Moore NH, Costa LG, Shaffer SA, Goodlett DR, Guizzetti M. Shotgun proteomics implicates extracellular matrix proteins and protease systems in neuronal development induced by astrocyte cholinergic stimulation. *J. Neurochem.* 108: 891-908, 2009.

Mukhatyar VJ, Salmeron-Sanchez M, Rudra S, Mukhopadaya S, Barker TH, Garcia AJ, Bellamkonda RV. Role of fibronectin in topographical guidance of neurite extension on electrospun fibres. *Biomaterials* 32: 3958-3968, 2011.

Muscoli C, Fresta M, Cardile V, Palumbo M, Renis M, Puglisis G, Paolino D, Nistico S, Rotiroti D, Mollace V. Ethanol-induced injury in rat primary cortical astrocytes involves oxidative stress: effect of idebenone. *Neurosci. Lett.* 329: 21-24, 2002.

Nakamura M, Mie M, Mihara H, Nakamura M, Kobatake E. Construction of a multi-functional extracellular matrix protein that increases number of N1E-115 neuroblast cells having neurites. *J. Biomed. Res. Mater. Pt.B: Appl. Biomater.* 91B: 425-432, 2009.

Nguyen L, Rigo JM, Rocher V, Belachew S, Malgrange B, Rogister B, Leprince P, Moonen G. Neurotransmitters as early signals for central nervous system development. *Cell Tissue Res.* 305: 187-202, 2001.

Pascual M, Guerri C. The peptide NAP promotes neuronal growth and differentiation through extracellular signal-regulated protein kinase and Akt pathways, and protects neurons co-cultured with astrocytes damaged by ethanol. *J. Neurochem.* 103: 557-568, 2007.

Pfrieger FW. Role of glia in synapse development. *Cell. Mol. Life Sci.* 66: 2037-2047, 2009.

Porter JT, McCarthy KD. Astrocytic neurotransmitter receptors in situ and in vivo. *Prog. Neurobiol.* 51: 439-455, 1997.

Previtali SC, Malaguti MC, Riva N, Scarlato M, Dacci P, Dina G, Triolo D, Porrello E, Lorenzetti I, Fazio R, Comi G, Bolino A, Quattrini A. The extracellular matrix affects axonal regeneration in peripheral neuropathies. *Naurology* 71: 322-331, 2008.

Price J, Hynes RO. Astrocytes in culture synthesize and secrete a variant form of fibronectin. *J. Neurosci.* 5: 2205-2211, 1985.

Qiao D, Seidler FJ, Slotkin TA. Oxidative mechanisms contributing to the developmental neurotoxicity of chlorpyrifos. *Toxicol. Appl. Pharmacol.* 206: 17-26, 2005.

Ren LQ, Garrett DK, Syapin M, Syapin PJ. Differential fibronectin expression in activated C6 glioma cells treated with ethanol. *Mol. Pharmacol.* 58: 1303-1309, 2000.

Riley EP, Infante MA, Warren KR. Fetal alcohol spectrum disorders: an overview. *Neuropsychol. Rev.* 21: 73-80, 2011.

Singh P, Carraher C, Schwarzbauer JE. Assembly of fibronectin extracellular matrix. *Annu. Rev. Cell Dev. Biol.* 26: 397-419, 2010.

Schmidt H. Muscarinic acetylcholine receptor in chick limb bud during morphogenesis. *Histochemistry* 71: 89-98, 1981.

Trentin AG. Thyroid hormone and astrocyte morphogenesis. *J. Endocrinol.* 189: 189-197, 2006.

Trentin AG, Mendes-de-Aguiar CBN, Garcez RC, Alvarez-Silva M. Thyroid hormone modulates the extracellular matrix organization and expression in cerebellar astrocyte: effects on astrocyte adhesion. *Glia* 42: 359-369, 2003.

VanDeMark KL, Guizzetti M, Giordano G, Costa LG. The activation of M1 muscarinic receptor signaling induces neuronal differentiation in pyramidal hippocampal neurons. *J. Pharmacol. Exp. Ther.* 329: 532-542, 2009.

Vassalli JD, Sappino AP, Belin D. The plasminogen activator/plasmin system. *J. Clin. Invest.* 88: 1067-1072, 1991.

Weiss ER, Maness P, Lauder JM. Why do neurotransmitters act like growth factors? *Prespect. Dev. Neurobiol.* 5: 323-335, 1998.

Wessler I, Reinheimer T, Klapproth H, Schneider FJ, Racke K, Hammer R. Mammalian glial cells in culture synthesize acetylcholine. *Naunyn Schmiedeberg's Arch. Pharmacol.* 356: 694-697, 1997.

Yao WD, Rusch J, Poo M, Wu CF. Spontaneous acetylcholine secretion from developing growth cones of Drosophila central neurons in culture: effects of cAMP-pathway mutations. *J. Neurosci.* 20: 2626-2637, 2000.

Yeh CH, Chang CK, Cheng MF, Lin HJ, Cheng JT. The antioxidative effect of bone morphogenetic protein-7 against high glucose-induced oxidative stress in mesangial cells. *Biochem. Biophys. Res. Commun.* 382: 292-297, 2009.

In: Fibronectin
Editors: James Beattie

ISBN 978-1-62257-298-4

Chapter IX

Evidence for Fibronectin Role on Uterine Natural Killer Cells Migration, Proliferation, Differentiation and Survival during Pregnancy

***Bruno Zavan*[1], *AndréaMollica do Amarante Paffaro*[1], *Paulo P. Joazeiro*[2], *Áureo. T.Yamada*[2] *and Valdemar A.Paffaro Jr*[1,*]**

[1]Integrative Biology Laboratory-LABAint, Biomedical Science Institute, UNIFAL-MG, Alfenas,MG. Brazil
[2]LCI/DHE/IB -UNICAMP, Campinas, SP.Brazil

Abstract

Fibronectin is an extracellular matrix component which has a crucial role in the establishment of a successful pregnancy. In this respectit plays an important function in cell migration. Uterine Natural Killer cells are the dominant lymphocytes found in the pregnant mammalian uterus. Studies have identified four differentiation stages of mouse uNK cells based on *Dolichos biflorus agglutinin* lectin cytochemistry. These differentiated cells are distributed in preferential regions in the uterus throughout pregnancy. One function of these cells is the release of cytokines such as IFN-γ leading to spiral artery dilation and the maintenance of decidual integrity.Therefore these cells are essential for the normal development of decidua and placenta. When integrin expression on uNK cells was investigated in mouse uterine sections from gestation day 6 to 17it was found thatα5, α6 and β7 integrins were expressed in cells from gestation day 8 to 10 but not after gd12. Lamininwas expressed only on the antimesometrial side. β7 integrinis an active receptor for VCAM-1 or MAdCAM-1 expressed by endothelial cells and this function promotes extra vasation of uNK cells through blood vessel walls. The absence of

* Email: paffaroj@unifal-mg.edu.br.

laminin closed to uNK cells suggests that these cells are not dependent on laminin and α6 integrin for their establishment within the extracellular matrix. However, fibronectin is present at gestation day 6 to 10 around uNK cells from the myometrium and the endometrial-myometrial junction. No staining for fibronectin was seen in the decidualized and non-decidualized endometrium near the placenta. Our results suggest that fibronectin appears to support uNK migration, proliferation, differentiation and survival in the uterus by binding toα5 integrin. The loss of α5 integrin ligation by the down regulation of fibronectin could inhibit these cellular events and perhaps unligated α5 can activelyinitiate apoptosis, possibly via a caspase 8-dependent pathway. This process has been described asintegrin-mediated cell death.

Introduction

The pregnant uterus is a dynamic environment wherechanges in physiology occur due to cell influx, migration and proliferation as well as alterations in expression of extracellular matrix molecules. A transient accumulation of uterine natural killer (uNK) cells is noted in pregnant rodent and human uterus [1-4]. These lynphocytes are the most abundant found in uterine mucosa tissues reaching 70% to 80% of total leukocytes in the first trimester of human gestation, then start to decline, and return to basal levels at the end of the pregnancy [5]. In rodents, influx of these cells starts after blastocyst implantation andare restricted to the mesometrial lymphoid aggregate of pregnancy (MLAp) and deciduas basalis within each implantation site [1][6-7]. Small non-granular lymphocytes are recognized as uNK cell precursors [8-11]. They proliferate and differentiate by accumulation ofsecretory lisosome granules content [7] confined to the mesometrial side of the pregnantuterus [4][9][12].

The major roles defined to date for mouse uNK cells are the decidual spiral arteries remodeling allowing dilation of these vessels to supply the blood flow into fetal placenta and maintaining decidual reaction[13-14] throughrelease of Interferon-γ (INF-γ)[15]. These findings, together with their distinct phenotype and gene expression profile [16], strongly suggest that uNK cells are pivotal in successful pregnancy outcomes.

Morphologicaland stereological studies have characterizedfour uNK cell subsets-associated to the maturation stages and their distribution within the pregnant mouse uterus (Figure 1). Thesesmall agranular immature forms are mainly distributed inthe MLAp, whilethe large granular mature forms are found preferentially in the deciduas Basalis (DB), were their incidences gradually increase until gestational day (gd) 12. [12] [17].

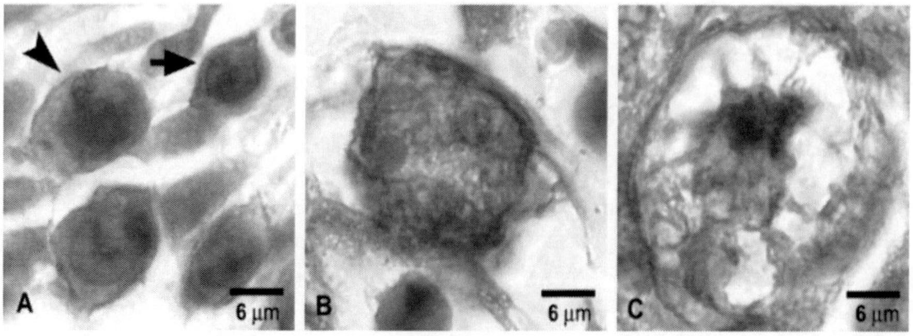

Figure 1.

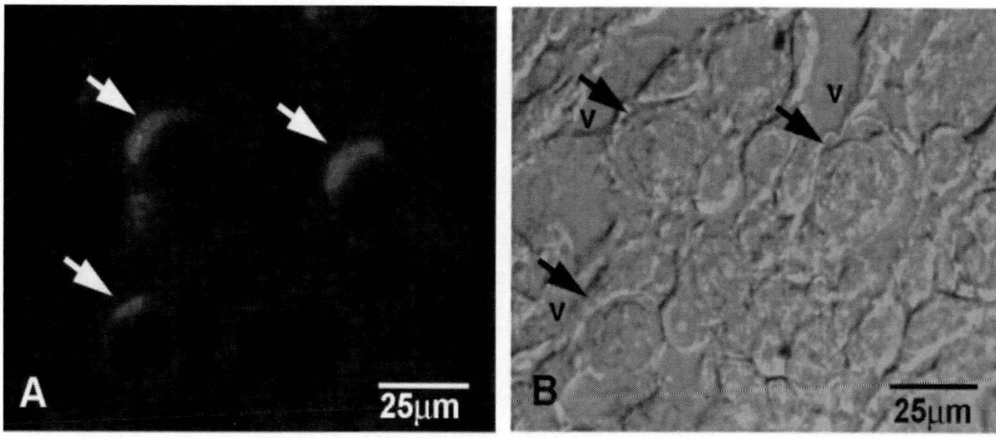

Figure 2.

Positive β7 immunocytochemicalstaining is predominantlyconcentrated on the uNK cell surfaceclose to the uterine blood vessels (Figure2), suggesting that the uNK cell can bind to blood vessels throughβ7 integrin which itself is probably associated α4 integrin subunit forming the VLA-4 molecule. Therefore, uNK cells could use α4β7 to bind endothelial VCAM-1 or MAdCAM-1 for infiltrationinto the uterus [17].

Uterine Natural Killer cells also express α6 integrin although nolaminin staining can be observed around these cells or at any location on the mesometrial aspect of the uterus. Therefore uNK cells could useα6 integrin in combination with the β1 or β4 (VLA-6)to bind to the laminin in other sitesas spleen, an important reservatory ofuNK precursors [18], beforemigration to the uterus.Once in the uterus, uNK would express VLA-6 to binding to another extracellular matrix compound instead to laminin.[17].

Fibronectin and α5 Integrins Ensure the Development of Healthy Pregnancy

Fibronectin has a heterogeneous distribution in pregnant mouse uterus as has its ligand α5 integrin on the uNK surface (Figure 3). This extracellular matrix component cannot be found in the most decidualized areas of the uterine wall. [19]. Therefore, the decidualization phenomenon that is characterized by the uterine fibroblast-like transformation to decidual cells[20], followed by reduction in the extracellular matrix content.[19], results in a decrease in fibronectin in the decidualized areas near the embryo.In the same region, fibronectin reduction is accompanied by loss of α5 integrin expression on the uterine Natural Killer cell surface until gd12. At the same period, uNK cells express α5 integrin in a rich fibronectin environment in nondecidualized mesometrial regions which are adjacent to decidualized tissues [17].Aftergd12, the amount of fibronectin decreases due to advancing decidualization and this phenomenon parallels a decrease in α5 integrin expression in uNK cells. Moreover, this is the period of pregnancy where the uNK cells are predominantly senescent and start to decrease in number in the pregnant uterus[17].These findings suggest that the uNK cells may use the α5β1(VLA-5) to bind tofibronectin and migrate in the uterus. Thisis supported by studieswhich showuNK cell viability in a fibronectin environment both *in vitro* and *in vivo*[3], and further studies which show uNK cell precursor migration from themyometrial region (fibronectin$^+$), through theendometrial area distant to the embryo (fibronectin$^+$) and arriving differentiated in endometrial regions close to the embryo (fibronectin-)[4].

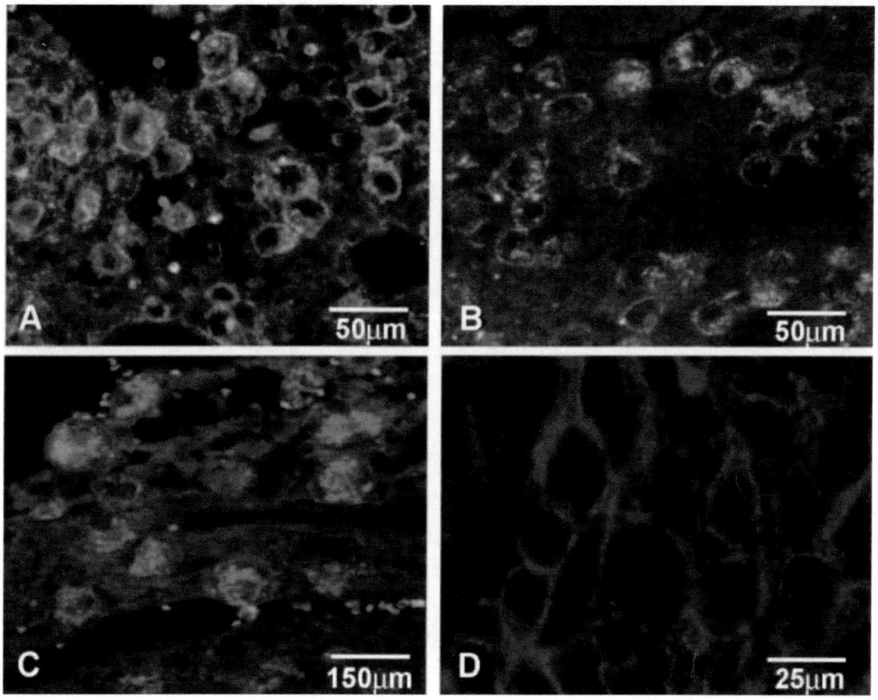

Figure 3.

The number of uNK cells decreases fromgd10 to the end of pregnancy[1][4] and in areas close to the embryo the senescent subtype of uNK cells predominate. While the pregnancy develops, the senescent uNK cell numbers increase in all mesometrial regions[4]. These findings are similar to the pattern of fibronectin expression in the uterus, with a decrease in expression of FN close to the site of embryo implantation. Similar expression patterns are seen in all areas of the mesometrial aspect during gd12-17. [17].

Notably, the concomitantly loss of α5 integrin and fibronectin expressionoccurs in a uterine region described as a senescent uNK cell rich area[4] while the over expression of unligated integrin α5 has been associated with apoptosis and reduced tumor cell growth in vitro and in vivo[21-24]promoting cell deathby a mechanismcharacterized as "integrin-mediated death".

Thus fibronectin and α5 integrins seem to be the key molecules for the establishment, proliferation, differentiation, migration and survival of uNK cells during pregnancy, and therefore, essential molecules to ensure a healthy pregnancy. [17].

References

[1] Stewart I and Peel S (1981) Granulated metrial gland cells in the virgin and early pregnant mouse uterus. *J. Anat.* 133: 535-541.

[2] King A and Loke YW (1991) On the nature and function of human uterine granular lymphocytes. *Immunol. Today*. Dec; 12(12):432-5.

[3] Croy BA; Kiso Y (1993) Granulated metrial gland cells, a natural subset of the pregnant murine uterus. *Microsc. Res. Tech.*, 25: 189 – 200.

[4] Paffaro Jr VA, Bizinotto MC, Joazeiro PP, Yamada AT (2003) Subset Classification Of Mouse Uterine Natural Killer Cells By DBA Lectin Reactivity. *Placenta.* 24: 479.

[5] Bulmer JN and Lash GE (2005) Human uterine natural killer cells: a reappraisal. *Mol. Immunol.*; 42:511–521.

[6] Dickson AD, Bulmer D (1971) The fine structure of cells in the rat metrial gland. *J. Anat.* 108: 123-133.

[7] Croy BA, Guimond MJ, Luross J, Hahnel A, Wang B, Van-Den Heuve M. (1997) terine Natural Killer Cells do not Require Interleukin for Differentiation or Maturation. *Am. J. Reprod. Immunol.* 37: 463.

[8] Stewart I (1983) An investigation into the differentiation of granulated metrial gland cells in the early pregnant mouse uterus. *J. Anat.* 137: 85-93.

[9] Peel S, Stewart I (1989) Rat granulated metrial gland cells differentiate in pregnant chimeric mice and may be cytotoxic for mouse trophoblast. *Cell Differ. Develop.* 28: 55-64.

[10] Parr EL, Parr MB, Zheng LM, Young JDE (1991) Mouse granulated metrial gland cells originated by local activation of uterine natural killer lymphocytes. *Biol. Reprod.* 44: 834-841.

[11] Van Den Heuvel M, McBey BA, Hahnel AC, Croy BA (1996) An analysis of the uterine lynphocyte-derived hobridoma cell line GWM 1-2 for expression of receptor for estrogen, progesterone and interleikin 2. *J. Reprod. Immunol.* 31: 37-50.

[12] Croy BA, Ashkar AA, Foster RA; Disanto JP, Magram J, Carson D, Gendler SJ, Grusby MJ, Wagner N, Muller W, Guimond MJ (1997) Histological studies of gene-ablated mice support important functional roles for natural killer cells in the uterus during pregnancy. *J. Reprod. Immunol.* 35: 111-133.

[13] Hunt JS, Miller L, Vassmer D, Croy BA (1997) Expression of the inducible nitric oxide synthase gene in mouse uterine leukocytes and potential relationships with uterine function during pregnancy. *Biol. Reprod.* 57:827.

[14] Ashkar AA, Croy BA (1999) Interferon-gamma contributes to the normalcy of murine pregnancy. *Biol. Reprod.* 61:493.

[15] Ashkar AA, Di Santo JP, Croy BA (2000) Interferon gamma contributes to initiation of uterine vascular modification, decidual integrity, and uterine natural killer cell maturation during normal murine pregnancy. *J. Exp. Med.* 192:259.

[16] Koopman LA, Kopcow HD, Rybalov B, Boyson JE, Orange JS, Schatz F, Masch R, Lockwood CJ, Schachter AD, Park PJ, Strominger JL (2003) Human decidual natural killer cells are a unique NK cell subset with immunomodulatory potential. *J. Exp. Med.*198:1201–1212, Oct 20.

[17] Zavan B, Paffaro AM, Joazeiro PP, Yamada AT, Paffaro VA Jr. 2010. Immunocytochemical studies of adhesion molecules on mouse uNK cells and their extracellular matrix ligands during mouse pregnancy.*Anat. Rec.* 293: 1081-1088.

[18] Zavan B, Carvalho CC, Rossi Jr WC, Paffaro AMA, Paffaro Jr VA. (2012). Splenectmy delays uterine natural killer cell recruitment to implantation sites and prolongs pregnancy in mice.*Anat. Rec.* . 295: 1221-1228.

[19] Rider V, Carlone DL, Witrock D, Cai C, Oliver N. (1992). Uterine fibronectin mRNA content and localization are modulated during implantation. *Dev. Dyn.* 195:1-14.

[20] Abrahamsohn PA and Zorn TMT (1993) Implantation and decidualization in rodents, *J. Exp. Zool.* 266: 603-28.

[21] Varner JA, Emerson DA, Juliano RL. 1995. Integrin alpha 5 beta 1 expression negatively regulates cell growth: reversal by attachment to fibronectin. *Mol. Biol. Cell.* 6:725-740.

[22] Kim S, Bell K, Mousa SA, Varner JA. 2000. Regulation of angiogenesis in vivo by ligation of integrin alpha5beta1 with the central cell-binding domain of fibronectin. *Am. J. Pathol.* 156:1345-1362.

[23] Plath T, Detjen K, Welzel M, von Marschall Z, Murphy D, Schirner M, Wiedenmann B, Rosewicz S. 2000. A novel function for the tumor suppressor p16(INK4a): induction of anoikis via upregulation of the alpha(5)beta(1) fibronectin receptor. *J. Cell Biol.* 150:1467-1478.

[24] Stupack DG, Puente XS, Boutsaboualoy S, Storgard CM, Cheresh DA. 2001. Apoptosis of adherent cells by recruitment of caspase-8 to unligated integrins. *J. Cell Biol.* 155:459-470.

In: Fibronectin
Editors: James Beattie

ISBN 978-1-62257-298-4

Chapter X

Fibronectin: Structure, Biosynthesis and Roles in Disease

Kátia da Silva Calabrese[1*], Luiz Otávio Pereira Carvalho[1], Tânia Zaverucha do Valle[1], Flávia de Oliveira Cardoso[1] and Ana Lúcia Abreu-Silva[2]

[1]Laboratório de Imunomodulação e Protozoologia, Instituto Oswaldo Cruz, FIOCRUZ. Rio de Janeiro, RJ, Brasil

[2]Departamento de Patologia, Universidade Estadual do Maranhão, Cidade Universitária Paulo VI, São Luís, MA, Brasil

Abstract

Fibronectin (FN) is a protein encoded by a single gene that, through alternative splicing, generates its two major forms: cellular FN and plasma FN. The cellular FN is an insoluble glycoprotein produced by some epithelial cells, fibroblasts, macrophages and endothelial cells, among others. It is involved in several cellular processes such as embryogenesis, tissue repair, cell migration and adhesion. It links the different components of the extracellular matrix (ECM) and helps to organize the cellular interaction with the matrix. Furthermore, it serves as an adhesion molecule, anchoring cells and pathogens. The plasma FN is a soluble form secreted by hepatocytes. The functional form of FN comprises two similar subunits of 220–250 kDa that together form a dimer maintained by antiparallel disulfide bonds. The protein is formed by a sequence of modular structures that are organized to form different binding sites for integrins, collagen, heparin, FN itself and other extracellular molecules. There are three modular structures FNI1, 2II and 3III, all formed by antiparallel beta sheets, but differing in size and primary structure. Both FN I and II can form disulfide bonds that stabilizes the structure, whereas FN III, which is the predominant module, cannot. In the last two decades, the concept of FN function has changed. Now it is known that FN as well as other components of the extracellular matrix areis involved in different cell functions

*Email:calabrese@ioc.fiocruz.br.

such as proliferation, differentiation, migration, and apoptosis. Therefore, FN plays an important role in different pathologies. In sickle cell disease, for instance, adhesion molecules expressed on the surface of the sickle erythrocyte bind to FN in the endothelium, provoking blood vessel occlusion. FN also plays an important role in some infectious diseases, especially by influencing tissue invasion. In *Leishmania* infection, the presence of the parasite influences the synthesis of FN which in turn interferes with the adherence, endocytosis and subsequent internalization of the parasites. FN also determines the increase of *Leishmania* migration at the infection site, facilitating their access to target cells and dissemination. Products of FN degradation recruit macrophages and stimulate the enhancement of parasite phagocytosis by these cells. Similarly, FN is also involved in the invasion mechanisms of *Yersinia pestis*, the etiological agent of plague. These bacteria express a protein that indirectly allows the cleavage of ECM proteins, such as FN and laminin, which are essential for the invasiveness and virulence of bacteria. In HIV infection FN fragments, which are produced by viral and host proteases in the course of inflammatory conditions, can stabilize and increase the infectivity of infected cells, helping the dissemination of the virus.

Structure

Fibronectin (FN) is a modular protein involved in several cellular processes. It is encoded by a single gene that, through alternative splicing, generates its two major forms: cellular FN and plasma FN. FN is one of the protein components of the extracellular matrix (ECM), where it helps to organize cellular interaction. It also participates in embryogenesis, tissue repair and blood clotting. In addition, it serves as an adhesion molecule, anchoring cells and pathogens.

FN is present in the blood plasma (300μg/mL) and in other body fluids in a soluble form and in the ECM in an insoluble multimeric form [1]. The molecular structure of FN is composed of two non-identical chains of 230 to 270 kDa, covalently linked by C-terminal disulphide bonds. Each chain is composed of a series of repeating modules or domains: 12 type I modules (1-12Fn1), 2 type II (1-2Fn2), 15 to 18 (depending on splicing) type III (15-18Fn3), and a variable (V) or IIICS region that is not homologous to other parts of FN (Figure 1).

Each module is encoded by a separate exon [2]. The modules type I, II, ED-A and ED-B correspond to single exons, type III corresponds to two exons (except for III9). The IIICS is combined with the first half of 15FN IIIn3 in one exon [(1]). Although FN molecules are the product of a single gene, found in the human chromosome 2 (at locus p14-16 or q34-36), which contains approximately 50 exons [3],.Tthe resulting protein can exist in multiple forms in consequence ofresulting from alternative splicing of a single pre-mRNA that can generate as many as 20 variants in human FN [4, 5]. Three regions of FN are subject to alternative splicing and in general the matrix form of the molecule has a higher content of these segments than the soluble form. The 2 type III modules that are subject to alternative splicing are called ED-A (ED for "extra domain") and ED-B. Plasma FN is synthesized in the liver by hepatocytes and contains neither ED-A nor ED-B, whereas so-called "cellular" FN (synthesized locally in tissues) contains variable amounts of either or both ED-A and ED-B [(1]). Furthermore, in plasma FN dimers, only 1 of the subunits contains the V region, whereas almost all cellular FN subunits contain this region. Each domain is linked by flexible polypeptide segments.

Different splice variants have specific cellular effects. Studies on chondrogenesis have shown the ED-A domain to be essential at the level of cellular condensation [6]. In an experimental rat hepatic fibrosis model, there was a marked increase in ED-A–positive FN expressed in sinusoidal endothelial cells, and the matrixes deposited by these cells stimulated the conversion of liver lipocytes to myofibroblasts [7]. This stimulation was blocked by antibodies specific to the ED-A domain, whereas recombinant FN peptides containing the ED-A domain mimicked the effect.

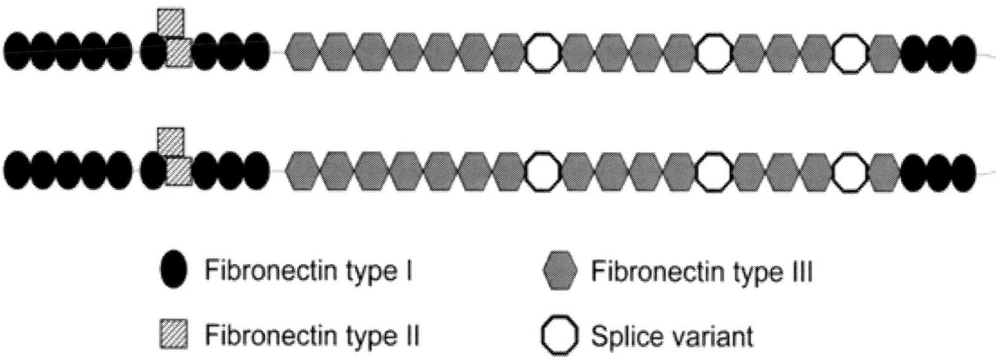

Figure 1. A schematic diagram of the modular, multidomain architecture of a cellular fibronectin dimer.

Fibrosarcoma cells expressing recombinant ED-A–domain have increased adhesion and spreading when compared to ED-A–negative FN cells [8]. Unlike the effects on liver lipocytes, this effect cannot be blocked by anti–ED-A antibodies but is blocked by Arg-Gly-Asp (RGD)-containing peptides or antibodies against α5 and ß1 integrin subunits. Furthermore, purified α5ß1 binds more avidly to ED-A–containing FN than ED-A–negative FN, suggesting that the effect of the ED-A exon splicing is to improve binding affinity to integrin.

FN binds to α5ß1 and αIIbß3 integrins through a cell adhesive site comprising modules III8-III10. The most critical site is the RGD sequence in III10 [9]. The FN molecule also has many other adhesive sites for various substances, including fibrin, heparin, collagen, and α4ß1 and α4ß7 integrins (Figure 1). A Gln residue close to the amino terminus serves as a transglutamination site for activated factor XIII (fXIIIa; plasma transglutaminase), cross-linking FN to various other proteins, including fibrin, fibrinogen, and FN itself. Cross-linking by fXIIIa is important for incorporation of FN into fibrin clots during the clotting process [10, 11]. The α4ß1 and α4ß7 integrins recognize a Leu-Asp-Val sequence in the V region of FN [12].

Type I Module Structure

The FNIn1 modules contains ~45 amino acids, with four cysteines forming two disulphide bonds in 1-3 and 2-4 configurations, and is found in the amino-terminal and carboxyl terminal regions of the full-length protein [13]. Two anti-parallel beta strands make the top sheet, which folds over a bottom sheet, composed of three antiparallel beta strands. These sheets interact through hydrophobic bonding that considerably stabilizes the module. Each FN1 I module is stabilized by two disulphide bonds, one between opposite sheets and one between beta strands of the same sheet [14].

Interactions between adjacent modules are imperative for the structure. For instance, the linker between the fourth and fifth module pair of the FnN1 I relies on hydrophobic interactions with the turns between strand B and C and the turns between D and E of I5F1. Such interactions avoid rotation between these FNIn1 modules.

The first five FN In1 modules (1-5F1) of the N-terminal domain are important in ECM assembly. Deletion of any of these modules prevents FN from contributing to the ECM. The 4-5 Fn1N I modules, along with the 10-12 F1N I modules, are involved in fibrin binding, an important event during the formation of blood clots. Finally, the F1N I modules that surround the two F2N II modules assist in collagen binding [15].

Type II Module Structure

Only two FN IIn2 modules occur in FN as a single tandem repeat. The Fn2N II module is highly conserved and is found in several proteins.,Hhowever, theits crystal structure of the Fn2 module of FN has not yet been determined. Similar to Fn1N I, Fn2N II modules contain approximately 49 residues and are stabilized by two disulphide bonds in 1-3 and 2-4 configurations [16]. The Fn2N II modules hasve four beta strands in its core.

Two antiparallel beta strands make the top sheet, which folds perpendicularly over the two antiparallel beta strands of the bottom sheet. Again, the folding of these beta sheets is stabilized by two disulfide linkages [17]. A cavity in the side of the module made of aromatic and hydrophobic residues, in conjunction with a charged serine, is thought to specifically bind leucine and/or isoleucine sequences in collagen [18].

Type III Module Structure

The largest of the modules, it contains ~ 90 amino acids in its sequence. In fact, over 67 proteins have been identified as containing F3 modules and several tertiary structures of this module are known. The structure of the III10 module is typical of the family. It has seven β-strands forming two antiparallel β-sheets of three and four strands, which overlap to form a hydrophobic core [19]. These substantial hydrophobic core interactions stabilize the module structure; they do not require stabilizing intrachaindisulphide bonds, which gives some mobility to the structure. This characteristic may be important in fibrillogenesis, as well as in determining the mechanical properties of the fibrillar matrix [20-22].

The main roles of FN IIIn3 modules are to mediate interactions between proteins and to act as spacers, positioning the protein in functional sites [23]. The III10 modules contain the RGD sequence, the FN receptor (integrin) binding motif. The III10 module alone does not bind to the integrin receptor with the same affinity as intact FN. Synergistic sites on III8 and III9 modules participate in RGD binding to integrin. FN IIIn3 modules also play a role in matrix assembly and heparin binding [24, 25].

Most F3N III modules are encoded by two exons, whereas F1N I and FN II2 modules are encoded by one exon each. Their structure is modulated through alternative pre-mRNA splicing producing multiple FN mRNAs from a single gene. Splicing occurs primarily in three regions: ED-B (also called EIIIB or EDII), ED-A (also called EIIIA or EDI) and IIICS (also called V or variable).

Depending on the species and tissue specificity the IIICS region is spliced through a complex mechanism, while, ED-B and ED-A can each be spliced through exon skipping [4]. A good example of how FN is regulated by alternative splicing is found during embryonic chondrogenesis or the formation of cartilage during development. In pre-cartilaginous mesenchymal cells, FN containing the ED-A exon is expressed. However, as these cells differentiate into cartilage, ED-A is removed through mRNA splicing. Only minus-ED-A fibronectin is found in mature chondrocytes [26].

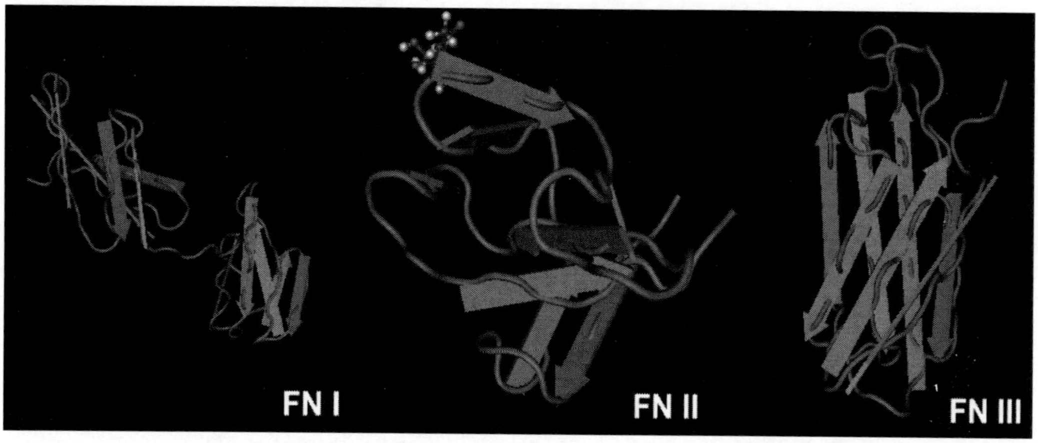

Figure 2. Simplified representations of the three dimensional structures of individual structural domains of fibronectin; FN nI (27) (PDB:1qgb), FnN II (28) (PDB:2fn2)and FnN III (19) (PDB:1ttf).

Biosynthesies

FN recognition by specific integrin receptors is necessary for FN deposition to take place. Several studies show that the main receptor responsible for the assembly of FN in the matrix is the α5β1 integrin, which mediates its unfolding in a process known as fibrillogenesis. Fibronectin deposition is characterized by the formation of specialized ECM–cell contact structures called fibrillar adhesion points, which are generated through an organized interplay between integrins, cytosolic proteins, and actin cytoskeleton [29, 30].

In accordance to Singh et al. [31], conformational changes expose additional FN-binding sites to promote further FN interactions, associations between fibrils, and matrix insolubility. Experimental studies show that the dynamics of FN secretion and formation of a fibrillar matrix occur preferentially in the presence of high concentrations of OH groups [32] and the kindling-2 protein can play a role in FN deposition probably due to interactions with integrins and phosphoinositides [33].

Roles in Disease

FN has been shown to be involved in the pathogenesis of multiple diseases, both infectious and non infectious. As a target for adhesive molecules present on the surface of cells and pathogens, FN acts as a substrate for different cells that possess specific FN ligands. One example of a non-infectious disease in which FN is involved is sickle-cell disease, or sickle-cell anemia. This is an autosomal recessive genetic blood disorder characterized by red blood cells that assume an abnormal sickle shape. This shape reduces cell flexibility, resulting in serious disease. The vaso-occlusive crisis is one of the causes for lethality among sickle-cell disease patients. This crisis is caused by the sickle-shaped red blood cells that obstruct capillaries and restrict blood flow to an organ, often resulting in organ damage. In comparison to normal red-blood cells, the sickle-erythrocytes present a higher number of adhesion molecules, including the α4β1 integrin, which is a ligand for plasma FN [34]. Therefore, sickle-cells have higher chance to interact and adhere to the endothelium, which favors the vaso-occlusive crisis. In addition to the endothelium, sickle-red-blood cells also seem to bind to monocytes and platelets that circulate in the peripheral blood. In this binding FN would act as a bridge, linking the two cell types together [35]. Patients with sickle-cell disease present reduced levels of FN in the plasma [36]. This fact may be linked to a higher consumption of these molecules in cell aggregates [35].

Despite the importance of FN in a series of non-pathogenic diseases, this chapter will focus on FN role during infection. When a pathogen infects its host, the pathogen needs to interact with the host in order to replicate and sustain infection. Intracellular pathogens, for instance, need to be in contact with their host cells before itthey can invade or be phagocytosed by the host cell. Extracellular pathogens need to adhere to their hosts to avoid being washed of. Therefore, each pathogen has developed different strategies to allow this kind of interaction. As we will show, several of these strategies pass by the interaction with host FN.

Staphylococcus aureus was the first bacterium that was shown to interact with host FN [37]. *S. aureus* is responsible for several pathologies including abscesses, pneumonia, osteomyelitis, meningitis, endocarditis and sepsis, as well as several skin and soft tissue infections such as impetigo, folliculitis, cellulitis and infected ulcers and wounds. The emergence of resistant strains has complicated the treatment of these infections [38].

Two proteins known as FnBPA and FnBPB were identified as being responsible for the interaction of *S. aureus* with host FN. [39]. They have been shown to have multiple high-affinity binding repeats, which can bind to consecutive FN In1 modules in the N-terminal region of FN [40]. As a facultative intracellular bacterium, *S. aureus* can use its FnBPs to adhere to the FN in ECM and also to attach itself to the surface of some cell types, such as fibroblasts, osteoblasts, endothelial and epithelial cells [1]. In order to invade the host cells, the FnBP from the bacteria forms a bridge with the cellular α5β1 integrin [41]. The binding of FnBPA to the FN on the surface of endothelial cells promotes bacterial movements and induces a cyclic recruitment of cellular actin and Rab5, which eventually will allow bacteria endocytosis. However, once inside the host cell, *S. aureus* is often eliminated by lysosomal effector mechanisms, unless they exert a high cellular toxicity before uptake [42]. Nevertheless, the bacterium has evolved mechanisms to escape destruction. It has been proposed that interaction between FnBPA/FN slows both the assembly of actin phagocytic cup and the formation of early phagosomes.That strategy would purposely delay cell invasion, allowing the time for the bacteria to produce high amounts of toxins before entering the cell [43].

On the other hand, an *S. aureus* invasion of endothelial cells is associated with the production of pro-inflammatory cytokines, such as interleukinIL-6 and interleukinIL-1β, which are important mediators of immunity against bacteria [44]. A double mutant lacking both FnBPA and FnBPB genes induce lower levels of cytokine production [45]. Therefore, FnBPs are not only important in establishing *S. aureus* infection, but also in activating the cellular response against infection. Some research has been done using *S. aureus*FnBPs to induce immunity. A recombinant fusion protein including an FnBP and a collagen binding protein was able to induce the production of antibodies that were able to bind to the surface of the bacterium, inhibiting the adhesion to FN and collagen. Mice immunized with the fusion protein survived significantly longer following the challenge with *S. aureus* than non-immunized mice [46].

After the description of *S. aureus* FN binding proteins, similar proteins from several other bacteria have also been described. Today we can find FN receptors in most patient isolates, which proves that these proteins are important virulent factors.

Yersinia pestis, for instance, has been described to have two kinds of FN receptors. This bacillus is the causative agent of plague, which is transmitted to humans by flea bite or by human to human contact. When transmitted by infected fleas, the bacteria will cause a systemic infection, known as bubonic plague. When acquired by the respiratory route, the bacillus will cause pulmonary plague, which is even more lethal than the former. Although many consider plague as an ancient infection, much attention has been brought to this bacterium with the description of a resistant strain and due to its potential as a biological weapon [47].

The plasminogen activator Pla is an outer membrane protease coded by a 9.5kb plasmid of *Y. pestis*. Its proteolytic activity is capable of activating plasminogen into plasmin, which is a potent serine protease that cleaves fibrin clots and non-collagenous proteins of the ECM, such as laminin and FN [48]. It has been thought that this activity could cause damage to the host tissues barriers, explaining the pathogenic function of this protein [47]. It has also been demonstrated that Pla can cleave the complement C3 protein, interfering with the activation of the complement system [49]. Further studies have shown that Pla is involved in a mechanism for producing host-derived activity that can degrade components of the ECM, such as laminin and FN. Pla has been described to enhance bacterial adherence to human ECM. Furthermore, a recombinant *E. coli* expressing Pla is able to adhere to the ECM efficiently [47].This activity seems to be essential for the bacteria invasiveness and virulence.

Another key virulence factor of *Y. pestis* is the outer membrane protein named Ail (Attachment Invasion Locus; gene *y1324*), which has a role in colonization and pathogenesis. Ail associates with FN in the ECM creating a bridge to the host cell that mediates bacterial adhesion. This interaction is important to allow the delivery of toxic components by *Y. pestis*[50].

In addition to bacteria, viruses also seem to interact with FN and use it in their favor. The Human Immunodeficient virus (HIV) replicates inside cells from the host immunological system, especially $CD4^+$ lymphocytes. The virus gp120 glycoprotein - present in its envelope – interacts with the host CD4 receptor, allowing the virus to invade the host cells. In the early stages of infection, HIV establishes a persistent infection within lymphoid tissues, where the virus continues to replicate and infect naive cells. The ECM of lymph nodes contains abundant FN (FN), particularly in the interfollicular (T cell-rich) compartment. The viral gp120 binds to a recombinant fragment derived from the first type III repeat of FN and results in increased viral adhesion to lymphocytes [51]. There is also *in vitro*evidence showing that when bound to the ECM, the viral particles remain infectious for a prolonged period [52]. The authors hypothesized that the trapping of the virus particles by matrix proteins might increase local concentration of particles and facilitate the binding of the virus to its receptors.[52].

In spite of binding to the ECM, proteases released in the course of HIV infection have been reported to cause 110-120 kDa FN fragments (FNf) to appear in the blood of many patients. These fragments seem to enhance transendothelial migration of HIV-infected cells, which may explain the dissemination of infected cells into cardiac, renal and central nervous system tissues [53]. Also the persistent presence of these fragments in the plasma of the patients might be a prognostic marker [54]. Finally, an increase in the expression of FN has been detected in the lymph nodes of AIDS patients [52].

As was reported earlier, FN is involved in infectious diseases acting as a substrate for cells that have specific FN ligand. This protein plays a significant role in some parasitic diseases like Leishmaniasis, Chagas disease and Toxoplasmosis, where the parasite binds to host FN facilitates parasite association with infected cells.

Brittingham*et al.* [55] proposed the idea that parasites can interact with ECM and basement membrane proteins such as FN within the skin, leading to enhanced invasion. Host FN is used by *Leishmania* spp. and *Trypanosomacruzi* to bridge their association with monocytes and macrophage host cells [56]. Invasive *Leishmania*sp has a surface protein,:leishmanolysin, observed in both promastigotes and amastigotes, which facilitates the binding of this parasite to FN. Promastigotes as well as amastigotes of different *Leishmania* species have been shown to degrade FN. After FN degradation by leishmanolysin, multiple fragments with capacity to down-regulate reactive-oxygen intermediate production in infected host macrophages are released. Based on these results Kulcani*et al.* [57] suggested that *Leishmania* species interact with FN through a specific receptor protein and that parasites extensively degrade FN using two protease systems. This promotes local ECM invasion and production of FN degradation products that can influence the activation state of host macrophages. Macrophages, for example, can interact with different FN domains via various receptors.

Proteolytic degradation of FN produces peptide fragments which may influence macrophage activation and function. The binding of macrophages to peptide fragments containing the FN interconnecting segment (ICS) domain can decrease macrophage expression of some cytokines such as interferon gamma, interleukin 12, monocyte chemotactic protein 1 and beta transforming growth faction [58]. However, an increase of tumor necrosis factor alpha secretion can be induced by interaction of macrophages with intact FN [59].

Abreu-Silva *et al.* [60] showed that *L. amazonensis* infection caused changes in ECM and proposed that these changes are related to the genetic background of the host. BALB/c - highly susceptible, C57BL/6 - intermediate susceptibility, and DBA/2 - resistant mice, display distinct susceptibilities to the cutaneous infection by *L. amazonensis*. Results showed that in these mice, a decrease of the FN expression in tissues was more marked in areas of high parasite burden. Based on this, the authors suggested that amastigotes could be directly involved in the degradation of ECM components as suggested by McGwire *et al.* [61].

Fibronectin expression was also studied in the liver of dogs naturally infected by *L. chagasi*. According to the authors, adhesive FN was expressed more highly in the livers of symptomatic dogs than of asymptomatic dogs [62].

Fibronectin can also attach to *Leishmania*promastigotes, *in vitro*, and act favoring their association with mononuclear phagocytes [63]. Using human retinal pigment epithelium (ARPE-19) cells infected by *L. amazonensis*, *in vitro*, FN production was evaluated and the results provide clear evidence that the promastigotes modulated FN production by ARPE-19 cells. An FN expression reduction in ARPE-19 cells was observed up to 24 hours after *L. amazonensis* infection, when the quantity of adherent and internalized parasites was high. However, the increase of this protein expression and parasite load decrease were observed up to 72 hours after infection.

It is known that FN mediates *Trypanosoma cruzi*-host cell adhesion *in vitro* [64]. *In vivo*, alterations in ECM have been described in *T. cruzi* infection in the murine model, in both acute and chronic phases. An increased expression of soluble FN in serum of infected mice [64] and the presence of FN in cardiac tissues has been reported to appear correlated with the presence of inflammatory cells [65].

Extracellular alterations have also been observed in central nervous systems colonized by *T. cruzi*; these alterations are associated with inflammatory lesions. The participation of FN and its receptor, the VLA-4 molecule, is associated the inflammatory cell migration into the central nervous system [66]. Moreover, the pattern of FN expression is similar to that observed in the central nervous system of other inflammatory disease like autoimmune diseases and tumors [67, 68].

References

[1] Henderson B, Nair S, Pallas J, Williams MA. Fibronectin: a multidomain host adhesin targeted by bacterial fibronectin-binding proteins. *Fems Microbiology Reviews*. 2011 Jan;35(1):147-200.

[2] Patel RS, Odermatt E, Schwarzbauer JE, Hynes RO. Organization of the fFibronectin gGene Pprovides Eevidence for Eexon Sshuffling Dduring Eevolution. *Embo Journal*. 1987 SEP;6(9):2565-72.

[3] Kornblihtt AR, Umezawa K, Vibepedersen K, Baralle FE. Primary Sstructure of Hhuman Ffibronectin - Ddifferential Ssplicing Mmay Ggenerate at Lleast 10 Ppolypeptides from a Ssingle Ggene. *Embo Journal*. 1985;4(7):1755-9.

[4] Ffrench-Constant C. Alternative Ssplicing of Ffibronectin - Mmany Ddifferent Pproteins but Ffew Ddifferent Ffunctions. *Experimental Cell Research*. 1995 DEC;221(2):261-71.

[5] Kosmehl H, Berndt A, Katenkamp D. Molecular variants of fibronectin and laminin: Structure, physiological occurrence and histopathological aspects. *Virchows Archiv: -an International Journal of Pathology*. 1996 DEC;429(6):311-22.

[6] Gehris AL, Stringa E, Spina J, Desmond ME, Tuan RS, Bennett VD. The region encoded by the alternatively spliced exon IIIA in mesenchymalfibronectin appears essential for chondrogenesis at the level of cellular condensation. *Developmental Biology*. 1997 OCT 15;190(2):191-205.

[7] Jarnagin WR, Rockey DC, Koteliansky VE, Wang SS, Bissell DM. Expression of vVariant Ffibronectins in Wwound- Hhealing: - Ccellular Ssource and Bbiological-Aactivity of the EIIIAiiia Ssegment in Rrat Hhepatic Ffibrogenesis. *Journal of Cell Biology*. 1994 DEC;127(6):2037-48.
[8] Manabe R, Ohe N, Maeda T, Fukuda T, Sekiguchi K. Modulation of cell-adhesive activity of fibronectin by the alternatively spliced EDA segment. *Journal of Cell Biology*. 1997 OCT 6;139(1):295-307.
[9] Liao YF, Gotwals PJ, Koteliansky VE, Sheppard D, Van de Water L. The EIIIA segment of fibronectin is a ligand for integrins alpha(9)beta(1) and alpha(4)beta(1) providing a novel mechanism for regulating cell adhesion by alternative splicing. *Journal of Biological Chemistry*. 2002 APR 26;277(17):14467-74.
[10] Hynes RO. *Fibronectins*. New York: Springer-VerlagInc; 1990.
[11] Mosher DF. *Fibronectin*. San Diego: Academic Press; 1989.
[12] Lobb RR, Hemler ME. The Ppathophysiologic Rrole of Aalpha-4 Iintegrins *in -Vvivo*. *Journal of Clinical Investigation*. 1994 NOV;94(5):1722-8.
[13] Bork P, Downing AK, Kieffer B, Campbell ID. Structure and distribution of modules in extracellular proteins. *Quarterly Reviews of Biophysics*. 1996 MAY;29(2):119-67.
[14] Vakonakis I, Staunton D, Ellis IR, Sarkies P, Flanagan A, Schor AM, et al. Motogenic Ssites in Hhuman Ffibronectin Aare Mmasked by Llong Rrange Iinteractions. *Journal of Biological Chemistry*. 2009 JUN 5;284(23):15668-75.
[15] Baron M, Norman D, Willis A, Campbell ID. Structure of the Ffibronectin Ttype 1 Mmodule. *Nature*. 1990 JUN 14;345(6276):642-6.
[16] Pickford AR, Smith SP, Staunton D, Boyd J, Campbell ID. The hairpin structure of the (6)F1(1)F2(2)F2 fragment from human fibronectin enhances gelatin binding. *Embo Journal*. 2001 APR 2;20(7):1519-29.
[17] Constantine KL, Madrid M, Banyai L, Trexler M, Patthy L, Llinas M. Refined Ssolution Sstructure and Lligand-Bbinding Pproperties of PDCdc-109 Ddomain -Bb. - aA Ccollagen-Bbinding Ttype-IiI Ddomain. *Journal of Molecular Biology*. 1992 JAN 5;223(1):281-98.
[18] Constantine KL, Ramesh V, Banyai L, Trexler M, Patthy L, Llinas M. Sequence-sSpecific 1H-1- NMRmr Aassignments and Sstructural Ccharacterization of Bbovine Sseminal Ffluid Pprotein PdcDC-109 Ddomain -Bb. *Biochemistry*. 1991 FEB 12;30(6):1663-72.
[19] Main AL, Harvey TS, Baron M, Boyd J, Campbell ID. The 3-Ddimensional Sstructure of the 10th tType-IIIii Mmodule of Ffibronectin: - an Iinsight into RgdGD-Mmediated iInteractions. *Cell*. 1992 NOV 13;71(4):671-8.
[20] Gao M, Craig D, Lequin O, Campbell ID, Vogel V, Schulten K. Structure and functional significance of mechanically unfolded fibronectin type III1 intermediates. *Proceedings of the National Academy of Sciences of the United States of America*. 2003 DEC 9;100(25):14784-9.
[21] Jones EY. The Iimmunoglobulin Ssuperfamily. *Current Opinion in Structural Biology*. 1993 DEC;3(6):846-52.

[22] Oberhauser AF, Fernandez-Badilla C, Fernandez JM. The mechanical conformations of a fibronectin module captured with force spectroscopy. *Biophysical Journal.* 2002 JAN;82(1):41a-a.
[23] Campbell ID, Spitzfaden C. Building pProteins with Ffibronectin Ttype-IIIii Mmodules. *Structure.* 1994 MAY 15;2(5):333-7.
[24] Obara M, Kang MS, Yamada KM. Site-Ddirected Mmutagenesis of the Ccell-Bbinding Ddomain of Hhuman Ffibronectin: - Sseparable, Ssynergistic Ssites Mmediate Aadhesive Ffunction. *Cell.* 1988 MAY 20;53(4):649-57.
[25] Danen EHJ, Aota SI, Vankraats AA, Yamada KM, Ruiter DJ, Vanmuijen GNP. Requirement for the Ssynergy Ssite for Ccell-Aadhesion to Ffibronectin Ddepends on the Aactivation sState of Iintegrin aAlpha-5-Bbeta-1. *Journal of Biological Chemistry.* 1995 SEP 15;270(37):21612-8.
[26] Gehris AL, Oberlender SA, Shepley KJ, Tuan RS, Bennett VD. Fibronectin mRNA alternative splicing is temporally and spatially regulated during chondrogenesis *in vivo* and *in vitro*. *Developmental Dynamics.* 1996 JUN;206(2):219-30.
[27] Potts JR, Bright JR, Bolton D, Pickford AR, Campbell ID. Solution structure of the N-terminal F1 module pair from human fibronectin. *Biochemistry.* 1999 JUN 29;38(26):8304-12.
[28] Sticht H, Pickford AR, Potts JR, Campbell ID. Solution structure of the glycosylated second type 2 module of fibronectin. *Journal of Molecular Biology.* 1998 FEB 13;276(1):177-87.
[29] Pankov R, Yamada KM. Fibronectin at a glance. *Journal of Cell Science.* 2002 OCT 15;115(20):3861-3.
[30] Schwarzbauer JE, Mao Y. Fibronectin fibrillogenesis, a cell-mediated matrix assembly process. *Matrix Biology*. 2005 SEP;24(6):389-99.
[31] Singh P, Carraher C, Schwarzbauer JE. Assembly of Fibronectin Extracellular Matrix. *Annual Review of Cell and Developmental Biology,* Vol 26. 2010;26:397-419.
[32] Llopis-Hernandez V, Rico P, Ballester-Beltran J, Moratal D, Salmeron-Sanchez M. Role of Ssurface Cchemistry in Pprotein Rremodeling at the Ccell-Mmaterial Iinterface. *Plos One*. 2011 MAY 9;6(5):e19610.
[33] Qu H, Tu YZ, Shi XH, Larjava H, Saleem MA, Shattil SJ, et al. Kindlin-2 regulates podocyte adhesion and fibronectin matrix deposition through interactions with phosphoinositides and integrins. *Journal of Cell Science.* 2011 MAR 15;124(6):879-91.
[34] Joneckis CC, Ackley RL, Orringer EP, Wayner EA, Parise LV. Integrin-Aalpha-4-Bbeta-1 and Gglycoprotein-IVv (Cd36) Aare Eexpressed on Ccirculating Rreticulocytes in Ssickle-Ccell-Aanemia. *Blood.* 1993 Dec 15;82(12):3548-55.
[35] Brittain JE, Knoll CM, Ataga KI, Orringer EP, Parise LV. Fibronectin bridges monocytes and reticulocytes via integrin Î±4Îalpha4beta˙ 1. *British Journal of Haematology.* 2008;141(6):872-81.
[36] Bolarin DM, Adenuga AO. Plasma Ffibronectin in Ssickle-Ccell Ddisease. *Acta Haematologica*. 1986;75(4):209-10.
[37] Kuusela P. Fibronectin Bbinds to *Staphylococcus-Aaureus*. *Nature.* 1978;276 (5689):718-20.

[38] Miller LS, Cho JS. Immunity against *Staphylococcus aureus* cutaneous infections. *Nature Reviews Immunology*. 2011 Aug;11(8):505-18.

[39] Jönsson K, Signäs C, Müller H-P, Lindberg M. Two different genes encode fibronectin binding proteins in *Staphylococcus aureus*. *European Journal of Biochemistry*. 1991;202(3):1041-8.

[40] Bingham RJ, Rudino-Pinera E, Meenan NAG, Schwarz-Linek U, Turkenburg JP, Hook M, et al. Crystal structures of fibronectin-binding sites from *Staphylococcus aureus* FnBPA in complex with fibronectin domains. *Proceedings of the National Academy of Sciences of the United States of America*. 2008 Aug 26;105(34):12254-8.

[41] Sinha B, Francois PP, Nusse O, Foti M, Hartford OM, Vaudaux P, et al. Fibronectin-binding protein acts as *Staphylococcus aureus* invasin via fibronectin bridging to integrin alpha(5)beta(1). *Cellular Microbiology*. 1999 Sep;1(2):101-17.

[42] Schröder A, Kland R, Peschel A, von Eiff C, Aepfelbacher M. Live cell imaging of phagosome maturation in andlt;iandgt;*Staphylococcus aureus*andlt;/iandgt; infected human endothelial cells: small colony variants are able to survive in lysosomes. *Medical Microbiology and Immunology*. 2006 2006-12-01;195(4):185-94.

[43] Schröder A, Schröder B, Roppenser B, Linder S, Sinha B, Fassler R, et al. *Staphylococcus aureus* fibronectin binding protein-A induces motile attachment sites and complex actin remodeling in living endothelial cells. *Molecular Biology of the Cell*. 2006 Dec;17(12):5198-210.

[44] Yao L, Bengualid V, Lowy FD, Gibbons JJ, Hatcher VB, Berman JW. Internalization of *Staphylococcusa-Aureus* by Eendothelial- Ccells Iinduces Ccytokine Ggene-Eexpression. *Infection and Immunity*. 1995 May;63(5):1835-9.

[45] Soderquist B, Alriksson I, Kallman J, Kihlstrom E. The influence of adhesive and invasive properties of *Staphylococcus aureus* defective in fibronectin-binding proteins on secretion of interleukin-6 by human endothelial cells. *Apmis*. 2006 Feb;114(2):112-6.

[46] Zhou H, Xiong ZY, Li HP, Zheng YL, Jiang YQ. An immunogenicity study of a newly fusion protein Cna-FnBP vaccinated against *Staphylococcus aureus* infections in a mice model. *Vaccine*. 2006 May 29;24(22):4830-7.

[47] Lahteenmaki K, Virkola R, Saren A, Emody L, Korhonen TK. Expression of plasminogen activator Pla of *Yersinia pestis* enhances bacterial attachment to the mammalian extracellular matrix. *Infection and Immunity*. 1998 DEC;66(12):5755-62.

[48] Mignatti P, Rifkin DB. Biology and Biochemistry of Pproteinases in Ttumor Iinvasion. *Physiological Reviews*. 1993 JAN;73(1):161-95.

[49] Sodeinde OA, Subrahmanyam YVBK, Stark K, Quan T, Bao YD, Goguen JD. A Ssurface Pprotease and the Iinvasive Ccharacter of Pplague. *Science*. 1992 NOV 6;258(5084):1004-7.

[50] Plesniak LA, Mahalakshmi R, Rypien C, Yang Y, Racic J, Marassi FM. Expression, refolding, and initial structural characterization of the *Y. pestis* Ail outer membrane protein in lipids. *Biochimica Eet Biophysica Acta-Biomembranes*. 2011 Jan 2011;1808(1):482-9.

[51] Tellier MC, Greco G, Klotman M, Mosoian A, Cara A, Arap W, et al. Superfibronectin, a Multimeric Form of Fibronectin, Increases HIV Infection of Primary CD4+ T Lymphocytes *The Journal of Immunology* 2000; March 15, 2000 164 (6):3236-45

[52] Greco G, Pal S, Pasqualini R, Schnapp LM. Matrix fibronectin increases HIV stability and infectivity. *Journal of Immunology*. 2002 Jun 1;168(11):5722-9.

[53] Birdsall HH, Porter WJ, Green DM, Rubio J, Trial JA, Rossen RD. Impact of fibronectin fragments on the transendothelial migration of HIV-infected leukocytes and the development of subendothelial foci of infectious leukocytes. *Journal of Immunology*. 2004 Aug 15;173(4):2746-54.

[54] Rossen RD, Rubio JA, Porter WJ, Trial J, Orson FM, Rodriguez-Barradas MC, et al. Monocyte CD49e and 110-120 kDafibronectin fragments: HIV prognostic indicators independent of viral load and CD4 T-cell counts. *Aids*. 2009 Nov 13;23(17):2247-53.

[55] Brittingham A, Chen G, McGwire BS, Chang KP, Mosser DM. Interaction of *Leishmania* gp63 with cellular receptors for fibronectin. *Infection and Immunity*. 1999 SEP;67(9):4477-84.

[56] Wyler DJ. Fibronectin in Parasitic Diseases. *Reviews of Infectious Diseases*. 1987 JUL-AUG;9:S391-S9.

[57] Kulkarni MM, Jones EA, McMaster WR, McGwire BS. Fibronectin binding and proteolytic degradation by *Leishmania* and effects on macrophage activation. *Infection and Immunity*. 2008 APR;76(4):1738-47.

[58] Korom S, Hancock WW, Coito AJ, Kupiec-Weglinski JW. Blockade of very late antigen-4 integrin binding to fibronectin in allograft recipients - II. Treatment with connecting segment-1 peptides prevents chronic rejection by attenuating arteriosclerotic development and suppressing intragraft T cell and macrophage activation. *Transplantation*. 1998 MAR 27;65(6):854-9.

[59] Beezhold DH, Personius C. Fibronectin Ffragments Sstimulate Ttumor- Nnecrosis-Ffactor Ssecretion by Hhuman Mmonocytes. *Journal of Leukocyte Biology*. 1992 JAN;51 (1):59-64.

[60] Abreu-Silva AL, Calabrese KS, Mortara RA, Tedesco RC, Cardoso FO, Carvalho LOP, et al. Extracellular matrix alterations in experimental murine *Leishmania (L.) amazonensis* infection. *Parasitology*. 2004 APR;128:385-90.

[61] McGwire BS, Chang KP, Engman DM. Migration through the extracellular matrix by the parasitic protozoan *Leishmania* is enhanced by surface metalloprotease gp63. *Infection and Immunity*. 2003 FEB;71(2):1008-10.

[62] Melo FA, Moura EP, Ribeiro RR, Alves CF, Caliari MV, Tafuri WL, et al. Hepatic extracellular matrix alterations in dogs naturally infected with *Leishmania (Leishmania) chagasi*. *International Journal of Experimental Pathology*. 2009 OCT;90(5):538-48.

[63] Wyler DJ, Sypek JP, Mcdonald JA. *In vitro* Pparasite-Mmonocyte Iinteractions in Hhuman Lleishmaniasis: - Ppossible Rrole of Ffibronectin in Pparasite Aattachment. *Infection and Immunity*. 1985;49(2):305-11.

[64] Truyens C, Rivera MT, Ouaissi A, Carlier Y. High Ccirculating Llevels of Ffibronectin and Aantibodies against Iits RgdGD Aadhesion Ssite Dduring Mmouse *Trypanosoma -cCruzi* iInfection: - Rrelation to Ssurvival. *Experimental Parasitology*. 1995 MAY;80(3):499-506.

[65] Andrade SG, Grimaud JA, Stockerguerret S. Sequential- Cchanges of the Cconnective Mmatrix Ccomponents of the Mmyocardium (Ffibronectin and Llaminin) and Eevolution of Ccardiac Ffibrosis in Mmice Iinfected with *Trypanosoma -Ccruzi*. *American Journal of Tropical Medicine and Hygiene*. 1989 MAR;40(3):252-60.

[66] Silva AA, Roffe E, Marino AP, dos Santos PVA, Quirico-Santos T, Paiva CN, et al. Chagas' disease encephalitis: Intense CD8(+) lymphocytic infiltrate is restricted to the acute phase, but is not related to the presence of Trypanosomacruzi antigens. *Clinical Immunology*. 1999 JUL;92(1):56-66.

[67] Shin TY, Kojima T, Tanuma N, Ishihara Y, Matsumoto Y. The sSubarachnoid Sspace as a Ssite for Pprecursor T-Ccell Pproliferation and Eeffector T-Ccell Sselection in Eexperimental Aautoimmune Eencephalomyelitis. *Journal of Neuroimmunology*. 1995 FEB;56(2):171-8.

[68] Pilkington GJ. The role of the extracellular matrix in neoplastic glial invasion of the nervous system. *Brazilian Journal of Medical and Biological Research*. 1996 SEP;29(9):1159-72.

In: Fibronectin
Editors: James Beattie

ISBN 978-1-62257-298-4

Chapter XI

Fibronectin in Renal Diseases and Peritoneal Dialysis

Susan Yung and Tak Mao Chan*
Department of Medicine, University of Hong Kong, Hong Kong

Abstract

The extracellular matrix (ECM) provides structural support to all organs and influences cellular functions such as cell adhesion, migration, proliferation and differentiation, which are involved in biological processes such as inflammation, wound healing, and fibrosis. Fibronectin is a large adhesive glycoprotein and is one of the most well characterized proteins of the ECM and basement membranes. It is encoded by a single gene on chromosome 2, but due to post-transcriptional alternative splicing of its mRNA and post-translational modification, numerous isoforms of fibronectin have been identified.

In the normal kidney, fibronectin is found in the mesangium of the glomerulus and in the interstitium, where it plays a critical role in the structural integrity of the kidney, maintenance of normal cell morphology and cell adhesion. Fibronectin is present in abundant quantities in glomerulosclerotic lesions and within the interstitium in various chronic kidney diseases often irrespective of the underlying aetiology and thus represents part of a final common pathway. Increased fibronectin expression results from a disturbed balance between its synthesis and degradation, mediated in part through an increase in TGF-β1 bio-activation, induction of intracellular signaling pathways and altered expression of matrix metalloproteinases. In immune-mediated renal disease such as lupus nephritis, anti-dsDNA antibodies have been shown to induce both soluble and insoluble fibronectin, which have distinct roles during inflammation and fibrotic processes.

Calcineurin inhibitors, used in the prevention of renal transplant rejection and the treatment of immune-mediated diseases, can induce TGF-β1 expression and renal

* Correspondence and reprint requests to: Dr Susan Yung and Professor Tak Mao Chan. Department of Medicine, University of Hong Kong, Queen Mary Hospital, Pokfulam Road, Hong Kong. Tel: (852) 2255 4542. Fax: (852) 2872 5828. Email: ssyyung@hku.hk.

fibrogenesis, whilst mycophenolic acid has been shown to possess anti-fibrotic properties. This review will discuss the biology and synthesis of fibronectin, its role in the kidney and the peritoneal membrane in the healthy state and in renal diseases or after peritoneal dialysis.

Keywords: Fibronectin, renal disease, epithelial-to-mesenchymal transdifferentiation, fibrosis, diabetic nephropathy, lupus nephritis, peritoneal dialysis

Introduction

The extracellular matrix (ECM) was once considered to function solely as a scaffold to maintain the structural integrity of tissues and organs, but there is now compelling evidence to demonstrate that components of the ECM can regulate many features of cell behavior such as cell proliferation, migration, differentiation and morphology.

The ECM comprises an intricate compilation of structural proteins and proteoglycans that provide a framework for tissue architecture and a substratum for cells to adhere. The ability of the ECM to meticulously regulate cellular functions is dependent on its specific composition, which varies between organs and tissues. ECM turnover and remodeling is a dynamic process that occurs during development and tissue repair, and the deposition and replenishment of ECM components is essential to preserve the structural and functional integrity of tissues. Excessive accumulation of matrix proteins in tissues and organs is often observed in various pathological conditions, which perturbs tissue structure resulting in fibrosis and organ dysfunction. Although the mechanisms that control ECM organization during pathological conditions remain to be fully elucidated, studies have demonstrated that an imbalance between matrix protein synthesis and degradation, induction of fibrogenic mediators and phenotypic changes in resident cells can contribute to aberrant matrix remodelling and organ fibrosis.

Fibronectin is a major component of the ECM and is also present in biological fluids [1]. It is the product of a single gene, but due to alternative splicing, up to 20 variants have been identified. Fibronectin exists in two forms, as a soluble dimer and also as an insoluble fibrillar network [1, 2]. The predominant source of soluble fibronectin is plasma where it can reach concentrations of up to 400 μg/ml, although it is also found in other biological fluids, albeit at lower concentrations [3]. Insoluble fibronectin is often found in embryonic and mesenchymal tissues. It is a prominent component of the kidney [4, 5] and is synthesized by resident renal cells such as the interstitial fibroblasts, mesangial cells, epithelial cells and endothelial cells [6-9]. The ability of fibronectin to polymerize is critical for the structural integrity of the kidney. Increased expression of fibronectin is observed in glomerulosclerotic lesions and within the interstitium in numerous chronic kidney diseases such as lupus nephritis and diabetic nephropathy [8, 10-13]. If left untreated, increased fibrogenesis will lead to renal fibrosis and end-stage renal failure, whereby patients require either kidney transplantation or dialysis to sustain life. This review will describe the role of fibronectin in the kidney under physiologic and pathologic conditions with particular emphasis on diabetic nephropathy and lupus nephritis, on the underlying mechanisms of its increase during chronic renal disease and therapeutic approaches that may prevent excessive synthesis of fibronectin and thus preserve the structural and functional integrity of the kidney.

Fibronectin Synthesis and Polymerization

Fibronectin is encoded by a single gene on chromosome 2, which generates fibronectin subunits with molecular weights ranging from 230-270 kDa [14-16]. Each subunit comprises 12 type I modules, 2 type II modules, 15 to 17 type III modules and a variable sequence (Figure 1) [17]. Due to post-transcriptional alternative splicing of the fibronectin gene and post-translational modification of the protein, various isoforms have been identified [18, 19] and to date, 20 isoforms of fibronectin have been identified in man. Alternative splicing can occur in 2 type III modules, called extradomain (ED) A^+ or EDB^+ and in the variable (V) region, the latter located between the penultimate and last repeats of the type III modules (Figure 1) [17]. Alternative splicing at the V region of the fibronectin mRNA can result in 5 variants of the fibronectin protein. Given that the V region contains recognition sites for integrins, changes to the V region can influence cell attachment, migration and proliferation.

The functional attributes of EDA^+ and EDB^+ domains to the fibronectin molecule remains to be fully elucidated but given that their expression is differentially modulated during embryogenesis and wound healing [20, 21], it is plausible to suggest that these spliced forms of fibronectin have specific physiological roles in the tissue. The inclusion of EDA^+ and EDB^+ domains into the fibronectin structure has been shown to modulate the regulation of cell attachment and cell-matrix interactions [22, 23]. Furthermore, the insertion of EDB^+ between III_7 and III_8 may alter inter-domain interfaces and interactions, resulting in conformational changes in the orientation of the domain thereby exposing new functional surfaces and otherwise cryptic sequences [23]. These spliced variants are selectively expressed during development and disease, both spatially and temporally. The alternative splicing of EDA^+ and EDB^+ domains is dynamically regulated *in vivo*. They are present during embryogenesis and are only observed in adults during pathological conditions when tissues are undergoing remodelling, wound healing or artherosclerosis [21, 24].

These spliced variants are sometimes referred to as oncofoetal isoforms of fibronectin to highlight their prominence during embryogenesis and tumourigenesis. It is therefore not unexpected that both spliced variants are either absent or present in diminutive amounts in tissues of normal adult subjects.

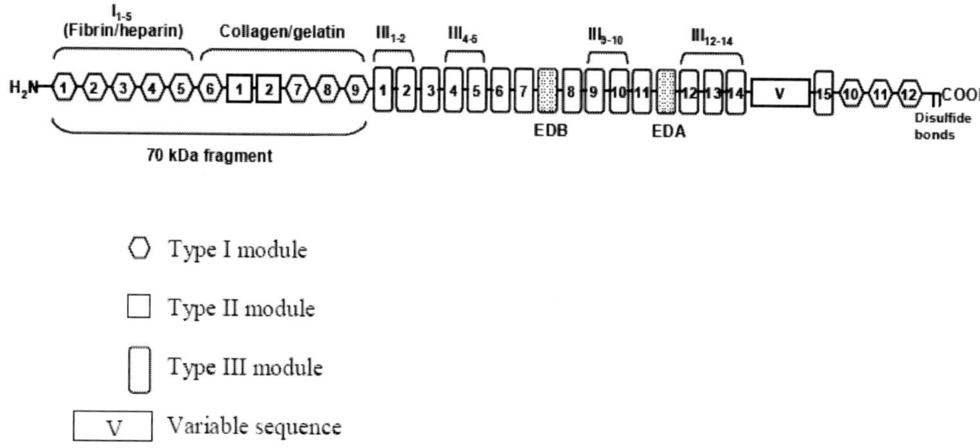

Figure 1. Structure of fibronectin.

Two subunits of fibronectin are linked together by a pair of disulphide bonds at the C terminus to form fibronectin molecules with molecular weights in excess of 450 kDa [17]. The inability of fibronectin subunits to polymerize into fibrils in the absence of cysteine molecules highlights the importance of the covalent bond in the polymerization of fibronectin and subsequent assembly into the ECM [25]. Polymerization of fibronectin and its assembly into the ECM is a complex cell-mediated process that involves several domains of the fibronectin molecule. The process is initiated when soluble fibronectin dimers bind to the cell surface through its 70 kDa region adjacent to the N-terminal of the protein. The 70 kDa region is essential for the formation of fibronectin fibrils since partial or complete deletion of the first five type I repeats ($I_{1\text{-}5}$) in recombinant fibronectin renders it incompetent to assemble fibrils [25]. Once fibronectin dimers are bound to the cell surface, cells then initiate fibronectin polymerization through the binding of $\alpha5\beta1$ integrins to the Arg-Gly-Asp (RGD) cell-binding domain located at III_{10} and to the synergy site at III_9. Although III_9 domain is not critical for fibronectin binding to the cell surface, this domain together with the RGD domain is critical for the initiation of fibril formation [26]. Fibronectin fibrils assembled into the ECM are not inert and rigid, but provide a dynamic and flexible extracellular milieu that accommodates cell migration, proliferation and differentiation.

Whilst an in-depth review of fibronectin assembly and its molecular interactions fall outside the scopeop of this review, readers are referred to the following articles [2, 27, 28].

Fibronectin and the Kidney

Resident Cells of the Kidney

The kidney functions to maintain fluid, electrolytes and pH balance in the body. Each kidney contains approximately 1 million nephrons, which are the basic filtering units of the kidney required for the production of urine. Each nephron consists of a renal corpuscle (glomerulus and Bowman's capsule) and the renal tubules. The glomerulus contains four cells types, namely the parietal epithelial cells, visceral epithelial cells (podocytes), endothelial cells and the mesangial cells (Figure 2).

Changes to one cell type will concomitantly affect cellular behaviour of the remaining 3 cell types. Mesangial cells are embedded in the mesangial matrix, which contains fibronectin, collagen type IV, laminin and proteoglycans. Its composition differs from that of the glomerular basement membrane (GBM) [29]. The quality and quantity of the mesangial matrix is tightly regulated by mesangial cells in the normal kidney and is noticeably altered during glomerular disease [30]. Accumulation of matrix proteins in the mesangium is the cornerstone of glomerulosclerosis and occurs when matrix synthesis exceeds matrix degradation [31]. While renal parenchymal diseases may start with distinct glomerular or tubulo-interstitial lesions, matrix deposition and progressive fibrosis affecting both the glomerular and tubulo-interstitial compartments are common features that lead to chronic renal failure. Furthermore, the degree of tubulo-interstitial injury and fibrosis strongly correlates with renal prognosis [32, 33].

The tubulo-interstitium comprises the tubular epithelium, vascular structure and the interstitium where fibroblasts reside. It accounts for 85 - 90% of the total kidney volume.

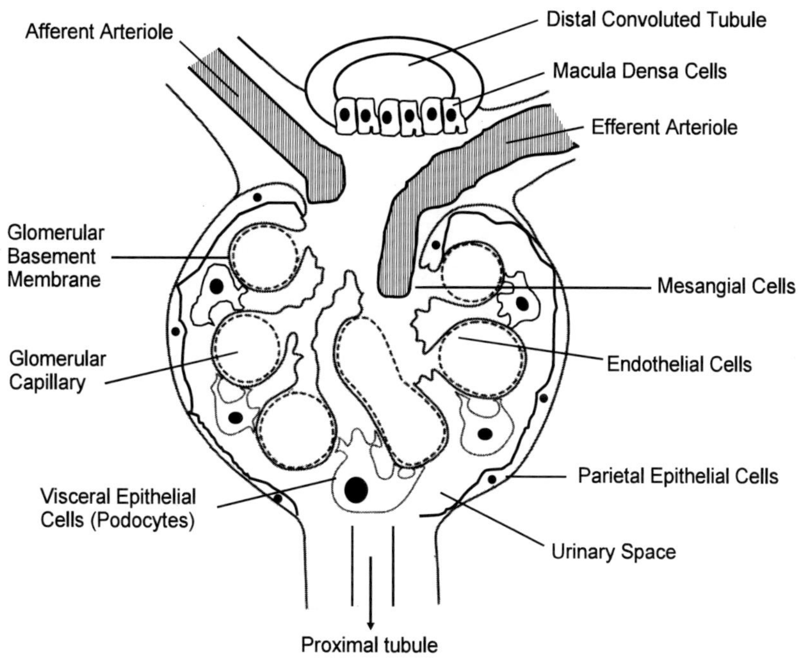

Figure 2. Schematic diagram of the kidney glomerulus.

Proximal tubular epithelial cells are the predominant cells of the renal cortex and play a pivotal role in the selective re-absorption of molecules from the glomerular filtrate. Proximal tubular epithelial cells have a high proliferative potential and contribute to the regeneration of the tubular epithelium in response to acute tubular injury through their ability to synthesis various growth factors and matrix proteins. Stimulation of proximal tubular epithelial cells with pro-fibrotic mediators will induce epithelial-to-mesenchymal transdifferentiation (EMT) and this phenotypic change has been shown to contribute to renal fibrogenesis [34]. EMT is characterized by the loss of epithelial cell adhesion, cell activation with actin reorganization and *de novo* synthesis of α-smooth muscle actin, disruption of the underlying basement membrane and increased cell migration and invasion [35-37]. Resident interstitial fibroblasts constitute a second cell type in the tubulo-interstitium. They are derived from the mesenchyme in the embryonic kidney although there is some support that a proportion of these cells may also originate from the bone marrow [38]. Interstitial fibroblasts are activated during chronic renal disease and acquire a myofibroblastic phenotype. There is compelling evidence that the presence of myofibroblasts is a predictor of fibrotic progression in animal models and human renal disease [39-44].

Fibronectin Expression in the Kidney Under Physiological and Pathological Conditions

A. Fibronectin Expression Under Physiological Conditions

Fibronectin is essential to maintain renal homeostasis and for cellular functions such as cell adhesion, migration, proliferation, differentiation and survival [16, 45, 46]. Fibronectin isoforms are differentially expressed during normal kidney development. In the human foetal kidney, EDA^+-containing fibronectin is widely distributed in the glomerular and tubular basement membranes, mesangial matrix and within the interstitium [47]. The distribution of EDB^+-containing fibronectin follows a similar distribution to that observed with EDA^+-containing fibronectin in the foetal kidney albeit at lower levels [47]. In the normal adult kidney, fibronectin is expressed in the mesangial matrix, glomerular and tubular basement membranes and the Bowman's capsule [13]. EDA^+-containing fibronectin is present in small amounts in the mesangium, whereas EDB^+-containing fibronectin is negligible [48].

Ultrastructural studies of the normal adult rat kidney have demonstrated that fibronectin localization is predominantly in the mesangial matrix, in particular at the interface between mesangial and endothelial cells, where it forms an extensive 3-dimensional network that is thought to function to stabilize the architecture of the glomerular tuft. The observation that fibronectin is also detected in the Golgi cisternae of mesangial cells provides evidence that these cells contribute to the synthesis of fibronectin in the kidney [4]. Fibronectin is also present at the basolateral aspect of endothelial and epithelial cells where they face the GBM, and also in the laminae rarae of the GBM [4]. The presence of fibronectin at the base of the endothelial and epithelial cells, within the GBM together with its ability to interact with collagen, laminin and heparan sulphate proteoglycans in the GBM highlights the role of fibronectin in the maintenance of glomerular capillary integrity and in the provision of a substrate onto which resident glomerular cells can adhere to the lamina densa.

B. Fibronectin Expression in Disease States

Injury to the kidney can result in a loss of nephrons and their replacement with fibrous tissue, resulting in glomerulosclerosis, tubulo-interstitial fibrosis and subsequent end-stage renal disease, which will necessitate either kidney transplantation or life-long dialysis. Although most patients are diagnosed well before they reach end-stage renal disease, there is currently no effective treatment that can completely impede the progressive decline in renal function. Glomerulosclerosis is defined as a loss of normal, functional glomerular tissue and its replacement with matrix proteins [31]. It is observed in renal disorders such as chronic glomerulonephritis, obstructive uropathy and retrovirus infection. Furthermore, patients with steroid-unresponsive nephritic syndrome have also been shown to develop glomerulosclerosis. Whilst the pathogenesis of glomerulosclerosis remains to be fully elucidated, it is most likely that all cell types within the glomerulus participate in fibrotic processes. There is emerging evidence to suggest that accumulation of matrix proteins within the glomerulus is initiated in the mesangium, which results in mesangial expansion that infringe into the capillary space thereby reducing the surface area for glomerular filtration. Studies have demonstrated that fibronectin is the predominant ECM component present in glomerulosclerotic lesions and is also the first matrix component to be deposited during tubulo-interstitial fibrosis [40, 49]. Despite the initial injury to the glomerulus, whether immune-mediated, inflammatory or metabolic, most chronic renal diseases will show a loss of containment of this injury in the glomerulus, which will invariably extend into the tubulo-interstitium. Evidently, it is the degree of tubular injury and interstitial fibrosis rather than glomerulosclerosis that best correlates with progression of renal insufficiency [32]. Animal models of chronic renal disease rarely progress in the absence of tubulo-interstitial lesions, whilst tubulo-interstitial fibrosis often precedes end-stage renal disease [44]. In an anti-Thy 1.1 rat model of glomerular disease, Yamamoto *et al* elegantly demonstrated that glomerulonephritis and matrix accumulation was transient and complete regression was possible in the absence of interstitial changes, whereas renal injury and matrix accumulation progressed to end-stage renal disease in the presence of tubulo-interstitial lesions [44].

Progression of renal fibrosis involves the activation of both infiltrating and resident renal cells in both the glomerular and tubulo-interstitial compartments. Three phases of fibrogenesis have been described for the kidney [50], namely an induction phase whereby inflammatory cells infiltrate the renal parenchymal and release pro-inflammatory and pro-fibrotic mediators. This initial event is followed by the activation of resident renal cells and studies to date have demonstrated that mesangial cells contribute significantly to glomerulosclerosis, whilst proximal tubular epithelial cells and interstitial fibroblasts both play pivotal roles in tubulo-interstitial fibrosis. Activation of mesangial cells, proximal tubular epithelial cells and interstitial fibroblasts results in their conversion to myofibroblasts. As disease progresses, mesangial cell proliferation and vascular collapse ensues with the activation of the fibrogenic phase and the subsequent destruction of the normal renal parenchyma and their replacement with fibrous scarring and ultimately renal failure [41].

Fibronectin expression is significantly increased in both the glomerular and tubulo-interstitial compartments of the kidney in various renal diseases such as diabetic nephropathy, lupus nephritis, IgA nephropathy and focal segmental glomerulosclerosis, and is observed within glomerulosclerotic lesions and in areas of interstitial fibrosis [8, 13, 48]. Irrespective of the aetiology of renal disease, a number of sliced variants can be observed within the glomerulus of patients with renal disease and these include EDB^- and EDB^+-containing fibronectin, and EDA^+-containing fibronectin, whereas increased fibronectin in the interstitium is predominantly attributed to EDA^+-containing fibronectin [48]. It has been suggested that splice variants of fibronectin may induce distinct biologic functions in tissues during pathology. In this respect, alternative splicing at the EDA exon can up-regulate the integrin binding affinity of fibronectin with subsequent induction of cell migration and proliferation, transduction of TGF-β1 signalling and conversion of fibroblasts to myofibroblasts. Cell-matrix interactions regulate gene expression, signalling pathways, intracellular pH and calcium levels. Therefore, perturbation of fibronectin synthesis in the kidney can have a significant effect on cellular behaviour in the kidney. EDB^+-containing fibronectin molecules are capable of forming a temporary scaffold to support the surrounding cells during tissue injury and can attract inflammatory cells to sites of injury [48]. EDB^+-containing fibronectin however, has been shown to be more susceptible to proteolytic attack, with the generation of fibronectin fragments that have distinct biologic roles from its parent molecule. Fragments of matrix proteins have been shown to induce cytokine secretion, activate signal transduction pathways and regulate the apoptotic response of immune and non-immune cells [51-53]. Fibronectin accumulation in the kidney is also a chemoattractant for fibroblasts in an animal model of glomerular basement membrane disease [40, 49]. These findings underscore the importance of fibronectin in the progression of kidney pathology. Furthermore, elevated urine levels of fibronectin have been suggested to predict poor outcome in patients with chronic glomerulonephritis [54].

Although progressive kidney deterioration and induction of fibrogenesis is common to all renal diseases, in the next section we will focus on the mechanisms of fibrogenesis in diabetic nephropathy and lupus nephritis since diabetic nephropathy is the most common cause of end-stage renal disease worldwide, whereas lupus nephritis is common in this locality and our research focuses on the pathogenesis of lupus nephritis.

Diabetic Nephropathy

Much attention has focused on the accumulation of matrix proteins in diabetic nephropathy. Mesangium expansion and thickening of the glomerular and tubular basement membranes are a hallmark of diabetic nephropathy and results from the excessive accumulation of matrix proteins such as fibronectin and collagen type IV within the glomerulus. The degree of mesangial expansion strongly correlates with clinical manifestations of diabetic nephropathy such as glomerular filtration rate and proteinuria [55, 56].

Numerous studies have documented a close resemblance of the effect of TGF-β1 and elevated glucose concentrations on the synthesis of matrix proteins in renal cells [57]. There is now compelling evidence to show that elevated glucose concentrations increase glucose uptake in cells, facilitated by glucose transporter GLUT1. The subsequent increase in glucose metabolic flux results in the activation of the polyol pathway, generation of reactive oxygen species, increased *de novo* synthesis of diacylglycerol, PKC activation and TGF-β1 bio-activation [58-66]. It is the autocrine up-regulation and activation of TGF-β1 by elevated glucose concentrations that mediates the accumulation of matrix proteins in renal cells as as observed in *in vitro* and animal studies of diabetic nephropathy. The ability of TGF-β1 neutralizing antibodies to attenuate increased synthesis of matrix proteins in mice with Type I diabetic nephropathy corroborated these findings. The role of TGF-β1 *per se* in mediating fibrogenesis in the kidney is further discussed in the below section. Although there is compelling evidence to show that elevated glucose concentrations mediates its fibrotic effects through increased bio-activation of TGF-β1, recent evidence hasve also highlighted a novel mechanism in mesangial cells whereby increased fibronectin synthesis occurs that can induce fibronectin synthesis in mesangial cells through the over-expression of GLUT1 in a TGF-β1-independent manner [67].

Oxidative stress is a critical pathogenic factor in the development of diabetic nephropathy [68-71]. TGF-β1 induction of fibronectin in mesangial cells occurs at least in part, through calcium influx and calmodulin/calcineurin-dependent activation of nuclear factor of activated T cell in a reactive oxygen species dependent manner [72]. Accumulating evidence suggests that Nox4 NAD(P)H oxidase is an essential enzyme for the generation of reactive oxygen species in fibrotic diseases. Gorin and co-workers recently demonstrated that activation of Nox4 NAD(P)H oxidase in mesangial cells mediated glomerular hypertrophy and increased fibronectin expression in the diabetic kidney [73]. These researchers also demonstrated that elevated glucose concentrations increased Nox4 NAD(P)H oxidase activity in cultured mesangial cells, which mediated increased fibronectin synthesis [73]. This study highlights the role of oxidative stress in the accumulation of matrix proteins in the glomerulus and development of glomerulosclerosis. It has been suggested that Nox4 functions as an upstream activator of ERK1/2, AKT/PKB and TGF-β1 signalling pathways that mediate renal hypertrophy and matrix accumulation [74, 75].

MicroRNAs are short, non-encoding RNA sequences comprising 18 to 25 nucleotides that regulate gene expression of target molecules by their ability to inhibit mRNA translation or induce mRNA degradation [76, 77]. Wang *et al* demonstrated an increase in intra-renal expression of microRNA-377 in diabetic mice and also in mesangial cells stimulated with elevated glucose concentrations, which was associated with increased synthesis of fibronectin [78]. These researchers demonstrated that microRNA-377 indirectly increased fibronectin synthesis in diabetic nephropathy through the prior down-regulation of p21-activated kinase and superoxide dismutase [78]. p21-activated kinase has been shown to activate members of the MAPK family and also NFκB-inducing kinase [79]. It is possible that a decrease in p21-activated kinase could result in altered kinase activity, thereby resulting in aberrant synthesis of fibronectin. Superoxide dismutases are a group of enzymes that catalyze the reduction of reactive oxygen species to less toxic components and regulate oxidative stress [80]. Studies have demonstrated that the activities and concentrations of these enzymes are decreased in diabetic patients [81] and furthermore, diabetic nephropathy is accelerated in mice deficient in superoxide dismutase [82]. The observation that anti-oxidants can significantly reduce elevated glucose induction of fibronectin mRNA in mesangial cells substantiates a role of oxidative stress in renal fibrosis in the pathogenesis of diabetic nephropathy [83]. Hyperglycaemia may therefore induce oxidative stress in the kidney, which increases microRNA-377 expression that mediates downstream intra-renal accumulation of matrix protein synthesis [78].

Glomerular hypertension has been shown to contribute to the development and progression of diabetic nephropathy. An increase in glomerular capillary pressure may result in an increase in capillary wall tension and subsequent expansion of the capillary lumen, thereby exposing mesangial cells to mechanical stretch [84]. Numerous studies have demonstrated that mechanical stretch can significantly increase fibronectin synthesis in cultured mesangial cells mediated in part through the activation of JNK/SAPK and p38 MAPK [85-89]. Nichio *et al* demonstrated that intracellular cAMP inhibited mechanical stretch-induced fibronectin in cultured mesangial cells through the suppression of MAPK activation, suggesting that agents that are capable of increasing cAMP levels in the kidney may be able to suppress renal fibrosis [90].

Much of the earlier studies that investigated the mechanisms through which elevated glucose concentrations induced matrix protein synthesis in the kidney were undertaken in mesangial cells since their involvement in ECM production is more apparent *in vivo* and *in vitro* compared to the other renal cell types. There is also data to highlight the importance of podocytes and endothelial cells in the development of glomerulosclerosis in diabetic nephropathy and that bi- and tri-directional communication between glomerular cells is critical for disease progression [29]. Emerging evidence suggests that podocyte injury contribute to glomerulosclerosis by directly stimulating matrix protein synthesis or suppressing their degradation in mesangial cells [91-93]. Podocytes have also been shown to up-regulate matrix protein synthesis indirectly through hemodynamic factors [94]. Furthermore, endothelial dysfunction has also been shown to induce glomerulosclerosis in animal models of diabetic nephropathy, underscoring the bi-directional cross-talk between endothelial and mesangial cells, leading to progressive accumulation of matrix protein in the glomerulus [95, 96]. Although much of the research in the field of diabetic nephropathy has focused on the fibrotic processes initiated by resident renal cells during pathogenesis of disease, there is evidence that infiltration of macrophages and inflammatory processes also play an important role in the development of glomerulosclerosis in diabetic nephropathy [97].

Although mesangial expansion and podocyte loss are present in the early development of diabetic nephropathy, it is tubulo-interstitial injury and subsequent tubulo-interstitial fibrosis that is critical for the progression to end-stage renal disease. Accumulating evidence suggests that the conversion of proximal tubular epithelial cells to myofibroblasts i.e. EMT plays a critical role in increased matrix protein synthesis in the tubulo-interstitium. We and others have demonstrated that elevated glucose induction of TGF-β1 can increase synthesis of matrix proteins in proximal tubular epithelial cells [98-103]. Induced in high glucose-1 (IHG-1) is a novel evolutionary transcript that is increased in the tubules of patients with diabetic nephropathy [101]. Murphy *et al* demonstrated that over-expression of IHG-1 in HK-2 cells amplified TGF-β1 expression in these cells, which was associated with a subsequent increase in CTGF and fibronectin [101].

The ability of glucose to mediate its fibrogenic effect through TGF-β1 appears to be cell-specific. Lam *et al* investigated the effect of elevated glucose concentration on fibronectin synthesis in cultured human renal fibroblasts and demonstrated that glucose-induced fibronectin synthesis in these cells occurred through a mechanism that was independent of TGF-β1 since the addition of TGF-β1 neutralizing antibody had no effect on glucose-induced fibronectin accumulation, and fibroblasts expressing a dominant negative TGF-β1 type II receptor were shown to synthesize increased levels of fibronectin following elevated glucose stimulation [104]. It is intriguing that glucose can induce fibronectin synthesis through distinct mechanisms in distinct cell populations from the same organ, and this may relate to the different ECM profiles in different locations of the kidney. It is also noteworthy that the fibrogenic effect of glucose is not specific to fibronectin since elevated glucose concentrations can also increase other matrix proteins such as collagen type III and laminin [105, 106].

Lupus Nephritis

Lupus nephritis is a severe organ manifestation of systemic lupus erythematosus, a prototype autoimmune disease that affects up to 60% of the SLE population [107]. This disease is characterized by the production of anti-double stranded (ds) DNA antibodies, their deposition along the glomerular and tubular basement membranes, complement activation and local production of inflammatory and fibrotic mediators [107, 108]. Histological features include the infiltration of inflammatory cells, mesangial cell proliferation, injury to the kidney and replacement of the normal kidney parenchyma with fibrous matter. Abnormalities in the mesangial area precede lesions in the glomerular capillary loop [109]. Whilst the levels of anti-dsDNA antibodies often correlate with disease activity [108, 110, 111], their role in pathogenesis remains to be defined. Fibronectin expression is increased in the expanded mesangium in immune-mediated glomerulonephritis [13, 48, 112, 113]. We have recently demonstrated that fibronectin expression is increased in the glomerulus of patients with active lupus nephritis, which co-localized with IgG deposition [8]. This finding suggests that a close relationship exists between autoantibody deposition and increased matrix protein accumulation in the renal parenchyma during pathogenesis of lupus nephritis. We have also shown that fibronectin synthesis is increased in cultured human mesangial cells following their exposure to human polyclonal anti-dsDNA antibodies, mediated through the prior activation of PKC-α, PKC-βI and PKC-βII and increased TGF-β1 secretion [8]. The role of TGF-β1 in mediating increased synthesis of matrix proteins during the progression of lupus nephritis is debatable. Yamamoto *et al* observed an increase in TGF-β1 expression in renal samples obtained from lupus patients and also in lupus-prone mice, which was associated with increased fibronectin expression [44, 114, 115]. In contrast, Peterson *et al* suggested that the development of renal fibrosis in patients with lupus nephritis occurred through MCP-1 rather than TGF-β1 since microarray analyses of isolated glomeruli from patients with lupus nephritis exhibited decreased TGF-β1 expression, whereas MCP-1 expression clustered with gene related to fibrogenesis [116]. That MCP-1 possesses fibrogenic properties has also been demonstrated in animal models of diabetic nephropathy [117, 118]. Baelde *et al* suggested that increased fibronectin expression in renal specimens obtained from lupus patients and experimental models of lupus nephritis was not a result of increased synthesis [10], but may in part be a result of specific trapping of plasma fibronectin from the circulation [119], mediated through the binding of heparin II binding domain in the fibronectin molecule to α5β1 integrin [12]. Increased levels of plasma fibronectin and anti-fibronectin antibodies are associated with disease activity in SLE patients [120-122]. Fibronectin has also been shown to mediate the internalization of nephritogenic auto-antibodies into glomerular endothelial cells and their deposition into electron dense deposits in the subendothelial region of glomeruli in SCID mice [123], suggesting that fibronectin contributes to the development of glomerular injury in the pathogenesis of lupus nephritis. The importance of fibronectin in the development of lupus nephritis is further highlighted in studies whereby intra-peritoneal administration of either native dimeric or soluble fibronectin into non-autoimmune mice resulted in humoral autoimmune response with increased levels of circulating immune complexes and IgG deposition in the glomeruli [124].

TGF-β1 and Fibronectin

Whilst much attention has been given to the role of TGF-β1 in the pathogenesis of diabetic nephropathy, this pro-fibrotic peptide has also been implicated in matrix protein accumulation in most renal diseases and is a key initiator of EMT. The TGF-β family consists of three isoforms, namely TGF-β1, TGF-β2 and TGF-β3, which are encoded by distinct genes. Whilst all three isoforms have been shown to induce matrix protein synthesis [125], most studies have concentrated on the role of TGF-β1. TGF-β1 is a multi-functional peptide that can influence ECM production, modulate inflammatory cell function and regulate growth inhibition and differentiation. Its importance in biological functions has been highlighted in TGF-β1 knockout mice, which develop diffuse mononuclear cell infiltration that is lethal within a few weeks of birth [126, 127]. TGF-β1 is synthesized as an inactive 25 kDa homodimer that is non-covalently associated with a latency-associated protein. The latency-associated protein inhibits TGF-β1 binding to its receptor and its dissociation is mediated by various factors that include an acidic environment, heat, shear force, thrombospondin-1 and proteolysis [128, 129]. TGF-β1 has been shown to initiate fibrogenesis through four distinct, but simultaneous pathways, namely (1) its direct ability to stimulate matrix protein synthesis, (2) inhibit degradation of matrix proteins by decreasing the expression of matrix metalloproteinases (MMPs), whilst increasing their inhibitors, (3) enhance cell-matrix interactions by increasing integrin expression and thus facilitating matrix deposition and (4) self-induction that results in the amplification of fibrotic processes [130, 131]. TGF-β1 mediates its fibrotic effects through the phosphorylation of Smad-2 and Smad-3, which associates with Smad-4 and this complex is then translocated to the nucleus where it up-regulates gene expression of target molecules [132]. TGF-β1 can also induce fibronectin synthesis in renal cells through mechanisms that are independent of Smad activation. In this respect, Peng *et al* noted that TGF-β1 induced fibronectin synthesis in mesangial cells was mediated through caveolae-dependent RhoA activation [133]. Although it may be tempting to target TGF-β1 for therapeutic intervention in order to inhibit renal fibrosis, studies have demonstrated that it is not as straightforward as one would assume since effective silencing of TGF-β1 in proximal tubular epithelial cells unexpectedly increased fibronectin synthesis consequent to a compensatory mechanism that induced the synthesis of TGF-β2 and TGF-β3, which contributed to pathological matrix accumulation [134].

Apart from TGF-β1, other mediators such as CTGF, PDGF, IL-1β, TNF-α and angiotensin II have been shown to increase fibronectin in mesangial cells, epithelial cells and endothelial cells and synergistic accumulation of fibronectin has been observed in mesangial cells when co-cultures with TGF-β1 and either IL-1β or TNF-α [135, 136]. These pro-fibrotic mediators were shown to increase fibronectin synthesis through the activation of signalling pathways such as PKC, p838 MAPK, ERK, JNK, up-regulation of α5β1 integrin activation or increased TIMP-1 expression [65, 135-139].

Glomerular infiltration of macrophages is a common feature of renal diseases. Accumulating evidence suggests that macrophages may also play a key role in the induction of glomerulosclerosis. Activated macrophages that infiltrate into the glomerulus secrete a plethora of inflammatory mediators that alter the secretory phenotype of resident mesangial cells, and how these cells respond to the secretory repertoire of macrophages will determine whether the outcome will result in the resolution of tissue injury or progressive scarring leading to renal fibrosis. Although TGF-β1 is regarded as a pro-fibrotic mediator, it also possesses immunosuppressive properties and is a potent negative regulator of monocytes/macrophages. Pawluczyk and Harris noted that conditioned medium obtained from TGF-β1 stimulated macrophages could suppress mesangial cell fibronectin synthesis compared to unstimulated macrophages, and this was mediated in part through the increased synthesis of transin/stromelysin, a neutral metalloproteinase [140]. These researchers suggested that TGF-β1 has a dual role in fibrogenesis and its role in the reparative processes should not be ignored.

Matrix Proteinases and Fibrosis

There is evidence to suggest that renal fibrosis is not only attributed to an increase in matrix protein accumulation within the renal parenchyma, but is also coupled with impairment in processes relating to matrix remodelling and degradation. The kidney synthesizes various proteases that include MMPs, serine proteinases and lysosomal cysteine proteinases [40, 141-146]. Each enzyme differs in their substrate specificity. The MMPs and members of the plasmin-dependent pathway have been extensively studied in the context of renal fibrosis [142, 146-148]. It is a misconception that proteinases played only a beneficial role in the degradation of matrix proteins since they have also been shown to cleave non-matrix substrates to release pro-fibrotic growth factors that contribute to renal fibrogenesis [149]. Furthermore, the ability of MMPs to degrade matrix protein components is dependent on the stabilization and binding of the ECM to the cell membrane.

MMPs are a family of zinc-dependent endopeptidases with differing substrate specificities [150]. Their activities are regulated by tissue inhibitors of metalloproteinases (TIMPs) [150]. Within the kidney, numerous studies have focused on MMP-2 and MMP-9 since they are able to degrade collagen type IV present in basement membranes. MMP-2 has also been shown to degrade fibronectin and laminin, whereas MMP-9 can also degrade aggrecan and elastin. Due to their overlapping activity and substrate specificities, it is difficult to dissect out their distinct role *in vivo*.

Studies whereby researchers have inhibited the activity of either an individual or a combination of MMP(s) *in vivo* either through gene silencing or pharmaceutical inhibition have revealed confusing results. Apart from MT1-MMP-deficient mice, which developed fibrosis of the soft tissues consequent to the ablation of the collagenolytic activity [151], gene silencing of other MMPs in mice do not develop tissue fibrosis [152]. Furthermore, the over-expression of MMP-2 in transgenic mice is associated with tubulo-interstitial fibrosis and EMT [153]. MMP-9 is normally considered to possess anti-fibrotic properties since it can degrade interstitial collagens, but it may also be considered a fibrotic mediator since it activates TGF-β1 and subsequent EMT in various cells [154, 155]. MMP-9 therefore appears to have a dual role in matrix protein synthesis and degradation. The role of MMPs in tissue fibrosis has been further confounded by studies demonstrating that mice deficient in TIMP-3 spontaneously develop interstitial fibrosis [156].

Emerging evidence suggests that the plasmin-dependent cascade also plays a key role in renal fibrosis [157]. Active plasmin is derived from the proteolytic cleavage of plasminogen by tissue-type plasminogen activator (tPA) or urokinase-type plasminogen activator (uPA), and their activities are inhibited by PAI-1 [158, 159]. In addition to its ability to degrade fibrin, plasmin has been shown to degrade fibronectin and laminin [160-162]. Plasmin has also been shown to activate latent MMPs and indirectly increase matrix protein degradation, whilst promoting EMT [163]. tPA knockout mice with ureteral obstruction have reduced tissue expression of MMP-9, which is accompanied by decreased interstitial fibrosis and EMT with the concomitant preservation of the tubular basement membrane [164]. Readers are referred to the following reviews for further information regarding the role of proteinases in the pathogenesis of renal fibrosis [149, 150, 165, 166].

Epithelial-to-Mesenchymal Transdifferentiation (EMT) in Kidney Diseases

Regardless of the initial cause of renal disease, renal fibrosis is characterized by the presence of interstitial myofibroblasts [167, 168]. It was not until almost two decades ago that it was conceptually accepted that differentiated epithelial cells with specialized functions could alter their phenotype and acquire mesenchymal properties, a process known as EMT [35, 169, 170]. The ability of proximal tubular epithelial cells to undergo phenotypic changes under physiological and pathological conditions underscores the plasticity property of these cells. EMT was first recognized in the early 1980s as an important mechanism of embryogenesis and development [171]. EMT is characterized by a loss of cytokeratin and E-cadherin expression, and dissolution of tight junctions resulting in a loss of cell-cell and cell-matrix interactions and the subsequent detachment of the epithelial cells from the underlying basement membrane. Increased synthesis of matrix metalloproteinases (MMPs) then digest the basement membrane permitting epithelial cells to migrate into the interstitium. EMT is accompanied by cell activation, increased synthesis of matrix proteins and up-regulation of Snail, Smads, vimentin, α-smooth muscle actin and fibroblast specific protein-1 [172].

Compelling evidence has highlighted the pivotal role of TGF-β1 in the induction of EMT in proximal tubular epithelial cells [173, 174]. The clinical relevance of EMT has been confirmed by Rastaldi *et al* who observed an increase in mesenchymal markers and a concomitant decrease in epithelial markers in renal biopsies obtained from patients with lupus nephritis, diabetic nephropathy, IgA nephropathy and primary focal segmental glomerulosclerosis [175]. The number of tubular epithelial cells with mesenchymal properties was associated with the degree of interstitial injury and renal impairment. These data therefore indicate that tubular epithelial cells are active participants in interstitial events in both immune-mediated and non-immune-mediated renal diseases [175].

The importance of tubular epithelial cell conversion to myofibroblasts in renal fibrosis was highlighted by Iwano *et al* who elegantly demonstrated in a murine model of unilateral ureteral obstruction that 36% of interstitial cells stained positive for fibroblast specific protein-1 were derived from the labelled tubular epithelium and these cells were found to contribute to collagen type I deposition in the tubulo-interstitium [38]. The conversion of tubular epithelial cells to myofibroblasts appears to occur during the latter stages of renal fibrosis in various animal models of renal fibrosis suggesting that tubular epithelial cells do not normally undergo EMT unless there is sustained, chronic injury. In this respect, Yang and Liu noted that in the presence of TGF-β1, α-smooth muscle actin expression was not induced in proximal tubular epithelial cells until 36 h after stimulation [39], whereas TGF-β1 induction of α-smooth muscle actin expression in interstitial fibroblasts was observed after 8 h [36]. Furthermore, in a murine model of unilateral ureteral obstruction, induction of α-smooth muscle actin expression in tubular epithelial cells was not noted until day 7 after surgery and was associated with a concomitant decrease in E-cadherin, whereas conversion of interstitial fibroblasts to myofibroblasts occurred after 3 days of renal injury [39], thereby corroborating findings that EMT in tubular epithelial cells is not an early event in the pathogenesis of renal fibrosis. Although numerous studies have focused on the synthesis of collagen type I in tubular cells undergoing EMT, it is noteworthy that fibronectin is also increased during EMT processes. Recently, Freire-de-Lima *et al* demonstrated that O-glycosylation of fibronectin containing EDA^+ and EDB^+ domains induced EMT in human prostate epithelial cell lines [176]. It is possible that a phenotypic change in tubular epithelial cells may be associated with the appearance of the oncofoetal isoforms of fibronectin that are O-glycosylated, although further studies are warranted to confirm this.

Fibronectin and the Peritoneal Membrane

Effect of Peritoneal Dialysis on the Peritoneal Membrane

Peritoneal dialysis (PD) is an established form of renal replacement therapy that is currently used by 11% of the global dialysis population. It involves the instillation of 1.0 to 2.0 litre(s) of PD fluid into the patient's abdomen where it is allowed to dwell for 4 to 8 h, during which time solutes and wastes are removed from the circulation and enter the PD solution. After the dwell time, the PD fluid is drained and a fresh bag is installed. PD patients require 3 to 4 exchanges a day, every day and in one year the peritoneum will be exposed to almost 3000 litres of PD fluid. Conventional PD solutions contain elevated concentrations of glucose at concentrations 15 to 40 times greater than the physiological concentration to provide the osmotic drive. Even after intra-peritoneal equilibration, glucose concentrations remain 5-16 times that of physiological concentrations. Heat sterilization of glucose-based PD solutions and subsequent storage gives rise to the formation of glucose degradation products (GDPs) and promote the formation of advanced glycation end-products (AGEs) [177, 178]. It is therefore not unexpected that constant exposure of the peritoneal membrane to un-physiological glucose concentrations in PD solutions is associated with diabetic-like modification of its structure.

The peritoneal membrane consists of a monolayer of mesothelial cells underneath which is the submesothelium that contains fibroblasts, collagen fibrils and capillaries [179]. Peritoneal mesothelial cells are specialized epithelial cells that line the peritoneal cavity and play critical roles in peritoneal homeostasis, maintenance of peritoneal membrane integrity and wound healing. Although mesothelial cells are derived from the mesoderm, they possess distinct characteristics of epithelial cells such as a polygonal, cobblestone morphology with surface microvilli, express cytokeratin and form a polarized monolayer that permits the translocation of molecules to the apical or basolateral sector of the cells [180-184]. Mesothelial cells also possess mesenchymal characteristics and express vimentin, α-smooth muscle actin and increased synthesis of matrix proteins when activated [180, 185, 186]. Mesothelial cells contribute to peritoneal homeostasis by virtue of their ability to synthesize a plethora of cytokines, growth factors and matrix proteins such as collagen, laminin and fibronectin, and when activated are major instigators of peritoneal fibrosis [185, 187-191].

Many patients on long-term PD exhibit re-duplication of the mesothelial and endothelial basement membranes, increased synthesis and deposition of matrix proteins within the submesothelium and progressive subendothelial hyalinization, with narrowing or obliteration of the vascular lumen [179, 192-194]. Vascular and interstitial changes become more apparent with progressive use of PD, thus demonstrating a temporal relationship between peritoneal fibrosis, vasculopathy and time on PD [192]. These structural changes are exacerbated by refractory or repeated episodes of peritonitis and are associated with a loss of ultrafiltration and solute clearance that leads to technique failure and adverse clinical outcomes. Peritoneal fibrosis can be detected in 50% and 80% of PD patients within one and two years respectively on PD [195-197]. Independent researchers have demonstrated significant changes to mesothelial cells following their exposure to PD, which include cell activation, cell hypertrophy, increased vacuolation, partial or complete loss of microvilli, dissolution of cell-cell contacts and alterations in the number of endoplasmic reticulum and micropinocytotic vesicles [179, 192, 194, 198]. Some degree of mesothelial denudation is invariably observed in PD patients and this is associated with thickening of the submesothelium and vasculopathy [192].

Pathophysiological Processes Leading to Peritoneal Fibrosis After Peritoneal Dialysis

The mesothelium provides the first line of defence against bacteria, chemical or surgical insult. It is therefore critical that following injury, restoration of the mesothelial monolayer quickly takes place. To maintain stable peritoneal homeostasis and a constant number of mesothelial cells lining the peritoneum, a fine balance between cell proliferation and cell death is required. Under normal conditions, mesothelial cells exhibit limited cell proliferation, with only 0.1-0.5% of the mesothelial cell population undergoing mitosis at any one time [182]. We and others have demonstrated that elevated glucose levels and GDPs can reduce mesothelial cell proliferation [199, 200], which would compromise remesothelialization. Although the underlying mechanism through which denudation of the mesothelium can assist in the development of peritoneal fibrosis remains to be fully elucidated, it is plausible to suggest that the inability to restore the mesothelium would allow the bio-incompatible PD fluid to seep into the submesothelium to activate interstitial fibroblasts and increase their fibrogenic ability.

An alternative explanation by which mesothelial cells can contribute to peritoneal fibrosis is their ability to change phenotype. In the presence of PD solutions, studies have demonstrated that mesothelial cells can undergo accelerated senescence or EMT *in vivo* and *in vitro* [201-204]. Both phenotypes can be induced in part, by the pro-fibrotic peptide TGF-β1, but the mechanism that dictates which phenotypic pathway mesothelial cells takes is currently unknown. Mesothelial cells with a senescent phenotype are hypertrophic, have a low proliferative potential and invariably possess multiply vacuoles and nuclei [201, 205]. Ksiazek *et al* demonstrated that elevated glucose concentrations mediated cell senescence though increased oxidative stress and/or TGF-β1 bio-activation [202, 203]. It is plausible to speculate that given their low proliferative potential and strong adherent property, their presence in the peritoneum would contribute to the inability of the mesothelium to undergo normal reparative processes and remesothelialization. Mesothelial cells that have undergone accelerated senescence have been shown to synthesize almost three times the amount of fibronectin than that synthesized by normal mesothelial cells and this increase would inevitably contribute to peritoneal fibrosis [202, 206].

We and others have demonstrated that elevated glucose concentrations detected in PD fluids can induce fibronectin synthesis in mesothelial cells, mediated through the prior activation of PKC-α and TGF-β1 bio-activation [64, 187]. Increased synthesis of pro-inflammatory cytokines during peritoneal inflammation has also been shown to induce fibronectin synthesis in mesothelial cells [207]. Angiotensin II is a potent vasoactive peptide that can regulate inflammation and fibrosis and its role in renal fibrosis has been extensively studied [208]. Recently, Kiribayashi *et al* demonstrated that mesothelial cells expressed angiotensin II type 1 and 2 receptors, and that angiotensin II can induce gene and protein expression of fibronectin in mesothelial cells through prior activation of ERK 1/2 and p38 MAPK, a finding also noted in renal interstitial fibroblasts [209, 210]. Mesothelial cells have been shown to express RAGE, the receptor for AGEs [211]. Using diabetic mice, De Vriese *et al* provided compelling evidence that the interaction of AGEs with RAGE in the mesothelium and sub-mesothelium resulted in increased TGF-β1 synthesis and its subsequent induction of fibronectin accumulation and neoangiogenesis [212]. Collectively, these studies highlight the importance of mesothelial cells in instigating peritoneal fibrosis.

Much attention has focused on the contributing role of mesothelial cells in fibrogenesis and the fibrotic and inflammatory mediators that induce synthesis of matrix proteins in mesothelial cells. Fibroblasts have been shown to play critical roles in tissue fibrosis in various organs including the kidney, but they have received little attention in the peritoneum. Ogata *et al* demonstrated that bFGF secretion was increased in mesothelial cells when exposed to elevated glucose concentrations, which subsequently promoted proliferation and secretion of fibronectin in peritoneal fibroblasts [213]. This study elegantly demonstrates the bi-directional communication between mesothelial cells and fibroblasts in the peritoneal membrane and their collaboration to mediate peritoneal fibrosis during PD. Studies have shown that fibroblasts play a direct role in tissue fibrosis in the kidney, lung and skin. It is thus possible that peritoneal fibroblasts can directly contribute exert a direct contribution to peritoneal fibrosis during PD although this warrants further investigation.

Effluent-derived mesothelial cells cultured *ex vivo* have been shown to possess a fibroblastic appearance [204]. In the clinical setting, mesothelial cells that have undergone EMT are destined for migration into the submesothelium where they contribute to peritoneal fibrosis. In an *in vitro* 3-dimensional migration model, Yamaguchi *et al* demonstrated that mesothelial cells that had undergone EMT following their exposure to TGF-β1 required a fibronectin gradient before they were able to migrate into a collagen gel, and this migration was mediated through the interaction of α5β1 integrin on mesothelial cells to the RGDS receptor on the fibronectin molecule [214]. The clinical relevance of EMT in peritoneal fibrosis is however controversial. Whilst Yanez-Mo *et al* demonstrated the presence of elongated mesothelial cells that are positive for cytokeratin and ICAM-1 embedded in the peritoneal submesothelium [204], Williams *et al* was unable to detect mesothelial cells that had undergone EMT in human peritoneal biopsy samples [215].

Drugs and Fibrosis

Calcineurin Inhibitors

The introduction of calcineurin inhibitors such as cyclosporine A (CsA) and tacrolimus has significantly reduced the incidence of acute rejection following renal transplantation [216]. The immunosuppressive action of these agents relates to their ability to inhibit lymphocyte activation and proliferation, but calcineurin inhibitors are nephrotoxic and induce both inflammatory and fibrotic processes in the kidney leading to chronic allograft nephropathy [217, 218]. Although various side-effects have been associated with CsA, we will focus only on their effect on fibrogenesis.

In vitro and *in vivo* studies have demonstrated that CsA can induce TGF-β1 secretion in proximal tubular epithelial cells, interstitial fibroblasts and T cells [219, 220]. TGF-β1 possesses both fibrotic and immunosuppressive properties [130, 131, 221, 222]. In a rat model of chronic CsA nephropathy, Shihab *et al* observed an increase in EDA^+-containing fibronectin in the glomeruli, tubulo-interstitium and blood vessels after 7 days treatment compared to untreated rats, which was mediated through increased TGF-β1 production [223]. The pathological role of CsA-induced TGF-β1 in mediating fibrogenesis was further highlighted by the observation that administration of anti-TGF-β1 neutralizing antibody to rats with CsA nephrotoxicity was accompanied by reduced intra-renal gene expression of matrix proteins [224]. Although CsA is a calcineurin inhibitor that can induce TGF-β1 synthesis, it is intriguingly that exogenous TGF-β1 can activate calcineurin in a dose and time-dependent manner in rat mesangial cells, which then mediates matrix protein accumulation [72]. Inhibition of calcineurin before the addition of TGF-β1 suppressed matrix protein accumulation thus confirming the pathological role of calcineurin in TGF-β1 mediated induction of fibrogenesis. The ability of TGF-β1 to activate calcineurin is attributed to the sustained increase in intracellular calcium since chelation of extracellular calcium abrogated TGF-β1-mediated influx of calcium and calcineurin activity [72]. There is also evidence that CsA can induce EMT in human proximal tubular epithelial cells, which is accompanied by actin polymerization, loss of adherens junctions and increased synthesis of fibronectin and collagen type IV, mediated in part through the prior activation of PKC-β isoforms [225, 226]. Using affymetrix gene microarray analysis, these researchers also demonstrated that in addition to TGF-β1, CsA is capable of inducing other profibrotic mediators, oncogenes and transcription regulators [226], which may cooperate with TGF-β1 in the process of EMT and matrix accumulation [225]. The mechanism through which CsA induces TGF-β1 levels in renal transplant patients remains to be fully elucidated, but effective and sustained inhibition of calcineurin with CsA may suppress TGF-β1 mediated induction of EMT and renal fibrosis.

Baboolal *et al* investigated the key determinants and mediators of renal injury and fibrosis in cadaveric renal transplant recipients who were treated with either CsA or tacrolimus. These researchers utilized serial protocol biopsies to assess the mRNA levels of TGF-β1, fibronectin and thrombospondin-1 at the time of engraftment and 3, 6 and 12 months post-transplantation and demonstrated an early and progressive increased in the gene expression of all three fibrotic markers in both treatment groups, which was associated with interstitial fibrosis and chronic allograft nephropathy [227]. Moreover, these researchers also demonstrated that changes in the levels of TGF-β1 mRNA predicted renal impairment. Comparison of CsA and tacrolimus treatment in the induction of TGF-β1 mRNA revealed a ten-fold increased in TGF-β1 mRNA in patients treated with CsA compared to tacrolimus after 6 months treatment [227]. In contrast, Khanna *et al* demonstrated that tacrolimus treatment significantly increased intra-renal expression of TGF-β1 and mRNA levels for fibronectin, collagen, MMP-2, MMP-9 and TIMP-2 in renal transplant patients compared to CsA treatment [228]. This discrepancy between both studies may be attributed to the time when biopsy samples were taken after renal transplantation and the dose and duration of CsA and tacrolimus treatment.

Independent researchers have also confirmed that tacrolimus can induce renal fibrosis in renal transplant patients [229]. In an animal model of tacrolimus-induced nephrotoxicity, accumulation of matrix proteins was observed predominantly in the tubulo-interstitium and blood vessels, and was mediated through increased production of TGF-β1 and PAI-1 [230]. Tacrolimus has also been shown to induce TGF-β1 and fibronectin in cultured rat mesangial cells [231], which may contribute to glomerulosclerosis.

Mychophenolic Acid (MPA)

Mycophenolate mofetil (MMF) is the pro-drug of MPA, which exerts its immunosuppressive effect by inhibiting inosine monophosphate dehydrogenase, the key enzyme in *de novo* purine synthesis that is essential for lymphocyte proliferation. MMF has proven efficacy in the prevention of organ transplant rejection and treatment of lupus nephritis [232-235]. Unlike calcineurin inhibitors, MMF does not increase TGF-β1 expression [232, 236] and does not induce fibrogenesis in the kidney. Recently, Dell'Oglio *et al* demonstrated that the gene and protein expression of neutral endopeptidase, an enzyme that can degrades angiotensin II, was induced in kidney transplant recipients converted to MMF treatment compared with others continued with azathioprine [237]. These researchers demonstrated that gene expression of neutral endopeptidase inversely correlated with proteinuria, whilst glomerular and tubular expression of neutral endopeptidase inversely correlated with glomerulosclerosis and tubulo-interstitial fibrosis respectively [237]. Given that angiotensin II is a key player in the progression of glomerular and tubulo-interstitial injury in patients with chronic renal disease [238], it is likely that MMF may inhibit progression of renal injury in kidney transplant patients. In an animal model of kidney transplantation, Djamali *et al* demonstrated that administration of MMF to rats delayed allograft fibrosis by the inhibition of TGF-β1-mediated induction of Nox-2 activation, NFκB phosphorylation and myofibroblast conversion in tubular cells [239].

In rat models of glomerulonephritis, MMF has been shown to reduce renal myofibroblast infiltration, matrix protein deposition and glomerular crescent formation [240]. There is also accumulating data to demonstrate that MMF may also elicit beneficial effects on nonlymphoid cells. *In vitro* studies have shown that MPA can suppress proliferation in tubular epithelial cells, mesangial cells, endothelial cells and fibroblasts and also reduce fibronectin synthesis in human mesangial cells stimulated with either FCS [241] or anti-dsDNA antibodies through the down-regulation of PKC phosphorylation and TGF-β1 bio-activation [8]. We have recently demonstrated that MMF can significantly reduce intra-renal expression of fibronectin in lupus-prone mice with active disease and this was associated with decreased PKC phosphorylation and improvement in clinical and serological parameters of disease [8]. MPA has also been shown to reduce EMT in proximal tubular epithelial cells [35].

Proliferation Signal Inhibitors

Sirolimus (rapamycin) and everolimus belong to the family of proliferation signal inhibitors, immunosuppressive agents that suppress T cell proliferation and expansion by inhibiting the mammalian target of rapamycin (mTOR) signalling cascade [242]. Unlike calcineurin inhibitors, experimental and clinical studies have demonstrated that sirolimus exhibits minimal nephrotoxicity at clinically relevant doses, although this group of drugs increases the intracellular level of calcineurin inhibitors and thus may enhance the nephrotoxic effect of the latter. The effect of sirolimus on renal fibrosis is controversial. Pontrelli *et al* observed that sirolimus could inhibit PAI-1 expression and reduce glomerulosclerosis and tubulo-interstitial fibrosis in patients with chronic allograft nephropathy [243], but Servais *et al* did not observe any significant reduction in renal fibrosis after one year when patients were converted from CsA to sirolimus 12 weeks after renal transplantation [244].

Treatment of lupus-prone mice with sirolimus has also provided encouraging results and there is emerging data to show that as with MMF, sirolimus may also elicit beneficial effects on resident renal cells [245-248]. Sirolimus can reduce cell proliferation in mitogen-stimulated mouse proximal tubular epithelial cells although cells become less sensitive to the anti-proliferative nature of sirolimus with prolonged exposure [249]. Sirolimus has also been shown to inhibit TGF-β1-mediated induction of EMT in human proximal tubular epithelial cells, collagen type I synthesis in human fibroblasts and tubulo-interstitial fibrosis in an animal model of unilateral ureteral obstruction-induced renal fibrosis [35, 250, 251]. Despite the ability of sirolimus to inhibit the fibrotic effects of TGF-β1, Swinford *et al* noted that sirolimus can increase TGF-β1 mRNA levels in tubular epithelial cells [252]. We have demonstrated that sirolimus can reduce TGF-β1-mediated induction of fibronectin in human mesangial cells and also glomerular and tubulo-interstitial expression of TGF-β1 and fibronectin in an animal model of active lupus nephritis (Yung, Zhang and Chan, unpublished data). It is thus possible that the effect of sirolimus on fibrogenesis may be cell-specific.

In an animal model of diabetic nephropathy, treatment with sirolimus was shown to suppress gene and protein expression of TGF-β1, fibronectin and laminin, an observation that was not noted in diabetic mice treated with CsA [253]. Sirolimus has also demonstrated beneficial effects in suppressing EMT in peritoneal mesothelial cells [254]. In an animal model of peritoneal fibrosis, Sekiguichi *et al* observed that rapamycin did not directly affect TGF-β1-mediated induction of peritoneal fibrosis and angiogenesis, but mediated its beneficial effects through hypoxia inducible factor-1α [255].

Pirfenidone

Pirfenidone (5-methyl-N-phenyl-2-(1H)-pyridone) is a low molecular weight synthetic molecule that possesses both anti-fibrotic and anti-inflammatory properties [256-259]. Its anti-fibrotic effect was first observed in 1995 in an animal model of bleomycin-induced pulmonary fibrosis where it was shown to significantly reduce collagen synthesis [256]. The mechanism through which pirfenidone inhibits matrix protein accumulation is currently unknown, but emerging evidence suggests that it is through its ability to inhibit TGF-β1 secretion, phosphorylation of Smad-2 and production of reactive oxygen species [259, 260]. Numerous studies have demonstrated that pirfenidone can reduce both glomerular and tubulo-interstitial fibrosis in animal models of diabetic nephropathy, unilateral ureteral obstruction and anti-glomerular basement membrane glomerulonephritis [258, 259, 261]. Furthermore, Miric *et al* noted that pirfenidone reduced cardiac, renal and plasma levels of fibronectin in diabetic rats after 4 weeks treatment, but had no effect on renal function [262]. In a separate study, Park *et al* investigated the effect of pirfenidone on renal fibrosis and function in a murine model of focal segmental glomerulosclerosis (FSGS). These researchers demonstrated that after 3 months treatment, pirfenidone significantly reduced glomerulosclerosis and tubulo-interstitial fibrosis compared to control mice, which was associated with an improvement in renal function [263].

In a recent randomized, double-blind, placebo-controlled clinical study, Sharma *et al* investigated the effect of pirfenidone on renal function in type I and type II diabetic patients with established diabetic nephropathy [264]. These investigators observed a marked improvement in the estimated glomerular filtration rate (eGFR) in patients treated with pirfenidone 6 months after treatment was initiated and this improvement was maintained until the end of the study. Baseline plasma levels of inflammatory and fibrotic biomarkers significantly correlated with baseline eGFR but did not predict the patients' response to pirfenidone [264]. Urine TGF-β1 levels, which reflects ongoing renal production, increased with time in the placebo and treatment groups, but there was no significant difference between the placebo and treatment groups.

Although the data suggest that pirfenidone may improve renal function through suppression of renal fibrosis, the observation that it did not alter the levels of pro-fibrotic mediators could be related to the small sample size and inter-patient variability.

In a separate study, Cho *et al* investigated the effect of pirfenidone on renal function in patients with FSGS and observed that pirfenidone slowed renal function decline [265]. Although the mechanism through which pirfenidone mediated its beneficial effect in these studies was not investigated, it is possible that it inhibited TGF-β1 activity and subsequent renal fibrosis.

Bone Morphogenetic Protein 7 (BMP-7) and Hepatocyte Growth Factor (HGF)

In addition to the aforementioned drugs that have demonstrated efficacy in the treatment of renal fibrosis, there is increasing data on the anti-fibrotic effects of HGF and BMP-7 in the pathogenesis of renal fibrosis. Animal and *in vitro* studies have demonstrated that both HGF and BMP-7 can inhibit EMT and accumulation of matrix proteins in resident renal and peritoneal cells [266-270].

The ability to induce endogenous anti-fibrotic mediators may be a promising therapeutic strategy in the prevention of renal and peritoneal fibrosis. A detailed discussion of the anti-fibrotic properties of HGF and BMP-7 falls outside the scopeop of this review and readers are referred to the following papers [271-274].

Conclusion

Chronic renal disease is characterized by a progressive accumulation of matrix proteins and deposition of fibrous tissue in the glomerular and tubuloe-interstitial compartments of the kidney, and progressive deterioration of kidney function.

The deposition of fibronectin is prominent in this process, and is accompanied by an increased expression of oncofoetal fibronectin isoforms. There is increasing data to show that fibronectin is not just an inert structural substrate. In contrast, it participates actively in many cellular functions including cell adhesion, migration, proliferation and differentiation. Soluble fibronectin has been shown to assist in the recruitment of infiltrating cells to sites of injury and also mediate the migration of myofibroblasts into the interstitium. Insoluble fibronectin can regulate the polymerization of other matrix proteins and thus play a pivotal role in tissue fibrosis. It may not be possible to translate or extrapolate data from *in vitro* or animal studies to the clinical setting.

Nevertheless, elucidation of the molecular and biochemical processes that initiate or aggravate fibronectin matrix assembly during physiological and pathological conditions, and its role in cell-matrix interaction and EMT, both in the kidney and the peritoneal membrane, is prerequisite to devising effective strategies to prevent or suppress organ fibrosis which portends functional failure.

Acknowledgments

Part of this work was supported by a General Research Fund Incentive Award (HKU 786111M), the Wai Hung Charitable Foundation, a donation from the Estate of the late Mr Chan Wing Hei and the UGC Matching Grant Scheme (Phase IV and V). Dr S Yung is partially supported by the Endowment Fund established for the 'Yu Chiu Kwong Professorship Chair in Nephrology' Medicine' Endowment Fund awarded to Tak Mao Chan.

References

[1] Hynes RO, Yamada KM. Fibronectins: multifunctional modular glycoproteins. *J. Cell Biol.* 1982; 95: 369-377.

[2] Singh P, Carraher C, Schwarzbauer JE. Assembly of fibronectin extracellular matrix. *Annu. Rev. Cell Dev. Biol.* 2010; 26: 397-419.

[3] Siri A, Carnemolla B, Raffanti S, Castellani P, Balzano E, Zardi L. Fibronectin concentrations in pleural effusions of patients with malignant and non-malignant diseases. *Cancer Lett.* 1984; 22: 1-9.

[4] Courtoy PJ, Kanwar YS, Hynes RO, Farquhar MG. Fibronectin localization in the rat glomerulus. *J. Cell Biol.* 1980; 87: 691-696.

[5] Del Prete D, Ceol M, Anglani F, Vianello D, Tiralongo E, Valente M, Graziotto R, Bonfante L, Scaparrotta G, Furian L, Rigotti P, Gambaro G, D'Angelo A. Early activation of fibrogenesis in transplanted kidneys: a study on serial renal biopsies. *Exp. Mol. Pathol.* 2009; 87: 141-145.

[6] Burger A, Wagner C, Viedt C, Reis B, Hug F, Hansch GM. Fibronectin synthesis by human tubular epithelial cells in culture: effects of PDGF and TGF-beta on synthesis and splicing. *Kidney Int.* 1998; 54: 407-415.

[7] Vesey DA, Cheung CW, Cuttle L, Endre ZA, Gobe G, Johnson DW. Interleukin-1beta induces human proximal tubule cell injury, alpha-smooth muscle actin expression and fibronectin production. *Kidney Int.* 2002; 62: 31-40.

[8] Yung S, Zhang Q, Zhang CZ, Chan KW, Lui SL, Chan TM. Anti-DNA antibody induction of protein kinase C phosphorylation and fibronectin synthesis in human and murine lupus and the effect of mycophenolic acid. *Arthritis Rheum.*2009;60:2071-2082.

[9] Weston BS, Wahab NA, Roberts T, Mason RM. Bacitracin inhibits fibronectin matrix assembly by mesangial cells in high glucose. *Kidney Int.* 2001; 60: 1756-1764.

[10] Baelde HJ, Eikmans M, van Vliet AI, Bergijk EC, de Heer E, Bruijn JA. Alternatively spliced isoforms of fibronectin in immune-mediated glomerulosclerosis: the role of TGFbeta and IL-4. *J. Pathol.* 2004; 204: 248-257.

[11] Dixey J, Moss J, Woodrow DF, Shore I, Maini RN. SLE nephritis: an ultrastructural immunogold study to evaluate the relationship between immune complexes and the basement membrane components type IV collagen, fibronectin and heparan sulphate proteoglycans. *Clin. Nephrol.* 1990; 34: 95-102.

[12] van Vliet AI, van Alderwegen IE, Baelde HJ, de Heer E, Bruijn JA. Fibronectin accumulation in glomerulosclerotic lesions: self-assembly sites and the heparin II binding domain. *Kidney Int.* 2002; 61: 481-489.

[13] Dixon AJ, Burns J, Dunnill MS, McGee JO. Distribution of fibronectin in normal and diseased human kidneys. *J. Clin. Pathol.* 1980; 33: 1021-1028.
[14] Prowse KR, Tricoli JV, Klebe RJ, Shows TB. Assignment of the human fibronectin structural gene to chromosome 2. *Cytogenet. Cell Genet.* 1986; 41: 42-46.
[15] Zardi L, Cianfriglia M, Balza E, Carnemolla B, Siri A, Croce CM. Species-specific monoclonal antibodies in the assignment of the gene for human fibronectin to chromosome 2. *EMBO J.* 1982; 1: 929-933.
[16] Hynes RO. Fibronectins: a family of complex and versatile adhesive glycoproteins derived from a single gene. *Harvey Lect.* 1985; 81: 133-152.
[17] Magnusson MK, Mosher DF. Fibronectin: structure, assembly, and cardiovascular implications. *Arterioscler. Thromb. Vasc. Biol.* 1998; 18: 1363-1370.
[18] Kornblihtt AR, Umezawa K, Vibe-Pedersen K, Baralle FE. Primary structure of human fibronectin: differential splicing may generate at least 10 polypeptides from a single gene. *EMBO J.* 1985; 4: 1755-1759.
[19] Kornblihtt AR, Vibe-Pedersen K, Baralle FE. Human fibronectin: cell specific alternative mRNA splicing generates polypeptide chains differing in the number of internal repeats. *Nucleic Acids Res.* 1984; 12: 5853-5868.
[20] Ffrench-Constant C, Hynes RO. Alternative splicing of fibronectin is temporally and spatially regulated in the chicken embryo. *Development* 1989; 106: 375-388.
[21] Ffrench-Constant C, Van de Water L, Dvorak HF, Hynes RO. Reappearance of an embryonic pattern of fibronectin splicing during wound healing in the adult rat. *J. Cell Biol.* 1989; 109: 903-914.
[22] Manabe R, Ohe N, Maeda T, Fukuda T, Sekiguchi K. Modulation of cell-adhesive activity of fibronectin by the alternatively spliced EDA segment. *J. Cell Biol.* 1997; 139: 295-307.
[23] Carnemolla B, Leprini A, Allemanni G, Saginati M, Zardi L. The inclusion of the type III repeat ED-B in the fibronectin molecule generates conformational modifications that unmask a cryptic sequence. *J. Biol. Chem.* 1992; 267: 24689-24692.
[24] Ting KM, Rothaupt D, McCormick TS, Hammerberg C, Chen G, Gilliam AC, Stevens S, Culp L, Cooper KD. Overexpression of the oncofetal Fn variant containing the EDA splice-in segment in the dermal-epidermal junction of psoriatic uninvolved skin. *J. Invest. Dermatol.* 2000; 114: 706-711.
[25] Schwarzbauer JE. Identification of the fibronectin sequences required for assembly of a fibrillar matrix. *J. Cell Biol.* 1991; 113: 1463-1473.
[26] Sechler JL, Corbett SA, Schwarzbauer JE. Modulatory roles for integrin activation and the synergy site of fibronectin during matrix assembly. *Mol. Biol. Cell* 1997; 8: 2563-2573.
[27] Pankov R, Yamada KM. Fibronectin at a glance. *J. Cell Sci.* 2002; 115: 3861-3863.
[28] Wierzbicka-Patynowski I, Schwarzbauer JE. The ins and outs of fibronectin matrix assembly. *J. Cell Sci.* 2003; 116: 3269-3276.
[29] Schlondorff D, Banas B. The mesangial cell revisited: no cell is an island. *J. Am. Soc. Nephrol.* 2009; 20: 1179-1187.
[30] Couchman JR, Beavan LA, McCarthy KJ. Glomerular matrix: synthesis, turnover and role in mesangial expansion. *Kidney Int.* 1994; 45:328-335.
[31] Fogo AB. Mesangial matrix modulation and glomerulosclerosis. *Exp. Nephrol.* 1999; 7: 147-159.

[32] Nath KA. The tubulointerstitium in progressive renal disease. *Kidney Int.* 1998; 54: 992-994.

[33] Nath KA. Tubulointerstitial changes as a major determinant in the progression of renal damage. *Am. J. Kidney Dis.* 1992; 20: 1-17.

[34] Burns WC, Kantharidis P, Thomas MC. The role of tubular epithelial-mesenchymal transition in progressive kidney disease. *Cells Tissues Organs* 2007; 185: 222-231.

[35] Copeland JW, Beaumont BW, Merrilees MJ, Pilmore HL. Epithelial-to-mesenchymal transition of human proximal tubular epithelial cells: effects of rapamycin, mycophenolate, cyclosporin, azathioprine, and methylprednisolone. *Transplantation* 2007; 83: 809-814.

[36] Liu Y. Epithelial to mesenchymal transition in renal fibrogenesis: pathologic significance, molecular mechanism, and therapeutic intervention. *J. Am. Soc. Nephrol.* 2004; 15: 1-12.

[37] Wynn T. Cellular and molecular mechanisms of fibrosis. *J. Pathol.* 2008; 214: 199-210.

[38] Iwano M, Plieth D, Danoff TM, Xue C, Okada H, Neilson EG. Evidence that fibroblasts derive from epithelium during tissue fibrosis. *J. Clin. Invest.* 2002; 110: 341-350.

[39] Yang J, Liu Y. Dissection of key events in tubular epithelial to myofibroblast transition and its implications in renal interstitial fibrosis. *Am. J. Pathol.* 2001; 159: 1465-1475.

[40] Eddy AA. Molecular insights into renal interstitial fibrosis. *J. Am. Soc. Nephrol.* 1996; 7: 2495-2508.

[41] Eddy AA. Molecular basis of renal fibrosis. *Pediatr. Nephrol.* 2000; 15: 290-301.

[42] Liu Y. New insights into epithelial-mesenchymal transition in kidney fibrosis. *J. Am. Soc. Nephrol.* 2010; 21: 212-222.

[43] Powell DW, Bertram CC, Cummins TD, Barati MT, Zheng S, Epstein PN, Klein JB. Renal tubulointerstitial fibrosis in OVE26 type 1 diabetic mice. *Nephron. Exp. Nephrol.* 2009; 111: e11-19.

[44] Yamamoto T, Noble NA, Miller DE, Border WA. Sustained expression of TGF-beta 1 underlies development of progressive kidney fibrosis. *Kidney Int.* 1994; 45: 916-927.

[45] Hynes RO. The dynamic dialogue between cells and matrices: implications of fibronectin's elasticity. *Proc. Natl. Acad. Sci. U S A* 1999; 96: 2588-2590.

[46] Hynes RO. Cell-matrix adhesion in vascular development. *J. Thromb. Haemost.* 2007; 5 Suppl 1: 32-40.

[47] Laitinen L, Vartio T, Virtanen I. Cellular fibronectins are differentially expressed in human fetal and adult kidney. *Lab. Invest.* 1991; 64: 492-498.

[48] Van Vliet A, Baelde HJ, Vleming LJ, de Heer E, Bruijn JA. Distribution of fibronectin isoforms in human renal disease. *J. Pathol.* 2001; 193: 256-262.

[49] Gharaee-Kermani M, Wiggins R, Wolber F, Goyal M, Phan SH. Fibronectin is the major fibroblast chemoattractant in rabbit anti-glomerular basement membrane disease. *Am. J. Pathol.* 1996; 148: 961-967.

[50] Strutz F, Neilson EG. New insights into mechanisms of fibrosis in immune renal injury. *Springer Semin. Immunopathol.* 2003; 24:459-476.

[51] Ren L, Blanchette JB, White LR, Clark SA, Heffner DJ, Tibbles LA, Muruve DA. Soluble fibronectin induces chemokine gene expression in renal tubular epithelial cells. *Kidney Int.* 2005; 68: 2111-2120.

[52] Lopez-Armada MJ, Gonzalez E, Gomez-Guerrero C, Egido J. The 80-kD fibronectin fragment increases the production of fibronectin and tumour necrosis factor-alpha (TNF-alpha) in cultured mesangial cells. *Clin. Exp. Immunol.* 1997; 107: 398-403.

[53] Wright SD, Meyer BC. Fibronectin receptor of human macrophages recognizes the sequence Arg-Gly-Asp-Ser. *J. Exp. Med.* 1985; 162: 762-767.

[54] Idasiak-Piechocka I, Oko A, Pawliczak E, Kaczmarek E, Czekalski S. Elevated urinary fibronectin excretion predicts poor outcome in patients with primary chronic glomerulonephritis. *Nephron. Clin. Pract.* 2010; 116: c47-52.

[55] Mauer SM, Steffes MW, Ellis EN, Sutherland DE, Brown DM, Goetz FC. Structural-functional relationships in diabetic nephropathy. *J. Clin. Invest.* 1984; 74: 1143-1155.

[56] Steffes MW, Osterby R, Chavers B, Mauer SM. Mesangial expansion as a central mechanism for loss of kidney function in diabetic patients. *Diabetes* 1989; 38: 1077-1081.

[57] Sharma K, Ziyadeh FN. Hyperglycemia and diabetic kidney disease. The case for transforming growth factor-beta as a key mediator. *Diabetes* 1995; 44: 1139-1146.

[58] Brosius FC, Heilig CW. Glucose transporters in diabetic nephropathy. *Pediatr. Nephrol.* 2005; 20: 447-451.

[59] Kashihara N, Haruna Y, Kondeti VK, Kanwar YS. Oxidative stress in diabetic nephropathy. *Curr. Med. Chem.* 2010; 17: 4256-4269.

[60] Whiteside CI, Dlugosz JA. Mesangial cell protein kinase C isozyme activation in the diabetic milieu. *Am. J. Physiol. Renal. Physiol.* 2002; 282: F975-980.

[61] Kapor-Drezgic J, Zhou X, Babazono T, Dlugosz JA, Hohman T, Whiteside C. Effect of high glucose on mesangial cell protein kinase C-delta and -epsilon is polyol pathway-dependent. *J. Am. Soc. Nephrol.* 1999; 10: 1193-1203.

[62] Kikkawa R, Umemura K, Haneda M, Arimura T, Ebata K, Shigeta Y. Evidence for existence of polyol pathway in cultured rat mesangial cells. *Diabetes* 1987; 36:240-243.

[63] Daniel C, Schaub K, Amann K, Lawler J, Hugo C. Thrombospondin-1 is an endogenous activator of TGF-beta in experimental diabetic nephropathy in vivo. *Diabetes* 2007; 56: 2982-2989.

[64] Ha H, Yu MR, Lee HB. High glucose-induced PKC activation mediates TGF-beta 1 and fibronectin synthesis by peritoneal mesothelial cells. *Kidney Int.* 2001; 59: 463-470.

[65] Lin S, Sahai A, Chugh SS, Pan X, Wallner EI, Danesh FR, Lomasney JW, Kanwar YS. High glucose stimulates synthesis of fibronectin via a novel protein kinase C, Rap1b, and B-Raf signaling pathway. *J. Biol. Chem.* 2002; 277: 41725-41735.

[66] Wahab NA, Harper K, Mason RM. Expression of extracellular matrix molecules in human mesangial cells in response to prolonged hyperglycaemia. *Biochem. J.* 1996; 316 (Pt 3): 985-992.

[67] Weigert C, Brodbeck K, Brosius FC, 3rd, Huber M, Lehmann R, Friess U, Facchin S, Aulwurm S, Haring HU, Schleicher ED, Heilig CW. Evidence for a novel TGF-beta1-independent mechanism of fibronectin production in mesangial cells overexpressing glucose transporters. *Diabetes* 2003; 52: 527-535.

[68] Jo F, Morimoto S, Nakahigashi M, Kusabe M, Someya K, Morita T, Jo H, Imada T, Kosaki A, Toyoda N, Nishikawa M, Iwasaka T. Olmesartan Induces Renoprotective Effects by Stimulating Angiotensin Type 2 Receptors and Reducing Oxidative Stress in Diabetic Nephropathy. *Kidney Blood Press Res.* 2011; 34: 418-423.

[69] Kitada M, Kume S, Imaizumi N, Koya D. Resveratrol improves oxidative stress and protects against diabetic nephropathy through normalization of Mn-SOD dysfunction in AMPK/SIRT1-independent pathway. *Diabetes* 2011; 60: 634-643.

[70] Noh H, Ha H. Reactive oxygen species and oxidative stress. *Contrib. Nephrol.* 2011; 170: 102-112.

[71] Stanton RC. Oxidative stress and diabetic kidney disease. *Curr. Diab. Rep.* 2011; 11: 330-336.

[72] Gooch JL, Gorin Y, Zhang BX, Abboud HE. Involvement of calcineurin in transforming growth factor-beta-mediated regulation of extracellular matrix accumulation. *J. Biol. Chem.* 2004; 279: 15561-15570.

[73] Gorin Y, Block K, Hernandez J, Bhandari B, Wagner B, Barnes JL, Abboud HE. Nox4 NAD(P)H oxidase mediates hypertrophy and fibronectin expression in the diabetic kidney. *J. Biol. Chem.* 2005; 280: 39616-39626.

[74] Gorin Y, Kim NH, Feliers D, Bhandari B, Choudhury GG, Abboud HE. Angiotensin II activates Akt/protein kinase B by an arachidonic acid/redox-dependent pathway and independent of phosphoinositide 3-kinase. *FASEB J.* 2001; 15: 1909-1920.

[75] Gorin Y, Ricono JM, Kim NH, Bhandari B, Choudhury GG, Abboud HE. Nox4 mediates angiotensin II-induced activation of Akt/protein kinase B in mesangial cells. *Am. J. Physiol. Renal. Physiol.* 2003; 285: F219-229.

[76] Akkina S, Becker BN. MicroRNAs in kidney function and disease. *Transl. Res.* 2011; 157: 236-240.

[77] Jiang X, Tsitsiou E, Herrick SE, Lindsay MA. MicroRNAs and the regulation of fibrosis. *FEBS J.* 2010; 277: 2015-2021.

[78] Wang Q, Wang Y, Minto AW, Wang J, Shi Q, Li X, Quigg RJ. MicroRNA-377 is up-regulated and can lead to increased fibronectin production in diabetic nephropathy. *FASEB J.* 2008; 22: 4126-4135.

[79] Slack-Davis JK, Eblen ST, Zecevic M, Boerner SA, Tarcsafalvi A, Diaz HB, Marshall MS, Weber MJ, Parsons JT, Catling AD. PAK1 phosphorylation of MEK1 regulates fibronectin-stimulated MAPK activation. *J. Cell Biol.* 2003; 162: 281-291.

[80] Buettner GR. Superoxide dismutase in redox biology: the roles of superoxide and hydrogen peroxide. *Anticancer Agents Med. Chem.* 2011; 11: 341-346.

[81] Uchimura K, Nagasaka A, Hayashi R, Makino M, Nagata M, Kakizawa H, Kobayashi T, Fujiwara K, Kato T, Iwase K, Shinohara R, Kato K, Itoh M. Changes in superoxide dismutase activities and concentrations and myeloperoxidase activities in leukocytes from patients with diabetes mellitus. *J. Diabetes Complications* 1999; 13: 264-270.

[82] DeRubertis FR, Craven PA, Melhem MF. Acceleration of diabetic renal injury in the superoxide dismutase knockout mouse: effects of tempol. *Metabolism* 2007; 56: 1256-1264.

[83] Catherwood MA, Powell LA, Anderson P, McMaster D, Sharpe PC, Trimble ER. Glucose-induced oxidative stress in mesangial cells. *Kidney Int.* 2002; 61: 599-608.

[84] Cortes P, Riser BL, Zhao X, Narins RG. Glomerular volume expansion and mesangial cell mechanical strain: mediators of glomerular pressure injury. *Kidney Int. Suppl.* 1994; 45: S11-16.

[85] Gruden G, Zonca S, Hayward A, Thomas S, Maestrini S, Gnudi L, Viberti GC. Mechanical stretch-induced fibronectin and transforming growth factor-beta1 production in human mesangial cells is p38 mitogen-activated protein kinase-dependent. *Diabetes* 2000; 49:655-661.

[86] Hirakata M, Kaname S, Chung UG, Joki N, Hori Y, Noda M, Takuwa Y, Okazaki T, Fujita T, Katoh T, Kurokawa K. Tyrosine kinase dependent expression of TGF-beta induced by stretch in mesangial cells. *Kidney Int.* 1997; 51: 1028-1036.

[87] Ingram AJ, James L, Ly H, Thai K, Scholey JW. Stretch activation of jun N-terminal kinase/stress-activated protein kinase in mesangial cells. *Kidney Int.* 2000; 58: 1431-1439.

[88] Yasuda T, Kondo S, Homma T, Harris RC. Regulation of extracellular matrix by mechanical stress in rat glomerular mesangial cells. *J. Clin. Invest.* 1996;98:1991-2000.

[89] Ishida T, Haneda M, Maeda S, Koya D, Kikkawa R. Stretch-induced overproduction of fibronectin in mesangial cells is mediated by the activation of mitogen-activated protein kinase. *Diabetes* 1999; 48: 595-602.

[90] Nishio T, Haneda M, Koya D, Inoki K, Maeda S, Kikkawa R. Cyclic AMP inhibits stretch-induced overexpression of fibronectin in glomerular mesangial cells. *Eur. J. Pharmacol.* 2002; 437: 113-122.

[91] Smeets B, Kuppe C, Sicking EM, Fuss A, Jirak P, van Kuppevelt TH, Endlich K, Wetzels JF, Grone HJ, Floege J, Moeller MJ. Parietal epithelial cells participate in the formation of sclerotic lesions in focal segmental glomerulosclerosis. *J. Am. Soc. Nephrol.* 2011; 22: 1262-1274.

[92] Zhu L, Jiang R, Aoudjit L, Jones N, Takano T. Activation of RhoA in podocytes induces focal segmental glomerulosclerosis. *J. Am. Soc. Nephrol.* 2011; 22: 1621-1630.

[93] Wendt TM, Tanji N, Guo J, Kislinger TR, Qu W, Lu Y, Bucciarelli LG, Rong LL, Moser B, Markowitz GS, Stein G, Bierhaus A, Liliensiek B, Arnold B, Nawroth PP, Stern DM, D'Agati VD, Schmidt AM. RAGE drives the development of glomerulosclerosis and implicates podocyte activation in the pathogenesis of diabetic nephropathy. *Am. J. Pathol.* 2003; 162: 1123-1137.

[94] Stieger N, Worthmann K, Schiffer M. The role of metabolic and haemodynamic factors in podocyte injury in diabetes. *Diabetes Metab. Res. Rev.* 2011; 27: 207-215.

[95] Fogo AB, Kon V. The glomerulus - --a view from the inside - --the endothelial cell. *Int. J. Biochem. Cell Biol.* 2010; 42: 1388-1397.

[96] Kelly DJ, Buck D, Cox AJ, Zhang Y, Gilbert RE. Effects on protein kinase C-beta inhibition on glomerular vascular endothelial growth factor expression and endothelial cells in advanced experimental diabetic nephropathy. *Am. J. Physiol. Renal. Physiol.* 2007; 293: F565-574.

[97] Ninichuk V, Khandoga AG, Segerer S, Loetscher P, Schlapbach A, Revesz L, Feifel R, Khandoga A, Krombach F, Nelson PJ, Schlondorff D, Anders HJ. The role of interstitial macrophages in nephropathy of type 2 diabetic db/db mice. *Am. J. Pathol.* 2007; 170: 1267-1276.

[98] Okada H, Kikuta T, Kobayashi T, Inoue T, Kanno Y, Takigawa M, Sugaya T, Kopp JB, Suzuki H. Connective tissue growth factor expressed in tubular epithelium plays a pivotal role in renal fibrogenesis. *J. Am. Soc. Nephrol.* 2005; 16: 133-143.

[99] Yung S, Lee CY, Zhang Q, Lau SK, Tsang RC, Chan TM. Elevated glucose induction of thrombospondin-1 up-regulates fibronectin synthesis in proximal renal tubular epithelial cells through TGF-beta1 dependent and TGF-beta1 independent pathways. *Nephrol. Dial. Transplant.* 2006; 21: 1504-1513.

[100] Wei JL, Peng YM, Liu F. Connective tissue growth factor and fibronectin secretion in renal tubular epithelial cells induced by TGF-beta1: suppressive effects of troglitazone. *Cell Biol. Int.* 2007; 31: 30-34.

[101] Murphy M, Docherty NG, Griffin B, Howlin J, McArdle E, McMahon R, Schmid H, Kretzler M, Droguett A, Mezzano S, Brady HR, Furlong F, Godson C, Martin F. IHG-1 amplifies TGF-beta1 signaling and is increased in renal fibrosis. *J. Am. Soc. Nephrol.* 2008; 19: 1672-1680.

[102] Zeng R, Han M, Luo Y, Li C, Pei G, Liao W, Bai S, Ge S, Liu X, Xu G. Role of Sema4C in TGF-beta1-induced mitogen-activated protein kinase activation and epithelial-mesenchymal transition in renal tubular epithelial cells. *Nephrol. Dial. Transplant.* 2011; 26: 1149-1156.

[103] Chen S, Hoffman BB, Lee JS, Kasama Y, Jim B, Kopp JB, Ziyadeh FN. Cultured tubule cells from TGF-beta1 null mice exhibit impaired hypertrophy and fibronectin expression in high glucose. *Kidney Int.* 2004; 65: 1191-1204.

[104] Lam S, Verhagen NA, Strutz F, van der Pijl JW, Daha MR, van Kooten C. Glucose-induced fibronectin and collagen type III expression in renal fibroblasts can occur independent of TGF-beta1. *Kidney Int.* 2003; 63: 878-888.

[105] Ayo SH, Radnik RA, Garoni JA, Glass WF, 2nd, Kreisberg JI. High glucose causes an increase in extracellular matrix proteins in cultured mesangial cells. *Am. J. Pathol.* 1990; 136: 1339-1348.

[106] Ayo SH, Radnik RA, Glass WF, 2nd, Garoni JA, Rampt ER, Appling DR, Kreisberg JI. Increased extracellular matrix synthesis and mRNA in mesangial cells grown in high-glucose medium. *Am. J. Physiol.* 1991; 260: F185-191.

[107] Cameron JS. Lupus nephritis. *J. Am. Soc. Nephrol.* 1999; 10: 413-424.

[108] Hahn BH. Antibodies to DNA. *N. Engl. J. Med.* 1998; 338: 1359-1368.

[109] Schwartz MM. The pathological classification of lupus nephritis. in: *Lupus Nephritis*, Lewis EJ, Schwartz MM, Korbet SM(eds) Oxford University Press: New York, 1999, pp 126-158.

[110] Chan TM, Leung JK, Ho SK, Yung S. Mesangial cell-binding anti-DNA antibodies in patients with systemic lupus erythematosus. *J. Am. Soc. Nephrol.* 2002; 13: 1219-1229.

[111] Isenberg DA, Manson JJ, Ehrenstein MR, Rahman A. Fifty years of anti-ds DNA antibodies: are we approaching journey's end? *Rheumatology* 2007; 46: 1052-1056.

[112] Alexander JJ, Zwingmann C, Jacob A, Quigg R. Alteration in kidney glucose and amino acids are implicated in renal pathology in MRL/lpr mice. *Biochim. Biophys. Acta* 2007; 1772: 1143-1149.

[113] Altunkova I, Minkova V, Belovezhdov N. Role of fibronectin in immune glomerulonephritis. *Nephron.* 1993; 63: 438-444.

[114] Yamamoto T, Noble NA, Cohen AH, Nast CC, Hishida A, Gold LI, Border WA. Expression of transforming growth factor-beta isoforms in human glomerular diseases. *Kidney Int.* 1996; 49: 461-469.

[115] Yamamoto K, Loskutoff DJ. Expression of transforming growth factor-beta and tumor necrosis factor-alpha in the plasma and tissues of mice with lupus nephritis. *Lab. Invest.* 2000; 80: 1561-1570.

[116] Peterson KS, Huang JF, Zhu J, D'Agati V, Liu X, Miller N, Erlander MG, Jackson MR, Winchester RJ. Characterization of heterogeneity in the molecular pathogenesis of lupus nephritis from transcriptional profiles of laser-captured glomeruli. *J. Clin. Invest.* 2004; 113: 1722-1733.

[117] Giunti S, Tesch GH, Pinach S, Burt DJ, Cooper ME, Cavallo-Perin P, Camussi G, Gruden G. Monocyte chemoattractant protein-1 has prosclerotic effects both in a mouse model of experimental diabetes and in vitro in human mesangial cells. *Diabetologia* 2008; 51: 198-207.

[118] Awad AS, Kinsey GR, Khutsishvili K, Gao T, Bolton WK, Okusa MD. Monocyte/macrophage chemokine receptor CCR2 mediates diabetic renal injury. *Am. J. Physiol. Renal. Physiol.* 2011; 301: F1358-1366.

[119] Bergijk EC, Baelde HJ, De Heer E, Killen PD, Bruijn JA. Specific accumulation of exogenous fibronectin in experimental glomerulosclerosis. *J. Pathol.*1995;176:191-199.

[120] Carsons S. High levels of fibronectin fragments in the plasma of a patient with active systemic lupus erythematosus. *J. Rheumatol.* 1987; 14: 1052-1054.

[121] Nishinarita S, Yamamoto M, Takizawa T, Hayakawa J, Karasaki M, Sawada S. Increased plasma fibronectin in patients with systemic lupus erythematosus. *Clin. Rheumatol.* 1990; 9: 214-219.

[122] Atta MS, Powell RJ, Hopkinson ND, Todd I. Human anti-fibronectin antibodies in systemic lupus erythematosus: occurrence and antigenic specificity. *Clin. Exp. Immunol.* 1994; 96: 20-25.

[123] Fujii H, Nakatani K, Arita N, Ito MR, Terada M, Miyazaki T, Yoshida M, Ono M, Fujiwara T, Saiga K, Ota T, Ohtani H, Lockwood M, Sasaki T, Nose M. Internalization of antibodies by endothelial cells via fibronectin implicating a novel mechanism in lupus nephritis. *Kidney Int.* 2003; 64: 1662-1670.

[124] Murphy-Ullrich JE, Oberley TD, Mosher DF. Serologic and pathologic studies of mice immunized with homologous fibronectin. *Am. J. Pathol.* 1986; 125: 182-190.

[125] Yu L, Border WA, Huang Y, Noble NA. TGF-beta isoforms in renal fibrogenesis. *Kidney Int.* 2003; 64: 844-856.

[126] Kulkarni AB, Huh CG, Becker D, Geiser A, Lyght M, Flanders KC, Roberts AB, Sporn MB, Ward JM, Karlsson S. Transforming growth factor beta 1 null mutation in mice causes excessive inflammatory response and early death. *Proc. Natl. Acad. Sci. U S A* 1993; 90:770-774.

[127] Shull MM, Ormsby I, Kier AB, Pawlowski S, Diebold RJ, Yin M, Allen R, Sidman C, Proetzel G, Calvin D, et al. Targeted disruption of the mouse transforming growth factor-beta 1 gene results in multifocal inflammatory disease. *Nature*1992;359: 693-699.

[128] Brown PD, Wakefield LM, Levinson AD, Sporn MB. Physicochemical activation of recombinant latent transforming growth factor-beta's 1, 2, and 3. *Growth Factors* 1990; 3: 35-43.

[129] Murphy-Ullrich JE, Poczatek M. Activation of latent TGF-beta by thrombospondin-1: mechanisms and physiology. *Cytokine Growth Factor Rev.* 2000; 11: 59-69.

[130] Border WA, Noble NA. TGF-beta in kidney fibrosis: a target for gene therapy. *Kidney Int.* 1997; 51: 1388-1396.
[131] Branton MH, Kopp JB. TGF-beta and fibrosis. *Microbes Infect.* 1999; 1: 1349-1365.
[132] Schnaper HW, Hayashida T, Poncelet AC. It's a Smad world: regulation of TGF-beta signaling in the kidney. *J. Am. Soc. Nephrol.* 2002; 13: 1126-1128.
[133] Peng F, Zhang B, Wu D, Ingram AJ, Gao B, Krepinsky JC. TGFbeta-induced RhoA activation and fibronectin production in mesangial cells require caveolae. *Am. J. Physiol. Renal. Physiol.* 2008; 295: F153-164.
[134] Qi W, Chen X, Holian J, Mreich E, Twigg S, Gilbert RE, Pollock CA. Transforming growth factor-beta1 differentially mediates fibronectin and inflammatory cytokine expression in kidney tubular cells. *Am. J. Physiol. Renal. Physiol.* 2006; 291: F1070-1077.
[135] Kreisberg JI, Kreisberg SH. High glucose activates protein kinase C and stimulates fibronectin gene expression by enhancing a cAMP response element. *Kidney Int. Suppl.* 1995; 51: S3-11.
[136] Pawluczyk IZ, Harris KP. Cytokine interactions promote synergistic fibronectin accumulation by mesangial cells. *Kidney Int.* 1998; 54:62-70.
[137] Goldberg H, Whiteside C, Fantus IG. O-linked beta-N-acetylglucosamine supports p38 MAPK activation by high glucose in glomerular mesangial cells. *Am. J. Physiol. Endocrinol. Metab.* 2011; 301: E713-726.
[138] Wilmer WA, Dixon CL, Hebert C. Chronic exposure of human mesangial cells to high glucose environments activates the p38 MAPK pathway. *Kidney Int.* 2001; 60:858-871.
[139] Isono M, Cruz MC, Chen S, Hong SW, Ziyadeh FN. Extracellular signal-regulated kinase mediates stimulation of TGF-beta1 and matrix by high glucose in mesangial cells. *J. Am. Soc. Nephrol.* 2000; 11: 2222-2230.
[140] Pawluczyk IZ, Harris KP. Transforming growth factor-beta suppresses macrophage-induced mesangial cell fibronectin expression. *Kidney Int.* 2001; 60: 533-542.
[141] Lelongt B, Ronco P. Role of matrix metalloproteinases in kidney development and glomerulopathy: lessons from transgenic mice. *Nephrol. Dial. Transplant.* 2002; 17 Suppl 9: 28-31.
[142] Lelongt B, Legallicier B, Piedagnel R, Ronco PM. Do matrix metalloproteinases MMP-2 and MMP-9 (gelatinases) play a role in renal development, physiology and glomerular diseases? *Curr. Opin. Nephrol. Hypertens.* 2001; 10: 7-12.
[143] Davies M, Martin J, Thomas GJ, Lovett DH. Proteinases and glomerular matrix turnover. *Kidney Int.* 1992; 41: 671-678.
[144] Zhang G, Eddy AA. Urokinase and its receptors in chronic kidney disease. *Front Biosci.* 2008; 13: 5462-5478.
[145] Zheng G, Harris DC. Plasmin in renal interstitial fibrosis: innocent or guilty? *Kidney Int.* 2004; 66: 455-456.
[146] Martin J, Eynstone L, Davies M, Steadman R. Induction of metalloproteinases by glomerular mesangial cells stimulated by proteins of the extracellular matrix. *J. Am. Soc. Nephrol.* 2001; 12: 88-96.
[147] Tveita AA, Rekvig OP, Zykova SN. Increased glomerular matrix metalloproteinase activity in murine lupus nephritis. *Kidney Int.* 2008.
[148] Thrailkill KM, Clay Bunn R, Fowlkes JL. Matrix metalloproteinases: their potential role in the pathogenesis of diabetic nephropathy. *Endocrine* 2009; 35: 1-10.

[149] Ronco P, Lelongt B, Piedagnel R, Chatziantoniou C. Matrix metalloproteinases in kidney disease progression and repair: a case of flipping the coin. *Semin. Nephrol.* 2007; 27: 352-362.
[150] Visse R, Nagase H. Matrix metalloproteinases and tissue inhibitors of metalloproteinases: structure, function, and biochemistry. *Circ. Res.* 2003; 92: 827-839.
[151] Holmbeck K, Bianco P, Caterina J, Yamada S, Kromer M, Kuznetsov SA, Mankani M, Robey PG, Poole AR, Pidoux I, Ward JM, Birkedal-Hansen H. MT1-MMP-deficient mice develop dwarfism, osteopenia, arthritis, and connective tissue disease due to inadequate collagen turnover. *Cell* 1999; 99: 81-92.
[152] Wang X, Zhou Y, Tan R, Xiong M, He W, Fang L, Wen P, Jiang L, Yang J. Mice lacking the matrix metalloproteinase-9 gene reduce renal interstitial fibrosis in obstructive nephropathy. *Am. J. Physiol. Renal. Physiol.* 2010; 299: F973-982.
[153] Cheng S, Pollock AS, Mahimkar R, Olson JL, Lovett DH. Matrix metalloproteinase 2 and basement membrane integrity: a unifying mechanism for progressive renal injury. *FASEB J.* 2006; 20: 1898-1900.
[154] Ronco P, Chatziantoniou C. Matrix metalloproteinases and matrix receptors in progression and reversal of kidney disease: therapeutic perspectives. *Kidney Int.* 2008; 74: 873-878.
[155] Tan TK, Zheng G, Hsu TT, Wang Y, Lee VW, Tian X, Cao Q, Harris DC. Macrophage matrix metalloproteinase-9 mediates epithelial-mesenchymal transition in vitro in murine renal tubular cells. *Am. J. Pathol.* 2010; 176: 1256-1270.
[156] Kassiri Z, Oudit GY, Kandalam V, Awad A, Wang X, Ziou X, Maeda N, Herzenberg AM, Scholey JW. Loss of TIMP3 enhances interstitial nephritis and fibrosis. *J. Am. Soc. Nephrol.* 2009; 20: 1223-1235.
[157] Edgtton KL, Gow RM, Kelly DJ, Carmeliet P, Kitching AR. Plasmin is not protective in experimental renal interstitial fibrosis. *Kidney Int.* 2004; 66: 68-76.
[158] Aya N, Yoshioka K, Murakami K, Hino S, Okada K, Matsuo O, Maki S. Tissue-type plasminogen activator and its inhibitor in human glomerulonephritis. *J. Pathol.* 1992; 166: 289-295.
[159] Eddy AA. Serine proteases, inhibitors and receptors in renal fibrosis. *Thromb. Haemost.* 2009; 101: 656-664.
[160] Cosgrove D. A complex, tissue-specific role for plasmin and its regulators in modulating fibrogenic activity. *Am. J. Physiol. Renal. Physiol.* 2007; 293: F10-11.
[161] Nakagami Y, Abe K, Nishiyama N, Matsuki N. Laminin degradation by plasmin regulates long-term potentiation. *J. Neurosci.* 2000; 20: 2003-2010.
[162] Uemura A, Nakamura M, Kachi S, Nishizawa Y, Asami T, Miyake Y, Terasaki H. Effect of plasmin on laminin and fibronectin during plasmin-assisted vitrectomy. *Arch. Ophthalmol.* 2005; 123: 209-213.
[163] Hahn-Dantona E, Ramos-DeSimone N, Sipley J, Nagase H, French DL, Quigley JP. Activation of proMMP-9 by a plasmin/MMP-3 cascade in a tumor cell model. Regulation by tissue inhibitors of metalloproteinases. *Ann. N. Y. Acad. Sci.* 1999; 878: 372-387.
[164] Yang J, Shultz RW, Mars WM, Wegner RE, Li Y, Dai C, Nejak K, Liu Y. Disruption of tissue-type plasminogen activator gene in mice reduces renal interstitial fibrosis in obstructive nephropathy. *J. Clin. Invest.* 2002; 110: 1525-1538.

[165] Catania JM, Chen G, Parrish AR. Role of matrix metalloproteinases in renal pathophysiologies. *Am. J. Physiol. Renal. Physiol.* 2007; 292: F905-911.

[166] Zeisberg M, Neilson EG. Mechanisms of tubulointerstitial fibrosis. *J. Am. Soc. Nephrol.* 2010; 21: 1819-1834.

[167] Tang WW, Ulich TR, Lacey DL, Hill DC, Qi M, Kaufman SA, Van GY, Tarpley JE, Yee JS. Platelet-derived growth factor-BB induces renal tubulointerstitial myofibroblast formation and tubulointerstitial fibrosis. *Am. J. Pathol.* 1996; 148: 1169-1180.

[168] Essawy M, Soylemezoglu O, Muchaneta-Kubara EC, Shortland J, Brown CB, el Nahas AM. Myofibroblasts and the progression of diabetic nephropathy. *Nephrol. Dial. Transplant.* 1997; 12: 43-50.

[169] Kalluri R, Neilson EG. Epithelial-mesenchymal transition and its implications for fibrosis. *J. Clin. Invest.* 2003; 112: 1776-1784.

[170] Li Y, Dai C, Wu C, Liu Y. PINCH-1 promotes tubular epithelial-to-mesenchymal transition by interacting with integrin-linked kinase. *J. Am. Soc. Nephrol.* 2007; 18: 2534-2543.

[171] Hay ED. Interaction of embryonic surface and cytoskeleton with extracellular matrix. *Am. J. Anat.* 1982; 165: 1-12.

[172] Guarino M, Tosoni A, Nebuloni M. Direct contribution of epithelium to organ fibrosis: epithelial-mesenchymal transition. *Hum. Pathol.* 2009; 40: 1365-1376.

[173] Zheng G, Lyons JG, Tan TK, Wang Y, Hsu TT, Min D, Succar L, Rangan GK, Hu M, Henderson BR, Alexander SI, Harris DC. Disruption of E-cadherin by matrix metalloproteinase directly mediates epithelial-mesenchymal transition downstream of transforming growth factor-beta1 in renal tubular epithelial cells. *Am. J. Pathol.* 2009; 175: 580-591.

[174] Yeh YC, Wei WC, Wang YK, Lin SC, Sung JM, Tang MJ. Transforming growth factor-{beta}1 induces Smad3-dependent {beta}1 integrin gene expression in epithelial-to-mesenchymal transition during chronic tubulointerstitial fibrosis. *Am. J. Pathol.* 2010; 177: 1743-1754.

[175] Rastaldi MP, Ferrario F, Giardino L, Dell'Antonio G, Grillo C, Grillo P, Strutz F, Muller GA, Colasanti G, D'Amico G. Epithelial-mesenchymal transition of tubular epithelial cells in human renal biopsies. *Kidney Int.* 2002; 62: 137-146.

[176] Freire-de-Lima L, Gelfenbeyn K, Ding Y, Mandel U, Clausen H, Handa K, Hakomori SI. Involvement of O-glycosylation defining oncofetal fibronectin in epithelial-mesenchymal transition process. *Proc. Natl. Acad. Sci. U S A* 2011; 108: 17690-17695.

[177] Witowski J, Bender TO, Gahl GM, Frei U, Jorres A. Glucose degradation products and peritoneal membrane function. *Perit. Dial. Int.* 2001; 21: 201-205.

[178] Tauer A, Zhang X, Schaub TP, Zimmeck T, Niwa T, Passlick-Deetjen J, Pischetsrieder M. Formation of advanced glycation end products during CAPD. *Am. J. Kidney Dis.* 2003; 41: S57-60.

[179] Dobbie J. Ultrastructure and pathology of the peritoneum in peritoneal dialysis. in: *Textbook of peritoneal dialysis*, Gokal R.(ed) Kluwer Academic Publishers: Dordrecht, 1994, pp 17-44.

[180] Yung S, Li FK, Chan TM. Peritoneal mesothelial cell culture and biology. *Perit. Dial. Int.* 2006; 26: 162-173.

[181] Yung S, Chan TM. Mesothelial cells. *Perit. Dial. Int.* 2007; 27 Suppl 2: S110-115.

[182] Mutsaers SE. The mesothelial cell. *Int. J. Biochem. Cell Biol.* 2004; 36: 9-16.

[183] Mutsaers SE. Mesothelial cells: their structure, function and role in serosal repair. *Respirology* 2002; 7: 171-191.

[184] Mutsaers SE, Whitaker D, Papadimitriou JM. Changes in the concentration of microvilli on the free surface of healing mesothelium are associated with alterations in surface membrane charge. *J. Pathol.* 1996; 180: 333-339.

[185] Stylianou E, Jenner LA, Davies M, Coles GA, Williams JD. Isolation, culture and characterization of human peritoneal mesothelial cells. *Kidney Int.* 1990; 37:1563-1570.

[186] Aroeira LS, Aguilera A, Sanchez-Tomero JA, Bajo MA, del Peso G, Jimenez-Heffernan JA, Selgas R, Lopez-Cabrera M. Epithelial to mesenchymal transition and peritoneal membrane failure in peritoneal dialysis patients: pathologic significance and potential therapeutic interventions. *J. Am. Soc. Nephrol.* 2007; 18: 2004-2013.

[187] Chan TM, Leung JK, Tsang RC, Liu ZH, Li LS, Yung S. Emodin ameliorates glucose-induced matrix synthesis in human peritoneal mesothelial cells. *Kidney Int.* 2003; 64: 519-533.

[188] Ferriola PC, Stewart W. Fibronectin expression and organization in mesothelial and mesothelioma cells. *Am. J. Physiol.* 1996; 271: L804-812.

[189] Fang CC, Yen CJ, Chen YM, Shyu RS, Tsai TJ, Lee PH, Hsieh BS. Pentoxifylline inhibits human peritoneal mesothelial cell growth and collagen synthesis: effects on TGF-beta. *Kidney Int.* 2000; 57: 2626-2633.

[190] Ha H, Cha MK, Choi HN, Lee HB. Effects of peritoneal dialysis solutions on the secretion of growth factors and extracellular matrix proteins by human peritoneal mesothelial cells. *Perit. Dial. Int.* 2002; 22: 171-177.

[191] Yung S, Davies M. Response of the human peritoneal mesothelial cell to injury: an in vitro model of peritoneal wound healing. *Kidney Int.* 1998; 54: 2160-2169.

[192] Williams JD, Craig KJ, Topley N, Von Ruhland C, Fallon M, Newman GR, Mackenzie RK, Williams GT. Morphologic changes in the peritoneal membrane of patients with renal disease. *J. Am. Soc. Nephrol.* 2002; 13: 470-479.

[193] Honda K, Nitta K, Horita S, Yumura W, Nihei H. Morphological changes in the peritoneal vasculature of patients on CAPD with ultrafiltration failure. *Nephron.* 1996; 72: 171-176.

[194] Dobbie JW. Morphology of the peritoneum in CAPD. *Blood Purif.* 1989; 7: 74-85.

[195] Garosi G, Di Paolo N. Morphological aspects of peritoneal sclerosis. *J. Nephrol.* 2001; 14 Suppl 4: S30-38.

[196] Garosi G, Di Paolo N. Pathophysiology and morphological clinical correlation in experimental and peritoneal dialysis-induced peritoneal sclerosis. *Adv. Perit. Dial.* 2000; 16: 204-207.

[197] Schneble F, Bonzel KE, Waldherr R, Bachmann S, Roth H, Scharer K. Peritoneal morphology in children treated by continuous ambulatory peritoneal dialysis. *Pediatr. Nephrol.* 1992; 6: 542-546.

[198] Di Paolo N, Sacchi G, De Mia M, Gaggiotti E, Capotondo L, Rossi P, Bernini M, Pucci AM, Ibba L, Sabatelli P, et al. Morphology of the peritoneal membrane during continuous ambulatory peritoneal dialysis. *Nephron.* 1986; 44: 204-211.

[199] Chan TM, Leung JK, Sun Y, Lai KN, Tsang RC, Yung S. Different effects of amino acid-based and glucose-based dialysate from peritoneal dialysis patients on mesothelial cell ultrastructure and function. *Nephrol. Dial. Transplant.* 2003; 18: 1086-1094.

[200] Witowski J, Wisniewska J, Korybalska K, Bender TO, Breborowicz A, Gahl GM, Frei U, Passlick-Deetjen J, Jorres A. Prolonged exposure to glucose degradation products impairs viability and function of human peritoneal mesothelial cells. *J. Am. Soc. Nephrol.* 2001; 12: 2434-2441.

[201] Yung S, Liu ZH, Lai KN, Li LS, Chan TM. Emodin ameliorates glucose-induced morphologic abnormalities and synthesis of transforming growth factor beta1 and fibronectin by human peritoneal mesothelial cells. *Perit. Dial. Int.* 2001; 21 Suppl 3: S41-47.

[202] Ksiazek K, Korybalska K, Jorres A, Witowski J. Accelerated senescence of human peritoneal mesothelial cells exposed to high glucose: the role of TGF-beta1. *Lab. Invest.* 2007; 87: 345-356.

[203] Ksiazek K, Breborowicz A, Jorres A, Witowski J. Oxidative stress contributes to accelerated development of the senescent phenotype in human peritoneal mesothelial cells exposed to high glucose. *Free Radic. Biol. Med.* 2007; 42: 636-641.

[204] Yanez-Mo M, Lara-Pezzi E, Selgas R, Ramirez-Huesca M, Dominguez-Jimenez C, Jimenez-Heffernan JA, Aguilera A, Sanchez-Tomero JA, Bajo MA, Alvarez V, Castro MA, del Peso G, Cirujeda A, Gamallo C, Sanchez-Madrid F, Lopez-Cabrera M. Peritoneal dialysis and epithelial-to-mesenchymal transition of mesothelial cells. *N. Engl. J. Med.* 2003; 348: 403-413.

[205] Gotloib L, Shostak A, Wajsbrot V, Kushnier R. High glucose induces a hypertrophic, senescent mesothelial cell phenotype after long in vivo exposure. *Nephron* 1999; 82: 164-173.

[206] Ksiazek K, Mikula-Pietrasik J, Korybalska K, Dworacki G, Jorres A, Witowski J. Senescent peritoneal mesothelial cells promote ovarian cancer cell adhesion: the role of oxidative stress-induced fibronectin. *Am. J. Pathol.* 2009; 174: 1230-1240.

[207] Yang WS, Kim BS, Lee SK, Park JS, Kim SB. Interleukin-1beta stimulates the production of extracellular matrix in cultured human peritoneal mesothelial cells. *Perit. Dial. Int.* 1999; 19: 211-220.

[208] Ruster C, Wolf G. Angiotensin II as a morphogenic cytokine stimulating renal fibrogenesis. *J. Am. Soc. Nephrol.* 2011; 22: 1189-1199.

[209] Kiribayashi K, Masaki T, Naito T, Ogawa T, Ito T, Yorioka N, Kohno N. Angiotensin II induces fibronectin expression in human peritoneal mesothelial cells via ERK1/2 and p38 MAPK. *Kidney Int.* 2005; 67: 1126-1135.

[210] Sekine S, Nitta K, Uchida K, Yumura W, Nihei H. Possible involvement of mitogen-activated protein kinase in the angiotensin II-induced fibronectin synthesis in renal interstitial fibroblasts. *Arch. Biochem. Biophys.* 2003; 415: 63-68.

[211] Boulanger E, Wautier MP, Wautier JL, Boval B, Panis Y, Wernert N, Danze PM, Dequiedt P. AGEs bind to mesothelial cells via RAGE and stimulate VCAM-1 expression. *Kidney Int.* 2002; 61: 148-156.

[212] De Vriese AS, Flyvbjerg A, Mortier S, Tilton RG, Lameire NH. Inhibition of the interaction of AGE-RAGE prevents hyperglycemia-induced fibrosis of the peritoneal membrane. *J. Am. Soc. Nephrol.* 2003; 14: 2109-2118.

[213] Ogata S, Yorioka N, Kohno N. Glucose and prednisolone alter basic fibroblast growth factor expression in peritoneal mesothelial cells and fibroblasts. *J. Am. Soc. Nephrol.* 2001; 12: 2787-2796.

[214] Yamaguchi Y, Ishigaki T, Sano K, Miyamoto K, Nomura S, Horiuchi T. Three-dimensional invasion of epithelial-mesenchymal transition-positive human peritoneal mesothelial cells into collagen gel is promoted by the concentration gradient of fibronectin. *Perit. Dial. Int.* 2011; 31: 477-485.

[215] Williams JD, Craig KJ, von Ruhland C, Topley N, Williams GT. The natural course of peritoneal membrane biology during peritoneal dialysis. *Kidney Int. Suppl.* 2003; 64: S43-49.

[216] Halloran PF. Immunosuppressive drugs for kidney transplantation. *N Engl. J. Med.* 2004; 351: 2715-2729.

[217] Pallet N, Legendre C. Deciphering calcineurin inhibitor nephrotoxicity: a pharmacological approach. *Pharmacogenomics* 2010; 11: 1491-1501.

[218] Morales JM. Immunosuppressive treatment and progression of histologic lesions in kidney allografts. *Kidney Int. Suppl.* 2005: S124-130.

[219] Li B, Sehajpal PK, Khanna A, Vlassara H, Cerami A, Stenzel KH, Suthanthiran M. Differential regulation of transforming growth factor beta and interleukin 2 genes in human T cells: demonstration by usage of novel competitor DNA constructs in the quantitative polymerase chain reaction. *J. Exp. Med.* 1991; 174: 1259-1262.

[220] Wolf G, Thaiss F, Stahl RA. Cyclosporine stimulates expression of transforming growth factor-beta in renal cells. Possible mechanism of cyclosporines antiproliferative effects. *Transplantation* 1995; 60: 237-241.

[221] Border WA, Noble NA. Fibrosis linked to TGF-beta in yet another disease. *J. Clin. Invest.* 1995; 96: 655-656.

[222] Li MO, Wan YY, Sanjabi S, Robertson AK, Flavell RA. Transforming growth factor-beta regulation of immune responses. *Annu. Rev. Immunol.* 2006; 24: 99-146.

[223] Shihab FS, Andoh TF, Tanner AM, Noble NA, Border WA, Franceschini N, Bennett WM. Role of transforming growth factor-beta 1 in experimental chronic cyclosporine nephropathy. *Kidney Int.* 1996; 49: 1141-1151.

[224] Khanna AK, Cairns VR, Becker CG, Hosenpud JD. Transforming growth factor (TGF)-beta mimics and anti-TGF-beta antibody abrogates the in vivo effects of cyclosporine: demonstration of a direct role of TGF-beta in immunosuppression and nephrotoxicity of cyclosporine. *Transplantation* 1999; 67: 882-889.

[225] McMorrow T, Gaffney MM, Slattery C, Campbell E, Ryan MP. Cyclosporine A induced epithelial-mesenchymal transition in human renal proximal tubular epithelial cells. *Nephrol. Dial. Transplant.* 2005; 20: 2215-2225.

[226] Slattery C, Campbell E, McMorrow T, Ryan MP. Cyclosporine A-induced renal fibrosis: a role for epithelial-mesenchymal transition. *Am. J. Pathol.* 2005;167: 395-407.

[227] Baboolal K, Jones GA, Janezic A, Griffiths DR, Jurewicz WA. Molecular and structural consequences of early renal allograft injury. *Kidney Int.* 2002; 61: 686-696.

[228] Khanna A, Plummer M, Bromberek C, Bresnahan B, Hariharan S. Expression of TGF-beta and fibrogenic genes in transplant recipients with tacrolimus and cyclosporine nephrotoxicity. *Kidney Int.* 2002; 62: 2257-2263.

[229] Miura Y, Satoh S, Saito M, Numakura K, Inoue T, Obara T, Tsuruta H, Narita S, Horikawa Y, Tsuchiya N, Komatsuda A, Kagaya H, Miura M, Habuchi T. Factors increasing quantitative interstitial fibrosis from 0 hr to 1 year in living kidney transplant patients receiving tacrolimus. *Transplantation* 2011; 91: 78-85.

[230] Shihab FS, Bennett WM, Tanner AM, Andoh TF. Mechanism of fibrosis in experimental tacrolimus nephrotoxicity. *Transplantation* 1997; 64: 1829-1837.

[231] Waiser J, Dell K, Kreutzkamp J, Bohler T, Budde K, Peters H, Neumayer HH. FK506, transforming growth factor-beta1 and mesangial matrix synthesis: parallels and differences compared with cyclosporine A. *Cytokine* 2006; 33: 59-65.

[232] Allison AC, Eugui EM. Mechanisms of action of mycophenolate mofetil in preventing acute and chronic allograft rejection. *Transplantation* 2005; 80: S181-190.

[233] Chan TM, Li FK, Tang CS, Wong RW, Fang GX, Ji YL, Lau CS, Wong AK, Tong MK, Chan KW, Lai KN. Efficacy of mycophenolate mofetil in patients with diffuse proliferative lupus nephritis. Hong Kong-Guangzhou Nephrology Study Group. *N. Engl. J. Med.* 2000; 343: 1156-1162.

[234] Chan TM, Tse KC, Tang CS, Mok MY, Li FK. Long-term study of mycophenolate mofetil as continuous induction and maintenance treatment for diffuse proliferative lupus nephritis. *J. Am. Soc. Nephrol.* 2005; 16: 1076-1084.

[235] Ginzler EM, Dooley MA, Aranow C, Kim MY, Buyon J, Merrill JT, Petri M, Gilkeson GS, Wallace DJ, Weisman MH, Appel GB. Mycophenolate mofetil or intravenous cyclophosphamide for lupus nephritis. *N. Engl. J. Med.* 2005; 353: 2219-2228.

[236] Shihab FS, Bennett WM, Yi H, Choi SO, Andoh TF. Combination therapy with sirolimus and mycophenolate mofetil: effects on the kidney and on transforming growth factor-beta1. *Transplantation* 2004; 77: 683-686.

[237] Dell'Oglio MP, Zaza G, Rossini M, Divella C, Pontrelli P, Verrienti R, Rutigliano M, Ditonno P, Stifanelli P, Ancona N, Schena FP, Grandaliano G. The anti-fibrotic effect of mycophenolic acid-induced neutral endopeptidase. *J. Am. Soc. Nephrol.* 2010; 21: 2157-2168.

[238] Kobori H, Nangaku M, Navar LG, Nishiyama A. The intrarenal renin-angiotensin system: from physiology to the pathobiology of hypertension and kidney disease. *Pharmacol. Rev.* 2007; 59: 251-287.

[239] Djamali A, Vidyasagar A, Yagci G, Huang LJ, Reese S. Mycophenolic acid may delay allograft fibrosis by inhibiting transforming growth factor-beta1-induced activation of Nox-2 through the nuclear factor-kappaB pathway. *Transplantation* 2010; 90: 387-393.

[240] Badid C, Vincent M, McGregor B, Melin M, Hadj-Aissa A, Veysseyre C, Hartmann DJ, Desmouliere A, Laville M. Mycophenolate mofetil reduces myofibroblast infiltration and collagen III deposition in rat remnant kidney. *Kidney Int.*2000;58:51-61.

[241] Dubus I, Vendrely B, Christophe I, Labouyrie JP, Delmas Y, Bonnet J, Combe C. Mycophenolic acid antagonizes the activation of cultured human mesangial cells. *Kidney Int.* 2002; 62: 857-867.

[242] Geissler EK, Schlitt HJ. The potential benefits of rapamycin on renal function, tolerance, fibrosis, and malignancy following transplantation. *Kidney Int.* 2010; 78: 1075-1079.

[243] Pontrelli P, Rossini M, Infante B, Stallone G, Schena A, Loverre A, Ursi M, Verrienti R, Maiorano A, Zaza G, Ranieri E, Gesualdo L, Ditonno P, Bettocchi C, Schena FP, Grandaliano G. Rapamycin inhibits PAI-1 expression and reduces interstitial fibrosis and glomerulosclerosis in chronic allograft nephropathy. *Transplantation* 2008; 85: 125-134.

[244] Servais A, Meas-Yedid V, Toupance O, Lebranchu Y, Thierry A, Moulin B, Etienne I, Presne C, Hurault de LB, Le Pogamp P, Le Meur Y, Glotz D, Hayem C, Olivo Marin

JC, Thervet E. Interstitial fibrosis quantification in renal transplant recipients randomized to continue cyclosporine or convert to sirolimus. *Am. J. Transplant.* 2009; 9: 2552-2560.

[245] Lui SL, Yung S, Tsang R, Zhang F, Chan KW, Tam S, Chan TM. Rapamycin prevents the development of nephritis in lupus-prone NZB/W F1 mice. *Lupus* 2008; 17:305-313.

[246] Lui SL, Tsang R, Chan KW, Zhang F, Tam S, Yung S, Chan TM. Rapamycin attenuates the severity of established nephritis in lupus-prone NZB/W F1 mice. *Nephrol. Dial. Transplant.* 2008; 23: 2768-2776.

[247] Ramos-Barron A, Pinera-Haces C, Gomez-Alamillo C, Santiuste-Torcida I, Ruiz JC, Buelta-Carrillo L, Merino R, de Francisco AL, Arias M. Prevention of murine lupus disease in (NZBxNZW)F1 mice by sirolimus treatment. *Lupus* 2007; 16: 775-781.

[248] Warner LM, Adams LM, Sehgal SN. Rapamycin prolongs survival and arrests pathophysiologic changes in murine systemic lupus erythematosus. *Arthritis Rheum.* 1994; 37: 289-297.

[249] Lieberthal W, Fuhro R, Andry C, Patel V, Levine JS. Rapamycin delays but does not prevent recovery from acute renal failure: role of acquired tubular resistance. *Transplantation* 2006; 82: 17-22.

[250] Shegogue D, Trojanowska M. Mammalian target of rapamycin positively regulates collagen type I production via a phosphatidylinositol 3-kinase-independent pathway. *J. Biol. Chem.* 2004; 279: 23166-23175.

[251] Wu MJ, Wen MC, Chiu YT, Chiou YY, Shu KH, Tang MJ. Rapamycin attenuates unilateral ureteral obstruction-induced renal fibrosis. *Kidney Int.* 2006; 69: 2029-2036.

[252] Swinford RD, Pascual M, Diamant D, Tang SS, Ingelfinger JR. Rapamycin increases transforming growth factor-beta mRNA expression in immortalized rat proximal renal tubular cells. *Transplantation* 2002; 73: 319-320.

[253] Wittmann S, Daniel C, Stief A, Vogelbacher R, Amann K, Hugo C. Long-term treatment of sirolimus but not cyclosporine ameliorates diabetic nephropathy in the rat. *Transplantation* 2009; 87: 1290-1299.

[254] Aguilera A, Aroeira LS, Ramirez-Huesca M, Perez-Lozano ML, Cirugeda A, Bajo MA, Del Peso G, Valenzuela-Fernandez A, Sanchez-Tomero JA, Lopez-Cabrera M, Selgas R. Effects of rapamycin on the epithelial-to-mesenchymal transition of human peritoneal mesothelial cells. *Int. J. Artif. Organs* 2005; 28: 164-169.

[255] Sekiguchi Y, Zhang J, Patterson S, Liu L, Hamada C, Tomino Y, Margetts PJ. Rapamycin inhibits transforming growth factor beta induced peritoneal angiogenesis by blocking the secondary hypoxic response. *J. Cell Mol. Med.* 2011.

[256] Iyer SN, Wild JS, Schiedt MJ, Hyde DM, Margolin SB, Giri SN. Dietary intake of pirfenidone ameliorates bleomycin-induced lung fibrosis in hamsters. *J. Lab. Clin. Med.* 1995; 125: 779-785.

[257] Shihab FS, Bennett WM, Yi H, Andoh TF. Pirfenidone treatment decreases transforming growth factor-beta1 and matrix proteins and ameliorates fibrosis in chronic cyclosporine nephrotoxicity. *Am. J. Transplant.* 2002; 2: 111-119.

[258] Shimizu T, Kuroda T, Hata S, Fukagawa M, Margolin SB, Kurokawa K. Pirfenidone improves renal function and fibrosis in the post-obstructed kidney. *Kidney Int.* 1998; 54: 99-109.

[259] RamachandraRao SP, Zhu Y, Ravasi T, McGowan TA, Toh I, Dunn SR, Okada S, Shaw MA, Sharma K. Pirfenidone is renoprotective in diabetic kidney disease. *J. Am. Soc. Nephrol.* 2009; 20: 1765-1775.

[260] Cho ME, Kopp JB. Pirfenidone: an anti-fibrotic therapy for progressive kidney disease. *Expert Opin. Investig. Drugs* 2010; 19: 275-283.

[261] Leh S, Vaagnes O, Margolin SB, Iversen BM, Forslund T. Pirfenidone and candesartan ameliorate morphological damage in mild chronic anti-GBM nephritis in rats. *Nephrol. Dial. Transplant.* 2005; 20: 71-82.

[262] Miric G, Dallemagne C, Endre Z, Margolin S, Taylor SM, Brown L. Reversal of cardiac and renal fibrosis by pirfenidone and spironolactone in streptozotocin-diabetic rats. *Br. J. Pharmacol.* 2001; 133: 687-694.

[263] Park HS, Bao L, Kim YJ, Cho IH, Lee CH, Hyun BH, Margolin SB, Park YH. Pirfenidone suppressed the development of glomerulosclerosis in the FGS/Kist mouse. *J. Korean Med. Sci.* 2003; 18: 527-533.

[264] Sharma K, Ix JH, Mathew AV, Cho M, Pflueger A, Dunn SR, Francos B, Sharma S, Falkner B, McGowan TA, Donohue M, Ramachandrarao S, Xu R, Fervenza FC, Kopp JB. Pirfenidone for diabetic nephropathy. *J. Am. Soc. Nephrol.* 2011; 22: 1144-1151.

[265] Cho ME, Smith DC, Branton MH, Penzak SR, Kopp JB. Pirfenidone slows renal function decline in patients with focal segmental glomerulosclerosis. *Clin. J. Am. Soc. Nephrol.* 2007; 2: 906-913.

[266] Dendooven A, van Oostrom O, van der Giezen DM, Leeuwis JW, Snijckers C, Joles JA, Robertson EJ, Verhaar MC, Nguyen TQ, Goldschmeding R. Loss of endogenous bone morphogenetic protein-6 aggravates renal fibrosis. *Am. J. Pathol.* 2011; 178: 1069-1079.

[267] Lund RJ, Davies MR, Hruska KA. Bone morphogenetic protein-7: an anti-fibrotic morphogenetic protein with therapeutic importance in renal disease. *Curr. Opin. Nephrol. Hypertens.* 2002; 11: 31-36.

[268] Yu MA, Shin KS, Kim JH, Kim YI, Chung SS, Park SH, Kim YL, Kang DH. HGF and BMP-7 ameliorate high glucose-induced epithelial-to-mesenchymal transition of peritoneal mesothelium. *J. Am. Soc. Nephrol.* 2009; 20: 567-581.

[269] Chatziantoniou C, Dussaule JC. Insights into the mechanisms of renal fibrosis: is it possible to achieve regression? *Am. J. Physiol. Renal. Physiol.* 2005; 289: F227-234.

[270] Matsuoka T, Maeda Y, Matsuo K, Naiki Y, Tamai Y, Sakaguchi M, Hasegawa H, Funauchi M, Kanamaru A. Hepatocyte growth factor prevents peritoneal fibrosis in an animal model of encapsulating peritoneal sclerosis. *J. Nephrol.* 2008; 21: 64-73.

[271] Liu Y. Hepatocyte growth factor and the kidney. *Curr. Opin. Nephrol. Hypertens.* 2002; 11: 23-30.

[272] Liu Y. Hepatocyte growth factor in kidney fibrosis: therapeutic potential and mechanisms of action. *Am. J. Physiol. Renal. Physiol.* 2004; 287: F7-16.

[273] Liu Y, Yang J. Hepatocyte growth factor: new arsenal in the fights against renal fibrosis? *Kidney Int.* 2006; 70: 238-240.

[274] Yanagita M. Modulator of bone morphogenetic protein activity in the progression of kidney diseases. *Kidney Int.* 2006; 70: 989-993.

In: Fibronectin
Editors: James Beattie

ISBN 978-1-62257-298-4

Chapter XII

Role of Fibronectin in Tissue Remodeling during Kidney Disease Development

Séin O'Connell, Stephen Nolan, Craig Slattery, Michael P Ryan and Tara McMorrow*

Renal Disease Research Group, UCD School of Biomolecular and Biomedical Science, UCD Conway Institute, University College Dublin, Belfield, Dublin, Ireland

Abstract

Chronic kidney disease is a progressive disorder characterized by significant histopathological alterations ultimately resulting in end-stage-renal-failure. The primary histological hallmarks of chronic kidney disease are glomerulosclerosis and tubulointerstitial fibrosis. These processes are characterized by inappropriate deposition and accumulation of extracellular matrix components in renal tissue. Several factors and molecular mechanisms are implicated in disease initiation and progression.

However, mechanisms of extracellular homeostasis are a major research focus in chronic kidney disease research. In the healthy kidney, extracellular matrix performs a number of critical functions including providing structural support to the cells of the nephron, influencing cellular phenotype and polarity, and modulating the activity of growth factors and cytokines.

Maintenance of renal extracellular matrix homeostasis is a highly regulated process, the result of a balance between matrix protein production and secretion, and enzymatic degradation. Fibronectin is a key component of the renal extracellular matrix.

Under normal conditions, fibronectin plays a role in regulating renal cell adhesion and migration. Under pathophysiological conditions, fibronectin expression levels are

* Email: Tara.McMorrow@ucd.ie.

dramatically increased in the renal interstitium with increased influence of fibronectin on normal cell function and phenotype.

There are also qualitative alterations to fibronectin under disease conditions as fibronectin is subject to additional post-translational modifications. These post-translational modifications decrease the susceptibility of fibronectin to proteolytic degradation, significantly disrupting extracellular matrix homeostasis and promoting matrix accumulation. This review will outline the role of fibronectin in normal renal tissue structure and function. Furthermore, the significance of alterations in fibronectin will be discussed in the context of deterioration of global organ architecture and function observed during chronic kidney disease.

1. Introduction

Fibronectin plays a vital role in normal adult renal architecture and tissue remodeling during kidney disease development but it also has a key role in embryonic renal development. The human adult kidney derives from the interaction of two mesodermal derivatives, the metanephric mesenchyme and the ureteric bud. Following this interactionthe metanephric mesenchyme undergoes a series of morphogenetic events resulting in the formation of an epithelium which eventually generates most of the tubules of the mature nephron [1].

Presently, there is a growing list of potential signaling molecules that includes many secreted factors and their receptors which are either expressed in the embryonic kidney or can affect development. Extracellular matrix components are in this category and fibronectin in particular plays a role in early renal development, as indicated by targeted mutations of fibronectin or either of its two major integrin receptors, which display distinct phenotypes, and are embryonic lethal [2].

2. Fibronectin and the Kidney

2.1. Structure and Function

Fibronectin is a large, multi-domain glycoprotein found in several segments of the nephron including the glomerular mesangial cells and renal interstitial cells [3]. The functional fibronectin molecule is a dimer consisting of two similar or identical subunits of ~220kDa that are held together by two disulphide bonds near their carboxyl-termini [4]. Each monomer is comprised of a combination of three different types of homologous repeating domains, termed Types I, II and III (Figure 1), [5, 6]. Fibronectin dimers are initially secreted in a soluble, compact, globular form and converted into a growing fibrillar fibronectin matrix through binding to specific cell surface integrin receptors (such as the $\alpha_5\beta_1$ integrin) [7].

This fibrillar fibronectin forms part of the normal renal ECM which is linked to the basolateral basement membrane via the integrinreceptors. Fibrillar fibronectin has been found to possess many biologically active domains and is thought to be particularly important in mediating renal cell adhesion to the ECM through its RGDS sequence in the cell binding domain of the FN-III$_{9\text{-}10}$ repeats [8, 9].

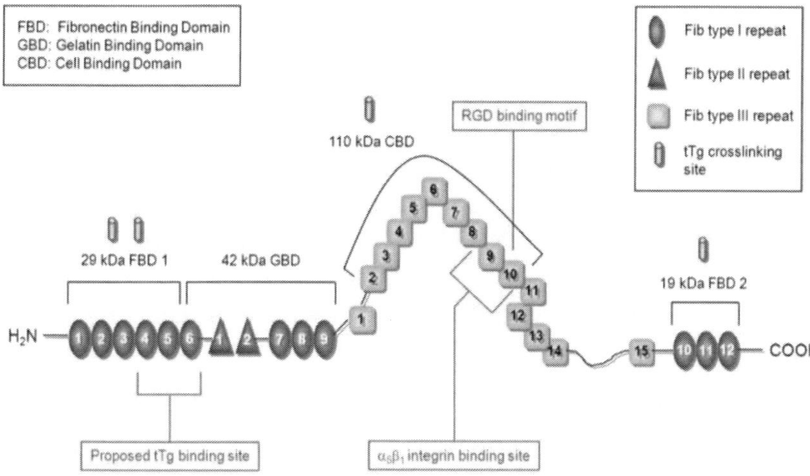

Figure 1. Proposed molecular structure of the fibronectin protein. Schematic diagram demonstrating the proposed order of repeats and domains in the fibronectin molecule. Binding sites for cell surface integrins and tissue transglutaminase are indicated. Proposed location of transglutaminase sensitive glutamine residues are also indicated.

2.2. Cell Adhesion

Cell adhesion is critical for the creation and organisation of both three-dimensional structure and physiological function of tissues, including the kidney. This adhesion is mediated through the expression of integrins and cell adhesion molecules (CAMs) and plays an integral role in many aspects of cell physiology. The integrin, cell surface superfamily of ECM receptors are the classical mediators of cell-matrix attachment. Integrin receptors are heterodimeric transmembrane proteins, possessing a large extracellular domain, a short membrane spanning segment and a small intracellular domain, formed by specific pairing between α and β subunits [10]. These integrins as well as cellular fibronectin secreted from the mesangial cells form part of the ECM [11].

Adhesion of renal epithelial cells to their underlying basement membrane has been demonstrated to have a profound influence on epithelial differentiation and cell polarity [12]. The main integrin implicated in epithelial collagen IV adhesion is the $\alpha_1\beta_1$ integrin [13, 14], whereas mesenchymal fibroblastic cells rely on $\alpha_5\beta_1$ engagement to mediate fibronectin-integrin adhesion [15]. Inhibition of integrin signaling in the kidney prevented fibronectin and other fibroblast marker expression increases. Blockade of integrin signaling prevented renal cell migration and halted renal disease progression [16].

This differing profile of integrin expression is reflective of not only the physiological environments of the respective cells but also their differing functional roles in the body. Integrins do not merely function to mediate physical cell attachment to the supporting extracellular matrix deposition (ECM) structures, but their ECM ligation also results in the recruitment of signaling and accessory proteins [17, 18]. The initiation of these intracellular signaling events in renal vascular smooth muscle cells has been shown following induction of extracellular stress [19].

2.3. Fibronectin Matrix in the Kidney

In addition to its role in cell adhesion, deposition of a fibronectin matrix is particularly important in wound healing where it acts as a provisional matrix for cell adhesion, survival and eventual re-epithelialisation [20, 21]. Cutaneous wound healing is divided into three phases: inflammation, proliferation, and maturation. These phases overlap, and their separation is somewhat arbitrary, but they help to understand the sequence of events that take place in the healing of skin wounds. The initial injury causes platelet adhesion and aggregation and the formation of a clot in the surface of the wound, leading to inflammation. In the proliferative phase there is formation of granulation tissue, proliferation and migration of connective tissue cells, and re-epithelialization of the wound surface. Maturation involves ECM deposition, tissue remodelling, and wound contraction [22], in which fibronectin is heavily involved.Given the analogies frequently drawn between wound healing and renal fibrotic development, aberrant fibronectin expression can be seen as a hallmark of tubulointerstitial fibrosis (TIF) [23].

However under normal circumstances fibronectin also plays an important role in the kidney. The glomerulus is the initial part of the nephron and is responsible for filtering blood to allow the passage of fluid into the tubular part of the nephron while still allowing the retention of necessary cellular constituents of the blood. The glomerulus is made up of a variety of different cells including endothelial, epithelial and mesangial cells each with their own specific function.

2.4. Fibronectin Matrix and Kidney Cells

Mesangial cells with their surrounding matrix compose the mesangium. Mesangial cells are irregularly shaped cells with elongated cytoplasmic processes in which filaments of contractile proteins (actin and myosin) are found. [24]. The mesangial matrix consists primarily of glycosaminoglycans, fibronectin and laminin. The glomerular mesangial cells contribute structural support and contractile and phagocytic properties to the glomerulus. Several cell surface receptors of the β-integrin family have been identified on the mesangial cells, including α1b1, α3b1, and the fibronectin receptor, α5b1 [25]. The glomerular basement membrane (GBM) surrounds the glomerulus and is composed primarily of collagen IV, laminin, entactin/nidogen, and sulphated proteoglycans but also fibronectin. Fibronectin is most abundant in the mesangial matrix but is also found in the GBM, and between the endothelial and epithelial cells which correlates with its known role in cell adhesion [26].

The proximal tubule begins at the urinary pole of the glomerulus and extends to the descending limb of the loop of Henle. It consists of a layer of epithelial cells covering the tubular basement membrane.Epithelial cells display highly defined apical-basal polarity perpendicular to the plane of the epithelium, and are linked to neighbouring cells through protein-protein interactions at apical junctional complexes, which ensures the mechanical integrity of the epithelial monolayer. The ECM is associated with an epithelial monolayer called the tubular basement membrane (TBM). The basement membrane normally underlying the epithelial monolayer of the proximal tubule is characterised by laminin 1, 10 and collagen IV [27, 28]. However, pathological interstitial matrix expansion is characterised by depositional profiles favouring fibronectin and collagen I and III, as is demonstrated in experimental animal models of renal disease progression [29-31] and will be discussed shortly.

2.5. Fibronectin in Kidney Basement Membranes

Both the TBM and GBM contain differing ECM components, and these in turn differ from the ECM found in the renal interstitium. The renal interstitium is composed of interstitial cells and extracellular matrix material consisting of sulphated and nonsulphated glycosaminoglycans [32]. The interstitial tissue in the cortex is mainly composed of collagen I, III and fibronectin [33]. It accounts for approximately 9% tissue volume and therefore tubules and capillaries are often directly opposed to each other. However, medullary interstitium is quite abundant and reaches approximately 30% to 40% tissue volume with interstitial cells in close contact with medullary vessels and tubules. The proliferation and matrix secreting capabilities of interstitial fibroblasts has been implicated in contributing to the development of interstitial fibrosis observed in many renal pathological conditions [34].

Fibroblasts have been demonstrated to produce extensive quantities of fibronectin, collagen I, III and V [35]. This production of ECM components passively provides structure and support to connective tissue within organs, however pathologically fibronectin deposition has been identified as a key constituent of the expanding renal interstitium which can contribute significantly to disease initiation and progression.

3. Kidney Disease Development

3.1. Acute Kidney Disease

Kidney disease is often classified into either acute or chronic. Acute kidney injury (AKI) is an extremely variable condition both in terms of aetiology and severity. It is characterised by rapid decline in glomerular filtration rate (GFR) and retention of waste products such as blood urea nitrogen (BUN) and creatinine [36]. Patients with AKI have been classified as being nonoliguric (urine output >400 mL/day), oliguric (urinary out-put <400 mL/day), or anuric (urinary output <100 mL/day) [37]. The aetiology and outcome of AKI is heavily influenced by the circumstances in which it occurs, such as whether it develops in the community or in the hospital. It is also important to determine whether AKI occurs as an isolated event or as part of a multiorgan syndrome. A quick prognosis of AKI remains a significant challenge in this area. Generally acute AKI is managed by monitoring GFR and upon identification, cessation and or treatment of the causative factor renal function will be restored.

3.2. Chronic Kidney Disease

AKI can progress to chronic kidney disease if it goes undetected. Chronic kidney disease (CKD), is defined as “the presence of kidney damage, manifested by abnormal albumin excretion or decreased kidney function, quantified by measured or estimated glomerular filtration rate (GFR) that persists for more than 3 months" [38]. CKD develops due to a number of initial etiologies including diabetes mellitus [39], obesity [40] and drug toxicity [41], among others. The molecular mechanisms underlying the progression of late stage CKD to end stage renal disease (ESRD) and eventual organ failure are poorly understood. Progression of CKD is hallmarked by a deformation of global organ architecture as a result of extensive, inappropriate deposition of extra cellular matrix (ECM) components. Fibrotic accumulation of ECM within the glomerulus and the tubulointerstitial space are referred to as glomerulosclerosis and TIF respectively, and are the primary histological lesions hallmarking CKD progression. The pathogenic mechanisms that lead to CKD seem to converge on this final common pathway of fibrotic accumulation [42].

3.3. Glomerulosclerosis

Glomerulosclerosis is an important contributor to end stage renal disease and involves a progressive fibrosing of the glomeruli concomitant with increasing albuminuria and declining renal function. Immune cell infiltration is thought to drive fibrotic progression in the glomerulus by modulating the function of resident glomerular cells. The release of pro-fibrotic cytokines from infiltrating immune cells results in cellular proliferation and release of ECM components resulting in impaired glomerular filtration. Examples of primary glomerular nephritis include focal segmental glomerulosclerosis [43], membranous nephritis [44] and focal and segmental glomerulosclerosis [45]. Glomerular mesangial cells also have the capabilities to produce significant amounts of profibrotic cytokines as well as fibronectin [46].

Morphological changes within the glomerular ultrastructure during glomerular damage include podocyte effacement, mesangial cell hypertrophy and glomerular basement membrane thickening. These changes combine to disrupt the filtering selectivity of the glomerulus, with resultant leakage of plasma proteins into the filtrate [47, 48]. This increased protein content in the tubular filtrate is believed to have a significant pathophysiological effect on the proximal tubular epithelial cells (PTECs). Overloading of protein re-uptake mechanisms in the PTECs results in the release of chemokines, which can drive interstitial inflammation [49-51].

3.4. Tubulointerstitial Fibrosis

The tubulointerstitium also plays a key role in renal disease progression. Together the tubules and interstitium account for 90% of renal volume [52]. Therefore it is not surprising that perturbations in interstitial architecture have a significant effect on renal function. Indeed, several studies have demonstrated the degree of tubulointerstitial expansion to hold an accurate correlation with declining GFR during renal disease progression [53-55]. This was most clearly demonstrated in a systemic morphometric analysis of over 1,700 renal biopsies of varying etiologies [56]. Although correlative data is evident, definitive functional association linking interstitial expansion and nephron loss is still lacking, although several theories persist. Tubulointerstitial expansion is thought to distort global organ architecture, blocking the tubular space, thus increasing the tubular filtration pressure. This increase in tubular filtration pressure is thought to be a contributing factor to overall nephron atrophy and loss.

Despite the diverse range of factors that can initiate and promote TIF, disease progression eventually reaches a point at which it can progress independently of the initiating factor [57, 58]. At this point it is thought that functional redundancy in the declining kidney is no longer capable of compensating for nephron loss and the resultant single nephron filtration pressure increases to the point of perpetuating further tubule damage.

Histologically, TIF progression is hallmarked by excessive interstitial deposition of extracellular matrix (ECM) components [59], a decrease in nephron [60] and peritubular capillary number [61], concomitant with an increase in inflammatory infiltrate [62] and α-smooth muscle actin (α-SMA) expressing interstitial fibroblasts [63]. It has been suggested that the progression of TIF can be likened to a maladapted wound healing response [64]. In normal situations this process would culminate in re-epithelialisation of the wound area and restoration of tissue function. However in the renal setting the chronic nature of the initiating insult could drive continued activation of the immune system and matrix deposition without facilitating resolution, thus promoting fibrotic development.

4. Extracellular Matrix in Kidney Disease

The fibrotic tubulointerstitial lesion associated with CKD has been demonstrated to be composed of a much broader molecular range of ECM molecules, identifying it as not merely a volumetric expansion, but also a compositional alteration of the interstitial matrix. Increased expression of fibronectin in particular [23], as well as collagen I, III, IV, V and VI [59, 65] were noted in the fibrotic interstitial space. The ECM environment does not just function to give support and spatial context to the surrounding cells, but can also directly and indirectly modulate the phenotype of the surrounding cells. The specific molecular composition of the ECM can directly modulate the phenotype of overlying cells through integrin mediated outside-in signaling [66, 67].

The ECM also functions as a reservoir for growth factors and cytokines which are actively secreted and then stored in the surrounding matrix [68, 69]. Upon proteolytic degradation of the matrix, or the cytokine associated matrix binding protein, these cytokines become biologically active, and in this way the ECM also holds an indirectly influence on the phenotype of embedded cells.

Mounting evidence suggests that interstitial ECM accumulation during CKD progression is not merely a product of increased matrix synthesis, but is exacerbated by dysregulation of proteins involved in post-translational matrix catabolism [70, 71]. The matrix metalloproteinases (MMPs) are a large, 26 member family of Zn^{2+} dependent proteolytic enzymes, primarily (although not exclusively) responsible for cleavage of ECM proteins. They are catagorised based upon their substrate preferences into collagenases, gelatinases, stromelysins and matrilysins among others [72-74]. Fibronectin is targeted by several MMPs, including MMP 2, 9 16 24 and 25 again indicating its importance in promoting a fibrotic phenotype.

The tissue inhibitors of metalloproteinases (TIMPs) are endogenous, physiological antagonists of the MMPs, carrying out their actions by binding to the catalytic site of MMPs forming 1:1 inhibitory complexes [75]. All members of the MMP family have been demonstrated to be inhibited by at least one of the four TIMP family members. The factors involved in matrix accumulation are governed by not only matrix synthesis, but also the post-depositional degradation of ECM proteins, which in turn is dictated by the local balance of MMP/TIMP expression. A local decrease in MMP activation, or increase in TIMP expression can thus lead to matrix accumulation independent of any alterations in the rate of matrix synthesis.

4.1. Role of Fibroblasts in ECM Deposition

Under normal physiological conditions the interstitial fibroblasts exist in a continuous network with dendritic cells throughout the renal interstitial space, where they function to regulate ECM production and turnover, thus providing structural support to the renal network of nephrons [76]. These quiescent fibroblasts exhibit little actin-associated cell-matrix adhesions, express no α-SMA and synthesise little ECM [77]. However, during the development of TIF, there is a significant increase in the interstitial population of activated, α-SMA expressing fibroblasts, termed myofibroblasts [57, 78]. It is these myofibroblasts that are believed to be the primary cell-type responsible for the interstitial expansion observed during CKD progression.

Fibroblasts exhibit extensive fibronectin associated focal adhesions which are linked to α-SMA-positive stress fiber assemblies [79]. Pathologically, excessive ECM deposition by activated α-SMA-positivemyofibroblasts leads to a disruption of organ form and function, this is highlighted by their demonstrated role in the fibrotic deterioration of the kidney. The molecular mechanisms of this fibrotic process have been the focus of much research. Several cytokine and profibrotic growth factors have been identified as important factors and contributors to TIF including transforming growth factor-β (TGF-β) [80], CTGF and bone morphogenic protein-7 (BMP-7) [81]. Alterations in these cytokines can promote TIF by altering the balance between matrix synthesis and degradation.

4.2 Role of TGF-β in ECM Deposition

TGF-β in particular is heavily involved in promoting TIF and ESRD. TGF-β is a member of a superfamily of molecules demonstrated to include TGF, BMPs, activins and inhibins [82]. TGF-β1 exerts its biological activity via distinct tyrosine-kinase-linked receptors termed TGF-β receptor type I and II. Once phosphorylated, TGF-β RI is activated and signals to downstream targets. The proteins of the SMAD signalling cascade are the best characterised of these downstream pathways. However, TGF-β has also been demonstrated to modulate several other pathways, including the Ras homolog (Rho) GTPases and the mitogen-activated protein kinases (MAPKs) [83, 84] in a variety of contexts.

TGF-β1 has also been demonstrated to regulate the proliferation and differentiation of mesenchymal cell types [85] and has also been demonstrated to play an important role in the regulation of ECM deposition, as well as the ECM constituents collagen I and fibronectin [86, 87]. Both CTGF and BMP-7 are downstream modulators of TGF-β1 signalling [88, 89]. CTGF is pro-fibrotic and is upregulated in many models of TIF [90-93]. In contrast, BMP-7 is anti-fibrotic and is decreased in models of renal fibrosis [94]. Fibroblasts have been demonstrated to secrete increased fibronectin in response to the cytokines TGF-β and CTGF [95, 96].

5. Fibronectin in Kidney Disease

Physiologically, in a wound setting, fibronectin provides an adhesive scaffold for cell migration, mediating further deposition of collagenous ECM and eventual wound re-epithelialisation [20, 97, 98]. Its expression diminishes rapidly upon re-establishment of the basement membrane and re-epithelialisation of the wound area [21]. However, in a chronic fibrotic disorder, such as ESRD, extensive, inappropriate and unchecked interstitial deposition of fibronectin persists and is an essential marker and mediator of interstitial fibrotic expansion and renal architectural deterioration (Figure 2).

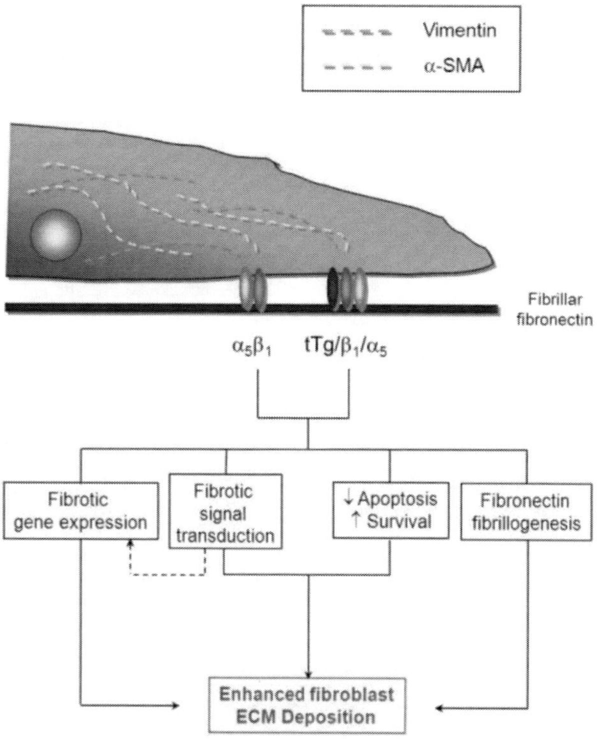

Figure 2. Implications of increased fibronectin adhension in renal fibrosis. Increased adhension of fibroblasts to fibronectin as mediated by increased tTg expression may have several implications in the progression of renal disease, including increased survival and fibrotic gene expression.

While dysregulation of matrix proteins and matrix modulating enzymes have been a constant hallmark of renal fibrotic disorders, as reviewed in [70, 71, 99]. It is also possible that altered spatial orientation or conformational changes within or between the fibronectin fibrils as a result of posttranslational modification and in particular enzymatic crosslinking may mask or impair access of MMPs to their cleavage sites and prevent degradation of the fibronectin and surrounding ECM further promoting renal fibrosis.

Fibronectin is an essential matrix glycoprotein expressed during the progression of renal tubulointerstitial fibrosis and functions as a provisional adhesive scaffold for cells and additional matrix deposition. Integrin binding of soluble globular fibronectin dimers is essential in their conversion to rigid fibrillar structures. Several ECM components have been identified as targets of enzyme mediated posttranslational modifications, including fibronectin. The $\alpha_5\beta_1$ integrin is an important integrin in mediating interactions with the extracellular glycoprotein fibronectin. It mediates fibroblast adhesion to fibronectin but also facilitates fibronectin fibrillogenesis through manipulation of soluble fibronectin dimmers into in-soluble fibrillar structures. $\alpha_5\beta_1$ expression may enhance the ability of fibroblasts to adhere to the fibronectin-rich interstitial matrix, promoting cell survival by preventing apoptosis which leads to prolonged production of ECM components and further renal fibrosis.

5.1. Post-Translational Modification of Fibronectin in Kidney Disease

Fibronectin (FN) is encoded by a single gene, but post-transcriptional alternative splicing and post-translational modifications give rise to multiple FN protein isoforms, whose production is regulated in a cell- and tissue-specific manner. It is known that the various isoforms of FN have different biological functions, and that it is not only a change in the amount of FN deposited at a specific location that may play a role in disease progression a change in the ratio between the various isoforms may also contribute to the development of fibrotic disease. Deposition of these FN isoforms can thus lead to a more efficient attraction of cells to the ECM in the kidney [23].

5.2. Tissue Transglutaminase and Fibronectin

One enzyme implemented in the post-translational modifications of fibronectin is tissue transglutaminase (tTg). tTg is a ~76kDa, 686 amino acid, member of the transglutaminase family of enzymes. To date 6 transglutaminases have been identified at the protein level with the ubiquitously expressed tTg being the first identified. Transglutaminases are also capable of modification of peptides in a number of ways, the most common of which is covalent protein-protein crosslinking through inter or intra-protein isopeptide bond formation. tTg is composed of four domains; an amino-terminal β-sandwich domain, a catalytic core, followed by 2 β-barrel domains [100]. The N-terminal domain of tTg contains a high affinity fibronectin binding site [101] which plays an important role in cell adhesion, ECM modulation and wound healing.

Normally tTg is found in a complex with fibronectin in the pericellular space. Extracellular tTg is intimately connected with the ECM network and is thought to be particularly important in tissue stabilisation during wound healing, however pathologically, tTg has been implicated in the progression of tissue fibrosis. The tTg catalytic cross-linking of fibronectin facilitates its multimerisation and stability [102], important in creation of the temporary matrix scaffold. Overexpression of tTg and its isopeptide bond product, ε-(γ-glutamyl)-lysine, has been associated with fibrotic progression in the kidney.

In a study performed by Johnson *et al.* [103], levels of tTg and its proteolytic cross-linking product, ε(γ-glutamyl)-lysine isopeptide bonds, were found to be elevated in biopsies of chronic renal disease. An accurate positive correlation was also found between the levels of tTg and the progression of renal tubulointerstitial fibrosis [104].

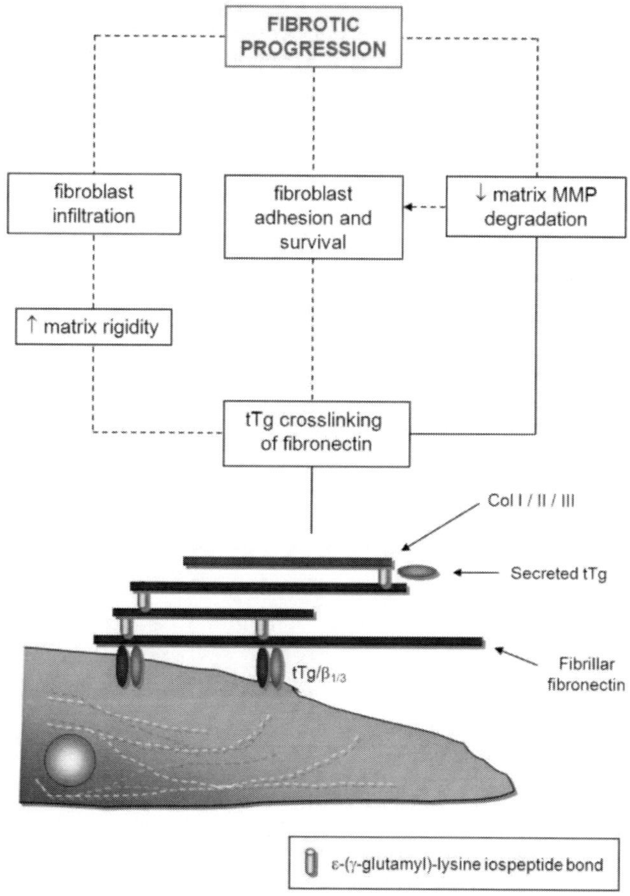

Figure 3. Hypothesised implications of increased tissue transglutaminase mediated interstitial matrix crosslinking. Possible implications of increased interstitial tTg mediated crosslinked fibronectin fibril formation on fibroblasts and the progression of interstitial fibrosis in chronic kidney disease.

Increased expression of tTg has been demonstrated in several experimental models of renal disease progression, such as the rat 5/6 nephrectomy and UUO models [105, 106]. tTg utilised in this model has been demonstrated to cross-link fibronectin which has been observed to alter the physical properties of the matrix by increasing its intrinsic rigidity,[107]. These post-translational modifications by tTg decrease the susceptibility of fibronectin to proteolytic degradation, significantly disrupting extracellular matrix homeostasis and promoting matrix accumulation.The implications of increased cross-linking of fibronectin in kidney disease are indicated in Figure 3.

Conclusion

Fibronectin is a major component of the ECM in the kidney and its modification and accumulation can contribute to the progression of renal injury. Recently fibronectin has also been proposed as a potential biomarker of renal disease. One study demonstrated an increase in fibronectin mRNA correlated with the development of renal cell carcinoma (RCC) [108]. In this study, 212 renal biopsies were examined and a 7 fold increase in fibronectin was observed in RCC patients. The authors propose that fibronectin may serve as prognostic marker for RCC aggressiveness. Another study demonstrated that increased urinary fibronectin correlated with a poor outcome in patients with glomerulonephritis [109]. Also the post-translational modifications of fibronectin by tTg identified in recent studies appears to significantly contribute to the deterioration of organ structure and function observed during chronic kidney disease development [107]. While further studies are warranted to fully determine the diagnostic potential of fibronectin it could be of significant diagnostic and prognostic value in chronic kidney disease.

Acknowledgments

Work detailed in this chapter was supported by the UCD Conway Institute of Biomolecular and Biomedical Research under the Programme for Research in Third Level Institutions (PRTLI) administered by the Higher Education Authority (HEA). This work was also supported by the European funded project, 'SysKID', Science Foundation Ireland, the Health Research Board of Ireland, Enterprise Ireland and by a research bursary from the Irish Nephrological Society and Amgen Ltd.

References

[1] Lechner, M.S. and G.R. Dressler, The molecular basis of embryonic kidney development. *Mech. Dev.*, 1997. 62(2): p. 105-20.

[2] George, E.L., et al., Defects in mesoderm, neural tube and vascular development in mouse embryos lacking fibronectin. *Development,* 1993.119(4): p. 1079-91.

[3] Castelletti, F., et al., Mutations in FN1 causeglomerulopathy with fibronectin deposits. *Proc. Natl. Acad. Sci. U S A, 2008*. 105(7): p. 2538-43.

[4] Hynes, R.O., Fibronectins. *Sci. Am.,* 1986. 254(6): p. 42-51.

[5] Hynes, R.O. and K.M. Yamada, Fibronectins: multifunctional modular glycoproteins. *J. Cell Biol.,* 1982.95(2 Pt 1): p. 369-77.

[6] Potts, J.R. and I.D. Campbell, Structure and function of fibronectin modules. *Matrix Biol.,* 1996.15(5): p. 313-20; discussion 321.

[7] Wierzbicka-Patynowski, I. and J.E. Schwarzbauer, The ins and outs of fibronectin matrix assembly. *J. Cell Sci.,* 2003.116(Pt 16): p. 3269-76.

[8] Feng, Y. and M. Mrksich, The synergy peptide PHSRN and the adhesion peptide RGD mediate cell adhesion through a common mechanism. *Biochemistry,* 2004.43(50): p. 15811-21.

[9] Fogerty, F.J., et al., Inhibition of binding of fibronectin to matrix assembly sites by anti-integrin (alpha 5 beta 1) antibodies. *J. Cell Biol.,* 1990.111(2): p. 699-708.

[10] Luo, B.H., C.V. Carman, and T.A. Springer, Structural basis of integrin regulation and signaling. *Annu. Rev. Immunol.,* 2007. 25: p. 619-47.

[11] Yong, J.L., et al., Fibronectin non-amyloid glomerulopathy. *Int. J. Clin. Exp. Pathol.,* 2009. 3(2): p. 210-6.

[12] Mackay, S., et al., Ultrastructural studies demonstrate that epithelial polarity is established in cultured mouse pre-Sertoli cells by extracellular matrix components. *J. Electron. Microsc.* (Tokyo), 1999.48(2): p. 159-65.

[13] Knight, C.G., et al., The collagen-binding A-domains of integrins alpha(1)beta(1) and alpha(2)beta(1) recognize the same specific amino acid sequence, GFOGER, in native (triple-helical) collagens. *J. Biol. Chem.,* 2000. 275(1): p. 35-40.

[14] Simon, E.E. and J.A. McDonald, Extracellular matrix receptors in the kidney cortex. *Am. J. Physiol.,* 1990. 259(5 Pt 2): p. F783-92.

[15] Midwood, K.S., et al., Modulation of cell-fibronectin matrix interactions during tissue repair. *J. Investig. Dermatol. Symp. Proc.,* 2006. 11(1): p. 73-8.

[16] Kang, Y.S., et al., Inhibition of integrin-linked kinase blocks podocyte epithelial-mesenchymal transition and ameliorates proteinuria. *Kidney Int.,* 2010.78(4): p. 363-73.

[17] Petit, V. and J.P. Thiery, Focal adhesions: structure and dynamics. *Biol. Cell,* 2000. 92(7): p. 477-94.

[18] Legate, K.R., et al., ILK, PINCH and parvin: the tIPP of integrin signalling. *Nat. Rev. Mol. Cell Biol.,* 2006.7(1): p. 20-31.

[19] Balasubramanian, L., et al., Integrin-mediated mechanotransduction in renal vascular smooth muscle cells: activation of calcium sparks. *Am. J. Physiol. Regul. Integr. Comp. Physiol.,* 2007. 293(4): p. R1586-94.

[20] Briggs, S.L., The role of fibronectin in fibroblast migration during tissue repair. *J. Wound Care,* 2005. 14(6): p. 284-7.

[21] Clark, R.A., Fibronectin matrix deposition and fibronectin receptor expression in healing and normal skin. *J. Invest. Dermatol.,* 1990. 94(6 Suppl): p. 128S-134S.

[22] Broughton, G., 2nd, J.E. Janis, and C.E. Attinger, The basic science of wound healing. *Plast. Reconstr. Surg.,* 2006. 117(7 Suppl): p. 12S-34S.

[23] Van Vliet, A., et al., Distribution of fibronectin isoforms in human renal disease. *J. Pathol.,* 2001.193(2): p. 256-62.

[24] Drenckhahn, D., et al., Ultrastructural organization of contractile proteins in rat glomerular mesangial cells. *Am. J. Pathol.,* 1990. 137(6): p. 1343-51.

[25] Cosio, F.G., D.D. Sedmak, and N.S. Nahman, Jr., Cellular receptors for matrix proteins in normal human kidney and human mesangial cells. *Kidney Int.,* 1990.38(5): p. 886-95.

[26] Courtoy, P.J., et al., Fibronectin localization in the rat glomerulus. *J. Cell Biol.,* 1980.87(3 Pt 1): p. 691-6.

[27] Sorokin, L.M., et al., Differential expression of five laminin alpha (1-5) chains in developing and adult mouse kidney. *Dev. Dyn.,* 1997. 210(4): p. 446-62.

[28] Butkowski, R.J., et al., Basement membrane collagen in the kidney: regional localization of novel chains related to collagen IV. *Kidney Int.,* 1989.35(5): p. 1195-202.

[29] Mozes, M.M., et al., Renal expression of fibrotic matrix proteins and of transforming growth factor-beta (TGF-beta) isoforms in TGF-beta transgenic mice. *J. Am. Soc. Nephrol.,* 1999. 10(2): p. 271-80.

[30] Uchio, K., et al., Abnormalities of extracellular matrices and transforming growth factor beta1 localization in the kidney of the hereditary nephrotic mice (ICGN strain). *J. Vet. Med. Sci.,* 1999. 61(7): p. 769-76.

[31] Lange-Sperandio, B., et al., Leukocytes induce epithelial to mesenchymal transition after unilateral ureteral obstruction in neonatal mice. *Am. J. Pathol.*, 2007. 171(3): p. 861-71.

[32] Lemley, K.V. and W. Kriz, Anatomy of the renal interstitium. *Kidney Int.,* 1991.39(3): p. 370-81.

[33] Mounier, F., J.M. Foidart, and M.C. Gubler, Distribution of extracellular matrix glycoproteins during normal development of human kidney. An immunohistochemical study. *Lab. Invest.*, 1986. 54(4): p. 394-401.

[34] Norman, J.T. and L.G. Fine, Progressive renal disease: fibroblasts, extracellular matrix, and integrins. *Exp. Nephrol.,* 1999. 7(2): p. 167-77.

[35] Obremski, V.J., P.M. Wood, and M.B. Bunge, Fibroblasts promote Schwann cell basal lamina deposition and elongation in the absence of neurons in culture. *Dev. Biol.,* 1993. 160(1): p. 119-34.

[36] Brady, H.R. and G.G. Singer, Acute renal failure. *Lancet,* 1995. 346 (8989): p. 1533-40.

[37] Klahr, S. and S.B. Miller, Acute oliguria. *N. Engl. J. Med.,* 1998. 338(10): p. 671-5.

[38] Levey, A.S., et al., Definition and classification of chronic kidney disease: a position statement from Kidney Disease: Improving Global Outcomes (KDIGO). *Kidney Int.,* 2005.67(6): p. 2089-100.

[39] Haroun, M.K., et al., Risk factors for chronic kidney disease: a prospective study of 23,534 men and women in Washington County, Maryland. *J. Am. Soc. Nephrol.,* 2003. 14(11): p. 2934-41.

[40] Hsu, C.Y., et al., Body mass index and risk for end-stage renal disease. *Ann. Intern. Med.,* 2006.144(1): p. 21-8.

[41] Schmitz, V., et al., Chronic renal dysfunction following liver transplantation. *Clin. Transplant.,* 2008. 22(3): p. 333-40.

[42] Bobadilla, N.A. and G. Gamba, New insights into the pathophysiology of cyclosporine nephrotoxicity: a role of aldosterone. *Am. J. Physiol. Renal. Physiol.,* 2007. 293(1): p. F2-9.

[43] Korbet, S.M., Primary focal segmental glomerulosclerosis. *J. Am. Soc. Nephrol.,* 1998. 9(7): p. 1333-40.

[44] Ronco, P. and H. Debiec, New insights into the pathogenesis of membranous glomerulonephritis. *Curr. Opin. Nephrol. Hypertens.* 2006. 15(3): p. 258-63.

[45] Noel, L.H., Morphological features of primary focal and segmental glomerulosclerosis.Nephrol Dial Transplant, 1999.14 Suppl 3: p. 53-7.

[46] Gruden, G., et al., Mechanical stretch-induced fibronectin and transforming growth factor-beta1 production in human mesangial cells is p38 mitogen-activated protein kinase-dependent. *Diabetes,* 2000.49(4): p. 655-61.
[47] Jefferson, J.A., S.J. Shankland, and R.H. Pichler, Proteinuria in diabetic kidney disease: a mechanistic viewpoint. *Kidney Int.*2008.74(1):p.22-36.
[48] D'Amico, G. and C. Bazzi, Pathophysiology of proteinuria. *Kidney Int.,* 2003.63(3): p. 809-25.
[49] Zoja, C., et al., Protein overload stimulates RANTES production by proximal tubular cells depending on NF-kappa B activation. *Kidney Int.,* 1998.53(6): p. 1608-15.
[50] Huang, Z., et al., Differential chemokine expression in tubular cells in response to urinary proteins from patients with nephrotic syndrome. *Cytokine,* 2008.42(2): p. 222-33.
[51] Eardley, K.S., et al., The relationship between albuminuria, MCP-1/CCL2, and interstitial macrophages in chronic kidney disease. *Kidney Int.,* 2006.69(7): p. 1189-97.
[52] Nath, K.A., The tubulointerstitium in progressive renal disease. *Kidney Int.,* 1998.54(3): p. 992-4.
[53] Nath, K.A., Tubulointerstitial changes as a major determinant in the progression of renal damage. *Am. J. Kidney Dis.,* 1992. 20 (1): p. 1-17.
[54] Mackensen-Haen, S., et al., Correlations between renal cortical interstitial fibrosis, atrophy of the proximal tubules and impairment of the glomerular filtration rate. *Clin. Nephrol.,* 1981. 15(4): p. 167-71.
[55] Mackensen-Haen, S., et al., The consequences for renal function of widening of the interstitium and changes in the tubular epithelium of the renal cortex and outer medulla in various renal diseases. *Clin. Nephrol.,* 1992. 37(2): p. 70-7.
[56] Bohle, A., et al., The long-term prognosis of the primary glomerulonephritides. A morphological and clinical analysis of 1747 cases. *Pathol. Res. Pract.,* 1992.188(7): p. 908-24.
[57] Rastaldi, M.P., et al., Epithelial-mesenchymal transition of tubular epithelial cells in human renal biopsies. *Kidney Int.*2002.62(1): p.137-46.
[58] Becker, G.J. and T.D. Hewitson, The role of tubulointerstitial injury in chronic renal failure. *Curr. Opin. Nephrol. Hypertens.,* 2000.9(2): p. 133-8.
[59] Eddy, A.A., Molecular insights into renal interstitial fibrosis. *J. Am. Soc. Nephrol.,* 1996. 7(12): p. 2495-508.
[60] Kriz, W. and M. LeHir, Pathways to nephron loss starting from glomerular diseases-insights from animal models. *Kidney Int.,* 2005. 67 (2): p. 404-19.
[61] Bohle, A., S. Mackensen-Haen, and M. Wehrmann, Significance of postglomerular capillaries in the pathogenesis of chronic renal failure. *Kidney Blood Press Res.,* 1996.19(3-4): p. 191-5.
[62] Strutz, F. and E.G. Neilson, The role of lymphocytes in the progression of interstitial disease. *Kidney Int.* Suppl., 1994. 45: p. S106-10.
[63] Qi, W., et al., The renal cortical fibroblast in renal tubulointerstitial fibrosis. *Int. J. Biochem. Cell Biol.,* 2006.38(1): p. 1-5.
[64] Mutsaers, S.E., et al., Mechanisms of tissue repair: from wound healing to fibrosis. *Int. J. Biochem. Cell Biol.,* 1997.29(1): p. 5-17.

[65] Vleming, L.J., et al., Progression of chronic renal disease in humans is associated with the deposition of basement membrane components and decorin in the interstitial extracellular matrix. *Clin. Nephrol.,* 1995. 44 (4): p. 211-9.
[66] Ruiz-Torres, M.P., et al., The accumulation of extracellular matrix in the kidney: consequences on cellular function. *J. Nephrol.,* 2005.18(3): p. 334-40.
[67] Dekkers, B.G., et al., Extracellular matrix proteins differentially regulate airway smooth muscle phenotype and function. *Am. J. Physiol. Lung Cell Mol. Physiol.,* 2007. 292(6): p. L1405-13.
[68] Schonherr, E. and H.J. Hausser, Extracellular matrix and cytokines: a functional unit. *Dev. Immunol.,* 2000. 7(2-4): p. 89-101.
[69] Macri, L., D. Silverstein, and R.A. Clark, Growth factor binding to the pericellular matrix and its importance in tissue engineering. *Adv. Drug Deliv. Rev.,* 2007. 59(13): p. 1366-81.
[70] Zaoui, P., et al., Role of metalloproteases and inhibitors in the occurrence and progression of diabetic renal lesions. *Diabetes Metab.,* 2000. 26 Suppl 4: p. 25-9.
[71] Catania, J.M., G. Chen, and A.R. Parrish, Role of matrix metalloproteinases in renal pathophysiologies. *Am. J. Physiol. Renal. Physiol.,* 2007. 292(3): p. F905-11.
[72] D'Armiento, J., Matrix metalloproteinase disruption of the extracellular matrix and cardiac dysfunction. *Trends Cardiovasc. Med.,* 2002.12(3): p. 97-101.
[73] Fu, X., W.C. Parks, and J.W. Heinecke, Activation and silencing of matrix metalloproteinases. *Semin. Cell Dev. Biol.,* 2008.19(1): p. 2-13.
[74] Ra, H.J. and W.C. Parks, Control of matrix metalloproteinase catalytic activity. *Matrix Biol.,* 2007.26(8): p. 587-96.
[75] Brew, K., D. Dinakarpandian, and H. Nagase, Tissue inhibitors of metalloproteinases: evolution, structure and function. *Biochim. Biophys. Acta,* 2000.1477(1-2): p. 267-83.
[76] Kaissling, B. and M. Le Hir, The renal cortical interstitium: morphological and functional aspects. *Histochem. Cell Biol.,* 2008. 130 (2): p. 247-62.
[77] Tomasek, J.J., et al., Myofibroblasts and mechano-regulation of connective tissue remodelling. *Nat. Rev. Mol. Cell Biol.,* 2002.3(5): p. 349-63.
[78] Zeisberg, M., F. Strutz, and G.A. Muller, Role of fibroblast activation in inducing interstitial fibrosis. *J. Nephrol.,* 2000.13 Suppl 3: p. S111-20.
[79] Hinz, B., Formation and function of the myofibroblast during tissue repair. *J. Invest. Dermatol.,* 2007. 127(3): p. 526-37.
[80] Annes, J.P., J.S. Munger, and D.B. Rifkin, Making sense of latent TGFbeta activation. *J. Cell Sci.,* 2003.116(Pt 2): p. 217-24.
[81] Gregory, K.E., et al., The prodomain of BMP-7 targets the BMP-7 complex to the extracellular matrix. *J. Biol. Chem.,* 2005. 280(30): p. 27970-80.
[82] Chang, H., C.W. Brown, and M.M. Matzuk, Genetic analysis of the mammalian transforming growth factor-beta superfamily. *Endocr. Rev.,* 2002. 23(6): p. 787-823.
[83] Moustakas, A. and C.H. Heldin, Non-Smad TGF-beta signals. *J. Cell Sci.,* 2005.118(Pt 16): p. 3573-84.
[84] Shi, Y. and J. Massague, Mechanisms of TGF-beta signaling from cell membrane to the nucleus. *Cell,* 2003.113(6): p. 685-700.
[85] Roberts, A.B., et al., Transforming growth factor-beta: multifunctional regulator of differentiation and development. *Philos. Trans. R Soc. Lond. B Biol. Sci.,* 1990. 327(1239): p. 145-54.

[86] Verrecchia, F., M.L. Chu, and A. Mauviel, Identification of novel TGF-beta /Smad gene targets in dermal fibroblasts using a combined cDNA microarray/promoter transactivation approach. *J. Biol. Chem.,* 2001. 276 (20): p. 17058-62.

[87] Niculescu-Duvaz, I., et al., The TGFbeta1-induced fibronectin in human renal proximal tubular epithelial cells is p38 MAP kinase dependent and Smad independent. *Nephron. Exp. Nephrol.,* 2007. 105(4): p. e108-16.

[88] Phanish, M.K., S.K. Winn, and M.E. Dockrell, Connective tissue growth factor-(CTGF, CCN2)--a marker, mediator and therapeutic target for renal fibrosis. *Nephron. Exp. Nephrol.,* 2010. 114(3): p. e83-92.

[89] Veerasamy, M., et al., Differential regulation of E-cadherin and alpha-smooth muscle actin by BMP 7 in human renal proximal tubule epithelial cells and its implication in renal fibrosis. *Am. J. Physiol. Renal. Physiol.,* 2009. 297(5): p. F1238-48.

[90] Gupta, S., et al., Connective tissue growth factor: potential role in glomerulosclerosis and tubulointerstitial fibrosis. *Kidney Int.,* 2000. 58 (4): p. 1389-99.

[91] Wang, S. and R. Hirschberg, BMP7 antagonizes TGF-beta -dependent fibrogenesis in mesangial cells. *Am. J. Physiol. Renal. Physiol.,* 2003. 284 (5): p. F1006-13.

[92] Yokoi, H., et al., Role of connective tissue growth factor in fibronectin expression and tubulointerstitial fibrosis. *Am. J. Physiol. Renal. Physiol.,* 2002. 282(5): p. F933-42.

[93] Ito, Y., et al., Expression of connective tissue growth factor in human renal fibrosis. *Kidney Int.,* 1998.53(4): p. 853-61.

[94] Turk, T., et al., BMP signaling and podocyte markers are decreased in human diabetic nephropathy in association with CTGF overexpression. *J. Histochem. Cytochem.,* 2009. 57(7): p. 623-31.

[95] Gao, X., et al., Connective tissue growth factor stimulates renal cortical myofibroblast-like cell proliferation and matrix protein production. *Wound Repair Regen.,* 2008. 16(3): p. 408-15.

[96] Chen, C.P., et al., Hypoxia and transforming growth factor-beta 1 act independently to increase extracellular matrix production by placental fibroblasts. *J. Clin. Endocrinol. Metab.*, 2005. 90(2): p. 1083-90.

[97] Greiling, D. and R.A. Clark, Fibronectin provides a conduit for fibroblast transmigration from collagenous stroma into fibrin clot provisional matrix. *J. Cell Sci.,* 1997.110 (Pt 7): p. 861-70.

[98] Grinnell, F., Fibronectin and wound healing. *J. Cell Biochem.,* 1984. 26 (2): p. 107-16.

[99] Marti, H.P., [Role of matrix metalloproteinases in the progression of renal lesions]. *Presse Med.,* 2000.29(14): p. 811-7.

[100] Casadio, R., et al., The structural basis for the regulation of tissue transglutaminase by calcium ions. *Eur. J. Biochem.*1999.262(3):p.672-9.

[101] Hang, J., et al., Identification of a novel recognition sequence for fibronectin within the NH2-terminal beta-sandwich domain of tissue transglutaminase. *J. Biol. Chem.,* 2005. 280(25): p. 23675-83.

[102] Gross, S.R., Z. Balklava, and M. Griffin, Importance of tissue transglutaminase in repair of extracellular matrices and cell death of dermal fibroblasts after exposure to a solarium ultraviolet A source. *J. Invest. Dermatol.,* 2003. 121(2): p. 412-23.

[103] Johnson, T.S., et al., Tissue transglutaminase and the progression of human renal scarring. *J. Am. Soc. Nephrol.,* 2003. 14(8): p. 2052-62.

[104] Ikee, R., et al., Involvement of transglutaminase-2 in pathological changes in renal disease. *Nephron. Clin. Pract.,* 2007. 105(3):p.c139-46.
[105] Johnson, T.S., et al., The role of transglutaminase in the rat subtotal nephrectomy model of renal fibrosis. *J. Clin. Invest.,* 1997. 99(12): p. 2950-60.
[106] Shweke, N., et al., Tissue transglutaminase contributes to interstitial renal fibrosis by favoring accumulation of fibrillar collagen through TGF-beta activation and cell infiltration. *Am. J. Pathol.,* 2008. 173(3): p. 631-42.
[107] Nelea, V., Y. Nakano, and M.T. Kaartinen, Size distribution and molecular associations of plasma fibronectin and fibronectin crosslinked by transglutaminase 2. *Protein J.,* 2008.27(4): p. 223-33.
[108] Waalkes, S., et al., Fibronectin 1 mRNA expression correlates with advanced disease in renal cancer. *BMC Cancer,* 2010.10: p. 503.
[109] Idasiak-Piechocka, I., et al., Elevated urinary fibronectin excretion predicts poor outcome in patients with primary chronic glomerulonephritis. *Nephron. Clin. Pract.,* 2010. 116(1): p. c47-52.

In: Fibronectin
Editors: James Beattie

ISBN 978-1-62257-298-4

Chapter XIII

Role of Fibronectin in Neoplasm

***Zhenggang Xiong*[1,*] *and Liqiong Liu*[2]**
[1]Department of Pathology and Laboratory Medicine,
Tulane University School of Medicine, New Orleans
[2]Department of Biochemistry and Molecular Biology, Louisiana
State University Health Science Center, New Orleans, US

Abstract

Fibronectin as an extracellular matrix glycoprotein is engaged in cell differentiation, growth, adhesion and migration. Alterations of its expression, degradation and organization are associated with neoplasm. Fibronectin plays an important role in tumor development, diagnosis, treatment and prognosis. Studies showed that increased fibronectin promotes growth, invasion and metastasis or spreading of lung cancer, breast cancer, glioma, renal cancer, leukemia, melanoma, squamous cell carcinoma, pancreatic cancer, etc. Molecular probes targeting fibronectin are shown being useful diagnostic tumor markers. Identification of fibronectin is an essential requisition for diagnosing fibroblastic neoplasm. A number of anticancer therapies targeting fibronectin were experimentally investigated over past years. Vaccines derived from fibronectin were demonstrated to reduce tumor size. Engineered fibronectin polypeptides could either interrupt the interaction of integrin $\alpha5\beta1$ with fibronectin or block epidermal growth factor receptor and insulin-like growth factor-I receptor. They could also activate cytokine-induced killer cells to execute anticancer activity. Several molecules were found to have therapeutic effects on cancer by either suppressing expression of fibronectin or impairing fibronectin assembly or acting as an antagonist against fibronectin receptor integrin. A recent study showed that an antisense oligonucleotide targeting fibronectin gene FN1 significantly decreases phosphorylation of FAK (Y397) and inhibits invasion of colon cancer cells. This illuminates a hope for gene therapy. Laboratory and clinical investigations revealed that fibronectin can be utilized to monitor and predict tumor response to treatment. High levels of fibronectin and its splice variant EDB-FN are found in tumors and their vasculatures, and monitoring these molecules can follow and predict

* Email: zxiong@tulane.edu.

the success of anticancer therapy. High FN1 expression in head and neck squamous cell carcinoma is associated with high intrinsic radiosensitivity, while alteration of extracellular matrix including fibronectin confers survival against chemotherapeutic agents in prostate cancer cells. Depending on the presence of fibronectin, the main binding protein for insulin-like growth factor IGFBP-3 can either inhibit or enhance EGF-mediated growth of mammary epithelial cells. The levels of fibronectin and IGFBP-3 within breast cancers can determine tumor response to treatments targeting EGFR. Fibronectin can also induce Mitomycin C resistance in human bladder cancer cells. A high level of fibronectin is correlated with lack of clinical response to high dose interleukin-2 therapy for tumors and decreases overall survival of patients. Quantification of oncofetal fibronectin (onfFN) mRNA can detect circulating thyroid cancer cells in patients who have been treated for papillary thyroid cancer with Levothyroxine, and this can be applied to clinical follow-up. This chapter discusses the role of fibronectin in oncogenesis, diagnosis of neoplasm, anticancer therapy and tumor prognosis as well as their rationales.

Fibronectin is a high molecular weight (approximately 440kD) glycoprotein. It exists as a dimer and consists of two almost identical monomers linked by a pair of disulfide bonds. [1] Fibronectin protein is produced from a single gene called FN1, but alternative splicing of its pre-mRNA results in several isoforms. Two types of fibronectin are present in vertebrates. Soluble fibronectin (also called cold-insoluble globulin) is produced in the liver by hepatocytes and is a major protein of blood plasma. Insoluble fibronectin is a soluble dimer initially and then is assembled into an insoluble matrix in a complex cellular process. It is major component of the extracellular matrix and is secreted primarily but not exclusively by fibroblasts.

Fibronectin binds to its receptors, membrane-spanning proteins called integrins. In addition, fibronectin also binds to other extracellular matrix components such as fibrin, collagen and heparan sulfate proteoglycans. Fibronectin plays an important role in cell differentiation, adhesion, growth and migration. Alteration of expression, degradation and organization of fibronectin and its receptors is associated with a number of diseases, especially neoplasm. [2] In this chapter, the roles of fibronectin in tumorigenesis, diagnosis of neoplasm, anticancer therapy and tumor prognosis are discussed.

Development and Progression of Neoplasm

Fibronectin is associated with development and progression of neoplasm, which has been suggested by numerous studies on a variety of tumors. The following neoplasms can serve as examples. Compared to normal renal tissue, fibronectin gene FN1 expression is higher in renal cell carcinoma. Mean tissue specific FN1 mRNA was found to be increased approximately seven fold in renal cell carcinoma over corresponding control renal tissue. Furthermore, tissue specific mean FN1 expression was increased close to eleven folds in clear cell renal cell carcinoma compared to papillary renal cell carcinoma. Patients with advanced tumor disease had higher FN1 expression when compared to organ-confined disease. [3]

Lymph node metastasis is more often seen in papillary thyroid carcinoma than in follicular thyroid carcinoma. Fibronectin is detected in 90% of papillary thyroid carcinoma, but only in a few cases of follicular thyroid carcinoma. An immunohistochemical study revealed that fibronectin expression in papillary thyroid carcinoma with lymph node metastasis was significantly higher than in papillary thyroid carcinoma without lymph node metastasis. [4]

Glioblastoma is the most common and malignant primary brain tumor. Its progression depends on the rapid proliferation of tumor cells. In an in vitro study, inhibition of fibronectin expression in glioma cells by short hairpin RNA-mediated silencing of gene expression delayed cell proliferation. This delay was partially explained by the reduction of fibronectin receptor integrin α5β1 expression observed in the same experimental system. In an animal model, mice with fibronectin-knockdown tumor had a mean survival advantage of 23 days over those with wild-type tumors. [5]

Integrin α9β1 as a receptor for fibronectin is also contributory to the aggressive behavior of tumors. Expression patterns of integrin α9β1 in 141 samples of mammary carcinoma and their relationship to patient outcome were studied. Effects of integrin α9β1 on tumor cell migration and invasion were also assessed in the same study using blocking antibody and gene transduction approaches. [6] A subset of these invasive carcinomas, mostly basal-like phenotype, exhibited expression of integrin α9β1 and a statistically significant association with reduced overall patient survival and distant-metastasis-free survival. Expression of integrin α9β1 was also screened in several mammary cancer cell lines, showing that the highly invasive basal-like cell line GI-101 expressed high levels of integrin α9β1. Migration and invasion of these GI-101 cells were greatly reduced by α9-blocking antibody and α9-knockdown with siRNA. In contrast, over-expression of integrin α9β1 enhanced migratory and invasive behavior of α9-negative MCF7 cells and α9-low MDA MB468 cells. This study demonstrated that integrin α9β1 promotes migration and invasion of mammary cancer.

Table 1. Tumors with Alterations of Fibronectin and Its Receptors

Tumor Origin	Alteration of Fibronectin	Alteration of Integrin	Number of References
Lung	Overexpression	Overexpression of α5β1	4
Ovary	Overexpression		3
Prostate	Overexpression	Overexpression of α5β1	2
Liver	Overexpression		4
Breast	Overexpression, Structural Alteration (Alternative Splicing)	Overexpression of α9β1	5
Glioma	Overexpression		3
Squamous Cell Carcinoma	Overexpression	Overexpression of α5β1	6
Gastro-intestine	Overexpression	Sialylation of α5β1	5
Lymphoma/ Leukemia	Overexpression, Structural Alteration (Alternative Splicing)		6
Bladder	Overexpression		2
Pancreas	Overexpression		2
Kidney	Overexpression		1
Thyroid	Overexpression		1

Studies on pancreatic cancer cell lines showed that tumor cell migration was strongly inhibited by treating the tumor cells with anti- integrin α5β1 antibodies. A high level expression of laminin, fibronectin and collagen type IV was found on the basement membrane of perineurium, which contributes to the affinity of pancreatic cancer cells for perineural tumor invasion. [7]

The tumors that have been associated with alterations of fibronectin and its receptors are summarized in the table 1.

Alterations of fibronectin and its receptors play their roles in tumorigenesis and progression through different pathways, depending on the cellular origin of the neoplasm. Understanding of these mechanisms helps us identifying early molecular changes and developing therapeutic drugs, which benefits the prevention and treatment of tumors.

1. FAK Signaling Pathway

In non-small cell lung cancer, this signaling pathway might be more commonly utilized by fibronectin. It was demonstrated that fibronectin activates the intracellular molecule, focal adhesion kinase (FAK). The effect of FAK signaling pathway in relation to the role of calpain-2 and RhoA in fibronectin-mediated migration and invasion of lung cancer cells was investigated. The study found that lung epithelial cell line A549 stimulated by fibronectin showed increased phosphorylation of FAK and its downstream targets, Src, extracellular signal regulated kinase (ERK1/2), phosphatidylinositol 3'-kinase (PI3K) and Akt. On the other hand, depletion of FAK by siRNA inhibited activity of Src, ERK1/2, PI3K and Akt. In addition, PP2 (Src inhibitor) blocked the phosphorylation of FAK, ERK1/2, PI3K and Akt. Conversely, inhibition of MEK1/2, the upstream kinase of ERK1/2, using PD98059 reduced the expression of matrix metalloproteinase-9 (MMP9) and calpain-2. The PI3K inhibitor, LY294002, also blocked the expression of MMP9 and RhoA. Inhibition of both MEK1/2 and PI3K caused reduced cell migration and invasion. These indicated that fibronectin promotes lung cancer invasion and metastasis through regulation of FAK, Src, ERK1/2, PI3K, Akt, MMP9/calpain-2 or MMP9/RhoA activity. [8]

In ovarian carcinoma, fibronectin binds to integrin α5β1, leading to a direct association of integrin α5β1 with c-Met, the receptor tyrosine kinase, and activation of c-Met. Subsequently, c-Met associates with and activates Src as well as FAK. Inhibition of integrin α5β1 decreases the phosphorylation of c-Met, FAK and Src in vitro and in vivo, resulting in inhibition of tumor cell invasion. This inhibition can be overcome by independent activation of c-Met by its native ligand, hepatocyte growth factor/scatter factor (HGF/SF), or overexpression of a constitutively active FAK in HeyA8 cells, indicating that integrin α5β1 is upstream of c-Met, Src and FAK. Inhibition of integrin α5β1 on cancer cells in two xenograft models of ovarian cancer metastasis resulted in a significant decrease of tumor burden, which was independent on the effect of integrin α5β1 on angiogenesis. These data suggest that fibronectin promotes invasion and metastasis of ovarian carcinoma through the integrin α5β1/c-Met/FAK/Src-dependent signaling pathway. [9]

Inhibition of fibronectin and integrin α5β1 expression in glial tumor cells results in delayed cell proliferation. Further study of the downstream signaling targets of integrin α5β1 demonstrated the reduced phosphorylation of Src kinase and STAT-3. These findings support the review that fibronectin plays its role in tumorigenesis of glial cells via the fibronectin/ integrin α5β1/Src kinase/STAT-3 signaling pathway. [5]

A study on the influence of fibronectin and tumor necrosis factor-alpha (TNF- α) on osteopontin expression in human head and neck squamous cell carcinoma cell line, HN-22, showed that the phosphorylation of ERK1/2 was strongly enhanced by TNF- α and fibronectin compared to the application of either one alone. This synergistic effect on ERK1/2 phosphorylation could also be induced by TNF- α and activating integrin β1 antibody. These data indicate that fibronectin coordinates TNF- α -mediated osteopontin induction in squamous cell carcinoma via integrin β1-dependent signaling mechanism that activates ERK1/2 and plays an important role during tumor progression. [10] A similar mechanism is also employed in cervical squamous cell carcinoma. Culture of human cervical cancer cell (SiHa) on fibronectin-coated surface showed that fibronectin activates integrin and subsequently FAK, ERK, PI-3K, integrin-linked kinase (ILK) and nuclear factor-kappa B (NF-kB). As the final destination, pro-MMP-9 and pro-MMP-2 are activated and promotes the cancer cell proliferation and migration. [11] This signaling mechanism through which fibronectin plays its role in neoplasm is also supported by another in vitro study with laryngeal carcinoma cell line, HEp-2. The study demonstrated that cultured cells in presence of extracellular matrix adhesion molecule fibronectin secrete matrix metalloproteinases including MMP-9 that facilitate cancer cell invasion. [12] The study reported that fibronectin induces MMP-9 mRNA and protein in HEp-2 cells and promotes gelatinolytic activity. This effect appears to be mainly mediated by integrin α5β1 receptor. Blocking integrin α5β1 attenuated the fibronectin-mediated stimulatory response on MMP-9. Further studies with siRNA suggested the involvement of FAK, PI-3K, ERK and NFκB in fibronectin-mediated MMP-9 induction. Immunocytochemical study revealed that ERK, PI-3K and NFκB are localized in nucleus. Immunoblotting demonstrated that the expression of p-FAK, p-PI-3K, p-ERK and NFκB are enhanced. It has been indicated that fibronectin-induced activation of the MMP-9 gene is enhanced by binding of NFκB, activator protein-1 (AP-1) and specificity protein-1 (Sp1) to the MMP-9 promoter.

Paxillin serves as one of the signal transducers for fibronectin. Studies on gastric cancer cells showed a dose dependence on fibronectin for cancer invasiveness and phosphorylation of paxillin (Y118). Cancer invasiveness and phosphorylation of paxillin in gastric cancer cells reached its peak when the concentration of fibronectin reached 100 nmol/L. siRNA targeting of paxillin significantly decreased the fibronectin-induced phosphorylation of paxillin compared with controls. Inhibition of FAK activity inhibited phosphorylation of paxillin induced by fibronectin, suggesting that fibronectin promotes paxillin (Y118) phosphorylation through FAK. [13] Another study on SW480 colon cancer cells achieved a similar finding. In that study, ionizing radiation of cells increased beta-galactoside alpha-(2, 6)-sialyltransferase (ST6Gal I) and sialylation of glycoprotein, especially for the key adhesion molecule integrin α5β1. The later, in turn, increased adhesion of SW480 colon cancer cells to fibronectin and contributed to cell survival through the activation of paxillin and AKT. On the other hand, knockdown of ST6Gal I or paxillin decreased the level of radiation-induced cell adhesion and increased the level of cell death. [14] Colon cancer cell line Colo320 also exhibits a dependence on fibronectin for its invasion and phosphorylation of FAK (Y397). Cancer invasion and phosphorylation of FAK in Colo320 reached their peak at 1 nmol/L fibronectin, and interestingly this effect can be reduced by a fibronectin antisense oligonucleotide. [15]

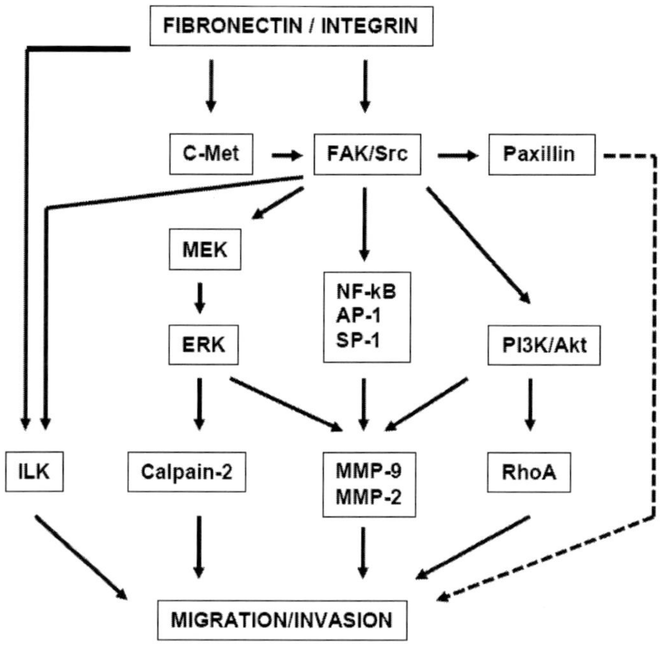

Figure 1. FAK Signaling Pathway.

The role of fibronectin in hematopoietic neoplasm may involve the FAK/PI-3K/ERK signaling pathway as well. Studies on the cultured human immortalized myelogenous leukemia line K562 cells showed that exposure of cells to fibronectin in serum free medium in the presence of fibronectin up-regulates the expression of pro-MMP-9 within 2 hours. Phosphorylation of FAK, ERK and PI-3K and nuclear translocation of EGFR and NF-kB were also found upon fibronectin binding to cells. These results support the suggestion that the FAK/PI-3K/ERK signaling pathway is involved in the fibronectin-integrin induced MMP-9 expression. [16]

In summary, fibronectin plays its role in tumorigenesis and tumor progression by binding to its receptor, integrin, and this complex, in turn, activates MMPs, Calpain-2 and RhoA through cascading activation of c-Met, FAK, paxillin, Src/STAT-3, MEK1/2, PI3K/Akt, ERK1/2, ILK and NF-kB. Furthermore, fibronectin-induced activation of MMP-9 is enhanced by binding of NFκB, AP-1 and Sp1 to the MMP-9 gene promoter. (Figure 1)

2. Cyclin Pathway

Fibronectin also promotes tumorigenesis and tumor progression via the cyclin pathway. Fibronectin has been shown to stimulate steroids interacting with androgen receptors. This interaction, in turn, controls the expression of cyclin D and other related genes involved in regulation of the cell cycle. Uncontrolled production of cyclin D following fibronectin overexpression affects amount of cyclin D-Cdk4 (cyclin-dependent kinase 4) complex which is formed and drives the cell through the G0/S and G1/S checkpoints with deregulation of cell cycling. Uncontrolled cell cycling is closely associated with cancer cell growth. [17] A recent in vitro study evaluating the influence of the receptor tyrosine kinase inhibitor AEE788 and the mammalian target of Rapamycin (mTOR) inhibitor RAD001 on adhesion and proliferation of renal cell carcinoma cells suggested that fibronectin alters the expression level of the cell cycle regulating proteins Cdk2, Cdk4, cyclin D1, cyclin E and p27, and, therefore, promotes tumor cell proliferation. [18]

During cell cycling, the concentration of cyclins varies in a cyclical fashion. Cyclins form complexes with Cdks to initiate phosphorylation and activation of Cdks. The resulting complexes (so-called maturation-promoting factors) activate other proteins through phosphorylation. These phosphorylated proteins, in turn, are responsible for specific events during cell division such as microtubule formation and chromatin remodeling. Different cyclins act in different cell cycle phases to drive cellular proliferation. G_1-phase cyclins bind to their Cdks and signal the cell to prepare for chromosome replication. S-phase cyclins bound to Cdk2 and enter the nucleus to regulate DNA synthesis. As DNA replication continues, cyclin E is destroyed, and the level of mitotic cyclins begins to rise in G_2 phase. M-phase cyclins and the M-phase Cdk (Cdk1) initiate assembly of the mitotic spindle, breakdown of the nuclear envelope, cessation of all gene transcription and condensation of the chromosomes. These events take cell to the metaphase of mitosis.

The cyclin pathway through which fibronectin plays its role in tumorigenesis and tumor progression is illustrated in figure 2.

It is worth mentioning that there is crosstalk between the FAK signaling pathway and the cyclin pathway. Cyclin D is regulated by the downstream pathway of mitogen receptors via MAP kinase, ERK and PI3K. ERK activates the downstream transcription factors Myc and AP-1 which, in turn, activate the transcription of Cdk4, Cdk6 and cyclin D genes. Rho family proteins and FAK activate the cyclin D gene in response to integrin. [19] More than one mechanism can be involved in the effects of fibronectin on a neoplasm. This is suggested by the observation that fibronectin stabilizes cyclin D1 and antagonizes Mitomycin C -induced G0/G1phase arrest in human bladder cancer cells, while fibronectin-mediated protection of human bladder cancer cells depends on the activation of the PI3-K/Akt signaling pathway. [20]

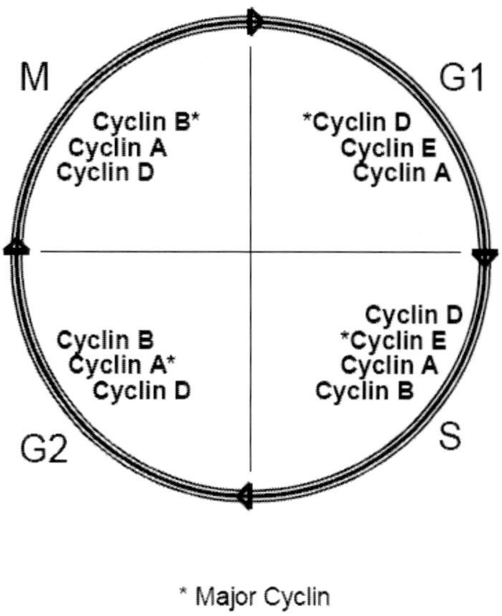

Figure 2. Cyclin Pathway.

3. Neovascularization

Neovascularization is a process of development and growth of new blood vessels from pre-existing vessels. Tumor angiogenesis is important in the carcinogenesis, growth and progression of human solid tumors.

Fibronectin plays a role in tumorigenesis also by facilitating tumor neovascularization to supply nutrients and oxygen for tumor growth. Migration-stimulating factor (extra domain B containing FN (EDB-FN)) is a potent oncofetal regulatory molecule. It is generated from the fibronectin gene by alternative splicing involving premature intra-intronic cleavage. EDB-FN is constitutively expressed by both epithelial and stromal cells during fetal development and in cancer. EDB-FN affects the behavior of a broad range of potential target cells including fibroblasts, endothelial cells and epithelial cells by stimulation of migration/invasion, matrix remodeling and induction of angiogenesis. It also functions as an autocrine survival factor for vascular endothelium. The significant association between elevated expression of EDB-FN and poor survival in patients with breast or oral cancer suggests that EDB-FN may function as a driver of tumor progression. Furthermore, EDB-FN is highly expressed in tumor vasculature. An in vitro study was carried out to investigate whether another alternatively spliced and tumor derived fibronectin EDA-FN (extra domain A containing FN) has lymphangiogenic potency on human lymphatic endothelial cells to facilitate tumor lymphatic metastasis. Lymphatic endothelial cells were cultured in three-dimensional culture system and treated with SW480 supernatant that is highly rich in EDA-FN. The results demonstrated that SW480 supernatant could facilitate tube-like formation of lymphatic endothelial cells, compared with controls. The same study also found co-localization of EDA-FN and integrin α9 on lymphatic endothelial cells as well as significant upregulation of integrin α9 in the SW480 supernatant treated group, indicating that Integrin α9 is a specific receptor for EDA-FN. Western blot and immunofluorescence revealed that EDA-FN had an important role in modulating the expression of some key regulators of lymphangiogenesis such as Prox1 and F-actin so as to promote motility and sprouting of lymphatic endothelial cells. In addition, all of these effects could be significantly inhibited by EDA-FN antibody, IST-9. These findings revealed that EDA-FN derived from tumor cells has an important role in facilitating lymphangiogenesis. [21]

Tissue engineering was utilized to investigate fibronectin-induced vascularization. Collagen scaffolds coated with fibronectin were shown to reduce short-term TUNEL staining by nearly 40%, while increasing long-term human umbilical vein endothelial cell survival by 30-45%, compared to collagen scaffolds without fibronectin. As a result, a 100% increase in the number of human umbilical vein endothelial cell -lined vessels was observed with fibronectin-coated scaffolds, as compared to collagen-only scaffolds, on day 7 and 14 after implantation. In addition, newly formed blood vessels appeared to be perfused with host erythrocytes by day 7, and vascular maturation and stabilization was apparent by day 14. [22]

Fibronectin facilitates tumor neovascularization by activating its integrin receptor(s). Several subunits of integrin are known to be involved in this process. The integrin β3 subunit forms a heterodimer with the integrin αv subunit and interacts with extracellular matrix components such as fibronectin. Integrin β3 subunit is not normally expressed in quiescent endothelial cells but is present in the vasculature of neoplastic and ischemic tissues.[23] Another study on human placental multipotent mesenchymal stromal cells (hPMSCs) isolated from term placenta showed that the hPMSCs express integrins α4, α5, αv, β1, β3 and β5, and they can be induced to differentiate into cells which express endothelial markers. An increase of cell surface integrin α5β1 accompanied endothelial differentiation. Fibronectin enhanced the adhesion and migration of differentiated hPMSCs through integrin α5β1, but not αvβ3 or αvβ5. This effect was augmented by vascular endothelial growth factor A. Antibodies to integrins α5 and β1 blocked formation of capillary-like structures in vitro from the differentiated cells. When hPMSCs were seeded onto chick chorioallantoic membrane (CAM), human von Willebrand factor-positive cells engrafted in the chick endothelium. CAMs transplanted with differentiated hPMSCs had a greater number of blood vessels containing human cells and more incorporated cells per vessel compared to CAMs transplanted with undifferentiated hPMSCs. These data suggest that differentiated hPMSCs contribute to blood vessel formation through interaction of fibronectin and integrin α5β1. [24]

Diagnosis of Neoplasm

Alterations in the expression of fibronectin and its integrin receptors, especially overexpression of fibronectin, are associated with neoplasm. This observation suggests that fibronectin may be used as a biomarker to aid diagnosis of neoplasm. Indeed, identification of fibronectin is an essential prerequisite for diagnosing a fibroblastic neoplasm. In fibroblastic tumors, cytoplasmic myofilaments are connected with extracellular filaments, fibronectin fibrils, through fibronexus at plasmalemmal plaque forming a myofilament-fibronexus-fibronectin complex that can be visualized under electron microscope. Electron microscopic examination has been considered a gold standard for diagnosis of myofibroblastoma or myofibrosarcoma, especially in the context of identification of fibronectin by immuno-histochemical study. [25]

Fibronectin is highly expressed in many other tumors, especially malignant tumors, and detection of this molecule can help in diagnosis of such tumors. Fibronectin is detected in approximately 90% of papillary thyroid carcinomas but very rarely in follicular thyroid carcinomas. Sialic acid is carried by fibronectin and can be targeted by monoclonal antibody JT-95. Immunohistochemical analysis with JT-95 can be used to distinguish papillary thyroid carcinoma from follicular thyroid carcinoma. [26]

Detection of urinary fibronectin and mutant p53 for diagnosis of bladder cancer was investigated. This study included 100 patients diagnosed with urothelial carcinoma, 93 patients with non-neoplastic urological disease and 47 healthy individuals. The urine supernatant was used for determination of fibronectin by ELISA, while urine sediment was used for cytology and detection of mutant p53 by PCR-SSCP followed by DNA sequencing. The results showed that the sensitivity and specificity of fibronectin assay were 82% and 84.3%, while the sensitivity and specificity were 59% and 91.4% for void urine cytology and 37% and 100% for detection of mutant p53. These data indicate that the fibronectin assay has the highest sensitivity compared to void urine cytology and mutant p53 detection in the diagnosis of bladder cancer. [27]

As indicated above, the spliced variant of fibronectin, EDB-FN, is highly expressed in tumor and fetal tissue vasculature. [28], [29] In contrast, normal adult tissue does not express EDB-FN. Molecular imaging of angiogenesis offers a serial non-invasive evaluation of both location and growth dynamics of tumors. Likewise, molecular probes targeting EDB-FN can be used as both early tumor markers and tools to monitor the success of anti-angiogenic cancer therapy. [30] A single chain antibody fragment (scFv) L19 with a high affinity (K_d = 54 pM) to EDB-FN was developed by Pini et al. [131] Radio iodinated (scFv) L19 showed specific accumulation around tumor neovasculature and stroma indicating high EDB-FN expression. [32] Recently, a bi-specific L19 was constructed by fusion of two (scFv) L19 to form a divalent scFv small immunoprotein (L19-SIP) with a molecular mass of ~80 kDa. Radiolabeled L19-SIP is being evaluated for imaging of neovasculature and radio-immunotherapy of cancer. Blood-based bioassays for breast cancer screening have been developed. In one study, differential gene expression assay in breast cancer and benign tumor samples was carried out to identify candidate biomarkers for blood-based screening. Two proteins, fibronectin and CXCL9, were identified from a set of gene expression data that included 120 breast cancer samples and 45 benign lesions. These two proteins were differentially expressed between cancer and benign lesion, and they were released into the extracellular medium and were stable in serum. The protein concentrations of fibronectin and CYCL9 in blood sampled from normal volunteers (n=119) and early breast cancer patients (n=133; 73% of patients had cT1-T2 tumor) were determined by ELISA. The results showed that the cancer patients had higher levels of CXCL9 and fibronectin than volunteers. CXCL9 mean concentration was 851 pg/ml and 635 pg/ml for patients and volunteers, respectively (P=0.013). Furthermore, CXCL9 concentration was significantly higher in patients with estrogen receptor-negative (999 pg/ml) compared to volunteers (P=0.003). Fibronectin concentration was 190 ug/ml for patients and 125 ug/ml for volunteers. A combined assay for both fibronectin and CXCL9 showed 53% sensitivity and 98% specificity. This study suggested that fibronectin and CXCL9 levels in serum can be used to detect early breast cancer, including ER-negative breast cancer. [33]

Treatment of Neoplasm

Cancer cells actively elaborate and remodel their extracellular matrix (ECM) to aid in survival and progression. Strategies to selectively enhance cancer sensitivity to radiation therapy are greatly beneficial to tumor treatment. It has been shown that inhibitory antibodies against integrin β1 can enhance the growth-inhibitory and apoptotic responses of human breast cancer cell lines to ionizing radiation, both in vitro and in vivo. A study showed that a specific alpha heterodimer of integrin β1 preferentially mediates a pro-survival signal in human breast cancer cells, which can be specifically targeted for therapy. Three-dimensional laminin-rich extracellular matrix (3D lrECM) culture conditions were used to compare expression of α5β1 integrin heterodimer in malignant and nonmalignant cell lines. In this study, the expression of integrin α5β1 was found to be up-regulated in malignant cells compared to nonmalignant breast cells. In addition, wild type and oncofetal splice variants of fibronectin were also markedly up-regulated in malignant cell lines compared with nonmalignant cells. Cells treated with a peptide that disrupts the interactions of integrin α5β1 with fibronectin promoted apoptosis of malignant cells, which further facilitated the apoptotic effects of radiation. To validate these results, gene expression array data from breast cancer patients were analyzed and revealed an association of high levels of integrin α5 expression with decreased survival. These findings indicated that fibronectin and integrin α5β1can be used as targets for breast cancer therapy. [34]

As the aforementioned, splice variant EDB-FN is highly expressed during angiogenesis in neoplastic and fetal tissue vasculatures, and a single chain antibody fragment (scFv) L19 with a high affinity to EDB-FN was developed. Radio iodinated (scFv) L19 showed specific accumulation around tumor neovasculature and stroma in which high EDB-FN expression occurs, indicating the therapeutic strategies targeting EDB-FN with combination of anti-angiogenic and cytotoxic drugs are feasible. [32]

Engineered domains of human fibronectin (Adnectins™) were generated to block epidermal growth factor receptor (EGFR) and insulin-like growth factor-I receptor (IGF-1R) which mediate both proliferative and survival cell signaling in cancer. Individual and optimized Adnectins specific for either EGFR or IGF-IR were engineered into a single protein (EI-Tandem Adnectin). In human cancer cell lines A431, H292, BxPC3 and RH41, the EI-Tandems inhibited phosphorylation of EGFR and IGF-IR and induced receptor degradation, which inhibits down-stream cell signaling and proliferation of cancer cells. An in vivo study also demonstrated that PEGylated EI-Tandem inhibits the growth of both EGFR and IGF-IR driven human tumor xenografts and induces degradation of EGFR. [35]

Fibronectin is a large glycoprotein with binding sites for many substances, including N-terminal and C-terminal heparin-binding domains. A study investigated the effects of highly purified rhFNHN29 and rhFNHC36 polypeptides cloned from the two heparin-binding domains of fibronectin on the adhesion and invasion of highly metastatic human hepatocellular carcinoma cells (MHCC97H). The MHCC97H cells which adhered to fibronectin in the presence of various concentrations of rhFNHN29 and rhFNHC36 polypeptides were quantified with crystal violet staining, and the effects of rhFNHN29 and rhFNHC36 on the invasion of MHCC97H cells were then detected using the Matrigel invasion assay as well as a lung-metastasis mouse model. The expression levels of integrins and FAK and the activity of matrix metalloproteinases and AP-1 were also analyzed. The results showed that both rhFNHN29 and rhFNHC36 inhibited adhesion and invasion of MHCC97H cells. Further study demonstrated that these inhibitory effects were mediated by integrin αvβ3. Polypeptides rhFNHN29 and rhFNHC36 decreased expression of integrins αv, β3 and β1, resulting in abrogated tyrosine phosphorylation of FAK and activation of AP-1 as well as the reduction of MMP-9 activity. These effects could be reversed by a protein tyrosine phosphatase inhibitor. It is reasonable to believe that polypeptides rhFNHN29 and rhFNHC36 could potentially be used as anti-adhesive and anti-invasive agents for human liver cancer. [4]

NK4, a fragment of hepatocyte growth factor (HGF), exerts a bifunctional action as an angiogenesis inhibitor and a competitive antagonist of HGF binding to its receptor c-Met. Studies on the anti-angiogenic mechanism of NK4 demonstrated that NK4 impairs fibronectin assembly. In cultured human endothelial cells, NK4 bound to perlecan. Perlecan is a multidomain heparan sulfate proteoglycan that interacts with fibronectin in the basement membrane and inhibits its assembly. Knockdown of perlecan expression by small interference RNA significantly reduced the inhibitory effect of NK4 on fibronectin assembly and cell spreading. In NK4-treated endothelial cells, phosphorylation of FAK and activation of Rac were decreased, while re-activation of Rac restored the DNA synthesis in these cells. These findings showed that the association between NK4 and perlecan impairs fibronectin assembly and, therefore, inhibits anchorage-dependent signaling and tumor growth. This indicates that fibronectin plays a critical role in the anti-angiogenic and anti-tumorigenic effects of NK4. [36]

Fibronectin and its interaction with integrins are necessary for pancreatic ductal adenocarcinoma (PDAC) progression. Endothelial monocyte-activating polypeptide II (EMAP II) was shown to inhibit PDAC progression through its ability to interfere with fibronectin-integrin angiogenesis signaling. [37] In heterotopic PDAC tumors, EMAP II reduced more than 65% of tumor size, more than 50% of microvascular density, and more than 40% of proliferative activity associated with the tumor. Cultured PDAC cells expressed integrins αvβ3 and α5β1. Treatment with EMAP II caused significant reduction of fibronectin expression in these tumor cells. These results indicated that antitumor activity of EMAP II against PDAC cells is at least partially mediated by interference with stromal fibronectin-integrin dependent PDAC cell proliferation.

Fibronectin is known to have two structural forms, globular FN (G-FN) and fibrillar FN (F-FN). G-FN plays a critical role in cell adhesion and growth, whereas F-FN results in apoptosis. Conversion G-FN into F-FN showed that F-FN promotes deactivation of Akt/GSK-3beta, induced phosphorylation of SHP-2, and activation of RhoA/Rho kinase (ROCK). As a result, F-FN inhibited the survival of SKOV-3 and MCF-7 cancer cells. SHP-2 inhibitor and anti-sense oligodeoxynucleotide decreased SHP-2 level and reversed the F-FN mediated apoptosis. ROCK inhibitor also modulated F-FN mediated apoptosis. These data demonstrated that F-FN induces cell apoptosis through phosphorylation of SHP-2, activation of RhoA/ROCK and deactivation of Akt/GSK-3beta. Fibrillar fibronectin may be one of the therapeutic choices against tumors. [38]

Oncofetal fibronectin (onfFN) is highly expressed in papillary thyroid cancer. Quantification of onfFN mRNA in blood will detect circulating thyroid cancer cells in patients. This methodology can be applied in the clinical follow-up of patients who have been treated for papillary thyroid cancer.

Cytokine-induced killer cells play an important role in adoptive cellular immunotherapy to neoplasm. An in vitro study revealed that fibronectin can activate cytokine-induced killer cells. The cytokine-induced killer cells were cultivated with recombinant human fibronectin, and the fibronectin-induced group had a statistically higher proliferation rate than control cultures. The proliferation rate of CD3+CD16+CD56+T cells in each group was thousands fold higher than the initial number. There were also a higher percentage of CD3+CD8+ T cells in fibronectin-induced group. These results indicate that fibronectin could be used to promote cellular immunity to tumors. [39]

Although monoclonal antibody based therapies have made an important contribution to current treatment strategies for cancer, they are generally not cost-effective. Therapeutic vaccination defined as induction of immunity against tumor-specific molecules (tumor antigens) has been an attractive alternative. A vaccine against EDB-FN was recently developed. A study demonstrated that it is possible to break self-tolerance and induce a strong antibody response against EDB-FN by vaccination. In this study, 19 of 20 vaccinated mice responded with production of anti-EDB-FN antibodies and a 70% reduction of tumor size, in contrast to those lacking anti-EDB-FN antibodies. Analysis of the tumor tissue revealed that immunization against EDB-FN induces several histopathological changes consistent with an attack by the immune system. These results indicate that vaccines against EDB-FN may be interesting candidate agents for treatment of tumors. [40]

Nm23 is an anti-metastatic tumor suppressor gene. The molecular mechanism of this anti-metastatic gene has previously remained unknown until a recent study. In this study, human hepatocellular carcinoma cell line H7721 cells were stably transfected with the plasmid pcDNA3/Nm23-H1. Cell adhesion, actin extension and wound-induced migration along with FAK phosphorylation and levels of integrins α5 and β1 were determined in transfected and control cells. N-glycosylation inhibitor tunicamycin was also used to deglycosylate the integrin β1 subunit. The results showed that overexpression of Nm23-H1 in H7721 cells reduced cell adhesion, migration and extension of actin stress fibers on dishes coated with fibronectin. Phosphorylation of FAK in Nm23-H1 transfected cells was also attenuated.

The surface expression of integrin β1 was down-regulated, although a partially glycosylated precursor of β1 was increased. These data suggest that Nm23-H1 down-regulates integrin β1 on the cell surface and may mediate intracellular signaling by modulating glycosylation of integrin β1, which subsequently suppresses cellular invasive processes. These results suggest a potential for Nm23 based gene therapy. [41]

Prognosis of Neoplasm

Fibronectin analysis can be used to monitor and predict tumor response to treatment and aid in forecasting the prognosis of a neoplasm.

To investigate whether cell adhesion to fibronectin induces drug resistance in human bladder cancer cells, the human bladder cancer cell line T24 was pre-coated with fibronectin and treated with Mitomycin C (MMC). Apoptosis induced by MMC was significantly reduced by fibronectin adhesion to T24 cells. Further study demonstrated that fibronectin provokes MMC-induced G0/G1-phase arrest by inactivating GSK-3beta to stabilize cyclin D1 expression and activates the PI3-K/Akt signaling pathway in T24 cells. These findings demonstrated that fibronectin protects T24 cells from MMC-induced cell death through development of chemoresistance to MMC. [20]

High dose interleukin-2 (IL-2) induces a durable therapeutic response in a subset of patients with metastatic melanoma and renal cell carcinoma. To identify predictive biomarkers of a clinical response to IL-2, sera from patients treated with high dose IL-2 were analyzed for soluble factors that are associated with immunity to tumor. Fibronectin and vascular endothelial growth factor were found as independent predictors of response. High levels of these proteins were correlated with lack of clinical response to IL-2 and decreased overall survival. [42]

Radiotherapy is widely used in head and neck cancer therapy, although some of these tumors are radio-resistant. To search for biomarkers that can predict the radiotherapy response of a tumor, microarray analysis was performed on head and neck squamous cell carcinoma cell lines that have various intrinsic radiosensitivity. The analysis identified 16 deregulated categories of genes. Of these genes, high FN1 expression was associated with high intrinsic radiosensitivity, suggesting that FN1 may be a potential biomarker for radiosensitivity of these tumors. [43] Although high FN1 expression was associated with high intrinsic radiosensitivity in head and neck squamous cell carcinoma, it seems that alteration of extracellular matrix including fibronectin confers survival only against chemotherapeutic agents but not radiotherapy in DU145 prostate cancer cells. Cells of prostate cancer cell line DU145 treated with Ceramide and Docetaxel as well as radiation showed that fibronectin promoted the resistance of DU145 prostate cancer cells to the apoptotic effects of Ceramide and Docetaxel but not to radiation. Furthermore, this survival effect involved the insulin-like growth factor and integrin β1. These findings indicated that the prostate tumors with increased fibronectin may have worse prognosis due to their resistance to chemotherapeutic agents. [44]

Progression of breast cancer is associated with a transition from estrogen dependence to EGF receptor (EGFR)/HER-2 dependence and alteration of the extracellular matrix, which is accompanied by an increase in the levels of the main binding protein for insulin-like growth factor, IGFBP-3.

Studies with breast epithelial cells demonstrated that IGFBP-3 inhibited EGF-induced proliferation and cell growth in presence of fibronectin but enhanced EGF-mediated growth of breast epithelial cells in absence of fibronectin. These findings showed that the levels of fibronectin and IGFBP-3 within breast cancers may determine tumor response to EGFR-targeting therapy. [45]

Reference

[1] Pankov R, Yamada KM . Fibronectin at a glance. *Journal of cell science* 2002; 115 : 3861–3.

[2] Williams CM, Engler AJ, Slone RD, Galante LL, Schwarzbauer JE. Fibronectin expression modulates mammary epithelial cell proliferation during acinar differentiation. *Cancer research* 2008; 68 (9): 3185–92.

[3] Waalkes S, Atschekzei F, Kramer MW, Hennenlotter J, Vetter G, Becker JU, Stenzl A, Merseburger AS, Schrader AJ, Kuczyk MA, Serth J. Fibronectin 1 mRNA expression correlates with advanced disease in renal cancer. *BMC Cancer*. 2010; 10:503.

[4] Tang NH, Chen YL, Wang XQ, Li XJ, Wu Y, Zou QL, Chen YZ. N-terminal and C-terminal heparin-binding domain polypeptides derived from fibronectin reduce adhesion and invasion of liver cancer cells. *BMC Cancer*. 2010;10:552.

[5] Sengupta S, Nandi S, Hindi ES, Wainwright DA, Han Y, Lesniak MS. Short hairpin RNA-mediated fibronectin knockdown delays tumor growth in a mouse glioma model. *Neoplasia*. 2010; 12(10):837-47.

[6] Allen MD, Vaziri R, Green M, Chelala C, Brentnall AR, Dreger S, Vallath S, Nitch-Smith H, Hayward J, Carpenter R, Holliday DL, Walker RA, Hart IR, Jones JL.Clinical and functional significance of α9β1 integrin expression in breast cancer: a novel cell-surface marker of the basal phenotype that promotes tumor cell invasion. *J. Pathol.* 2011; 223(5): 646-58.

[7] Ryschich E, Khamidjanov A, Kerkadze V, Büchler MW, Zöller M, Schmidt J. Promotion of tumor cell migration by extracellular matrix proteins in human pancreatic cancer. *Pancreas*. 2009; 38(7):804-10.

[8] Meng XN, Jin Y, Yu Y, Bai J, Liu GY, Zhu J, Zhao YZ, Wang Z, Chen F, Lee KY, Fu SB. Characterization of fibronectin-mediated FAK signaling pathways in lung cancer cell migration and invasion. *Br. J. Cancer*. 2009; 101(2):327-34.

[9] Mitra AK, Sawada K, Tiwari P, Mui K, Gwin K, Lengyel E. Ligand-independent activation of c-Met by fibronectin and α(5)β(1)-integrin regulates ovarian cancer invasion and metastasis. *Oncogene*. 2011; 30(13):1566-76.

[10] Kaomongkolgit R, Manokawinchoke J, Sanchavanakit N, Pavasant P, Sumrejkanchanakij P. Fibronectin supports TNF-alpha-induced osteopontin expression through beta1 integrin and ERK in HN-22 cells. *Arch. Oral Biol.* 2010; 55(2):101-7.

[11] Maity G, Fahreen S, Banerji A, Roy Choudhury P, Sen T, Dutta A, Chatterjee A. Fibronectin-integrin mediated signaling in human cervical cancer cells (SiHa). *Mol. Cell Biochem*. 2010; 336(1-2):65-74.

[12] Sen T, Dutta A, Maity G, Chatterjee A. Fibronectin induces matrix metalloproteinase-9 (MMP-9) in human laryngeal carcinoma cells by involving multiple signaling pathways. *Biochimie*. 2010; 92(10):1422-34.

[13] Li D, Ding J, Wang X, Wang C, Wu T. Fibronectin promotes tyrosine phosphorylation of paxillin and cell invasiveness in the gastric cancer cell line AGS. *Tumori.* 2009; 95 (6):769-79.

[14] Lee M, Park JJ, Lee YS. Adhesion of ST6Gal I-mediated human colon cancer cells to fibronectin contributes to cell survival by integrin beta1-mediated paxillin and AKT activation. *Oncol. Rep*. 2010; 23(3):757-61.

[15] Ding J, Li D, Wang X, Wang C, Wu T. Fibronectin promotes invasiveness and focal adhesion kinase tyrosine phosphorylation of human colon cancer cell. *Hepatogastroenterology*. 2008; 55(88):2072-6.

[16] Dutta A, Sen T, Chatterjee A. Culture of K562 human myeloid leukemia cells in presence of fibronectin expresses and secretes MMP-9 in serum-free culture medium. *Int. J. Clin. Exp. Pathol*. 2010; 3(3):288-302.

[17] Resnitzky D, Reed SI . Different roles for cyclins D1 and E in regulation of the G1-to-S transition. *Mol. Cell. Biol.* 1995; 15 (7): 3463–9.

[18] Juengel E, Engler J, Natsheh I, Jones J, Mickuckyte A, Hudak L, Jonas D, Blaheta RA. Combining the receptor tyrosine kinase inhibitor AEE788 and the mammalian target of rapamycin (mTOR) inhibitor RAD001 strongly inhibits adhesion and growth of renal cell carcinoma cells. *BMC Cancer*. 2009; 9:161.

[19] Assoian RK, Klein EA. Growth control by intracellular tension and extracellular stiffness. *Trends Cell Biol.* 2008,18 (7): 347–52.

[20] Pan CW, Shen ZJ, Wu TT, Tang XY, Wang M, Sun J, Shao Y. Cell adhesion to fibronectin induces mitomycin C resistance in bladder cancer cells. *BJU Int*. 2009; 104 (11):1774-9.

[21] Ou JJ, Wu F, Liang HJ. Colorectal tumor derived fibronectin alternatively spliced EDA domain exerts lymphangiogenic effect on human lymphatic endothelial cells. *Cancer Biol. Ther*. 2010; 9 (3): 186-91.

[22] Cooper TP, Sefton MV. Fibronectin coating of collagen modules increases in vivo HUVEC survival and vessel formation in SCID mice. *Acta Biomater*. 2011; 7(3): 1072-83.

[23] Hayashi H, Kume T. Foxc2 transcription factor as a regulator of angiogenesis via induction of integrin beta3 expression. *Cell Adh. Migr*. 2009; 3(1):24-6.

[24] Lee MY, Huang JP, Chen YY, Aplin JD, Wu YH, Chen CY, Chen PC, Chen CP. Angiogenesis in differentiated placental multipotent mesenchymal stromal cells is dependent on integrin alpha5beta1. *PLoS One*. 2009; 4(10):e6913.

[25] Matthew SW, Hoffmann AH, Xiong Z. Mammary myofibrosarcoma: case report and literature review. *Breast J.* 2011; 17(3):300-4.

[26] Takeyama H, Kyoda S, Okamoto T, Manome Y, Watanabe M, Kinoshita S, Uchida K, Sakamoto A, Morikawa T. The expression of sialic fibronectin correlates with lymph node metastasis of thyroid malignant neoplasm. *Anticancer Res.* 2011; 31(4):1395-8.

[27] Eissa S, Zohny SF, Zekri AR, El-Zayat TM, Maher AM. Diagnostic value of fibronectin and mutant p53 in the urine of patients with bladder cancer: impact on clinicopathological features and disease recurrence. *Med. Oncol.* 2010; 27(4):1286-94.

[28] Castellani P., Viale G., Dorcaratto A., Nicolo G., Kaczmarek J., Querze G., Zardi L. The fibronectin isoform containing the ED-B oncofetal domain: a marker of angiogenesis. *Int. J. Cancer.* 1994; **59**(5):612–8.

[29] Santimaria M., Moscatelli G., Viale G.L., Giovannoni L., Neri G., Viti F., Leprini A., Borsi L., Castellani P., Zardi L., Neri D., Riva P. Immunoscintigraphic detection of the ED-B domain of fibronectin, a marker of angiogenesis, in patients with cancer. *Clin. Cancer Res.* 2003; **9**(2): 571–9.

[30] Berndorff D., Borkowski S., Moosmayer D., Viti F., Muller-Tiemann B., Sieger S., Friebe M., Hilger C.S., Zardi L., Neri D., Dinkelborg L.M. Imaging of tumor angiogenesis using 99mTc-labeled human recombinant anti-ED-B fibronectin antibody fragments. *J. Nucl. Med.* 2006; **47**(10):1707–16.

[31] Pini A., Viti F., Santucci A., Carnemolla B., Zardi L., Neri P., Neri D. Design and use of a phage display library. Human antibodies with subnanomolar affinity against a marker of angiogenesis eluted from a two-dimensional gel. *J. Biol. Chem.* 1998; **273** (34): 21769–76.

[32] Tarli L., Balza E., Viti F., Borsi L., Castellani P., Berndorff D., Dinkelborg L., Neri D., Zardi L. A high-affinity human antibody that targets tumoral blood vessels. *Blood.* 1999; **94**(1):192–8.

[33] Ruiz-Garcia E, Scott V, Machavoine C, Bidart JM, Lacroix L, Delaloge S, Andre F. Gene expression profiling identifies Fibronectin 1 and CXCL9 as candidate biomarkers for breast cancer screening. *Br. J. Cancer.* 2010; 102(3):462-8.

[34] Nam JM, Onodera Y, Bissell MJ, Park CC. Breast cancer cells in three-dimensional culture display an enhanced radioresponse after coordinate targeting of integrin alpha5beta1 and fibronectin. *Cancer Res.* 2010; 70 (13): 5238-48.

[35] Emanuel SL, Engle LJ, Chao G, Zhu RR, Cao C, Lin Z, Yamniuk AP, Hosbach J, Brown J, Fitzpatrick E, Gokemeijer J, Morin P, Morse BA, Carvajal IM, Fabrizio D, Wright MC, Das Gupta R, Gosselin M, Cataldo D, Ryseck RP, Doyle ML, Wong TW, Camphausen RT, Cload ST, Marsh HN, Gottardis MM, Furfine ES. A fibronectin scaffold approach to bispecific inhibitors of epidermal growth factor receptor and insulin-like growth factor-I receptor. *MAbs.* 2011; 3(1):38-48.

[36] Sakai K, Nakamura T, Matsumoto K, Nakamura T. Angioinhibitory action of NK4 involves impaired extracellular assembly of fibronectin mediated by perlecan-NK4 association. *J. Biol. Chem.* 2009; 284(33): 22491-9.

[37] Schwarz RE, Awasthi N, Konduri S, Caldwell L, Cafasso D, Schwarz MA. Antitumor effects of EMAP II against pancreatic cancer through inhibition of fibronectin-dependent proliferation. *Cancer Biol. Ther.* 2010; 9(8):632-9.

[38] Huang CY, Liang CM, Chu CL, Peng JM, Liang SM. A fibrillar form of fibronectin induces apoptosis by activating SHP-2 and stress fiber formation. *Apoptosis.* 2010; 15(8):915-26.

[39] Wang S, Du W, Zhang H, Wulan T, Zhang Y, He Y, Yang Y, Liu S, Zhang Z, Wang J. Biological characteristics and antitumor activity of CIK cells activated by recombinant human fibronectin for human lung cancer cell lines in vitro. *Zhongguo Fei Ai Za Zhi.* 2010; 13(4):277-81.

[40] Huijbers EJ, Ringvall M, Femel J, Kalamajski S, Lukinius A, Abrink M, Hellman L, Olsson AK. Vaccination against the extra domain-B of fibronectin as a novel tumor therapy. *FASEB J.* 2010; 24(11):4535-44.

[41] She S, Xu B, He M, Lan X, Wang Q. Nm23-H1 suppresses hepatocarcinoma cell adhesion and migration on fibronectin (coated dishes) by modulating glycosylation of integrin beta1. *J. Exp. Clin. Cancer Res.* 2010; 29:93.

[42] Sabatino M, Kim-Schulze S, Panelli MC, Stroncek D, Wang E, Taback B, Kim DW, Deraffele G, Pos Z, Marincola FM, Kaufman HL.Serum vascular endothelial growth factor and fibronectin predict clinical response to high-dose interleukin-2 therapy. *J. Clin. Oncol.* 2009; 27 (16): 2645-52.

[43] Jerhammar F, Ceder R, Garvin S, Grénman R, Grafström RC, Roberg K. Fibronectin 1 is a potential biomarker for radioresistance in head and neck squamous cell carcinoma. *Cancer Biol. Ther.* 2011; 10(12): 1244-51.

[44] Thomas F, Holly JM, Persad R, Bahl A, Perks CM. Fibronectin confers survival against chemotherapeutic agents but not against radiotherapy in DU145 prostate cancer cells: involvement of the insulin like growth factor-1 receptor. *Prostate.* 2010; 70(8):856-65.

[45] McIntosh J, Dennison G, Holly JM, Jarrett C, Frankow A, Foulstone EJ, Winters ZE, Perks CM. IGFBP-3 can either inhibit or enhance EGF-mediated growth of breast epithelial cells dependent upon the presence of fibronectin. *J. Biol. Chem.* 2010; 285(50):38788-800.

In: Fibronectin
Editors: James Beattie

ISBN 978-1-62257-298-4

Chapter XIV

Fibronectin and Brain Tumors

Mariella Caffo, Gerardo Caruso*, Valeria Barresi, Marcello Passalacqua, Filippo F. Angileri and Francesco Tomasello
Neurosurgical Clinic, Department of Neurosciences
Psychiatry and Anaesthesiology, °Department of Human Pathology,
University of Messina School of Medicine, Messina, Italy

Abstract

Brain tumors are difficult to treat. Intracranial tumors infiltrate surrounding normal brain tissue, forming secondary foci which results in local compression and degradation of normal tissue. The extracellular matrix (ECM) plays a pivotal role in numerous cellular functions during normal and pathological processes, such as differentiation, apoptosis, tumor invasion and migration. Increased expression of several classes of proteases has been positively correlated with the ability of gliomas, meningiomas, and neurinomas to form such foci. Production of several ECM components has been shown to be altered in intracranial tumors. The unique BMs and ECMs of the tumor perivasculature prevent formation of brain tumor metastases to tissue outside the central nervous system (CNS). Proteolysis of cell-matrix interactions are tightly controlled by tumor cells, which must maintain a substratum upon which to move. Glioma invasion has been interpreted by numerous authors in terms of interaction between neoplastic cells and ECM. Critical factors include the synthesis and deposition of ECM components by tumour and mesenchymal cells, the release of ECM-degrading activities for remodelling interstitial spaces, the expression of adhesion molecules (matrix receptors on gliomas cell surfaces that specifically recognize and adhere to ECM components), and the effects of cell-matrix interactions on the behaviour of glioma cells. These phenomena result in changes in cell-cell and cell-matrix interactions, chemotaxis, and cellular migration. ECM modification aids the loss of contact inhibition that allows tumour cells to freely

* Address Correspondence to: Gerardo Caruso, M.D. Neurosurgical Clinic, Department of Neurosciences, Psychiatry and Anaesthesiology, A.O.U. Policlinico "G. Martino", Via Consolare Valeria, 1, 98125 Messina, Italy. Phone: ++39-090-2217167, fax: ++39-090-693714. E-mail: gcaruso@unime.it.

migrate and invade surrounding tissues. In particular, glioblastoma (GBM) cells follow defined anatomical structures, such as basement membrane of blood vessels and of the subependimal space, glia limitans externa, and/or parallel and intersecting nerve fiber tracts in the white matter and even at an early stage of tumour progression, colonies of neoplastic cells are frequently found in the perivascular spaces. In recent years several ECM components have been proposed as possible key molecules for the invasive phenotype of GBM, including fibronectin, laminin and collagen type IV. Fibronectin is a glycoprotein found in ECM as aggregates or fibrils. FN has many biological functions including cellular adhesion, migration and invasion and mediates a variety of adhesive events by binding to fibrinogen/fibrin, collagen, heparin sulphate, and hyaluronic acid. FN is found at the gliomesenchymal junction of tumours and in tumour-associated blood vessel, and focally within and around gliomas cells *in situ. In vitro* FN is expressed by GBM cell lines. Glioma cells express FN receptors, cluster around FN *in vivo,* and migrate in response to FN in vitro. Of several ECM tested, GBM cells migrated most efficiently on FN. Migration along FN positive mesenchymal cells may lead to gathering of gliomas cells in the perivascular regions and along the meninges. *In vitro* migration studies FN could be important for clarifying the processes underlying *in situ* glioma invasion along cerebral vessels and tumor spread beneath the glia limitans. A promising strategy could be an *in vivo* invasion model where glioma cells are able to diffusely infiltrate brain tissue and the expression of ECM components, matrix receptors, and other invasion-relevant proteins can be tightly and specifically regulated.

Abbreviations

BM	Basement Membrane
CNS	Central Nervous System
ECM	Extracellular matrix
FN	Fibronectin
GBM	Glioblastoma Multiforme
MMP	Matrix Metalloproteinases
RGD	Arg-Gly-Asp
VEC	Vascular Endothelial Cells

1. Introduction

The prognosis of patients with malignant glioma remains extremely poor despite surgery and improvements in radio- and chemo-therapies. Malignant gliomas account for 45% of all primary CNS tumors and 77% of all malignant primary CNS tumors (Luis et al. 2007). The median survival after diagnosis of GBM is still only 50 weeks, and less than 2% of patients survives three years post-diagnosis (Senger et al. 2003). Surgical treatment followed by adjuvant therapies such as radiotherapy and chemotherapy represents the classical strategy in brain tumors management. Aggressive treatment modalities have extended the median survival from 4 months to 1 year but the survival is often associated with significant impairment in the quality of life. Improvements in surgical technique including intra-operative mapping of the brain offer some benefits in prolonging survival. Almost all cases of malignant gliomas recur after initial therapy. In patients with recurrent GBM, the 6-months progression-free survival is only 21% after treatment with temozolomide (Yung et al. 2000). Most patients with recurrent malignant glioma do not survive for 1 year after diagnosis of recurrent disease (Wong et al. 1999). Cerebral gliomas show a unique pattern of invasion and with rare exceptions do not metastasize outside of the brain. Glioma invasion is well depicted by three passages such as adhesion of glioma cells to proteins of the surrounding ECM mediated by cell adhesion receptors, proteases secretion by glioma cells which locally degrade the ECM components, and migration of glioma cells into the newly created space through the ECM.

FN is a member of a family of glycoproteins that show many biological functions, including normal cell adhesion, growth, and migration. FN is involved in many cellular processes, including tissue repair, embryogenesis, blood clotting, and cell migration/adhesion. Both decreased expression and elevated degradation of FN have been shown to be responsible for some of the morphological changes observed in tumors and tumor-derived cell lines (Hynes 1990). As a peripheral protein, FN mainly acts as a bridge to link the cell surface and ECM. Therefore, the absence or reduction of FN in tumor cells may reduce the adhesion between tumor cells and matrix components, and decrease the matrix's control of cell differentiation, proliferation, and migration. As a major protein in blood and as component of the wound provisional matrix, plasma FN contributes to tissue repair and neuronal survival following cerebral ischemia (Sakai et al. 2001). FN exists in two main forms: 1) as an insoluble glycoprotein dimer that serves as a linker in the ECM, and 2) as a soluble disulphide linked dimer found in the plasma. The plasma form is synthesized by hepatocytes, and the ECM form is made by fibroblasts, chondrocytes, endothelial cells, macrophages, as well as certain epithelial cells. FN can also serve to organize cellular interaction with the ECM by binding to different components of the ECM and to membrane-bound FN receptors on cell surfaces. FN is found at the gliomesenchymal junction of tumours and in tumour-associated blood vessel, and focally within and around glioma cells in situ. Glioma cells express FN receptors, cluster around FN *in vivo*, and migrate in response to FN in vitro. FN may also combine with the fibrinogens which are exuded due to the vessel reactions with the surrounding cancer, to promote the formation of new matrix in tumor providing a condition for the seeding and proliferation of the invading tumor cells. Meanwhile, FN interacting with collagen may displace and release the metalloproteinase-2 bound to collagens, which may play a critical role in tumor invasion. In this chapter we will describe the structure, function and role of FN in brain tumors.

2. Glioma Progression and Extracellular Matrix

Cerebral gliomas show a unique pattern of invasion and with rare exceptions do not metastasize outside of the brain. How invasive glioma cells survive in the setting of invasion, evading immune detection, thwarting cytotoxic therapies, and deferring commitment to proliferation, remains unknown. Invading glioma cells normally migrate to distinct anatomical structures. These structures include the basement membrane (BM) of blood vessels, the subependymal space, the glial limitans externa, and parallel and intersecting nerve fiber tracts in the white matter. Glioma invasion is well depicted by three passages such as adhesion of glioma cells to proteins of the surrounding ECM mediated by cell adhesion receptors, proteases secretion by glioma cells which locally degrade the ECM components, and migration of glioma cells into the newly created space through the ECM. The proteolytic degradation of the BM and of ECM is mediated by proteases, such us the MMPs, secreted by tumor and stromal cells (Idbaih and Ducray 2008). MMPs play an important role in human brain tumor invasion, probably due to an imbalance between the production of MMPs and tissue inhibitor of metalloproteinases-1 (TIMP-1) by the tumor cells. MMP-1 is the crucial enzyme able to initiate breakdown of the interstitial collagen types; in this way it activates the other MMPs which allows the glioma cells to infiltrate normal brain tissue.

ECM is a complex network of different collagens, proteoglycans, hyaluronic acid, laminin, fibronectin, and many other glycoproteins, including proteolytic enzymes involved in degradation and remodelling of the ECM (Chintala et al. 1996). Extracellular matrix exists in two forms: interstitial matrix that fills in the intercellular space and the more specialized BM, which is a thin sheet of extracellular matrix underlying the epithelium. ECM provides the microenvironment for the cells and serves as a tissue scaffold, guiding cell migration during embryonic development and wound repair. Beyond that, it also functions as the repository and modulator of growth factors and cytokines, and therefore is responsible for transmitting environmental signals to the cells. ECM proteins such as FN, laminins or collagens form distinct protein networks that show tissue-specific variation in composition and architecture. Cell responses to contact with these networks depend on the cell's repertoire of ECM receptors. Connections from the matrix through these receptors determine the organization of cytoskeletal structures and the localization and activation of signaling molecules leading to unique tissue specific cell functions. Both decreased expression and elevated degradation of FN have been shown to be responsible for some of the morphological changes observed in tumors and tumor-derived cell lines. Critical factors, in glioma invasion, include the synthesis and deposition of ECM components by glioma and mesenchymal cells, the release of ECM-degrading activities for remodeling interstitial spaces, the presence of adhesion molecules. ECM modification aids the loss of contact inhibition allowing tumor cells to freely migrate and invade the surrounding tissues. Various ECM components, such us hyaluronan, vitronectin, fibronectin, tenascin-C, osteopontin are regulated within the tumor stroma and at the advancing edge of the tumor within brain parenchyma. Changes in these ECM components are felt to modulate brain tumor growth, proliferation and invasion, although specific interactions and exact mechanisms are unknown. Cell adhesion is the binding of the cells to each other and to the ECM through cell adhesion molecules such as integrins, selectins, cadherins, the Ig (immunoglobulin) superfamily and lymphocyte homing receptors. Cell adhesion mediates cell attachment, migration, and signalling to and from the extracellular matrix. Adhesion complexes include focal adhesions, adherens junctions, tight junctions, desmosomes, hemi-desmosomes, and gap junctions. The best characterized adhesion molecules are the integrins and the best characterized adhesion complexes are focal adhesions. The extracellular ligands that anchor these adhesions include laminin, fibronectin, vitronectin, and various collagens. Focal adhesions can be considered both as sensors of force and as sites that originate cytoskeletal forces through anchored actin-microfilament bundles. Focal adhesions have many SH2-containing components (Src kinases, PI3K, SHP-2), as well as many tyrosine-phosphorylated molecules (focal adhesion kinase paxillin, tensin, caveolin).

Therefore tyrosine phosphorylation of these sites, which is induced by integrin clustering, growth factor stimulation, or applied force, could potentially stimulate focal adhesion assembly. Integrins are heterodimers of α- and β - subunits that bind to the ECM through large extracellular domains and connect intracellularly to the actin cytoskeleton filaments. Integrins regulate many aspects of the cell behavior including survival, proliferation, migration and differentiation. Integrins of the β1 and αv classes are expressed on different cell types, including neurons, glial cells, meningeal and endothelial cells. β2 integrins are specifically expressed by leukocytes and they are found on microglia and on infiltrating leukocytes within the CNS, αvβ3 integrin is associated with elevated levels of its ligands and with vitronectin in human gliomas (Gladson 1996). Antibody block against the β1 subunit inhibits adhesion, motility and invasion of glioma cells *in vitro* (Rooprai et al. 1999). Furthermore, decreasing the in vivo expression of the β1 subunit by an antisense strategy in the intracranial C6 glioma model leads to an inhibition of glioma associated angiogenesis at the invasive edge of the tumors (Bello et al. 2001). Down-regulated *β*1 integrin protein levels in vivo probably affect interactions of glioma cells with ECM components, leading to reduced migration along vascular basement membranes.

3. Fibronectin Structure

Fibronectin structure is rod-like, composed of three different types of homologous, repeating modules, types I, II, and III. These modules comprise functional domains that mediate interactions with other ECM components, with cell surface receptors and with FN itself. Twelve type I modules make up the amino-terminal and carboxy-terminal region of the molecule, and are involved mainly in fibrin and collagen binding. Only two type II modules are found in FN. They are instrumental in binding collagen. The most abundant module in FN is type III, which contains the RGD-FN receptor recognition sequence along with binding sites for other integrins and heparin. Depending on the tissue type and/or cellular conditions, the FN molecule is made up of 15-17 type III modules. In addition, there is a module that does not fall into any of these categories, called IIICS. This module, along with EDB and EDA is regulated through alternative splicing of FN pre-mRNA. FN molecules can form two disulphide bridges at their carboxy-termini, producing a covalently-linked dimer.

The type I module of FN1 is made up of ~45 amino acids, and is found in the amino-terminal and carboxy terminal regions of the full-length protein. Each FN1 module, like all FN modules, is constructed of antiparallel β-sheets. Two antiparallel beta strands make up the top sheet, which folds over a bottom sheet, composed of three antiparallel β-strands. These sheets interact through hydrophobic bonding that considerably stabilizes the module. One such interaction can be visualized in the hydrophobic stacking of two highly conserved aromatic residues on opposite sheets. In addition to hydrophobic interactions between β-sheets, each F1 module is stabilized by two disulfide bridges, one between opposite sheets and one between β-strands of the same sheet. Interactions between adjacent modules are important to FN structure. Such interactions prevent rotation between these FN1 modules. The first five FN1 modules of the N-terminal domain are important in ECM assembly. The FN1 modules are involved in fibrin binding, an important event during the formation of blood clots. Finally, FN1 modules that surround the two FN2 modules assist in collagen binding.

The FN2 module is found in a wide array of proteins, and is highly conserved. So, though the crystal structure of the FN2 module of FN has not yet been determined, it is likely that it will be very similar to the FN2 module of PDC109b protein. The FN2 module includes four beta strands in its core and is ~60 amino acids in length. Two antiparallel beta strands make up the top sheet, which folds perpendicularly over the two antiparallel beta strands of the bottom sheet (Pickford et al. 1997). Again, the folding of these beta sheets is stabilized by two disulphide linkages. A cavity in the side of the module made up of aromatic and hydrophobic residues, in conjunction with a charged serine, is thought to bind specific leucine and/or isoleucine sequences in collagen.

Type III modules make up a large part of the FN protein with each module being ~90 amino acids in length. In fact, over 67 proteins have been identified as containing FN3 modules. Like other FN modules, type III module cores are made up of overlapping beta sheets. In FN3 modules, the top sheet contains four antiparallel beta strands and the bottom sheet is three-stranded (Dickinson et al. 1994). Unlike FN2 or FN1 modules, disulphide bridges do not stabilize FN3 structure. Instead, this occurs solely through hydrophobic interactions in the module core. The FN3 module contains the RGD, the FN receptor (integrin) binding motif. The RGD protrudes from the protein and is exposed to solution. The FN3 module alone does not bind to the integrin receptor with the same affinity as intact fibronectin. This observation is due to the fact that a synergistic site on the FN3 module participates in RGD binding to integrin. In addition to binding various forms of integrins, FN3 modules can participate in heparin binding. One of the most intriguing features of F3 modules in FN is that their structure is modulated through alternative pre-mRNA splicing. Whereas FN1 and FN2 modules are encoded by one exon each, two exons encode most FN3 modules. Through the alternative splicing of certain type III exons, multiple fibronectin mRNAs can arise from a single gene. Splicing occurs primarily in three regions: EDB (also called EIIIB or EDII)), EDA (also called EIIIA or EDI), and IIICS (also called V or variable) (Liao et al. 2002; Chang et al. 2004). EDB and EDA can each be spliced out through exon skipping, while IIICS is spliced through a more complex mechanism, which is species and tissue specific. In humans there are potentially 20 different forms of FN, most arising from alternative splicing of the IIICS module. The IIICS module encodes two integrin binding sites, and is involved in heparin binding, so by changing the structure of this molecule through alternative splicing, the function of FN can be changed.

3.1. Fibrillogenesis

FN matrix assembly is a complex process involving binding domains and repeating modules from all regions of FN which participate in interactions with cell surface receptors and with other FN molecules. ECM proteins such as FN, laminins, or collagens form distinct protein networks that show tissue specific variation in composition and architecture. Cell responses to contact with these networks depend on the type of matrix as well as on the cell's repertoire of ECM receptors. Connections from the matrix through these receptors determine the organization of cytoskeletal structures and the localization and activation of signalling molecules leading to unique tissue-specific cell functions. Many of the receptors that integrate matrix information belong to the integrin superfamily of transmembrane proteins (Sonnenberg 1993; Schwartz et al. 1995). At the cell surface, integrins are intimately involved in constructing fibrillar matrix networks from soluble secreted matrix proteins. Stepwise assembly of FN matrix proceeds from initiation through elongation to stabilization. Each step involves multiple FN domains for binding to cells, to other FN molecules, and to ECM components like collagen or heparin. Initiation depends on interactions between the cell-binding domain of FN and cell surface integrin receptors - usually α5β1 binding to the RGD and synergy site. As a dimeric ligand, FN can induce receptor clustering by binding to two intergrins. Integrin–FN clusters then create a relatively high local concentration of FN helping to promote self association and also to connect FN to the actin cytoskeleton through receptor cytoplasmic domain interactions. Although integrins other than α5β1 bind to FN, only a few of these have been shown to support FN matrix assembly (Yang and Hynes 1996; Wu 1997). Thus, multiple integrin-specific assembly mechanisms are available to the cell. Once initiated, fibrils are propagated by continual addition of FN dimers to growing multimers. Many of the FN binding sites involved in fibril elongation have been mapped and include the amino-terminal repeats I_{1-5}, the first one or two type III repeat, and the carboxyterminal heparin domain (Ingham et al. 1997; Hocking et al. 1994). The sites in repeat III_1 and the heparin domain appear to be cryptic as native III_1 does not bind FN, and fragments containing the heparin domain were not as effective as small heparin domain peptides in blocking assembly. Interestingly, all of the known FN binding sites interact with the amino-terminal domain of FN and it has been proposed that ternary complexes formed through co-operative interactions of repeats I_{1-5}, III_1, and either the cell or heparin domain (Bultmann et al. 1998) could be important during fibrillogenesis. The III_1 repeat is not essential in this process as recombinant FN lacking only this repeat is assembled into a perfectly normal fibrillar matrix. The presence of multiple binding sites for I_{1-5} along FN might provide alternative assembly mechanisms and fibril structures. As fibrils mature, they are gradually converted into a detergent-insoluble form yielding stable matrix. This model suggests that the compact form of soluble FN is an inactive state and that binding to integrins leads to conformational changes that allow FN self-association at the cell surface. The rate of formation of detergent-insoluble matrix is also dependent on the strength of FN–integrin interactions as α5β1 integrin contact with both the RGD sequence and the synergy site is needed for assembly. The mechanism for development of detergent insolubility is not understood but could be due to disulfide bond formation through FN's inherent disulfide isomerase activity or by formation of highly stable protein-protein interactions (Langenbach and Sottile 1999). The latter process could be similar to formation of amyloid fibrils and involve exchange of β-strands between the β-sheets of folded type III repeats in adjacent molecules. Litvinovich *et al.* showed that partial unfolding of repeat III_9 was sufficient to allow formation of thermostable multimers (Litvinovich et al. 1998). Perhaps this type of mechanism is used to form FN into the highly stable, detergent-insoluble structures characteristic of mature matrix.

FN matrix assembly is a cell-dependent process. Newly secreted FN dimers bind to transmembrane integrin receptors. Integrins link FN to the actin cytoskeleton through interactions between their cytoplasmic domains and cytoskeletal-associated proteins. On the outside of the cell, interactions between FN and integrins promote FN–FN association and fibril formation perhaps by inducing conformational changes in bound FN. The thickness of FN fibrils varies substantially from 10 to 1000 nm in diameter (Singer 1979). Extended FN molecules are about 3 nm in diameter (Leahy et al. 1996) indicating that fibrils probably range from a few to several hundred FN molecules across. The organization of FN within the fibrils is not well understood but is probably determined by intermolecular interactions involving some subset of the following domains. Function-blocking antibodies against either the cell binding domain of FN or α5β1 integrin inhibit FN fibril formation. Although the RGD sequence is required in the initiation steps of matrix assembly, later stages do not rely on this sequence. Mutant recombinant FN lacking the RGD sequence is able to co-assemble into fibrils with native, RGD+FN (Schwarzbauer 1991). The synergy site appears to play a modulatory role because matrix levels are drastically reduced but not ablated with FN lacking this sequence (Sechler et al. 1997). However, this defect can be overcome by stimulation of integrin activity with Mn2+. The effect of integrin activity levels may be explained by a recent study showing that the synergy site affects the III_{9-10} inter-domain interface which can modulate accessibility of the RGD motif (Altroff et al. 2004).

Along the FN molecule, at least four FN binding domains have been identified but only the site within the N-terminal assembly domain (I_{1-5}) is essential. FN–FN interactions are important for fibril formation but they can also participate in intramolecular interactions that keep soluble FN in a compact form, unable to form fibrils in solution. In fact, the interaction between III_{2-3} and III_{12-14} appears to function in this capacity. III_{12-14} also contains a heparin and syndecan proteoglycan binding site (Tumova et al. 2000). Fragments containing this domain inhibit FN assembly, possibly through dual blockade of FN–FN and FN–cell interactions (Bultmann et al. 1998). FN matrix assembly is reduced by antibodies that recognize the III_{1-2} region, by FN fragments and peptides derived from this region, and by recombinant FN lacking this domain (Chernousov et al. 1991; Sechler et al. 2001). Functionally, this small domain binds to proteoglycans and has at least two FN binding sites. In solid phase binding assays, native III_2 binds intact FN and 70 kDa fragment (Sechler et al. 2001). III_1 can also interact with FN and 70 kDa fragment but the interaction requires heat-denaturation of the module. Thus, the FN binding site in III_1 is cryptic but it is also dispensable since normal matrix amounts and organization are formed from a recombinant FN lacking III_1 (Sechler et al. 2001). In contrast, deletion of the entire III_{1-2} domain significantly reduced matrix assembly and the defect was primarily in formation of DOC-insoluble matrix (Sechler et al. 2001). Some insights into the role of the cryptic site in III_1 have come from studies of III_{1-C}, a fragment containing the Cterminal two-thirds of III_1 and also called anastellin (Morla and Ruoslahti 1992). Incubation of soluble FN with anastellin results in FN aggregation into fibrillar structures (Morla et al. 1994). Thus, this fragment can induce FN polymerization in the absence of cells, probably by increasing exposure of FN binding sites, although the precise mechanism has not been determined. FN polymers formed by incubation with anastellin have biological activity including enhanced adhesive properties (Morla et al. 1994). In vivo studies indicate that this polymeric FN can inhibit tumor growth, angiogenesis and metastasis in mouse models (Yi and Ruoslahti 2001).

3.2. Molecular Events in Fibrillogenesis

FN fibril assembly begins with FN binding to integrin receptors and this step initiates the molecular events required for subsequent FN-FN interactions and formation of multimeric structures. Conformational changes are induced during fibril assembly. Initially, integrin-bound FN is diffusely localized at the cell surface. Receptor clustering by dimeric FN helps to organize FN into short fibrils. The cytoplasmic domains of integrins become associated with the actin cytoskeletal network forming a connection that is essential for FN matrix assembly. The contractility of the cytoskeleton mediated by actin-myosin interactions contributes to FN fibril formation. The dynamic interactions beetween FN, integrins, and the cytoskeleton require intracellular adapter and signaling proteins (Miranti and Brugge 2002). Complexes containing a5h1 integrin, focal adhesion kinase (FAK), vinculin, and paxillin were found in the adhesion complexes that form during fibroblast attachment to a 3-dimensional ECM (Cukierman et al. 2001). Focal adhesions provide firm cell attachment to 2-dimensional substrates and contain avh3 integrin, paxillin and vinculin whereas fibrillar adhesions are enriched in tensin and integrin a5h1. Fibrillar adhesions are generated by movement of ligated FN and a5h1 integrin from the ends of stationary focal adhesions toward the cell center along actin stress fibers (Ohashi et al. 2002). This movement may stretch FN molecules and thus promote FN–FN interactions. Cell surface receptors also participate in FN matrix assembly. Concomitant binding of integrins and syndecan-4, a transmembrane heparan sulfate proteoglycan, to FN stimulates Rho GTPase and FAK activities (Wilcox-Adelman et al. 2002) initiating signals that enhance fibril assembly (Ilic et al. 2004).

Many different approaches are being applied to decipher the mechanisms for FN conformational changes during assembly. Recently, fluorescence resonance energy transfer has been used to monitor changes in FN folding during matrix assembly (Baneys et al., 2002). FN extension plays an important role in exposing FN binding sites and allowing intermolecular interactions required for FN incorporation into fibrils. FN binding sites are distributed along the length of the molecule and several sites have more than one binding partner which could lead to different alignments of FN molecules within the fibrils. FN–FN interactions initially give rise to short fibrils around the cell periphery. As assembly proceeds, longer and thicker fibrils are formed. Two types of interactions are involved in this process. Incorporation of FN requires interactions, but FN is also recruited from solution into existing fibrils through homophilic interaction tht are independent of the RGD integrin binding site. As the fibrils grow, they are converted from a detergent DOC-soluble form into DOC-insoluble material, an irreversible process. This conversion provides a useful method for monitoring and quantifying matrix assembly through analysis of DOC-insoluble FN. FN matrix assembly is regulated by intracellular signaling pathways. Binding to integrins, organization of the actin cytoskeleton, stimulation of cell contractility, and activation of kinase cascades all contribute to the assembly process (Wierzbicka-Patynowski and Schwarzbauer 2003). The extracellular environment also plays an important role in matrix regulation. Three-dimensional effects on cell morphology, composition of adhesion complexes, cell growth rates, and FN matrix assembly have been documented (Cukierman et al. 2001; Wenk et al. 2000). In addition, cell attachment to adhesive protein substrates other than FN can dramatically affect FN fibril formation (Bae et al. 2004). FN-null cells growing on FN or laminin substrates were able to assemble exogenously-supplied FN into fibrils. In contrast, a vitronectin substrate suppressed the ability of these cells to assemble FN matrix. The mechanism of this suppressive effect of vitronectin is unclear but most likely involves crosstalk among vitronectin and fibronectin receptors including a5h1 and avh3 integrins and/or heparan sulfate proteoglycans. Clearly, proper integration of extracellular signals with active intracellular pathways plays an important role in the initiation and progression of FN matrix assembly. FN matrix is relatively stable and its turnover under normal fibroblast culture conditions is very slow. Continual FN polymerization is critical for the deposition and maintenance of cell surface FN fibrils as well as the ECM proteins type I collagen and thrombospondin-1. Removal of FN or inhibition of polymerization results in rapid loss of FN matrix through a caveolin-1-dependent process (Sottile and Chandler 2005). Whether the receptors that assemble FN also participate in its internalization and turnover remains to be determined. FN matrix is not a static structure. Fibrils are significantly stretched and loss of contact with cells results in contraction to up to one-quarter the stretched length. In addition, cell movements deform the fibrils without breaking them. A lot of changes within fibrils that might provide elasticity have been postulated (Erickson 2002), but the exact mechanisms remain to be determined.

4. Function and Role

Fibronectin is found in blood and tissues of vertebrates and in cultures of variuos adherent cells such as fibroblasts, astroglial cells, myoblasts and certain epithelial cells and primitive mesenchymal cells. The fibronectins of cultured human fibroblasts, human plasma and human cerebrospinal fluid are immunologically and biochemically very similar. The biological role of FN is poorly understood, and the protein has been studied mostly under cell culture conditions. Cellular FN is assembled into the ECM, an insoluble network that separates and supports the organs and tissues of an organism. It plays a crucial role in wound healing. Along with fibrin, plasma fibronectin is deposited at the site of injury, forming a blood clot that stops bleeding and protects the underlying tissue. As repair of the injured tissue continues, fibroblasts and macrophages begin to remodel the area, degrading the proteins that form the provisional blood clot matrix and replacing them with a matrix that more resembles the normal, surrounding tissue. Fibroblasts secrete proteases, including matrix metalloproteinases, that digest the plasma fibronectin, and then the fibroblasts secrete cellular FN and assemble it into an insoluble matrix. Fragmentation of FN by proteases has been suggested to promote wound contraction, a critical step in wound healing. Fragmenting FN further exposes its V-region, which contains the site for α4β1 integrin-binding. These fragments of FN are believed to enhance α4β1 integrins-expression allowing cells to adhere to and forcefully contract the surrounding matrix. FN is necessary for embryogenesis, and inactivating the gene for FN results in early embryonic lethality. FN is important for guiding cell attachment and migration during embryonic development. In mammalian development, the absence of FN leads to defects in mesodermal, neural tube, and vascular development. Similarly, the absence of a normal FN matrix in developing amphibians causes defects in mesodermal patterning and inhibits gastrulation.

Normal adherent cells, unlike malignant cells, deposit fibronectin in a pericellular matrix form, and the protein seems to be involved in cell adhesion. Studies of FN in differentiating cultured cells and during embryogenesis suggest that its expression is closely linked to developmental changes. FN is thought to promote adhesion and spreading of cells by linking them to collagen substrates. The splicing segments of FN can also promote cell migration and angiogenesis. FN matrix could stimulate cell proliferation in several different FN-deficient cell lines (Sechler and Schwarzbauer 1998). Assembly of recombinant $FNDIII_{1-7}$ that exhibits a fibril organization inhibited cell growth by specifically blocking G1 to S phase progression through reduced activation of a focal adhesion kinase (FAK) signaling pathway. FN matrix can also be manipulated through the use of function altering peptides and antibodies (III_{1-C}) corresponding to part of the first type III repeat can reduce FN matrix and proliferation rates of several different cell types (Bourdoulous et al. 1998). Disruption of FN matrix corresponded with a reduction in activity of cyclin E–cdk2, which would directly affect cell cycle progression. Collaborations between FN matrix and growth factors can regulate cell differentiation and migration (Nagel and Winklbauer 1999). Intracellular signals can also impinge on FN assembly through integrin cytoplasmic domain interactions and via an unknown pathway involving the von Hippel Lindau tumor suppressor protein (Ohh et al. 1998). Fibronectin exists in both plasma and cellular forms, which have a similar amino acid composition and differ noticeably only in their solubility and apparent molecular weights. Each FN domain is responsible for one of fibronectin's binding functions. Alterations in adhesion and cytoskeletal organization have long been associated with the loss of fibronectin from oncogenically transformed cells. How complex biological events such as cellular proliferation, change in cell shape, and cellular differentiation are mediated through the interaction between ECM macromolecules and their specific receptors is speculative. Changes in FN expression are closely linked with cell differentiation and organogenesis (Salmeron-Sanchez et al. 2011). It has been proposed that FN has a structural role in positioning and anchoring cellls *in vivo* as well as in cell adhesion *in vitro.* FN matrix assembly is regulated by intracellular signaling pathways. Bindings to integrins, organization of the actin cytoskeleton, stimulation of cell contractility, and activation of kinase cascades all contribute to the assembly process. The extracellular environment also plays an important role in matrix regulation. Cell behaviour within a three-dimensional matrix differs from cells on a two-dimensional substrate indicating that matrix dimensionality influences cellular functions. Three-dimensional effects on cell morphiligy, composition of adhesion complexes, cell growth rates, and FN matrix assembly have been documented (Cukierman et al. 2001). Integration of extracellular signals with active intracellular pathways plays an important role in the initiation and progression of FN matrix assembly.

5. Fibronectin and Brain Tumor

5.1. Cerebral Gliomas

The ECM affects numerous functions and processes in the brain. During brain development, the ECM modulates the migration of glial and neuronal precursor cells, guides axonal growth cones, synapse formation and cell proliferation. In normal reparative processes, as well as in primary brain neoplasia, the ECM is present in increased amounts and undergoes remodeling. The roles played by the ECM in tissue repair and neoplastic transformation are complex and only now beginning to be uncovered. Evidence is accumulating that changes in ECM components modulate brain tumor growth, proliferation and invasion through a variety of different mechanisms. The ECM in the brain is in two-way communication with the cells located within it. Integrins on the cell surface create the primary link in this communication system, binding to and transducing signals between the cytoskeleton on the inside of the cell and the ECM on the outside of the cell. Ligation of cell surface receptors by the ECM can modulate cell motility, migration, proliferation, integrin expression patterns, and the production of both ECM itself and proteases, which modify the ECM. The ECM also binds to ands modulates the activity of several growth factors (Tysnes and Mahesparan 2001). ECM component, including fibrillary collagens, laminin anf fibronectin, are limited to vascular and perivascular areas within the brain (Gladson 1999).

The lack of fibrillar collagen and these other matrix components gives the brain a softer consistency compared with other systemic organs.

Fibronectin in brain tumor was first described by Sherbet et al (Sherbet et al. 1982) and has been primarily associated with cell migration upon binding with integrin receptors (Ohnishi et al. 1998). Migration along FN-positive mesenchymal cells may lead to glioma cells gathering in the perivascular region and around the tumor. FN is an important component in the ECM of gliomas, largely in the vessel wall (Lin et al. 2002; Caffo et al. 2004). Basement membrane of the blood vessels stained positively for type IV collagen in glioblastoma, anaplastic astrocytoma, low-grade gliomas and normal brain tissue. Along with staining in the basement membrane, intense FN staining was observed in the ECM of the glial cells in GBM and anaplastic astrocytoma sections with only a very weak staining was seen in low-grade gliomas and none in normal brain (Chintala et al. 1996). In order to confirm these in vivo results, they cultured established GBM cell lines and stained them for the various components. Interestingly, FN staining occurred in SNB19, U251, UWR1, and UWR2 cells. UWR1 cells showed intense staining for FN compared to other cell lines. In glioblastoma, anaplastic astrocytoma, low-grade gliomas, and normal brain, FN showed a positive staining that was mainly confined to the vessel walls. However, especially in GBM a fine and irregular network-like staining for FN was observed in the cells as well as in the ECM deposited by these cells. Finally they found that type IV collagen, FN, and laminin were localized mostly in the basement membrane of the vasculature in glioblastoma, anaplastic astrocytoma, low-grade gliomas, and in normal brain. However, positive staining was observed in the cells only in glioblastoma sections (Chintala et al. 1996). The immunohistochemical data obtained suggests that type IV collagen, FN and laminin are integral components of the basement membrane of the vasculature. In addition they observed positive staining for the ECM components in the cells as well as the autologous ECM, which strongly suggests that GBM cells produce type IV collagen, FN, and laminin both in vitro and in vivo. Expression of FN has been characterized in human GBM and in a number of astrocytoma and glioblastoma cell lines. In a study of pediatric GBM, a marked positive staining of FN in walls of small and medium size vessels was demonstrated. FN expression was seen in perivascular sheets, where it looked like a fine irregular network (Caffo et al. 2004). *In vitro* experiments have proven that components of ECM, LN and FN, can strongly stimulate the migration of glioma cells, which occurs after glioma cells express the relevant surface receptors (Mahesparan et al. 1999). Knott et al. found that when normal brain tissues were invaded by glioma, ECM components such as LN, FN, and collagen type IV may are available and that tumor cells may express specific integrins, depending on the change of the interior environment, to interact with these ECM components, and enhance tumor cell invasion (Knott et al. 1998). Although there are many *in vitro* and *in vivo* models for FN-promoted invasion and transmigration (Mahesparan et al. 2003), the uniqueness of these findings may be attributed to interaction between tumor cells and vascular endothelial cells (VECs), as changes of tumoral microenvironment molecules can affect tumor status and progression. Recently it has been demonstrated that FN expression is correlated with glioma migration and glioma malignancy (Lin et al. 2010). The expression of FN and other extracellular matrix components by glioma is directly linked to the expression of integrins on cell surface. Ritchie et al. showed that in glioma cells with induced ectopic expression of nuclear factor kB, there was an increase in the expression of both FN and integrins (Ritchie et al. 2000). More recently, FN has been reported to mantain extracellular matrix rigidity to promote structural rigidity, motility, and proliferation of established glioma cell lines *in vitro* (Ulrich et al. 2009). However, the functional mechanism of FN-mediated regulation of glioma progression has remained unknown. In a recent study it has been showed the mechanism that involves the maintenance of integrin ß1 FN receptors in glioma cells required by FN to induce a Src kinase-dependent survivin activity that promotes brain tumor proliferation. The FN binds to membrane-spanning integrin receptors and plays an important role in cell signaling, in defining cellular shape, in mobility, and in regulating the cell cycle. The inhibition of FN expression in glioma cells, using short hairpin RNA-mediated silencing of gene expression, delayed cell proliferation *in vitro*. This delayed growth is explained, in part, by the observed reduced expression of integrin ß1 FN receptor, which was restored by the inhibition of proteosomal activity (Sengupta et al. 2010). Interestingly, a lot of studies have shown that microglial adhesion and morphology vary greatly between different ECM substrates. In the light of this ECM regulation of microglial behavior and evidence showing that integrin expression in some cell types is regulated by the ECM, Milner et al. evaluated the influence of individual ECM molecules (FN, laminin. vitronectin) on microglial integrin expression. The authors have demonstrated, in cultured mouse tissue, microglial MHC class I expression was significantly increased by FN. They additionally showed high serum levels of FN and vitronectin, but not laminin, and therefore, breakdown of the blood-barrier associated with CNS inflammation or trauma will greatly increase the availability of these proteins within the CNS parenchyma, and thus influence the behaviour of microglial cells (Milner and Campbell 2003). These observation showed that FN and vitronectin promote microglial activation which results in the amoeboid morphology, increased expression of MHC class I, and increased of the α4ß1, α5ß1 and Mac-1 integrins. FN and vitronectin expression within the CNS is also increased in epilepsy, multiple sclerosis, and tumor formation. These observations suggest that regulation of the expression and distribution of FN and vitronectin within CNS is an important factor influencing the degree of microglial activation.

5.2. Meningiomas

Meningiomas are derived from the meninges, which in turn originate from prechordal plate mesenchyme, mesoderm, and neural crest elements. Meningiomas grow by expansion, leading to compression of adjacent structures. A major mesenchymal function of meningioma cells is their ability to elaborate extracellular matrix protein (Ng and Wong 1993). The ECM components is significant in meningiomas because they increase meningial dimensions through the increased production of ECM and, on the other hand, the proliferation of cells characteristically associated with neoplasia needs is controlled by signals and interactions with ECM. It is also possible those meningiomas may secrete their own ECM proteins, and that their invasive behavior in vitro is modulated, at least in part, by ECM proteins. Our studies have evidenced a significant immunoreactivity for extracellular matrix proteins, mainly localized at level of neoplastic cells, basement membrane of vessels, in surrounding fibrillar tissue, in perivascular sites and in micro capillaries and arterioles. These features are associated with a remarkable neoangiogenetic activity. The development of a network of blood vessels is a prerequisite for the local expansion of tumor colonies and, moreover, the presence of new capillaries around and within the growing tumor mass provides numerous entry sites to the circulation for disseminating neoplastic cells (Caffo et al. 2008). Although meningiomas are generally not considered malignant, they can show infiltration into the surrounding bone and dura mater. In a few cases local microinvasion may ever occur into the brain tissue. At microscopical level, outgrowths between the subdural spaces are often observed. Fingerlike extensions at the interface between tumor and normal brain might be an expression of invasion or an atypical aspect. Invasion and angiogenesis are the newest therapeutic targets in brain tumors, as they involve local endothelial cell invasion, proliferation and migration in response to many molecular substances produced by the brain tumors.

FN is expressed in 98-100% of the extracellular matrix of meningiomas, including vessel walls, whorls formations, and psammoma bodies (Kochi et al. 1983; Ng and Wong 1993; Caffo et al. 2008). FN expression in meningiomas seems to be rather complex, since this antigen has been described both in meningiomatous cells and in the ECM surrounding these cells. Kochi et al. reported FN immunohistochemical expression in the vessel walls and neoplastic cells in cases of meningioma, suggesting its potential role in the pathogenesis of meningioma (Kochi et al. 1983). Bellon et al. found that FN was present over the whole width of the vessel wall in meningiomas, in the surrounding fibrillar network and, in some cases, in the perivascular connective tissue. Moreover, Bellon found fine fibrillar deposits between meningiomatous cells in five cases of transitional meningioma (Bellon et al. 1985). FN immunohistochemical expression outside the lesion may suggest that the presence of FN precedes tumor development, as well as suggesting that this protein has a tendency to migrate (Caffo et al. 2008).

Caffo et al., in a series of secretory meningiomas, found that the vessel walls, the surrounding fibrillar tissue and the perivascular sites were markedly stained; moreover, they also observed immunopositivity in meningothelial cells, in pseudopsammoma bodies within the neoplastic proliferation and in the surrounding edematous parenchymal area (Caffo et al. 2008). In a recent study, FN was overexpressed in 86 cases of meningioma, with cytoplasmic staining, which was more intense in the whorl formations and psammomatous bodies than in the vessel walls. The percentage of benign meningiomas overexpressing high levels of FN was slightly less than the percentage of transitional meningiomas expressing high levels of FN. In general, the expression level of all markers examined in this study, tended to be higher in the atypical meningiomas than in the benign meningiomas (Das et al. 2003).

Conclusion

The treatment of gliomas represents one of the most challenging areas in neurosurgery and oncology. Malignant gliomas involve multiple aberrant signaling pathways and the blood brain barrier restricts the delivery of many chemotherapeutic agents. The effectiveness of the actual chemotherapeutic approach, multimodal targeted therapies, differently than the other malignant extra cerebral tumors, remains very modest in gliomas (Caruso et al. 2011). The cellular microenvironment, specified by cell-cell and ECM interactions, is crucial for regulating the development, maintenance, and functionality of cells within tissues. In cancer, the microenvironment plays a central role in disease progression. In cancers, histologically normal tissues adjacent to tumors have been found to harbor the same genetic abnormalities as the tumor cells (Forsti et al. 2001), further supporting the notion that the microenvironment contributes to oncogenic phenotypes. Therefore, to understand cancer progression, the effects of molecules that direct cell responses within the microenvironment must be identified. The ECM governs a wide variety of cellular processes, including proliferation, differentiation, and migration, and changes in the ECM microenvironment occur during both developmental and disease processes (Aszodi et al. 2006; Engbring and Kleinman 2003). Within the matrix the contributions of multiple components and the co-ordination of their assembly have yet to be deciphered. Proteases, ECM components, the adhesion cell molecules, and their related signalling pathways play an important role in glioma cells migration and invasion. Each component can be identified as potential target for glioma therapy. In addition, the increased expression of FN and vitronectin documented in various pathological states is probably caused by activated microglia. The ECM substrates, which might act in a feed-forward fashion to further activated neighbouring microglia and lead to a chronic inflammatory state within the CNS. In addition, laminin and FN induce different patterns of matrix metalloproteinase expression within cultured microglial cells (Milner et al. 2003), which is additional evidence that the ECM, via integrin signalling, is able to regulate gene expression and subsequent behaviour in microglial cells. Microglia could attach strongly to the basal lamina around blood vessels and then migrate along vessels to reach their target site of inflammation. Several line of evidence suggests that failure of effective regeneration in the CNS might be attributable to an inadequate microglial/macrophage response after injury (Perry et al. 1987). Taken together several evidences suggests that one way of promoting more effective therapeutic approach in the CNS would be to stimulate the macrophage/ microglia response.

The distribution of FN in the adult human central and peripheral nervous system, as well as in cerebral gliomas, meningiomas and Schwannomas has only recently been studied. The evidence indicates that mature human glial and neuronal cells do not express FN, but it is found in the human nervous tissue at sites of contact between the neuroectoderm and mesenchyme (Paetau et al. 1980). Recently it has been reported that expression of ECM proteins such as FN, laminin and various types of collagen has been augmented both *in vitro and in* vivo in human malignant gliomas compared to benign brain tumors (Kochi et al. 1983; Bellon et al. 1985; Chintala et al. 1996). In a recent study, the biological activity of laminins in neoplastic tissues was shown to promote cells motility and growth, angiogenesis and metastases.

The results showed that downregulation of laminin α4 chain expression inhibits gliomas invasion *in vitro* and *in vivo*, decreasing tumor cell motility and invasion *in vitro* and *in vivo*. Targeting the laminin α4 chain, using antisense oligonucleotide, might have novel potential as an anti-invasive therapy for patients with malignant gliomas (Nagato et al. 2005). Khazenzon et al. have demonstrated the effects of antisense inhibition of laminin-8 expression in glioma therapy through an in vitro model using human glioblastoma multiforme cell lines M059K and U-87MG co-cultured with normal human brain microvascular endothelial cells. During glioma progression the expression of capillary BM laminins containing α4 chain switches from the predominant laminin-9 into lanimin-8. Laminin-8 and its receptor, integrins α3β1 and α6β1, are important for the functioning of endothelial cell BMs. Laminin-8 (α4 and β1 chains) plays an important role in glioma cell invasiveness, in combination with other proteins associated with glioma progression, such as tenascin-C, MMP-2 and MMP-9. In fact, up-regulation of laminin-8 expression in both glioma cells and glioma-adjacent capillary endothelial cells may reduce glial cell adhesion and enhance migration, which is necessary for local tumor invasiveness. The result of this study was a significant reduction of invasion of co-cultures through Matrigel and through the use of morpholino oligos against α4 and β1 chains of laminin-8 (Khazenzon et al. 2003). Glioma cell invasion is a complex and multifactorial mechanism involving a large series of molecules and cell-cell and cell-ECM interactions.

These processes allow individual tumor cells to migrate and invade the surrounding brain parenchyma. ECM, proteases, cell adhesion molecules and their related signaling pathways show an important role in glioma cell migration and invasion. It is evident that the complexity and cross-talk between signal transduction pathways limits the potential efficacy of targeting a single receptor or molecule (Caruso et al. 2011).

New and improved culture systems and recombinant protein expression will probably be needed to dissect, in detail, the assembly of multicomponent networks. Identification of additional matrix components, possibly through genetic approaches, will help to characterize the extracellular environment. Considering the influence of ECM proteins on function and behaviour of tumor cells more physiological *in vitro* models are necessary to evaluate the functional mechanisms of migration. New and emerging technologies will help to solve the very intriguing problems of tissue-specific matrix formation and the effects of matrix composition and architecture on cell behaviour.

References

Altroff, H; Schlinkert, R; van der Walle, CF; et al. Interdomain tilt angle determines integrin-dependent function of the ninth and tenth FIII domains of human fibronectin. *J. Biol. Chem.,* 2004 279, 55995-56003.

Aszodi, A; Legate, KR; Nakchbandi, I; Fassler, R. What mouse mutants teach us about extracellular matrix function. *Annu. Rev. Cell Dev. Biol.,* 2006 22, 591–621.

Bae, E; Sakai, T; Mosher, DF. Assembly of exogenous fibronectin by fibronectin-null cells is dependent on the adhesive substrate. *J. Biol. Chem.,* 2004 279, 35749-35759.

Baneys, G; Baugh, L; Vogel, V. Fibronectin extension and unfolding within cell matrix fibrils controlled by cytoskeletal tension. *Proc. Natl. Acad. Sci. USA,* 2002 99, 5139-5143.

Bello, L; Francolini, M; Marthyn, P; et al. Alpha(v)beta3 and alpha(v)beta5 integrin expression in glioma periphery. *Neurosurgery,* 2001 49, 380-389.

Bellon, G; Caulet, T; Cam, Y; et al. Immunohistochemical localization of macromolecules of the basement membrane and extracellular matrix of human gliomas and meningiomas. *Acta Neuropathol.,* 1985 66, 245–52.

Bourdoulous, S; Orend, G; MacKenna, DA; Pasqualini, R; Ruoslathi, E. Fibronectin matrix regulates activation of RHO and CDC42 GTPases and cell cycle progression. *J. Cell Biol.,* 1998 143, 267-276.

Bultmann, H; Santas, AJ; Pesciotta Peters, DM. Fibronectin fibrillogenesis involves the heparin II binding domain of fibronectin. *J. Biol. Chem.,* 1998 273, 2601-2609.

Caffo, M; Caruso, G; Galatioto, S; et al. Immunohistochemical study of the extracellular matrix proteins laminin, fibronectin and type IV collagen in secretory meningiomas. *J. Clin. Neurosci.,* 2008 115, 806-811.

Caffo, M; Germanò, A; Caruso, G; et al. An immunohistochemical study of extracellular matrix proteins laminin, fibronectin and type IV collagen in paediatric glioblastoma multiforme. *Acta Neurochir.*,2004 146,1113-1118.

Caruso G, Caffo M, Raudino G, Alafaci C, Tomasello F (2011) New Therapeutic Strategies in Gliomas Treatment, Brain Tumors - Current and Emerging Therapeutic Strategies, Ana L. Abujamra (Ed.), ISBN: 978-953-307-588-4, InTech.

Chang, ML; Chen, JC; Alonso, CR; Kornblihtt, AR; Bissell, DM. Regulation of fibronectin splicing in sinusoidal endothelial cells from normal or injured liver. *Proc. Natl. Acad. Sci. USA,* 2004 101, 18093-18098.

Chernousov, MA; Fogerty, FJ; Koteliansky, VE; Mosher, DF. Role of the I-9 and III-1 modules of fibronectin in formation of an extracellular fibronectin matrix. *J. Biol. Chem.,* 1991 266, 10851-10858.

Chintala, SK; Sawaya, R; Gokaslan, ZL; et al. Immunohistochemical localization of extracellular matrix proteins in human glioma, both in vivo and in vitro. *Cancer Lett.,* 1996 101, 107-114.

Cukierman, E; Pankov, R; Stevens, DR; Yamada, KM. Taking cell-matrix adhesions to the third dimension. *Science,* 2001 294, 1708-1712.

Das, A; Tan, W; Smith, DR. Expression of extracellular matrix markers in benign meningiomas. *Neuropathology,* 2003 23:275-281.

Dickinson, CD; Gay, DA; Parello, J; Ruoslahti, E; Ely, KR. Crystals of the cell-binding module of fibronectin obtained from a series of recombinant fragments differing in length. *J. Mol. Biol.,* 1994 238:123-127.

Engbring, J.A. and Kleinman, H.K. (2003) The basement membrane matrix in malignancy. *J. Pathol. 200*, 465–470.

Erickson, H.P. (2002) Stretching fibronectin. *J. Muscle Res. Cell Motil. 23,* 575-580.

Forsti, A; Louhelainen, J; Soderberg, M; Wijkstrom, H; Hemminki, K. Loss of heterozygosity in tumour adjacent normal tissue of breast and bladder cancer. *Eur. J. Cancer,* 2001 37:1372–1380.

Khazenzon, NM; Ljubimova, AV; Lakhter, AJ; et al. Antisense inhibition of laminin-8 expression reduces invasion of human gliomas in vitro. *Mol. Cancer Ther.,* 2003 2:985-994.

Knott, JC; Mahesparan, R; Garcia-Cabrera, I; et al. Stimulation of extracellular matrix components in the normal brain by invading glioma cells. *Int. J. Cancer,* 1998 75:864-872.

Kochi, N; Tani, E; Morimura, T; Itagaki, T. Immunohistochemical study of fibronectin in human glioma and meningioma. *Acta Neuropathol.*, 1983 59: 119-126.

Gladson, C.L. (1996) Expression of integrin alpha v beta 3 in small blood vessels of glioblastoma tumors. *J. Neuropathol. Exp. Neurol. 55*, 1143-1149.

Gladson, C.L. (1999). The extracellular matrix of gliomas: Modulation of cell function. *J. Neuropathol. Exp. Neurol. 58,* 1029-1040.

Hocking, DC; Sottile, J; McKeown-Longo, PJ. Fibronectin's III-1 module contains a conformation-dependent binding site for the amino-terminal region of fibronectin. *J. Biol. Chem.,* 1994 269:19183-19191.

Hynes, R.O. (1999) The dynamic dialogue between cells and matrices: implications of fibronectin's elasticity. *Proc. Natl. Acad. Sci. USA 96*, 2588–2590.

Idbaih, A; Ducray, F; Sierra del Rio, M; et al. Therapeutic application of noncytotoxic molecular targeted therapy in gliomas: growth factor receptors and angiogenesis inhibitors. *Oncologist,* 2008 13:978-992.

Ilic, D; Kovacic, B; Johkura, K; et al. FAK promotes organization of fibronectin matrix and fibrillar adhesions. *J. Cell Sci.,* 2004 117:177-187.

Ingham, KC; Brew, SA; Huff, S; Litvinovich, SV. Cryptic self-association sites in type III modules of fibronectin. *J. Biol. Chem*., 1997 272:1718-1724.

Langenbach, K.J. and Sottile, J. (1999) Identification of protein-disulfide isomerase activity in fibronectin. *J. Biol. Chem. 274,* 7032-7038.

Leahy, DJ; Aukhil, I; Erickson, HP. A crystal structure of a four-domain segment of human fibronectin encompassing the RGD loop and synergy region. *Cell,* 1996 84:155-164.

Liao, YF; Gotwals, PJ; Koteliansky, VE; Sheppard, D; Van De Water, L. The EIIIA segment of fibronectin is a ligand for integrins alpha 9beta 1 and alpha 4beta 1 providing a novel mechanism for regulating Cell adhesion by alternative splicing. *J. Biol. Chem*., 2002 277: 14467-14474.

Lin, ZX; Zhang, PF; Jiang, CZ; et al. The exploration of relationship between immunohistochemistry of LN, FN, P53 and timorous invasion microecosystem in human brain glioma. *Acta Anatomica Sinica,* 2002 33:360-365.

Lin, ZX; Yang, LJ; Huang, Q; Fu, J. Activated vascular endothelia regulate invasion of glioma cells through expression of fibronectin. *Chin. Med. J.,* 2010 123:1754-1761.

Litvinovich, SV; Brew, SA; Aota, S; et al. Formation of amyloid-like fibrils by self-association of a partially unfolded fibronectin type III module. *J. Mol. Biol.,* 1998 280:245-258.

Luis DN, Ohgaki H, Wiestler OD, Cavenee WK (2007) World Health Organization Classification of Tumours of the Central Nervous System. IARC, Lyon.

Mahesparan, R; Tyseness, BB; Read, TA; et al. Extracellular matrix induced cell migration from glioblastoma biopsy specimens *in vitro*. *Acta Neuropathol*., 1999 97:231-239.

Mahesparan, R; Read, TA; Lund-Johansen, M; et al. Expression of extracellular matrix component in a highly infiltrative *in vivo* glioma model. *Acta Neuropathol*., 2003 105:49-57.

Milner, R. and Campbell, I.L. (2003) The extracellular matrix and cytokines regulate microglial integrin expression and activation. *J. Immunol. 170,* 3850-3858.

Miranti, C.K. and Brugge, J.S. (2002) Sensing the environment: a historical perspective on integrin signal transduction. *Nat. Cell Biol. 4,* E83-E90.

Morla, A. and Ruoslahti, E. (1992) A fibronectin self-assembly site involved in fibronectin matrix assembly: reconstruction in a synthetic peptide. *J. Cell Biol. 118*, 421-429.

Morla, A; Zhang, Z; Ruoslahti, E. Superfibronectin is a functionally distinct form of fibronectin. *Nature,* 1994 367:193-196.

Nagato, S; Nakagawa, K; Harada, H; et al. Downregulation of laminin α4 chain expression inhibits glioma invasion in vitro and in vivo. *Int. J. Cancer,* 2005 117:41-50.

Nagel, M. and Winklbauer, R. (1999) Establishment of substratum polarity in the blastocoels roof of the Xenopus embryo. *Development 126*, 1975-1984.

Ng, H.K. and Wong, A.T. (1993) Expression of epithelial and extracellular matrix protein markers in meningiomas. *Histopathology 22*, 113-125.

Ohashi, T; Kiehart, DP; Erickson, HP. Dual labeling of the fibronectin matrix and actin cytoskeleton with green fluorescent protein variant. *J. Cell Sci*., 2002 115:1221-1229.

Ohh, M; Yauch, RL; Lonergan, KM; et al. The von Hippel-Lindau tumor suppressor protein is required for proper assembly of an extracellular fibronectin matrix. *Mol. Cell,* 1998 1:959-968.

Ohnishi, T; Hiraga, S; Izumoto, S; et al. Role of fibronectin-stimulated tumor cell migration in glioma invasion *in vivo*: clinical significance of fibronectin and fibronectin receptor expressed in human glioma tissue. *Clin. Exp. Metastasis,* 1998 16:729-741.

Paetau, A; Mellstrom, K; Vaheri, A; Haltia, M. Distribution of a major connective tissue protein, fibronectin, in normal and neoplastic human nervous system. *Acta Neuropathol.*, 1980 51:47-51.

Perry, VH; Brown, MC; Gordon, S. The macrophage response to central and peripheral nerve injury. *J. Exp. Med.,* 1987 165:1218-1223.

Pickford, AR; Potts, JR; Bright, JR; Phan, I; Campbell, ID. Solution structure of a type 2 module from fibronectin: implications for the structure and function of the gelatin-binding domain. *Structure,* 1997 5:359– 370.

Ritchie, CK; Giordano, A; Khalili, K. Integrin involvement in glioblastoma multiforme: possible regulation by NF-kappaB. *J. Cell Physiol.,* 2000 184:214-221.

Rooprai, HK; Vanmeter, T; Panou, C; et al. The role of integrin receptors in aspects of glioma invasion in vitro. *Int. J. Dev. Neurosci.,* 1999 17:613-623.

Sakai, T; Johnson, KJ; Murozono, M; et al. Plasma fibronectin supports neuronal survival and reduces brain injury following transient focal cerebral ischemia but is not essential for skin-wound healing and hemostasis. *Nat. Med.,* 2001 7:324–330.

Salmeron-Sanchez, M; Rico, P; Moratal, D; et al. Role of material-driven fibronectin fibrillogenesis in cell differentiation. *Biomaterials,* 2001 32:2099-2105.

Schwartz, MA; Schaller, MD; Ginsberg, MH. Integrins: emerging paradigms of signal transduction. *Ann. Rev. Cell Dev. Biol.,* 1995 11:549-599.

Schwarzbauer, J.E. (1991) Identification of the fibronectin sequences require for assembly of a fibrillary matrix. *J. Cell Biol. 113*, 1463-1473.

Sechler, JL; Rao, H; Cumiskey, AM; et al. A novel fibronectin binding site required for fibronectin fibril growth during matrix assembly. *J. Cell Biol.,* 2001 154:1081-1088.

Sechler, J.L. and Schwarzbauer, J.E. (1997) Coordinated regulation of fibronectin fibril assembly and actin stress fiber formation. *Cell Adhes. Commun. 4,* 413-424.

Sechler, J.L. and Schwarzbauer, J.E. (1998) Control of cell cycle progression by fibronectin matrix architecture. *J. Biol. Chem. 273*, 25533-25536.

Senger, D; Cairncross, JG; Forsyth, PA. Long-term survivors of glioblastoma: statistical aberration or important unrecognized molecular subtype? *Cancer J.*, 2003 9:214-221.

Sengupta, S; Nandi, S; Hindi, ES; et al. Short hairpin RNA-mediated fibronectin knockdown delays tumor growth in a mouse glioma model. *Neoplasia,* 2010 12:837-847.

Sherbet, GV; Tidle, ME; Stidolph, S. Fibronectin in human astrocytomas grown in tissue culture. *Anticancer Res.*, 1982 2:251-254.

Singer, I.I. (1979) The fibronexus: a transmembrane association of fibronectin-containing fibers and bundles of 5 nm microfilaments in hamster and human fibroblasts. *Cell 16,* 675-685.

Sonnenberg, A. (1993) Integrins and their ligands. *Curr. Top Micro Immun. 184,* 7-35.

Sottile, J. and Chandler, J. (2005) Fibronectin matrix turnover occurs through a caveolin-1-dependent process. *Mol. Biol. Cell 16*, 757-768.

Tumova, S; Woods, A; Couchman, JR. Heparan sulfate chains from glypican and syndecans bind the Hep II domain of fibronectin similarly despite minor structural differences. *J. Biol. Chem.*, 2000 275:9410-9417.

Tysnes, B.B. and Mahesparan, R. (2001) Biological mechanisms of glioma invasion and potential therapeutic targets. *J. Neurooncol. 53*, 129-147.

Ulrich, TA; de Juan Pardo, EM; Kumar, S. The mechanical rigidity of the extracellular matrix regulates the structure, motility, and proliferation of glioma cells. *Cancer Res.,* 2009 69:4167-4174.

Wenk, MB; Midwood, KS; Schwarzbauer, JF. Tenascin-C suppresses Rho activation. *J. Cell Biol.*, 2000 150:913-920.

Wierzbicka-Patynowski, I. and Schwarzbauer, J.F. (2003) The ins and outs of fibronectin matrix assembly. *J. Biol. Chem. 116*, 3269-3276.

Wilcox-Adelman, SA; Denhez, F; Goetnick, PF. Syndecan-4 modulates focal adhesion kinasephosphorylation. *J. Biol. Chem., 2002* 277:32970-32977.

Wong, ET; Hess, KR; Gleason, MJ; et al Outcomes and prognostic factors in recurrent glioma patients enrolled onto phase II clinical trials. *J. Clin. Oncol.,* 1999 17:2572-78.

Wu, C. (1997) Roles of integrins in fibronectin matrix assembly. *Histol. Histopathol. 12*, 233-240.

Yang, J.T. and Hynes, R.O. (1996) Fibronectin receptor functions in embryonic cells deficient in $\alpha_5 v_1$ integrin can be replaced by αv integrins. *Mol. Biol. Cell 7*, 1737-1748.

Yi, M. and Ruoslahti, E. (2001) A fibronectin fragment inhibits tumor growth, angiogenesis, and metastasis. *Proc. Natl. Acad. Sci. USA 98*, 620-624.

Yung, WK; Albright, R; Olson, J; et al. A phase II study of temozolamide vs. procarbazine in patients with glioblastoma multiforme at first relapse. *Br. J. Cancer,* 2000 83:588-93.

Index

A

Abraham, 109
access, 214, 274
accessibility, 104, 160, 308
acetic acid, 145
acetone, 144, 145, 151
acetylcholine, 192, 193, 195, 199, 203, 205
acid, 5, 11, 22, 28, 35, 37, 38, 44, 46, 63, 65, 68, 98, 105, 108, 109, 115, 117, 144, 145, 146, 147, 159, 169, 177, 179, 188, 196, 200, 202, 228, 248, 252, 262, 291, 302, 304
acidic, 237
action potential, 193
activation state, 220
activity level, 308
acute leukemia, 18
acute renal failure, 263
additives, 8
adenocarcinoma, 16, 89, 128, 134, 135, 294
adhesions, 15, 40, 79, 82, 84, 85, 86, 96, 101, 105, 106, 107, 109, 111, 118, 135, 137, 140, 161, 163, 164, 167, 168, 172, 178, 180, 182, 272, 277, 304, 309, 317, 318
adhesive properties, 308
adipose, 96
adolescents, 18
ADP, 23, 156, 168
adsorption, 114
adults, 4, 198, 229
advancement, 9
advantages, 12
adverse conditions, 113
aetiology, 227, 233, 269
AFM, 108
age, 36
aggregation, 4, 11, 93, 117, 160, 176, 268, 308
aggressive behavior, 8, 285
aggressiveness, 131, 135, 276
agonist, 41, 60, 193
AIDS, 220
albumin, 269
albuminuria, 270, 279
alcoholic liver disease, 15
aldosterone, 278
allergy, 94
alters, 33, 80, 86, 201, 289
American Heart Association, 151, 175
amino, 3, 5, 22, 28, 35, 36, 38, 44, 46, 50, 60, 63, 66, 98, 105, 107, 108, 126, 130, 159, 169, 177, 179, 215, 216, 254, 259, 274, 277, 305, 306, 307, 311, 317
amino acid, 3, 28, 35, 36, 38, 44, 46, 50, 60, 63, 98, 126, 130, 159, 169, 177, 179, 215, 216, 254, 259, 274, 277, 305, 306, 311
amino groups, 22
ammonium, 142
amniotic fluid, 10
amoeboid, 87, 313
amphibians, 310
anchorage, 8, 100, 111, 157, 161, 166, 171, 174, 182, 185, 293
anchoring, 31, 213, 214, 311
androgen, 288
anemia, 218
angiogenesis, 4, 5, 23, 29, 37, 41, 56, 67, 87, 92, 94, 97, 98, 103, 104, 113, 114, 115, 118, 120, 121, 122, 125, 129, 136, 139, 141, 152, 153, 169, 181, 183, 184, 185, 186, 212, 246, 263, 286, 290, 291, 293, 294, 297, 298, 305, 308, 310, 314, 315, 318, 320
angiogenic process, 108
angiotensin II, 242, 245, 252, 260
annealing, 84
antibody, 8, 10, 11, 12, 18, 44, 57, 59, 60, 115, 144, 146, 147, 148, 149, 151, 153, 170, 176, 181, 193,

195, 236, 243, 248, 261, 285, 287, 290, 292, 293, 294, 298
anti-cancer, 97
anticancer activity, 283
antigen, 115, 116, 120, 225, 314
antioxidant, 198
antisense, 91, 283, 287, 305, 315
antisense oligonucleotides, 91
antitumor, 294, 299
apoptosis, vii, 6, 7, 13, 26, 32, 72, 73, 88, 89, 91, 95, 98, 100, 113, 115, 119, 133, 135, 137, 155, 167, 169, 173, 174, 179, 184, 187, 208, 210, 214, 274, 292, 294, 299, 301
apoptotic mechanisms, 88
aqueous humor, 6
ARC, 99
architecture, 88, 94, 100, 109, 111, 112, 157, 176
arrest, 133, 171, 289, 295
arrests, 263
ARS, 32
arteries, 208
arterioles, 116, 117, 313
artery, 207
arthritis, 37, 44, 104, 257
articular cartilage, 32, 65
ascites, 100
aspartic acid, 35
aspirate, 147
assessment, 9, 10, 193
astrocytes, 191, 192, 193, 194, 195, 196, 197, 198, 199, 200, 201, 202, 203, 204
astrocytoma, 128, 129, 203, 312
asymptomatic, 221
ATF, 165
atherosclerosis, 14, 75, 96, 157, 176
ATP, 131
atrophy, 271, 279
attachment, 5, 6, 8, 17, 26, 28, 30, 40, 131, 138, 141, 161, 166, 167, 168, 172, 177, 178, 182, 212, 224, 229, 267, 304, 309, 310
Austria, 152
authors, 23, 24, 63
autoimmune disease, 221, 236
automata, 115
autosomal recessive, 218
avian, 67

B

bacillus, 219
background, 47, 57, 60, 61
bacteria, 3, 7, 12, 49, 62, 214, 218, 219, 241
bacteriophage, 153
bacterium, 7, 218, 219
barriers, 186, 219
basal lamina, 16, 92, 278, 315
base, 11, 79, 111, 133, 232
basement membrane, 5, 17, 41, 66, 72, 90, 92, 94, 95, 99, 133, 142, 152, 169, 220, 227, 230, 231, 232, 233, 234, 236, 239, 241, 246, 248, 250, 257, 266, 267, 268, 270, 273, 280, 286, 293, 302, 303, 305, 312, 313, 316, 317
behaviors, 8
bending, 75, 80, 84
beneficial effect, 174, 245, 246, 247
benefits, 286, 302
benign, 292, 314, 315, 317
biochemical processes, 4, 247
biochemistry, 29, 31, 178, 179, 180, 257
biological activity, 28, 29, 46, 66, 102, 108, 118, 272, 308, 315
biological fluids, 10, 228
biological processes, 99, 227
biomarkers, 246, 292, 295, 298
biomaterials, 8
biomechanics, 111
biomolecules, 21
biopsy, 149, 150, 243, 244, 318
bladder cancer, 284, 289, 291, 295, 297, 298, 317
bleeding, 310
blood, vii, 2, 3, 6, 8, 10, 12, 16, 73, 75, 92, 97, 114, 133, 141, 160, 179, 207, 208, 209, 214, 216, 218, 220, 243, 244, 268, 269, 284, 290, 291, 292, 294, 298, 302, 303, 305, 310, 312, 313, 314, 317
blood clot, vii, 3, 8, 214, 216, 303, 305, 310
blood flow, 208, 218
blood plasma, 10, 73, 75, 141, 214, 284
blood pressure, 6
blood urea nitrogen, 269
blood vessels, 6, 97, 133, 160, 179, 209, 243, 244, 290, 291, 298, 302, 303, 312, 313, 315, 317
bloodstream, 7
body fluid, 2, 214
bonding, 216, 305
bonds, 3, 7, 22, 74, 75, 80, 109, 141, 159, 213, 214, 215, 216, 230, 266, 275, 284
bone, 25, 64, 74, 101, 152, 205, 231, 264, 272, 314
brain, 6, 15, 91, 96, 113, 114, 175, 192, 193, 195, 196, 198, 200, 201, 203, 301, 302, 303, 304, 311, 312, 314, 315, 316, 317, 319
brain tumor, 301, 302, 303, 304, 311, 312, 314, 315
branching, 91, 111, 114, 115
Brazil, 207
breakdown, 36, 40, 62, 63, 183, 289, 304, 313

breast cancer, 18, 24, 26, 32, 54, 65, 120, 129, 135, 185, 283, 292, 296, 298
breast carcinoma, 53, 96, 97, 121, 128
bronchial epithelial cells, 89, 113
bubonic plague, 219
budding, 91, 131, 134
Butcher, 121

C

Ca^{2+}, 115, 130, 179
calcium, 89, 113, 185, 233, 234, 244, 277, 281
calorimetry, 24
CAM, 135, 202, 291
cancer, 2, 8, 9, 10, 12, 16, 18, 24, 26, 32, 36, 37, 43, 44, 54, 62, 63, 65, 68, 90, 96, 97, 113, 114, 120, 121, 122, 125, 128, 129, 131, 132, 133, 134, 135, 136, 137, 138, 139, 140, 141, 147, 152, 156, 166, 174, 175, 176, 178, 181, 184, 185, 186, 187, 188, 189, 282, 283, 285, 286, 287, 288, 289, 290, 291, 292, 293, 294, 295, 296, 297, 298, 303, 314
cancer cells, 24, 26, 32, 54, 90, 113, 114, 120, 121, 129, 132, 133, 138, 147, 174, 187, 188, 286, 287, 289, 292, 293, 294, 296, 298
cancer progression, 125, 128, 136, 315
cancer therapy, 184, 189, 292, 293
cancerous cells, 22
capillary, 73, 88, 89, 94, 97, 109, 113, 117, 232, 235, 236, 271, 291, 315
capsule, 230, 232
carbohydrates, 72
carboxyl, 3, 27, 215, 266
carcinogenesis, 1, 290
carcinoma, 8, 31, 53, 61, 62, 66, 67, 68, 89, 96, 101, 113, 120, 121, 122, 128, 129, 135, 136, 137, 173, 176, 187, 284, 285, 286, 287, 291, 295, 297
cartilage, 32, 65, 217
cartilaginous, 217
cascades, 127, 134, 143, 165, 174, 185, 309, 311
casein, 17
catabolism, 271
catalytic activity, 65, 167, 280
category a, 266
CBD, 22, 23, 32, 35, 48, 51, 74, 77, 86
CBP, 165
CCN2/CTGF, 29
CD8+, 294
cDNA, 14, 281
cell biology, vii, 13, 16, 25, 27, 29, 30, 31, 152, 175, 176, 177, 178, 179, 180, 181, 182, 183, 185, 186, 187, 188
cell body, 132
cell culture, 2, 8, 9, 10, 24, 26, 77, 83, 89, 98, 102, 110, 115, 177, 258, 310
cell cycle, 88, 100, 108, 152, 157, 167, 171, 174, 175, 288, 289, 310, 313, 316, 319
cell death, 12, 30, 208, 210, 242, 281, 287, 295
cell differentiation, 9, 27, 29, 174, 196, 283, 284, 303, 311, 319
cell division, 91, 156, 168, 289
cell fate, 158
cell invasion, 16, 68, 95, 131, 133, 136, 174, 219, 286, 287, 296, 312, 314, 316
cell invasiveness, 297, 315
cell line, 8, 17, 23, 26, 50, 53, 54, 61, 62, 63, 64, 89, 90, 97, 135, 137, 178, 185, 211, 240, 285, 286, 287, 292, 293, 295, 297, 299, 302, 303, 304, 310, 312, 315
cell lines, 8, 17, 26, 53, 54, 62, 63, 89, 97, 135, 137, 185, 240, 285, 286, 292, 293, 295, 299, 302, 303, 304, 310, 312, 315
cell motility, 7, 24, 32, 35, 36, 37, 39, 40, 41, 44, 56, 57, 60, 61, 62, 63, 64, 66, 67, 86, 87, 115, 161, 166, 167, 182, 311, 315
cell movement, 5, 39, 64, 91, 193, 310
cell signaling, 73, 74, 75, 78, 80, 84, 87, 88, 127, 160, 293, 313
cell surface, vii, 1, 2, 3, 5, 6, 8, 9, 16, 21, 22, 35, 38, 39, 40, 43, 46, 50, 71, 73, 74, 76, 77, 79, 80, 81, 83, 93, 94, 96, 105, 125, 126, 129, 130, 132, 134, 136, 143, 156, 158, 159, 160, 178, 209, 230, 266, 267, 268, 291, 295, 301, 303, 305, 306, 308, 309, 311, 312
cellulitis, 218
central nervous system, 194, 201, 204, 220, 221, 301
central nervous system XE "nervous system" (CNS), 201, 301
ceramic, 149
cerebellar neurons, 192
cerebral cortex, 202
cerebrospinal fluid, 310
cervical cancer, 129, 132, 137, 287, 297
Chagas disease, 220
character, 15
chemical, 80, 160, 241
chemicals, 199
chemokine receptor, 255
chemokines, 270
chemotaxis, 37, 39, 40, 42, 48, 49, 50, 51, 52, 56, 60, 61, 64, 66, 67, 69, 301
chemotherapeutic agent, 284, 296, 299, 314
chemotherapy, 137, 302
Chicago, 191
chicken, 2, 14, 28, 30, 115, 249
children, 13, 198, 202, 259

chimera, 162
CHO cells, 26, 79, 89, 132, 134, 136, 139
choline, 193
chondroitin sulfate, 142, 192, 202
chromatography, 9, 55, 56
chromosome, 2, 13, 21, 157, 158, 214, 227, 229, 249, 289
chronic kidney XE "kidney" disease, 227, 228, 256, 265, 266, 269, 275, 276, 278, 279
chronic rejection, 225
chronic renal failure, 230, 279
cilium, 115
circulation, 5, 133, 158, 237, 240, 313
CKD, 269, 271, 272
classes, 199, 301, 305
classification, 254, 278
cleavage, 108, 126, 131, 132, 136, 138, 160, 214, 239, 271, 274, 290
clinical application, 1
clinical trials, 320
cloning, 44, 63
closure, 41
clustering, 5, 76, 78, 80, 81, 127, 130, 143, 161, 162, 167, 168, 181, 182, 304, 307, 308
clusters, 82, 91, 307
CNS, 27, 191, 202, 302, 305, 313, 315
coatings, 12, 84
cobalt, 200
coding, 14
codon, 44
collaboration, 243
colon, 16, 18, 121, 283, 287, 297
colon cancer, 16, 18, 283, 287, 297
colorectal cancer, 96, 120
communication, 141, 142, 196, 200, 235, 243, 311
community, 269
compaction, 94
comparative analysis, 121
compensation, 79
compilation, 228
complement, 3, 219, 236
complexity, 2, 4, 71, 75, 79, 81, 84, 85, 91, 157, 169, 316
compliance, 73
composition, 15, 36, 72, 85, 87, 95, 98, 101, 102, 111, 158, 159, 160, 163, 178, 228, 230, 271, 304, 306, 309, 311, 316
compounds, 36, 41, 174, 194, 199, 200
compression, 301, 313
condensation, 215, 221, 289
configurations, 7
connective tissue, 15, 72, 93, 94, 95, 118, 177, 257, 268, 269, 280, 281, 314, 319
constituents, 23, 26, 63, 268, 273
consumption, 218
contact dermatitis, 94, 109
control condition, 56
control group, 6
controversial, 243, 245
convergence, 91, 183
coordination, 4, 14, 92
correlation, 6, 8, 64, 109, 120, 137, 259, 270
cortex, 231, 269, 277, 279
cost, 294
covalent bond, 109, 230
covering, 91, 148, 268
creatinine, 269
cretinism, 197
crystal structure, 103, 126, 216, 305, 318
CT, 29, 107, 115, 123, 181
cues, 72, 81, 161, 169
culture, 1, 2, 8, 9, 10, 17, 23, 26, 36, 55, 57, 58, 59, 60, 61, 65, 73, 82, 84, 87, 88, 90, 94, 96, 98, 99, 102, 110, 112, 114, 115, 122, 124, 137, 177, 193, 199, 204, 205, 248, 259, 278, 290, 292, 297, 298, 309, 316, 319
culture conditions, 90, 94, 292, 309
culture medium, 297
cycling, 198, 288, 289
cyclins, 289, 297
cyclophosphamide, 143, 262
cyclosporine, 243, 261, 262, 263, 278
cyst, 115
cysteine, 230, 238
cytokines, 5, 11, 12, 41, 56, 57, 65, 67, 81, 95, 162, 165, 169, 207, 219, 220, 241, 242, 265, 270, 271, 272, 273, 280, 304, 318
cytokinesis, 131, 134, 138
cytology, 291
cytoplasm, 7, 12, 126
cytoplasmic tail, 134, 163, 166
cytoskeleton, 4, 39, 40, 64, 67, 78, 80, 82, 83, 84, 85, 105, 107, 117, 124, 127, 143, 160, 161, 163, 167, 168, 172, 180, 183, 186, 192, 217, 258, 305, 307, 308, 309, 311, 318
cytotoxic agents, 122
cytotoxicity, 198, 199

D

database, 129, 139
deaths, 133
defects, 2, 4, 71, 92, 116, 117, 126, 140, 158, 166, 170, 177, 179, 310
defence, 241

deficiency, 115, 119, 166, 170, 198, 202
deformation, 83, 270
degradation, 68, 81, 93, 95, 118, 122, 130, 137, 141, 144, 152, 161, 162, 165, 171, 194, 195, 214, 220, 225, 227, 228, 230, 234, 235, 237, 238, 239, 241, 257, 258, 260, 265, 266, 271, 272, 274, 275, 283, 284, 293, 301, 303, 304
dehydrate, 151
denaturation, 80, 308
dendritic cell, 272
Denmark, 180
dephosphorylation, 25
deposition, 6, 11, 33, 64, 74, 75, 81, 83, 84, 91, 92, 93, 94, 95, 96, 97, 100, 101, 102, 110, 111, 123, 142, 143, 152, 157, 160, 165, 174, 179, 217, 223, 228, 230, 236, 237, 240, 241, 245, 247, 262, 265, 267, 268, 269, 270, 271, 272, 273, 274, 277, 278, 280, 301, 304, 309
deposits, 13, 91, 237, 276, 314
depth, 230
deregulation, 288
derivatives, 266
dermatitis, 94, 109, 118
dermatology, 29, 30, 188
dermis, 94, 96
desensitization, 175
destruction, 218, 233
detachment, 40, 84, 239
detectable, 83, 96
detection, 10, 17, 144, 145, 146, 147, 148, 149, 150, 153, 291, 298, 303
developing brain, 192, 199, 202
developmental change, 310
developmental process, 91, 92
diabetes, 252, 253, 255, 270
diabetic kidney disease, 251, 252, 264, 279
diabetic nephropathy, 228, 233, 234, 235, 237, 240, 246, 251, 252, 253, 256, 258, 263, 264, 281
diabetic patients, 235, 246, 251
diacylglycerol, 234
dialysis, 228, 232, 240, 258, 259, 260, 261
diffusion, 97
digestion, 43, 55, 56, 59
dilation, 207, 208
dimensionality, 311
direct action, 195, 197
direct measure, 168
directionality, 160
disease activity, 236
disease progression, 36, 71, 91, 98, 235, 257, 267, 269, 270, 271, 274, 275, 314
diseases, 13, 35, 36, 66, 75, 118, 119, 120, 166, 174, 214, 218, 220, 227, 230, 232, 233, 234, 237, 238, 240, 248, 254, 256, 264, 279, 284
disintegrin, 24
disorder, 6, 218, 265, 273
dispersion, 90
displacement, 24
dissociation, 24, 164, 237
distilled water, 151
distribution, 2, 28, 55, 66, 67, 120, 125, 134, 152, 163, 172, 179, 197, 201, 204, 208, 209, 222, 232, 282, 313, 315
diversity, 105, 138, 159, 180
DNA, 133, 157, 165, 175, 182, 236, 248, 254, 261, 289, 291, 293
DNA repair, 133
DNA sequencing, 291
DOC, 308, 309
dogs, 221, 225
domain structure, 13, 38, 43, 74
down-regulation, 143, 161, 235, 245
dressings, 11
drinking water, 197, 203
Drosophila, 205
drug delivery, 1, 7, 12, 19
drug resistance, 128, 129, 135, 173, 187, 295
drug toxicity, 270
drugs, 11, 97, 129, 143, 144, 147, 245, 247, 261, 286, 293
dura mater, 314
dwarfism, 257
dynamics, 9, 27, 64, 101, 103, 105, 111, 116, 137, 138, 143, 160, 163, 164, 167, 168, 171, 178
dynamism, 85
dysplasia, 67, 107

E

E-cadherin, 8, 139, 174, 239, 240, 258, 281
ECM degradation, 81, 95, 162
ECM proteins, 2, 6, 9, 72, 74, 81, 84, 86, 94, 95, 142, 157, 162, 165, 169, 173, 214, 271, 272, 304, 306, 309, 313, 315, 316
ECs, 163, 169, 170, 173, 312
edema, 169
efficiency, 157, 162, 164
elastin, 142, 239
electron, 76, 102, 126, 237, 291
electrophoresis, 72, 142, 144, 145
ELISA, 10, 123, 143, 146, 148, 291, 292
ELISA method, 143, 148
elongation, 86, 91, 113, 143, 195, 278, 306
elucidation, 247

e-mail, 71
emboli, 117
embolization, 117
embryogenesis, 23, 74, 83, 99, 141, 157, 163, 201, 213, 214, 229, 303, 310
embryonic stem cells, 161
encephalitis, 226
encoding, 156, 174, 184, 234
end stage renal disease, 270
end stage renal disease (ESRD), 270
endocarditis, 218
endocrine, 132
endocrinology, 29, 32, 185
endothelial cells, 4, 5, 7, 16, 17, 22, 29, 37, 68, 72, 73, 78, 84, 86, 87, 88, 89, 90, 92, 93, 95, 97, 98, 109, 110, 111, 120, 121, 130, 140, 141, 142, 144, 147, 158, 159, 163, 166, 180, 200, 207, 213, 215, 218, 219, 224, 228, 230, 232, 235, 237, 238, 245, 253, 255, 290, 293, 297, 303, 305, 312, 315, 317
endothelial dysfunction, 235
endothelium, 214, 218, 290, 291
end-stage renal disease, 232, 233, 235, 278
energy transfer, 72, 79, 102, 309
engineering, 2, 8, 11, 64, 90, 97, 110, 122, 177, 178, 290
England, 175, 179, 182
entrapment, 90
environment, 21, 42, 84, 85, 93, 95, 98, 99, 142, 156, 160, 208, 209, 237, 271, 309, 311, 312, 316, 318
environments, 9, 18, 71, 73, 82, 87, 90, 94, 96, 97, 98, 99, 119, 256, 267
enzymatic activity, 166
enzyme, 10, 55, 56, 234, 238, 244, 274, 304
enzyme-linked immunosorbent assay, 10
enzymes, 40, 87, 118, 152, 198, 235, 273, 274
epidermis, 94
epilepsy, 313
epithelia, 72, 92, 134
epithelial cells, 6, 7, 11, 13, 16, 17, 83, 85, 87, 88, 89, 90, 96, 97, 110, 111, 113, 121, 163, 179, 181, 213, 218, 228, 230, 231, 232, 233, 235, 238, 239, 240, 241, 243, 245, 248, 250, 253, 254, 258, 261, 267, 268, 270, 279, 281, 284, 290, 296, 299, 303, 310
epithelial ovarian cancer, 137
epithelium, 63, 91, 128, 221, 230, 231, 240, 250, 253, 258, 266, 268, 279, 304
equilibrium, 5
erythrocytes, 179, 218, 290
ESRD, 272, 273
ester, 198
estrogen, 123, 211, 292, 296
ethanol, 56, 150, 151, 192, 193, 194, 195, 196, 200, 202, 203, 204
eukaryotic, 12, 39, 44, 173
eukaryotic cell, 39, 173
evidence, 7, 8, 17, 21, 22, 23, 25, 26, 33, 64, 68, 106, 117, 160, 163, 166, 169, 170, 171, 193, 195, 220, 221, 228, 231, 232, 234, 235, 238, 239, 240, 242, 244, 246, 271, 313, 315
evolution, 13, 38, 180, 280
excitotoxicity, 202
exclusion, 4, 75
excretion, 251, 270, 282
exertion, 156
exocytosis, 132, 136
exons, 2, 4, 21, 44, 214, 217, 306
experimental design, 56
exposure, 75, 79, 80, 93, 94, 122, 159, 160, 192, 193, 195, 196, 197, 198, 199, 200, 236, 241, 243, 245, 256, 260, 281, 288, 308
external environment, 98
extracellular glycosaminoglycans (GAGS), vii
extracellular matrix (ECM), 1, 2, 36, 72, 125, 142, 155, 157, 213, 214, 227, 228, 271, 292, 301
extravasation, 133
exudate, 94, 118

F

families, 22
family members, 167, 272
FAS, 195
fat, 144, 149
feedback, 78, 81, 94, 166, 168
fetal development, 203, 290
fiber, 108, 110, 160, 163, 168, 172, 187, 272, 299, 302, 303, 319
fibers, 73, 83, 84, 107, 109, 111, 142, 163, 168, 173, 180, 295, 309, 319
fibrin, vii, 3, 5, 16, 21, 28, 29, 35, 43, 56, 62, 75, 79, 87, 90, 92, 93, 94, 103, 104, 116, 117, 215, 216, 219, 239, 281, 284, 302, 305, 310
fibrinogen, 74, 101, 116, 117, 126, 160, 162, 215, 302, 303
fibroblast growth factor, 67, 148, 156, 197, 203, 260
fibroblast proliferation, 166, 174, 189
fibrogenesis, 10, 71, 118, 173, 189, 228, 231, 233, 234, 237, 238, 242, 243, 245, 246, 248, 250, 253, 255, 260, 281
fibronectin expression, 111, 120, 122, 123, 185, 192, 193, 194, 196, 198, 200, 203, 204, 210, 227, 234, 236, 252, 254, 256, 260, 265, 268, 281, 285, 294
fibrosarcoma, 16
fibrous tissue, 232, 247

filament, 87, 172, 185, 187
filtration, 55, 56, 73, 232, 234, 246, 269, 270, 271, 279
fish, 112
fixation, 51, 52, 60
flexibility, 37, 74, 104, 218
fluid, 10, 15, 22, 37, 69, 96, 118, 230, 240, 242, 268
fluorescence, 72, 79, 83, 87, 102, 148, 309
focal adhesion kinase (FAK), 36, 40, 84, 127, 310
focal segmental glomerulosclerosis, 233, 240, 246, 253, 264, 270, 278
foils, 148
folliculitis, 218
force, 9, 40, 73, 80, 82, 84, 103, 108, 109, 111, 118, 133, 156, 160, 168, 223, 237, 304
forecasting, 295
formula, 43
fragments, 2, 8, 10, 11, 18, 26, 28, 35, 37, 41, 43, 44, 45, 46, 47, 50, 51, 56, 63, 65, 81, 82, 94, 98, 113, 116, 138, 177, 214, 220, 225, 233, 255, 298, 307, 308, 310, 317
France, vi, 301
frost, 151
funding, 63, 151, 175
funds, 134
fusion, 31, 131, 132, 134, 135, 136, 137, 138, 139, 140, 162, 166, 201, 219, 224, 292

G

gastric mucosa, 128
gastric ulcer, 121
gastrulation, 84, 91, 99, 111, 166, 193, 310
GDP, 131, 156, 168
GEF, 156, 168
gel, 41, 42, 43, 44, 45, 46, 47, 48, 49, 53, 54, 55, 56, 57, 58, 59, 60, 61, 62, 68, 72, 81, 142, 144, 145, 149, 150, 160, 243, 261, 298
gelatin binding domain (GBD), 35, 43
gene expression, vii, 9, 17, 23, 26, 113, 119, 129, 134, 135, 137, 140, 161, 173, 176, 178, 208, 233, 234, 237, 243, 244, 245, 250, 256, 258, 273, 285, 292, 313, 315
gene promoter, 288
gene silencing, 132, 239
gene therapy, 188, 256, 283, 295
genes, 26, 95, 133, 134, 135, 156, 165, 169, 174, 219, 224, 237, 261, 288, 289, 295
genetic background, 220
genetic engineering, 2
genome, 39, 130
genotype, 137, 187
Georgia, 141, 151, 155, 175
Germany, 66
gestation, 10, 126, 207, 208
gland, 62, 64, 91, 211
glaucoma, 6
glia, 201, 204, 302
glial cells, 73, 191, 194, 202, 205, 287, 305, 312
glioblastoma, 128, 129, 137, 175, 178, 302, 312, 315, 317, 318, 319, 320
Glioblastoma, 285, 302
glioblastoma multiforme, 315, 317, 319, 320
glioma, 26, 86, 96, 112, 204, 283, 285, 296, 301, 302, 303, 304, 305, 312, 315, 316, 317, 318, 319, 320
glomerulonephritis, 232, 233, 236, 245, 246, 251, 254, 257, 276, 282
glomerulus, 73, 227, 230, 231, 232, 233, 234, 235, 236, 238, 248, 253, 268, 270, 277
glucose, 97, 123, 200, 203, 205, 234, 235, 236, 241, 242, 243, 248, 251, 254, 256, 259, 260, 264
GLUT4, 187
glutamate, 198
glutamic acid, 22
glutamine, 198, 267
glutathione, 198, 202
glycerol, 102
glycine, 35
glycogen, 64, 131, 139, 156, 185
glycol, 90
glycoproteins, 16, 36, 126, 177, 248, 249, 276, 278, 303, 304
glycosaminoglycans, vii, 36, 142, 268, 269
glycosylation, 22, 27, 130, 136, 240, 258, 295, 299
GNP, 223
grants, 201
granules, 208
graph, 45, 47, 49, 51, 52, 53, 55, 59
growth dynamics, 291
growth rate, 309, 311
GTPases, 108, 109, 131, 152, 156, 163, 164, 168, 171, 172, 180, 181, 186, 272, 316
Guangzhou, 262
guanine, 30, 168
guidance, 171, 204
guilty, 256

H

head and neck cancer, 295
healing, vii, 4, 5, 11, 12, 14, 15, 18, 29, 36, 37, 39, 41, 50, 54, 62, 63, 64, 65, 67, 75, 86, 92, 93, 94, 95, 98, 99, 101, 112, 114, 115, 116, 117, 118,

119, 120, 121, 123, 141, 152, 153, 157, 158, 166, 169, 175, 177, 184, 229, 259, 268, 277, 310
health, 185, 201
heart disease, 6, 15, 75, 104
heat shock protein, 7, 16
hematology, 186
hemostasis, 15, 114, 175, 319
hepatic fibrosis, 187, 215
hepatic stellate cells, 113
hepatitis, 187, 189
hepatocellular carcinoma, 96, 121, 293, 295
hepatocytes, 5, 8, 10, 17, 21, 73, 158, 213, 214, 284, 303
heterogeneity, 5, 138, 255
hippocampus, 201
HIV, 214, 219, 220, 224, 225
homeostasis, 36, 39, 63, 71, 98, 110, 139, 165, 176, 178, 185, 187, 231, 241, 265, 266, 275
Hong Kong, 227, 262
hormone, 193, 196, 201, 202, 205
hormones, 2, 8, 12, 192
host, 7, 11, 16, 72, 119, 214, 218, 219, 220, 221, 290
hot spots, 133
human brain, 304, 315, 318
human lung fibroblasts, 102, 122
human neutrophils, 186
human skin, 66, 104, 116
hybrid, 24, 25
hydrogels, 90, 92, 115
hydrogen, 60, 192, 199, 252
hydrogen peroxide, 192, 199, 252
hydrophobicity, 17
hyperglycaemia, 251, 260
hypertension, 235, 262
hypertrophy, 234, 241, 252, 254, 270
hypothesis, 199
hypothyroidism, 193, 196, 197, 201, 204
hypoxia, 246

I

ICAM, 243
ice, 144, 145, 151
ICS, 220
ID, 66, 68, 102, 103, 104, 108, 188, 222, 223, 319
identification, 35, 43, 180, 269, 291
idiopathic, 33, 118
IFN, 207
image, 135
image analysis, 135
images, 148, 149
immune response, 3, 261
immune system, 271, 294
immune-mediated renal disease, 227, 240
immunity, 219, 294, 295
immunization, 294
immunofluorescence, 6, 194, 290
immunogenicity, 224
immunoglobulin, 304
immunohistochemistry, 318
immunomodulatory, 211
immunoprecipitation, 24, 25
immunoreactivity, 313
immunosuppression, 261
immunosuppressive agent, 245
immunotherapy, 12, 292, 294
impacts, 71
impetigo, 218
improvements, 302
in vitro, 3, 8, 9, 10, 16, 17, 23, 41, 67, 68, 83, 87, 91, 93, 97, 98, 99, 104, 113, 121, 122, 139, 142, 148, 170, 174, 181, 192, 196, 197, 199, 204, 210, 220, 221, 223, 234, 235, 242, 243, 247, 255, 257, 259, 285, 286, 287, 288, 290, 291, 292, 294, 299, 302, 303,뫬305, 311, 312, 313, 315, 316, 317, 318, 319
in vivo, 1, 4, 8, 9, 12, 13, 17, 23, 27, 71, 73, 83, 85, 86, 87, 90, 91, 93, 94, 96, 98, 99, 114, 136, 142, 143, 153, 162, 169, 170, 181, 183, 184, 192, 199, 204, 210, 212, 223, 229, 235, 239, 242, 243, 251, 260, 261, 286, 292, 293, 297, 302, 303, 305, 311, 312, 315, 317, 318, 319
incidence, 8, 243
inclusion, 4, 14, 75, 80, 158
incompatibility, 172
indirect effect, 158
individuals, 291
induction, 32, 88, 94, 118, 165, 174, 177, 181, 182, 188, 200, 212, 227, 228, 233, 234, 235, 236, 237, 238, 240, 242, 244, 245, 246, 248, 254, 262, 267, 287, 290, 294, 297
infants, 13
infarction, 6, 11
infection, 189, 214, 218, 219, 220, 221, 225, 232
inflammation, 11, 39, 75, 96, 141, 157, 227, 242, 268, 270, 313, 315
inflammatory cells, 221, 233, 236
inflammatory disease, 221, 255
inflammatory mediators, 238, 242
inhibition, 25, 40, 54, 61, 64, 86, 87, 88, 96, 143, 152, 162, 168, 171, 172, 173, 195, 196, 199, 237, 239, 244, 245, 253, 285, 286, 298, 301, 304, 305, 309, 313, 315, 317
inhibitor, 44, 57, 60, 61, 62, 64, 65, 66, 80, 82, 88, 118, 149, 162, 167, 170, 173, 195, 197, 243, 257, 261, 286, 288, 293, 294, 295, 297, 304

initiation, 76, 77, 83, 159, 168, 178, 211, 230, 265, 267, 269, 306, 307, 309, 311
injury, 5, 6, 15, 91, 94, 114, 156, 163, 170, 175, 185, 201, 204, 230, 231, 232, 233, 235, 236, 238, 240, 241, 244, 245, 247, 248, 250, 252, 253, 255, 257, 259, 261, 268, 269, 276, 279, 310, 315, 319
inositol, 183
insecticide, 199
insertion, 229
insulin, vii, 30, 31, 32, 33, 64, 124, 172, 175, 185, 186, 283, 293, 296, 298, 299
integration, 31, 309
integrators, 136
iintegrity, 22, 68, 141, 142, 160, 163, 207, 211, 227, 228, 232, 241, 257, 268
intercellular contacts, 7
interface, 80, 122, 126, 232, 308, 314
interference, 191, 192, 293, 294
interferon, 220
intermolecular interactions, 76, 80, 160, 307, 309
internalization, 7, 12, 16, 117, 130, 170, 171, 214, 237, 309
interstitial nephritis, 257
interstitial pneumonia, 118
intervention, 11, 238, 250
intestine, 285
intima, 96
intracellular calcium, 244
intraocular, 6
intron, 44
ionizing radiation, 287, 292
ions, 60, 281
Ireland, 265, 276
ischaemic heart disease, 75
ischemia, 15, 114, 175, 303, 319
Islam, v, 35, 62
islands, 81, 84
isoflavone, 60
isolation, 124
isoleucine, 35, 216, 306
isozyme, 24, 251
issues, 128, 202, 220
Italy, 191, 301

K

Kentucky, 125
keratinocyte, 30, 69, 86, 93, 95, 112, 136
keratinocytes, 8, 17, 37, 66, 86, 93, 141, 166
kidney, 6, 10, 17, 72, 73, 87, 91, 95, 99, 100, 111, 119, 122, 128, 129, 182, 227, 228, 230, 231, 232, 233, 234, 235, 236, 238, 242, 243, 245, 247, 250, 251, 252, 254, 256, 257, 261, 262, 263, 264, 265, 266, 267, 268, 269, 271, 272, 274, 275, 276, 277, 278, 279, 280
kidney transplantation, 228, 232
kidneys, 248, 249
killer cells, 88, 89, 283, 294
kinase activity, 64, 127, 166, 183, 235
kinetics, 83, 164
Kinsey, 101, 255
Korea, 1

L

labeling, 107, 318
latency, 165, 237
lead, vii, 13, 36, 39, 41, 43, 46, 54, 63, 71, 72, 92, 160, 167, 175, 228, 230, 252, 270, 272, 274, 302, 309, 312, 315
Leahy, 31, 103, 307, 318
leakage, 143, 169, 270
lesions, 6, 8, 26, 95, 120, 174, 221, 227, 228, 230, 232, 233, 236, 248, 253, 261, 270, 280, 281, 292
leucine, 216, 306
leukemia, 18, 283, 288, 297
leukocytes, 4, 208, 211, 225, 252, 305
LFA, 134
ligament, 64, 122
ligand, 5, 12, 23, 25, 32, 39, 56, 57, 60, 78, 84, 87, 117, 126, 129, 140, 165, 175, 179, 209, 218, 220, 222, 286, 307, 318
light, 12, 80, 126, 128, 132, 147, 157, 158, 168, 171, 313
limitations, 27
lipids, 224
liver, 10, 14, 15, 174, 179, 185, 189, 214, 215, 221, 278, 284, 293, 296, 317
liver cancer, 293, 296
liver damage, 10
liver disease, 15
liver transplant, 278
liver transplantation, 278
localization, 27, 32, 109, 115, 117, 162, 163, 164, 166, 167, 168, 171, 172, 173, 174, 212, 232, 248, 277, 278, 290, 304, 306, 316, 317
locus, 214
Louisiana, 283
lumen, 235, 241
lung cancer, 16, 129, 134, 174, 178, 187, 283, 286, 296, 299
Luo, 254, 277
lupus, 227, 228, 233, 236, 240, 244, 245, 248, 254, 255, 256, 262, 263

lupus nephritis, 227, 228, 233, 236, 240, 244, 246, 254, 255, 256, 262
lymph, 220, 285, 298
lymph node, 220, 285, 298
lymphangiogenesis, 290
lymphatic system, 7
lymphocytes, 4, 207, 208, 211, 219, 279
lymphoid, 208, 219
lymphoid tissue, 219
lysine, 22, 275
lysis, 149

M

machinery, 73, 83, 111, 140
macromolecules, 41, 44, 311, 316
macrophages, 87, 112, 132, 159, 213, 220, 235, 238, 251, 253, 279, 303, 310
major histocompatibility complex, 30
majority, 26
malignancy, 262, 312, 317
malignant cells, 7, 292, 310
malignant melanoma, 96
malignant tumors, 96, 291
mammalian cells, 166
mammalian target of rapamycin (mTOR), 152, 156, 178, 245, 297
mammals, 2, 64, 165
man, 229
management, 302
manganese, 192, 193, 203
manipulation, 39, 274
MAPK/ERK, 187
mapping, 302
markers, 137
marrow, 25
Mars, 257
Maryland, 278
masking, 80
mass, 27, 80, 84, 96, 97, 105, 278, 313
mass spectrometry, 27
materials, 99, 110, 143
matrix metalloproteinase, 6, 65, 72, 157, 160, 176, 179, 185, 187, 227, 237, 239, 256, 257, 258, 271, 280, 281, 286, 287, 293, 297, 310, 315
matrix metalloproteinases, 65, 179, 185, 227, 237, 239, 256, 258, 271, 280, 281, 287, 293, 310
matrixes, 215
matter, iv, 236
MB, 101, 107, 113, 181, 182, 183, 187, 211, 255, 320
MCP, 187, 236, 279
MCP-1, 236, 279
measurement, 10, 168
mechanical properties, 71, 82, 85, 94, 160, 216
mechanical stress, 253
media, 8, 96, 144, 145, 171
median, 302
medical, 10, 99, 185
medicine, 13, 175, 176, 184, 187
medulla, 279
MEK, 60, 113, 185
melanoma, 8, 16, 43, 53, 54, 66, 68, 96, 120, 283, 295
melatonin, 198
mellitus, 270
membranes, 5, 10, 48, 63, 92, 99, 118, 131, 232
membranous glomerulonephritis, 278
memory, 203
meninges, 302, 313
meningioma, 313, 314, 317
meningitis, 218
mental retardation, 197
mesangial cells, 123, 168, 182, 200, 205, 228, 230, 232, 233, 234, 235, 236, 237, 238, 243, 244, 245, 246, 248, 251, 252, 253, 254, 255, 256, 262, 266, 267, 268, 270, 277, 279, 281
mesenchyme, 231, 266, 313, 315
mesoderm, 13, 44, 91, 114, 126, 136, 175, 241, 276, 313
mesothelioma, 259
mesothelium, 241, 242, 259, 264
metabolism, 29, 32, 184, 185, 201
metabolites, 199
metalloproteinase, 6, 72, 157, 160, 176, 238, 257, 258, 280, 303
metaphase, 289
metastasis, 7, 8, 9, 13, 16, 24, 54, 64, 87, 97, 121, 122, 125, 128, 133, 134, 139, 141, 152, 157, 173, 174, 186, 283, 284, 285, 286, 290, 293, 297, 298, 308, 320
metastatic cancer, 134
methanol, 145
methodology, 56, 294
methylprednisolone, 250
Mexico, 1
MHC, 313
mice, 2, 4, 14, 18, 41, 44, 74, 92, 93, 94, 95, 96, 97, 98, 100, 114, 115, 116, 119, 120, 140, 152, 156, 157, 158, 160, 162, 166, 169, 179, 182, 184, 211, 212, 219, 220, 221, 224, 234, 236, 237, 239, 242, 245, 246, 250, 253, 254, 255, 256, 257, 263, 278, 285, 294, 297
microenvironments, 73, 84, 134, 135
microRNA, 234
microscope, 41, 148, 193, 291

microscopy, 41, 76, 80, 83, 87, 102, 126, 148, 194
migration stimulating factor (MSF), 35, 43, 66, 68
mitogen, 32, 64, 128, 156, 175, 183, 188, 194, 203, 245, 253, 254, 260, 272, 279, 289
mitosis, 242, 289
MMP, 6, 8, 11, 18, 72, 81, 94, 109, 157, 238, 239, 244, 256, 257, 271, 272, 287, 288, 293, 297, 302, 304, 315
MMP-2, 18, 238, 239, 244, 256, 287, 315
MMP-3, 8, 95, 257
MMP-9, 8, 81, 94, 238, 239, 244, 256, 287, 288, 293, 297, 315
MMPs, 11, 162, 165, 174, 237, 238, 239, 271, 272, 274, 288, 304
modeling, 108
models, 2, 4, 39, 71, 73, 86, 92, 96, 122, 129, 182, 184, 231, 232, 235, 237, 240, 245, 246, 269, 273, 275, 279, 286, 308, 312, 316
modifications, 14, 41, 148, 249, 266, 274, 275, 276
modules, 3, 14, 22, 38, 43, 44, 46, 47, 48, 49, 52, 62, 68, 74, 75, 80, 81, 83, 88, 90, 100, 102, 103, 105, 108, 114, 116, 158, 159, 160, 214, 215, 216, 217, 218, 222, 229, 276, 297, 305, 306, 317, 318
molecular biology, 33, 189
molecular dynamics, 80
molecular mass, 105, 292
molecular structure, 214, 267
molecular weight, 1, 36, 37, 44, 55, 56, 63, 67, 80, 81, 93, 141, 229, 230, 246, 284, 311
molecules, 2, 4, 7, 8, 25, 36, 37, 38, 39, 40, 42, 43, 46, 56, 74, 76, 79, 81, 84, 88, 105, 125, 127, 130, 139, 143, 159, 162, 163, 173, 181, 192, 195, 202, 203, 208, 210, 212, 213, 214, 218, 230, 231, 233, 234, 237, 241, 251, 266, 267, 271, 272, 283, 294, 301, 304, 305, 306, 307, 309, 312, 315, 316
monoclonal antibody, 10, 11, 18, 115, 176, 291, 294
monolayer, 15, 17, 41, 42, 68, 82, 98, 147, 241, 268
monomers, 22, 141, 284
morphogenesis, 15, 17, 39, 68, 91, 99, 101, 104, 108, 109, 111, 112, 114, 115, 157, 202, 205
morphology, 1, 9, 17, 73, 85, 92, 98, 110, 111, 112, 142, 160, 171, 172, 227, 228, 241, 259, 309, 313
morphometric, 194, 270
mosaic, 159
Moses, 178, 188
motif, 3, 4, 43, 46, 48, 60, 62, 66, 99, 126, 159, 179, 216, 306, 308
MR, 66, 100, 101, 104, 106, 110, 119, 123, 183, 189, 203, 251, 254, 255, 264
MRI, 63
mRNA, 4, 8, 12, 14, 19, 21, 62, 66, 82, 91, 113, 119, 121, 128, 162, 178, 194, 195, 198, 200, 212, 214, 217, 223, 227, 229, 234, 244, 246, 249, 254, 263, 276, 282, 284, 287, 294, 296, 305, 306
mRNAs, 14, 103, 217, 306
MSF, 35, 36, 38, 43, 44, 45, 46, 47, 48, 49, 50, 51, 52, 53, 54, 55, 56, 57, 59, 60, 61, 62, 63, 65, 66, 67, 68
mucosa, 53, 54, 120, 128, 208
multicellular organisms, 36, 39
multidimensional, 111
multiple myeloma, 30
multiple sclerosis, 313
multipotent, 290, 297
muscarinic receptor, 191, 193, 195, 196, 197, 201, 203, 205
mutagenesis, 28, 177
mutant, 48, 49, 51, 52, 88, 93, 170, 219, 291, 298
mutation, 44, 46, 52, 77, 78, 92, 98, 157, 162, 170, 173, 255
mutations, 48, 49, 51, 52, 77, 133, 175, 179, 205, 266
myeloid cells, 179
myoblasts, 310
myocardial infarction, 11
myofibroblasts, 94, 95, 96, 98, 120, 121, 158, 174, 215, 231, 233, 235, 239, 240, 247, 272
myosin, 40, 76, 78, 80, 83, 85, 90, 107, 157, 167, 168, 171, 172, 183, 268, 308

N

NAD, 234, 252
nanomaterials, 19
nanowires, 11, 19
NAP, 204
National Academy of Sciences, 13, 19, 30, 32, 175, 176, 177, 179, 180, 181, 182, 184, 187, 222, 224
National Institutes of Health, 134, 151, 175
natural killer cell (NK), 89, 113, 211, 212
neck cancer, 129
necrosis, 113, 251
neoangiogenesis, 242
neoplasm, 283, 284, 286, 287, 288, 289, 291, 294, 295, 298
neoplastic tissue, 315
neovascularization, 87, 97, 115, 290
nephrectomy, 275, 282
nephritic syndrome, 232
nephritis, 10, 17, 227, 228, 232, 233, 236, 240, 244, 246, 248, 254, 255, 256, 262, 263, 264, 270
nephron, 230, 265, 266, 268, 270, 271, 279
nephropathy, 228, 233, 234, 235, 236, 240, 243, 244, 245, 253, 257, 261, 262
nephrotic syndrome, 279

nerve, 141, 192, 201, 302, 303, 319
nervous system, 195, 196, 201, 202, 221, 226, 301, 319
neuritogenesis, 191, 192, 193, 194, 195, 196, 197, 198, 200, 201, 202, 203
neuroblastoma, 28, 200, 201
neuronal cells, 91, 315
neurons, 26, 132, 138, 191, 192, 193, 194, 195, 196, 197, 198, 199, 200, 203, 204, 205, 278, 305
neuroscience, 33
neurosurgery, 314
neurotoxicity, 192, 198, 199, 200, 201, 202, 203, 204
neurotransmitter, 192, 193, 204
neurotransmitters, 201, 205
neutral, 238, 245, 262
neutrophils, 133, 137, 138, 186
NH2, 105, 281
nitric oxide, 211
nitrogen, 149, 151
NMR, 102
nodes, 84
normal development, 44, 63, 143, 166, 196, 207, 278
normal renal tissue structure, 266
nucleation, 5
nuclei, 148, 242
nucleotides, 234
nucleus, 39, 156, 165, 237, 280, 287, 289
null, 2, 23, 44, 74, 78, 79, 86, 91, 92, 93, 94, 96, 97, 105, 157, 158, 160, 162, 167, 173, 254, 255, 309, 316
nutrients, 97, 290
nutrition, 198

O

obesity, 270
obstruction, 239, 240, 246, 263, 278
obstructive uropathy, 232
occlusion, 214
oncogenes, 183, 244
oncogenesis, 284
oncoproteins, 174, 187
oral cancers, 62
organ, 6, 39, 93, 95, 198, 218, 228, 236, 244, 247, 258, 266, 270, 272, 276, 284
organelles, 129
organism, 7, 11, 39, 310
organize, 84, 85, 86, 87, 96, 213, 214, 303, 308
organizing, 88
organs, 8, 133, 143, 169, 227, 228, 242, 269, 310, 312
osteomyelitis, 218
ovarian cancer, 129, 131, 137, 139, 260, 286, 297
overlap, 76, 216, 268
overlay, 88, 145
overproduction, 253
ox, 71
oxidative stress, 192, 193, 198, 199, 200, 202, 204, 205, 234, 235, 242, 252, 260
oxygen, 97, 192, 199, 203, 220, 234, 252, 290
oxygen analog diazoxon, 192, 199

P

p53, 88, 100, 122, 133, 135, 137, 138, 163, 174, 179, 187, 188, 291, 298
pairing, 267
pancreatic cancer, 129, 129, 283, 286, 296, 298
paradigm, 152, 184
parallel, 41, 85, 94, 95, 97, 134, 215, 302, 303
parasite, 214, 220, 221
parasitic diseases, 220
parenchyma, 233, 236, 238, 304, 313, 316
participants, 240
partition, 22, 72, 137
pathogenesis, 62, 138, 152, 175, 176, 189, 218, 219, 232, 233, 235, 236, 237, 239, 240, 247, 253, 255, 256, 278, 279, 314
pathogens, 213, 214, 218
pathology, iv, 27, 33, 176, 182, 185, 233, 254, 258
pathophysiological, 156, 265, 270
pathophysiology, 278
pathways, vii, 4, 6, 9, 11, 22, 23, 24, 25, 26, 27, 36, 39, 40, 44, 54, 56, 60, 61, 62, 63, 64, 65, 67, 71, 81, 88, 107, 117, 126, 128, 129, 130, 134, 135, 136, 140, 156, 158, 160, 161, 163, 164, 165, 167, 168, 169, 171, 172, 173, 174, 176, 178, 181, 185, 187, 191, 194, 195, 197, 203, 204, 233, 234, 237, 238, 254, 272, 286, 309, 311, 315, 316
PCR, 291
PDGFR, 25
peptide, 8, 11, 16, 22, 26, 46, 48, 50, 61, 62, 63, 80, 97, 98, 105, 123, 126, 129, 165, 204, 220, 237, 242, 277, 292, 318
peptides, 4, 8, 11, 13, 18, 46, 48, 49, 50, 59, 62, 67, 77, 95, 98, 100, 106, 107, 119, 126, 215, 225, 274, 307, 308, 310
periodontal, 64, 122
peripheral blood, 218
peripheral nervous system, 315
peritoneal cavity, 241
peritoneum, 240, 242, 258, 259
peritonitis, 241
permeability, 6, 15, 153, 184
permit, 160

peroxide, 142
pH, 60, 144, 149, 230, 233
phage, 298
phagocytosis, 214
pharmaceutical, 239
pharmacology, 30, 31
phenotype, 4, 8, 17, 43, 64, 71, 98, 110, 115, 118, 123, 137, 162, 167, 173, 174, 177, 208, 231, 238, 239, 242, 260, 265, 266, 271, 280, 285, 296, 302
phenotypes, 16, 68, 96, 97, 174, 242, 266, 315
phosphate, 82, 88, 123, 142
phosphoinositides, 217, 223
phosphorylation, vii, 22, 23, 24, 25, 27, 36, 39, 59, 60, 61, 78, 88, 97, 120, 127, 156, 161, 165, 166, 167, 168, 169, 170, 171, 173, 179, 181, 182, 183, 186, 201, 237, 245, 246, 248, 252, 283, 286, 287, 289, 293, 294, 295, 297, 304
physical properties, 2, 275
Physiological, 91, 184, 185, 224, 231
physiology, 9, 21, 30, 178, 179, 181, 182, 183, 184, 185, 208, 255, 256, 262, 267
PI3K, 25, 40, 89, 157, 158, 161, 163, 164, 165, 166, 167, 168, 169, 170, 171, 172, 173, 174, 184, 185, 286, 288, 289, 304
placebo, 246
placenta, 10, 15, 207, 208, 290
plaque, 291
plasma levels, 10, 104, 174, 187, 246
plasma membrane, 125, 126, 130, 131, 132, 133, 134, 136, 163, 170
plasma proteins, 270
plasmid, 51, 52, 219, 295
plasminogen, 82, 123, 162, 187, 194, 203, 205, 219, 239, 257
plasticity, 178, 188, 203, 239
platelet aggregation, 11, 117, 176
platelets, 4, 93, 116, 117, 218
platform, 139, 167
pleural effusion, 248
pneumonia, 118, 218
point mutation, 44, 133
polarity, 92, 115, 163, 167, 172, 186, 265, 267, 268, 277, 318
polarization, 72, 85, 171
polyacrylamide, 72, 83, 85, 86, 97, 142, 144, 145, 149, 150, 153
polycystic kidney disease, 182
polymerase chain reaction, 261
polymerization, 15, 31, 77, 82, 85, 86, 87, 91, 94, 100, 101, 105, 106, 118, 145, 155, 159, 160, 162, 163, 172, 178, 230, 244, 247, 308, 309
polymers, 90, 308
polypeptide, 2, 3, 12, 28, 36, 159, 181, 215, 249, 294
polypeptides, 249, 283, 293, 296
polystyrene, 17
polyvinylidene fluoride, 142
population, 104, 236, 240, 242, 272
positive correlation, 275
potential benefits, 262
precipitation, 56, 144
precursor cells, 114, 311
predictive accuracy, 99
pregnancy, 13, 207, 208, 209, 210, 211, 212
preparation, iv, 62, 146
preservation, 239
prevention, 227, 244, 247, 286
primary brain tumor, 285
primary tumor, 7, 97, 137
probe, 148
progenitor cells, 8, 17, 96, 110
progesterone, 211
prognosis, 230, 269, 279, 283, 284, 295, 296, 302
pro-inflammatory, 11, 219, 233, 242
project, 276
proliferation, 7, 23, 24, 25, 30, 31, 33, 56, 67, 68, 72, 73, 88, 89, 90, 91, 92, 93, 94, 96, 97, 98, 107, 110, 112, 113, 117, 121, 125, 127, 128, 137, 142, 155, 157, 158, 160, 161, 166, 168, 169, 171, 174, 175, 183, 185, 187, 192, 193, 195, 203, 208, 210, 214, 227, 228, 229, 230, 231, 233, 236, 242, 243, 244, 245, 247, 268, 269, 270, 272, 281, 285, 286, 287, 289, 293, 294, 296, 298, 303, 304, 305, 310, 311, 313, 314, 315, 320
promoter, 281, 287
propagation, 165
prostate cancer, 10, 113, 129, 136, 137, 152, 174, 284, 296, 299
prostate carcinoma, 86
prostheses, 90
proteases, 18, 81, 93, 94, 112, 119, 160, 162, 174
protection, 30, 289
protective role, 6
protein components, 214, 238
protein folding, 44
protein kinase C, 30, 31, 60, 69, 162, 194, 203, 248, 251, 253, 256
protein kinases, 64, 183, 272
protein structure, 102
protein synthesis, 12, 158, 171, 228, 235, 237, 239
proteinase, 118
protein-protein interactions, 81, 268, 307
proteinuria, 234, 245, 277, 279
proteoglycans, 3, 5, 6, 25, 72, 78, 88, 192, 228, 230, 232, 248, 268, 284, 304, 308, 309
proteolysis, 2, 13, 65, 93, 112, 162, 194, 237
proteolytic enzyme, 87, 194, 271, 304

proteome, 27, 107
proteomics, 79, 139, 187, 204
prototype, 180, 236
proximal tubules, 279
pseudopodia, 91
PTEN, 39, 40, 66, 88, 89, 157, 161, 172, 174, 175, 188, 189
PVA, 226
P-value, 128
PVP, 51, 52
pyrophosphate, 149

Q

quality of life, 302
quantification, 263

R

Rab, 131, 133, 134, 138
radiation, 174, 287, 292, 296
radiation therapy, 292
radio, 292, 295, 302
radioresistance, 299
radiotherapy, 295, 299, 302
RANTES, 279
RCP, 133
reactions, 303
reactive oxygen, 199, 234, 235, 246
real time, 60
receptors, vii, 1, 4, 6, 7, 9, 11, 15, 21, 22, 23, 24, 35, 38, 39, 40, 43, 56, 60, 66, 74, 76, 77, 78, 79, 80, 84, 85, 86, 88, 93, 106, 111, 125, 143, 156, 158, 159, 160, 162, 164, 165, 191, 192, 193, 194, 195, 196, 197, 199, 202, 204, 217, 219, 220, 225, 242, 256, 257, 266, 267, 268, 272, 277, 284, 286, 288, 289, 291, 301, 303, 304, 305, 306, 307, 308, 309, 311, 312, 318, 319
reciprocity, 161, 163, 180
recognition, 3, 67, 75, 103, 106, 136, 153, 160, 170, 184, 217, 229, 281, 305
recombinant proteins, 44, 47, 50
reconstruction, 15, 105, 318
recovery, 263
recruiting, 168, 170
recurrence, 298
recycling, 129, 130, 131, 132, 133, 134, 137, 138, 139, 140, 169, 170, 185
red blood cells, 218
redistribution, 14, 106
redundancy, 271
regeneration, 37, 141, 192, 204, 231, 315
regression, 129, 179, 232, 264
rejection, 227, 243, 244, 262
relevance, 63, 175, 189, 240, 243
remodelling, 95, 118, 177, 228, 229, 238, 280, 301, 304
renal cell carcinoma, 137, 276, 284, 289, 295, 297
renal failure, 228, 233, 278
renal replacement therapy, 240
renin, 262
repair, 1, 4, 11, 13, 18, 32, 37, 71, 75, 95, 116, 119, 133, 155, 158, 163, 170, 176, 213, 214, 228, 257, 259, 277, 279, 280, 281, 303, 304, 310, 311
replacement, 12
replication, 289
repression, 165, 171
reproduction, 32
requirements, 8, 28, 108, 136
researchers, 2, 234, 238, 239, 241, 244, 245, 246
residues, 22, 24, 35, 43, 44, 60, 106, 107, 133, 159, 167, 169, 185, 216, 267, 305, 306
resistance, 6, 73, 83, 128, 129, 135, 173, 174, 187, 263, 284, 296, 297
resolution, 94, 238, 271
respect, 23, 52
response, 4, 14, 23, 26, 29, 31, 33, 35, 40, 41, 42, 45, 47, 48, 49, 50, 51, 52, 53, 54, 55, 56, 60, 61, 62, 65, 68, 73, 81, 82, 83, 85, 95, 101, 109, 110, 114, 121, 143, 144, 156, 164, 166, 169, 170, 174, 178, 181, 182, 185, 186, 219, 231, 233, 237, 246, 251, 255, 256, 263, 271, 273, 279, 283, 287, 289, 294, 295, 296, 299, 302, 303, 314, 315, 319
responsiveness, 15, 96
restoration, 170, 241, 271
retardation, 169, 184, 195
reticulum, 125, 130, 160, 241
retirement, 63
retrovirus, 232
rheumatoid arthritis, 37, 104
risk, 10, 18, 278
RNA, 234, 285, 293, 296, 313, 319
rodents, 208, 212
room temperature, 145, 146
rubber, 81

S

salivary gland, 62, 64, 91
salivary glands, 91
scar tissue, 6
scatter, 115, 286
science, 13, 14, 15, 175, 176, 178, 180, 181, 182, 183, 187, 277, 296
scleroderma, 143

sclerosis, 174, 232, 259, 264
SDS-PAGE, 59, 60, 72, 81, 144, 145, 149
secrete, 87, 97, 141, 142, 204, 238, 273, 287, 310, 313
secretion, 4, 17, 81, 96, 124, 141, 143, 144, 153, 156, 160, 162, 169, 170, 174, 193, 205, 217, 220, 224, 233, 236, 242, 243, 246, 254, 259, 265, 303, 304
sediment, 291
sedimentation, 77, 102
seeding, 17, 303
segmental glomerulosclerosis, 270, 278
selectivity, 160, 270
self-assembly, 96, 97, 105, 112, 178, 248, 318
self-organization, 86, 87
senescence, 137, 242, 260
sensing, 66, 110, 172, 186
sensitivity, 291, 292
sensors, 304
sepsis, 218
sequencing, 46
serine, 22, 24, 25, 61, 88, 93, 118, 131, 165, 169, 172, 184, 186, 195, 216, 219, 238, 306
Sertoli cells, 277
serum, 8, 10, 17, 19, 48, 59, 116, 137, 144, 145, 147, 150, 197, 221, 288, 292, 297, 313
shape, 13, 14, 39, 84, 97, 102, 106, 110, 152, 161, 183, 218, 311, 313
shear, 237
shock, 7, 16, 142
showing, 144, 168, 220, 285, 308, 313
sickle cell, 214
side chain, 79
signal peptide, 126
signal transduction, 7, 12, 40, 56, 59, 60, 62, 100, 134, 135, 138, 162, 164, 167, 172, 181, 182, 191, 193, 194, 195, 196, 203, 233, 316, 318, 319
signaling pathway, 4, 6, 9, 81, 88, 107, 127, 128, 143, 158, 160, 161, 163, 164, 165, 169, 171, 172, 173, 176, 179, 184, 185, 188, 195, 227, 251, 286, 287, 288, 289, 295, 296, 297, 309, 310, 314, 316
signalling, vii, 21, 22, 23, 24, 25, 26, 27, 29, 32, 35, 39, 40, 44, 54, 61, 63, 65, 66, 67, 71, 113, 136, 178, 182, 183, 185, 189, 233, 234, 238, 245, 272, 273, 277, 304, 306, 315
signals, 21, 66, 72, 73, 78, 86, 87, 88, 91, 107, 113, 124, 136, 139, 155, 156, 158, 160, 163, 164, 167, 168, 172, 175, 179, 181, 183, 185, 186, 192, 204, 272, 280, 304, 309, 311, 313
silica, 11, 19
single chain, 292, 293
SIP, 121, 292
siRNA, 79, 92, 132, 285, 286, 287
skeletal muscle, 64
skin, 14, 15, 26, 33, 53, 54, 66, 94, 104, 109, 112, 114, 115, 116, 118, 119, 122, 149, 150, 152, 169, 175, 177, 188, 218, 220, 243, 249, 268, 277, 319
slaves, 135
smooth muscle, 25, 30, 31, 32, 94, 101, 119, 128, 138, 231, 239, 240, 241, 248, 267, 271, 277, 280, 281
smooth muscle cells, 25, 101, 267, 277
SNAP, 131, 132, 137
sodium, 72, 109, 142, 149, 153
software, 148, 151
solid phase, 77, 308
solid tumors, 97, 103, 290
solidification, 145
solubility, 5, 311
solution, 75, 144, 145, 146, 147, 149, 150, 151, 240, 306, 308, 309
space, 10, 12, 27
spastic, 197
species, 4, 5, 12, 75, 93, 199, 203, 217, 220, 234, 235, 246, 252, 306
spectroscopy, 223
spindle, 85, 289
spleen, 179, 209
Spring, 115
sprouting, 290
squamous cell, 62, 66, 67, 96, 97, 98, 101, 120, 122, 135, 283, 287, 295, 299
squamous cell carcinoma, 62, 66, 67, 96, 97, 98, 101, 120, 122, 135, 283, 287, 295, 299
stability, 15, 46, 101, 102, 104, 116, 117, 160, 162, 178, 225, 274
stabilization, 28, 113, 168, 238, 290, 306
stasis, 11
state, 2, 6, 11, 80, 99, 127, 156, 160, 228, 307, 315
states, vii, 44, 78, 79, 127, 315
stem cells, 26, 90, 137, 161
sterile, 147
steroids, 288
stimulation, 25, 26, 40, 54, 61, 62, 63, 67, 157, 164, 165, 166, 170, 175, 178, 181, 197, 202, 204, 215, 236, 240, 256, 290, 304, 308, 309, 311
stimulus, 39, 40, 42
storage, 26, 241
strategy, 135
stress, 73, 82, 84, 90, 96, 97, 98, 107, 108, 109, 110, 160, 163, 168, 172, 173, 180, 187, 192, 199, 201, 202, 203, 234, 235, 251, 252, 253, 260, 267, 272, 295, 299, 309, 319
stretching, 84, 109
striatum, 198
stroke, 75, 104

stroma, 6, 62, 63, 83, 93, 96, 117, 121, 281, 292, 293, 304
stromal cells, 26, 32, 68, 290, 297, 304
structural changes, 241
structural gene, 249
structural protein, 78, 153, 228
structure, 2, 9, 13, 18, 24, 35, 36, 37, 38, 43, 66, 71, 72, 74, 76, 79, 83, 84, 85, 87, 95, 96, 97, 103, 104, 112, 118, 126, 140, 143, 152, 156, 158, 159, 163, 177, 182, 211, 213, 216, 217, 222, 223, 228, 229, 230, 241, 249, 257, 259, 266, 267, 269, 276, 277, 280, 303, 305, 306, 309, 319, 320
substrate, 23, 37, 72, 82, 83, 84, 85, 88, 97, 105, 110, 112, 138, 144, 146, 149, 150, 151, 160, 168, 172, 218, 220, 232, 238, 247, 271, 309, 311, 316
substrates, 9, 11, 73, 83, 84, 85, 87, 92, 96, 97, 98, 169, 173, 238, 309, 310, 313, 315
subtraction, 27
sulfate, 3, 107, 109, 117, 142, 144, 153, 284, 293, 309, 320
Sun, 33, 110, 122, 123, 138, 182, 184, 259, 297
suppression, 235, 247
surface area, 232
surface chemistry, 2, 19
surface component, 3
surface plasmon resonance, 24
surgical technique, 302
survey, 135
survival, 6, 9, 15, 23, 25, 29, 39, 87, 88, 89, 90, 92, 94, 97, 113, 114, 124, 125, 127, 128, 133, 137, 157, 158, 161, 163, 164, 166, 168, 169, 170, 174, 175, 179, 186, 188, 192, 193, 203, 208, 210, 231, 263, 268, 273, 274, 284, 285, 287, 290, 292, 293, 294, 295, 297, 299, 302, 303, 305, 319
survivors, 319
susceptibility, 36, 43, 220, 266, 275
swelling, 198
Switzerland, 149
symmetry, 112
synapse, 202, 204, 311
synaptic plasticity, 178
synaptogenesis, 192
syndrome, 199, 269
synergistic effect, 287
synthesis, 7, 12, 16, 36, 37, 54, 55, 56, 59, 61, 62, 63, 64, 65, 67, 68, 78, 94, 95, 100, 122, 123, 141, 143, 152, 157, 158, 161, 162, 165, 170, 174, 175, 176, 178, 183, 188, 214, 227, 228, 230, 231, 232, 233, 234, 235, 236, 237, 238, 239, 240, 241, 242, 243, 244,뫬245, 246, 248, 249, 251, 254, 259, 260, 262, 271, 272, 289, 293, 301, 304
synthetic polymers, 90
systemic lupus erythematosus, 236, 254, 255, 263
systemic sclerosis, 189

T

T cell, 7, 12, 220, 225, 234, 243, 245, 261, 294
T cell, 7, 12, 243, 261, 294
target, 8, 12, 24, 44, 54, 71, 98, 121, 125, 131, 136, 139, 152, 156, 157, 165, 174, 178, 185, 186, 187, 189, 195, 199, 214, 218, 234, 237, 245, 256, 263, 281, 289, 290, 297, 315
techniques, 2, 9, 24
technologies, 316
temperature, 145, 146
temporal lobe, 128
tensile strength, 36, 94, 118
tension, 78, 80, 91, 94, 102, 111, 160, 163, 167, 235, 297, 316
tensions, 73
testing, 10
tetanus, 132, 138
Tetanus, 136, 138
texture, 149
therapeutic approaches, 228
therapeutic effects, 283
therapeutic interventions, 259
therapeutic targets, 186, 314, 320
therapeutics, 12, 14, 18, 63, 158, 175
therapy, 98, 104, 121, 184, 262, 264, 284, 292, 295, 296, 299, 302, 315, 318
third dimension, 39, 109, 135, 317
threonine, 22, 24, 25, 88, 131, 165, 169, 184
thrombolytic therapy, 104
thrombosis, 31, 157, 177
thrombus, 116, 117
thymus, 153
thyroid, 89, 113, 124, 173, 185, 192, 197, 202, 203, 284, 291, 294, 298
thyroid cancer, 284, 294
Thyroid hormones (T3), 192
TIMP, 18, 238, 239, 244, 272, 304
TIMP-1, 18, 238, 304
tissue engineering, 8, 11, 90, 110, 122, 280
tissue homeostasis, 36, 71, 165
tissue remodelling, 118, 177, 268
TNF, 89, 238, 251, 287, 297
TNF-alpha, 251, 297
TNF-α, 89
toxicity, 174, 199, 204, 218
toxicology, 199, 202
toxin, 12, 19, 132, 136, 138
trafficking, 125, 126, 129, 130, 131, 132, 133, 134, 135, 136, 138, 139, 140, 169, 170, 180

transactions, 30, 33, 189
transcription, 16, 26, 39, 75, 88, 123, 133, 156, 158, 165, 169, 178, 179, 188, 244, 289, 297
transcription factors, 88, 165, 289
transducer, 167
transduction, 7, 12, 15, 40, 56, 59, 60, 62, 100, 119, 134, 135, 138, 140, 162, 164, 167, 172, 181, 182, 191, 202, 233, 285
transferrin, 133, 136
transformation, 13, 16, 96, 104, 119, 166, 174, 182, 184, 186, 187, 188, 209, 311
transforming growth factor, 65, 72, 102, 118, 123, 156, 177, 181, 182, 188, 251, 252, 253, 254, 255, 258, 260, 261, 262, 263, 272, 278, 279, 280, 281
translation, 39, 76, 130, 159, 161, 171, 174, 234
translocation, 79, 80, 105, 156, 160, 163, 173, 178, 241, 288
transmission, 118
transplant, 227, 244, 261, 263
transplant recipients, 244, 245, 261, 263
transplantation, 91, 114, 243, 244, 245, 261, 262
transport, 107, 130, 131, 132, 135, 138, 139, 170, 198, 201
trauma, 104, 313
traumatic brain injury, 91, 114
treatment, 10, 11, 23, 62, 79, 105, 158, 166, 173, 178, 197, 218, 227, 232, 243, 244, 246, 247, 261, 262, 263, 269, 283, 286, 292, 294, 295, 302, 314
triggers, 33, 39, 119, 164, 168
triiodothyronine, 196
trypsin, 57
tryptophan, 169
tumor, 7, 8, 9, 16, 18, 31, 32, 64, 65, 68, 87, 90, 96, 97, 98, 100, 107, 112, 113, 115, 120, 121, 122, 124, 128, 129, 133, 134, 136, 137, 143, 150, 152, 174, 179, 186, 188, 189, 210, 212, 220, 255, 257, 283, 284, 285, 286, 287, 288, 289, 290, 291, 292, 293, 294, 295, 296, 297, 298, 299, 301, 303, 304, 308, 311, 312, 313, 314, 315, 316, 319, 320
tumor cells, 32, 87, 96, 100, 112, 124, 174, 188, 285, 286, 290, 294, 301, 303, 304, 312, 314, 316
tumor development, 283, 314
tumor growth, 32, 98, 121, 122, 129, 133, 143, 290, 293, 296, 308, 319, 320
tumor invasion, 8, 9, 186, 286, 301, 303, 304
tumor metastasis, 7, 16, 64, 87, 121, 152
tumor necrosis factor, 113, 220, 255, 287
tumor progression, 121, 136, 137, 287, 288, 289, 290
tumorigenesis, 4, 71, 83, 97, 98, 135, 157, 173, 284, 286, 287, 288, 289, 290
tumors, 83, 96, 97, 98, 103, 110, 120, 129, 137, 138, 172, 221, 283, 284, 285, 286, 291, 294, 295, 301, 302, 303, 304, 305, 314, 317
tumours, 39, 44, 62, 64, 96, 97, 302, 303
turnover, 5, 12, 15, 40, 87, 109, 130, 139, 140, 160, 161, 162, 166, 167, 179, 185, 195, 228, 249, 256, 257, 272, 309, 319
tyrosine, 16, 22, 23, 24, 31, 60, 64, 84, 88, 120, 127, 128, 157, 164, 166, 167, 168, 181, 182, 184, 272, 286, 288, 293, 297, 304

U

UK, 21, 35, 63, 99
ulcer, 32, 94, 118, 121
ultrastructure, 259, 270
underlying mechanisms, 228
uniform, 8, 84, 91
United, 13, 19, 30, 32, 71, 175, 176, 177, 179, 180, 181, 182, 184, 187, 222, 224
United Kingdom, 71
United States, 13, 19, 30, 32, 175, 176, 177, 179, 180, 181, 182, 184, 187, 222, 224
urea, 16
urine, 10, 230, 233, 269, 291, 298
urine cytology, 291
urokinase, 187, 239
USA, 178, 201, 316, 317, 320
uterus, 207, 208, 209, 210, 211
uvula, 128

V

vaccine, 294
validation, 27
valve, 92, 101
variations, 54, 102
vascular endothelial growth factor (VEGF), 74
vascular system, 2, 126
vascularization, 143, 166, 290
vasculature, 2, 91, 259, 290, 291, 312
VCAM, 207, 209, 260
VEGF expression, 97
VEGFR, 22
vein, 17, 90, 200, 290
versatility, 159, 169, 177
vertebrates, 2, 284, 310
vesicle, 7, 131, 134, 136, 138, 139
vessels, 6, 92, 97, 121, 133, 160, 179, 208, 209, 269, 290, 302, 312, 313, 315
viruses, 219
VLA, 209, 221

W

walking, 81
Washington, 191, 278
waste, 269
water, 37, 145, 151
web, 142, 158
wells, 146
Western blot, 59, 60, 144, 148, 153, 194, 199, 290
white matter, 302, 303
wild type, 292
Wisconsin, 123
workers, 234
World Health Organization, 318
worldwide, 233
wound healing, vii, 4, 5, 11, 12, 14, 15, 18, 29, 36, 37, 39, 41, 50, 54, 62, 63, 64, 65, 75, 86, 92, 93, 94, 95, 98, 99, 101, 112, 114, 115, 116, 117, 118, 119, 120, 123, 141, 153, 157, 158, 166, 169, 175, 177, 184, 227, 229, 241, 249, 259, 268, 271, 274, 277, 279, 281, 310, 319

X

xenografts, 293
X-ray, 19

Y

yeast, 24, 137
yield, 126

Z

zinc, 238